T0387968

Microbiological Analysis of Foods and Food Processing Environments

Microbiological Analysis of Foods and Food Processing Environments

Osman Erkmen

Department of Food Engineering, Faculty of Engineering, University of Gaziantep, Gaziantep, Turkey

Academic Press is an imprint of Elsevier
125 London Wall, London EC2Y 5AS, United Kingdom
525 B Street, Suite 1650, San Diego, CA 92101, United States
50 Hampshire Street, 5th Floor, Cambridge, MA 02139, United States
The Boulevard, Langford Lane, Kidlington, Oxford OX5 1GB, United Kingdom

Notices
Knowledge and best practice in this field are constantly changing. As new research and experience broaden our understanding, changes in research methods, professional practices, or medical treatment may become necessary.

Practitioners and researchers must always rely on their own experience and knowledge in evaluating and using any information, methods, compounds, or experiments described herein. In using such information or methods they should be mindful of their own safety and the safety of others, including parties for whom they have a professional responsibility.

To the fullest extent of the law, neither the Publisher nor the authors, contributors, or editors, assume any liability for any injury and/or damage to persons or property as a matter of products liability, negligence or otherwise, or from any use or operation of any methods, products, instructions, or ideas contained in the material herein.

British Library Cataloguing-in-Publication Data
A catalogue record for this book is available from the British Library

Library of Congress Cataloging-in-Publication Data
A catalog record for this book is available from the Library of Congress

ISBN: 978-0-323-91651-6

Publisher: Nikki P. Levy
Acquisitions Editor: Nina Bandeira
Editorial Project Manager: Devlin Person
Production Project Manager: Vijayaraj Purushothaman
Cover Designer: Miles Hitchen

Typeset by MPS Limited, Chennai, India

Microbiological Analysis of Foods and Food Processing Environments

Osman Erkmen

Department of Food Engineering, Faculty of Engineering, University of Gaziantep, Gaziantep, Turkey

Academic Press is an imprint of Elsevier
125 London Wall, London EC2Y 5AS, United Kingdom
525 B Street, Suite 1650, San Diego, CA 92101, United States
50 Hampshire Street, 5th Floor, Cambridge, MA 02139, United States
The Boulevard, Langford Lane, Kidlington, Oxford OX5 1GB, United Kingdom

Notices
Knowledge and best practice in this field are constantly changing. As new research and experience broaden our understanding, changes in research methods, professional practices, or medical treatment may become necessary.

Practitioners and researchers must always rely on their own experience and knowledge in evaluating and using any information, methods, compounds, or experiments described herein. In using such information or methods they should be mindful of their own safety and the safety of others, including parties for whom they have a professional responsibility.

To the fullest extent of the law, neither the Publisher nor the authors, contributors, or editors, assume any liability for any injury and/or damage to persons or property as a matter of products liability, negligence or otherwise, or from any use or operation of any methods, products, instructions, or ideas contained in the material herein.

British Library Cataloguing-in-Publication Data
A catalogue record for this book is available from the British Library

Library of Congress Cataloging-in-Publication Data
A catalog record for this book is available from the Library of Congress

ISBN: 978-0-323-91651-6

For Information on all Academic Press publications
visit our website at https://www.elsevier.com/books-and-journals

Publisher: Nikki P. Levy
Acquisitions Editor: Nina Bandeira
Editorial Project Manager: Devlin Person
Production Project Manager: Vijayaraj Purushothaman
Cover Designer: Miles Hitchen

Typeset by MPS Limited, Chennai, India

Contents

Section III Isolation and counting of indicator and pathogenic microorganisms

Section V Identification of foods safety and quality

About the author

Osman Erkmen

Born in 1955 in Konya, Turkey, Osman Erkmen has been a professor of Food Microbiology in the Department of Food Engineering at Gaziantep University (Gaziantep, Turkey) since 2004. He received his BSc degree in Biology (1985) and MSc degree in Food Microbiology (1987) from the Middle East Technical University (Ankara, Turkey). He gained his PhD from the Department of Microbiology and Clinical Microbiology, Faculty of Medicine, Gaziantep University in 1994. He started his career as a research assistant at the Department of Food Engineering in 1985 and later became assistant Professor in 1994 and associate Professor of Food Microbiology in 1999. Since 2004, he has been working as a professor in this department. At the Department of Food Engineering, he expanded his research on the uses of nonthermal processes and natural antimicrobials in food preservation; in the production of fermented foods; in the microbial production of lycopene, thiamin, alcohol, and citric acid from industrial wastes; and in microbial inactivation and modeling. He has received funding for research from the University of Gaziantep Foundation, the Scientific and Technological Research Council, and the Republic of Turkey State Planning Organization. He has studied the combined effect of nonthermal processes, natural antimicrobials in the destruction of microbial cells and spores, their application in food preservation, and the characteristics of white and red wines production from Gaziantep Grapes. He teaches courses in the Food Engineering Department and the Nutrition and Dietetics Department. He teaches courses in General Microbiology, Food Microbiology, Food Sanitation, and Food Toxicology. Professor Erkmen has published over 150 research articles, reviews, book chapters, proceedings papers, and popular articles in the fields of Food Microbiology, General Microbiology, Food Toxicology, and Food Sanitation with more than 3500 citations. He is the editor of two books: "Gıda Mikrobiyolojisi" (Food Microbiology; 5th edition) and "Fermente Ürünler Teknolojisi ve Mikrobiyolojisi" (Fermented Products Technology and Microbiology) in the Turkish language; and is the author of three books: "Food Microbiology: Principles into Practice (two volumes)" (John Wiley and Sons, Ltd), "Laboratory Practice in Microbiology" (Academic Press), and "Food Safety (Nobel Academic Publishing). He is an advisor to PhD and MSc graduates. He has more than 10 patents, has organized more than 20 international scientific symposiums, and has participated more than 65 international symposiums. He has held administrative positions as Director of the Institute of Natural and Applied Sciences, Director of the Institute of Social Sciences, Director of the Technical Sciences Vocational School, and as Head of the Department of Food Sciences.

Preface

Food microbiology is a dynamic discipline covering many areas including microbiological analysisof foods, identification of foodborne pathogens, development of techniques for isolating and enumerating microorganisms, new food formulations, development of new food processing, distribution techniques, and carrying out important activities to ensure the quality and safety of food.

Conducting microbiological research to preserve the quality of foods has resulted in the development of many microbiological techniques. It is essential that the techniques used in microbiological analysis of foods should be the same or equivalent procedures. Otherwise, health food production in accordance with the standards is not possible and thus evaluation of the data obtained from the analysis of foods from different laboratories or research cannot be assured. Essentially, there is a need for a resource in the food industry where standard techniques for microorganisms and foods are offered for the isolation, enumeration, and identification of microorganisms. This book presents the most accurate and basic microbiological techniques for determining the microbiological quality of foods in accordance with the standards. Microbiological analyses in food and food production areas are divided into five parts in this book. In the first part, general food microbiology analysis techniques are given. Techniques for counting important microbial groups in foods are explained in the second chapter. Part three is concerned with techniques for the isolation, counting, and identification of indicator and pathogenic microorganisms. Part four consists of techniques for the detection of toxigenic fungi, viruses, and parasites. In the last part, microbiological analyses of foods are included in order to determine the microbiological qualities of foods. Flowcharts are provided for the isolation, counting, and identification of microorganisms, enabling the processes to be understood faster and easier. Visuals are also provided for the morphological appearances and test results of microorganisms that facilitate faster identification. In addition, the phenotypic characteristics of microorganisms are given in tables, providing easy determination of their characteristics and easy understanding of their distinctions. The book presents standard and applicable techniques that students, researchers, analysts, and teachers can apply in food microbiology courses, the food industry, the health field, and in analysis laboratories.

Corrections, technical questions, and other interests could be sent to: osmerkmen@gmail.com

Osman Erkmen
Department of Food Engineering, Faculty of Engineering,
University of Gaziantep, Gaziantep, Turkey

Acknowledgments

I always thank my students who like microbiology lessons. The desire of my students to learn and access information inspired me to write this book. I would like to thank my wonderful wife Assist. Prof. Dr. Ayşe Erkmen; my daughter, Disaster Management specialist, Ümran Erkmen; and my sons Dr. Barış Erkmen and Electric Electronic Engineer Hüseyin Erkmen for their support, sincere encouragement, and endurance.

Finally, I would like to thank the ubiquitous and invisible microorganisms, I wrote and published a book due to their existence, and also had to join the "real world."

Laboratory rules

The laboratory is an important part of microbiology. Laboratories are NOT optional. The microbiology laboratory involves working with living microorganisms. Proper techniques will be demonstrated and should always be used before microbiological analysis. Persons will be required to read the related part of this book and they should understand the inherent risks involved in working with living microorganisms.

General laboratory rules:

1. Wear appropriate clothes. In general, do not come to lab with a lot of skin exposed. Disposable paper is required. DO NOT transport lab coats back and forth/to and from lab.
2. NEVER eat, drink, or apply cosmetics in the laboratory.
3. Avoid touching your face while working with cultures.
4. Avoid contact of your hair with Bunsen burners and cultures.
5. Report all spills, accidents, or injuries immediately to the instructor.
6. Before you leave the lab for the day, disinfectant your working area. Show your working area to your instructor with regard to cleanliness before leaving.
7. Scrub your hands before you begin lab work (to help minimize contamination of your cultures) and scrub them before you leave to remove anything you may have picked up in the lab.
8. Old culture tubes are to be placed in the racks provided so they can be autoclaved.
9. Old Petri plates are to be placed in the provided autoclave bags.
10. Microscope slides which are no longer needed are to be placed in the provided glass disposal containers.
11. Do not remove cultures or other material from the lab.
12. ALWAYS use the proper techniques for handling cultures. These techniques will be demonstrated during the laboratory classes.
13. Read laboratory exercises before coming to the lab. It is a good idea to begin your write-ups before coming to class.

SECTION

General food microbiology analyzing practices

I

Rapid and conventional techniques are used in the microbiological analysis of foods. These techniques are widely used in the counting of microorganisms and detection of the microbiological quality of foods. They were developed over many years and are used as the official techniques in most food microbiology laboratories. These techniques are accepted internationally and recommended for the microbiological analysis of foods. Conventional approaches for the isolation of microorganisms involve various solid and liquid media. Solid and liquid culturing techniques are used either for the target microorganism detection (presence/absence test) or quantitation. The rapid direct microscopic count techniques use a small volume (e.g., 0.01 mL) of sample but membrane filter technique involves using samples of up to 100 mL or more. The plate count techniques have a maximum sample use of a volume of about 1 mL. Liquid culturing is commonly uses the most probable number technique. This technique allows flexible sample volume use. The conventional techniques used to isolate, count, and identify foodborne microorganisms are based on their cultural characteristics. Generally, plate count agar or other nonselective agar media can be used for nonspecific total viable count. Selective and/or differential media are used for counting and isolation of specific groups or species of microorganisms. Most of the standards for the indication of safety and hygiene conditions in food production involve counting indicator microorganisms (e.g., aerobic bacteria, coliforms, fecal coliforms, yeasts, and molds). This section describes the most commonly used food microbiology techniques in the isolation and counting of microorganisms.

PRACTICE

Sampling and sample preparation techniques

1

1.1 Introduction

The objective of this practice is to enable the analyzer to have representative samples in the laboratory that are microbiologically unchanged from the time of sampling and to prepare the samples for analysis. If samples are collected incorrectly or do not represent the entire product, the microbiological results will be meaningless. The person who collects the samples should able to apply an appropriate sampling plan, prevent contamination, and minimize microbial changes during transport, storage, and handling. The statistical sampling procedure should be performed during sampling according to the physical condition of the food (e.g., solid, semisolid, viscous, or liquid). A specific sampling procedure may also be used depending on the specific microorganisms, types of foods, and other factors. For such sampling, refer to the related technique in this book. Sampling, sample preparation, and all sampling techniques should be performed under aseptic conditions in duplicate unless otherwise specified. All solutions and equipment that may cause contamination of microorganisms must be sterilized or disinfected. Some of the equipment can be dangerous and inadequately sterilized by dipping into alcohol and flaming with flame. Such heat-labeled equipment can be disinfected using disinfectants.

1.2 Sampling plan

A sampling plan is a systematic way to assess the microbiological quality of foods. A sampling plan includes both the sampling procedure and the decision criteria. To examine a food for the presence of microorganisms, a representative sample must be examined by defined procedures. A quantity ("lot") of product is produced, handled, and stored within a limited time period under uniform conditions. It is impractical to examine all products. Instead of this, a certain number and size of sample units from the lot are analyzed. The samples should be taken from the lot independently and randomly. In developing a sampling plan, a number of factors should be considered: properties of food, production processes, storage conditions, associated risks, targeted consumers, and limitations. Each food product should be considered individually. A sampling plan includes the following elements:

1. Microorganism or group of microorganisms of concern;
2. Number of samples to be tested (n);
3. Testing technique(s);
4. Microbiological limit(s); m and M:
   a. Acceptable ($<$m);
   b. Marginally acceptable ($>m$ and $<M$);
   c. Unacceptable ($>M$);
5. Sampling procedure.

Microbiological Analysis of Foods and Food Processing Environments. DOI: https://doi.org/10.1016/B978-0-323-91651-6.00037-9

Two types of sampling plans can be used to indicate limit(s) for product: two-class sampling plan and three-class sampling plan.

A two-class sampling plan consists of the following specifications: *n, c, m.*

A three-class sampling plan consists of the following specifications: *n, c, m, M.*

where

n = number of sample units (packages, beef patties, and so forth) from a lot.

c = the maximum acceptable number or maximum allowable number of sample units. When this number of sample units exceeds the microbiological criterion, the lot is rejected.

m = maximum number of microorganisms (criteria) per g; values above this level are either marginally acceptable or unacceptable. It is used to separate acceptable from unacceptable foods in a two-class sampling plan or separate good quality from marginally acceptable quality foods in a three-class sample plan.

M = a quantity that is used to separate marginally acceptable quality from unacceptable food. It is used only in three-class sampling plans. Values above M for any sample are unacceptable, indicating health hazards, sanitary indicators, and spoilage potential.

In a two-class sampling plan, only one microbiological limit "*m*" is involved, therefore the two-class sampling plan attributes $<m$ and $>m$ by maximum allowable number of c sample(s). The lot will be accepted or rejected according to the two-class sampling plan as illustrated in the following diagram.

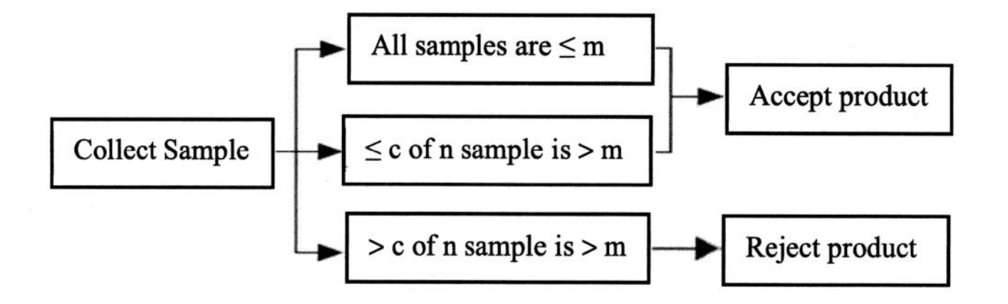

A two-class plan is used to accept or reject a batch (lot) of food in a presence/absence decision by a plan. A sampling plan for *Salmonella* is $n = 5$, $c = 0$, where $n = 5$ means that five individual samples of the lot are examined microbiologically for the presence of *Salmonella*, and $c = 0$ means that all five units must be free of the bacteria for the lot to be acceptable. If any unit is positive for *Salmonella*, the entire lot is rejected.

Samples may contain coliforms with a sampling plan $n = 5$, $c = 2$. By this plan, if three or more of the five-unit samples contained coliforms, the entire lot would be rejected. If up to 100 coliforms per g in two of the five units is allowed, the sampling plan would be $n = 5$, $c = 2$, $m = 10^2$. After the five units have been examined for coliforms, the lot is acceptable if no more than two of the five contain as many as 10^2 coliforms per g but is rejected if three or more of the five units contain 10^2 coliforms per g.

A three-class sampling plan is used to indicate acceptable/marginally acceptable/unacceptable foods. Assume that for a given food product, the standard plate count (SPC) shall not exceed 10^6 cfu per g^{-1} (*M*) or be higher than 10^5 colony forming unit (cfu) per g from three or more of five units examined. The specifications become $n = 5$, $c = 2$, $m = 10^5$, $M = 10^6$. If any of the five units exceeds 10^6 cfu per g, the lot is rejected (unacceptable). If any sample unit cannot give results above m, the lot is acceptable. If two of the five units exceeds m and do not exceed M, the lot is marginally accepted.

In a three-class sampling plan, two microbiological limits, *m* and *M*, are set. The microbiological limit *m* commonly reflects the upper limit of a good manufacturing practice. The criterion *M* marks the limit beyond which the level of contamination is hazardous or unacceptable. The lot will be evaluated according to a three-class sampling plan as shown in the following diagram:

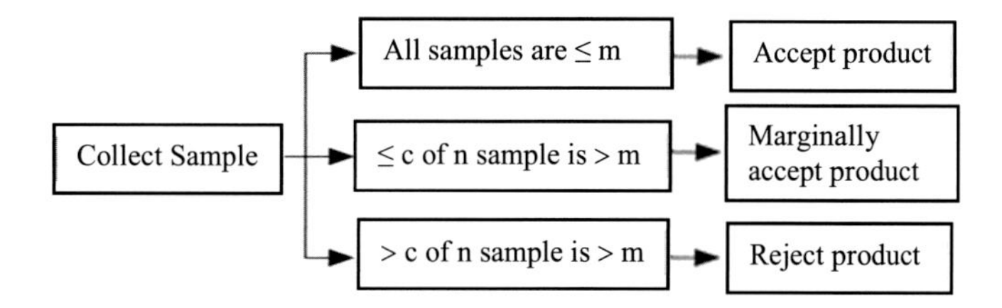

In general, a two-class sampling plan is preferred when the microorganism of concern is not permitted in food samples. If a number of microorganisms is allowable, a three-class sampling plan is usually prepared. To enhance food safety and improve food quality, more stringent microbiological limits (*m* and/or *M*) should be adopted.

1.3 Sampling

Containers used for placing samples should be leak-proof, wide-mouthed, clean, and dry, but their mouth must be suitable to protect against microbial contamination. Appropriate samples should be taken to represent the food or other sampling source. Laboratory results and interpretation are only valid when appropriate samples are analyzed. A sampling unit should be at least 100 g or mL. When liquid samples are removed, an additional liquid sample must be taken for a temperature control. The samples are collected under appropriate conditions. Samples obtained from foods or other sources should be delivered to the laboratory as soon as possible. Upon arrival at the laboratory, the time of arrival should be recorded. The samples are prepared at the start of microbiological analysis. During storage and transport of samples, microbial contamination should be prevented from the external environment, including the air, sample container, sampling device, and the shipment vehicle. The packaged food should only be opened in the laboratory. A form should be labeled with the following information: sample name, data conditions, place, time of collection, name of people, and temperature at the sampling time.

1.3.1 Liquid samples

Thermometers used in controlling the temperature of liquid food should be sanitized before use. Metal thermometers are preferred since the breakage of a mercury thermometer would contaminate the product. Before removing a sample, aseptically mix the liquid foods (e.g., milk, ice cream, fruit juice, water, and sirups) to homogenize the contents of the container as much as possible. In the laboratory, the liquid should be mixed thoroughly before removing the required samples for analysis. Sampling of liquid foods from in-process may involve the use of a sterile metal tube. A special line sampling technique involves the use of a disposable sterile hypodermic needle and syringe. The needle is inserted into a rubber closer of a stainless-steel nipple. Sampling cocks may be used on holding tanks and product pipelines. The temperature of the control sample should be checked during the transport, storage, and before analysis. The sample must be labeled with the time of delivery to the laboratory.

1.3.2 Solid samples

Sterile spoons, spatulas, or scalpels, depending on the nature of the foods, can be used to handle solid food samples. Samples must be taken from several parts of the food and protected from excessive humidity. Large samples should be mixed thoroughly in the laboratory before taking smaller samples for analysis. For large solid food (frozen or unfrozen), samples should be taken aseptically from several parts using sterile knifes and forceps, then mixed as a composite. Sampling of solid or semisolid line foods may be accomplished using the same equipment and procedures. Automatic sampling devices can be used for powdered products and others.

1.3.3 Semisolid samples

Aseptically transfer a representative sample to a sterile container from semisolid foods. Before removing a sample, aseptically mix the food to homogenize it as much as possible. If adequate mixing of the semisolid product is not possible, multiple samples should be removed from the semisolid product in the container from different areas. Perishable samples must be cooled to 0°C–4°C quickly. In addition, a sterile sample container should be used for temperature control.

1.3.4 Frozen samples

Frozen foods should be sampled aseptically from several areas of frozen food using sterile augers, bits, knives, forceps, or other sharp sampling instruments. These portions should be mixed as a composite to provide a sample representative of the whole food. They must be collected in prechilled containers. Containers should be placed in a freezer long enough to chill thoroughly. Frozen samples should be kept frozen until arrival at the laboratory and analysis. They should be transported with approved hard construction insulated containers to the laboratory. Thawing of frozen samples must be avoided during transport and storage.

1.3.5 Packaged samples

Packaged food should be sampled from the original unopened package or containers. When samples are removed from large packages, contamination of microorganisms should be prevented during opening from air, equipment, and package. From small packages, remove the whole package sample. The packaged foods should only be opened before analysis.

1.3.6 Special purpose samples

In some instances, samples are tested as part of a foodborne disease investigation or on the basis of a consumer complaint. In outbreak investigations, samples are removed from all perishable leftover foods or from suspect meals as soon as possible. Ingredients or raw items using in the preparation of food must be collected if available. The samples as well as foods from patients should be considered. Samples from patients may include stools, vomit, and serum. These samples should be collected in sterile leak-proof containers, and properly identified with the patient's name, type of sample, and date of collection.

1.3.7 Surface samples

Surface samples are transferred to a microbiological medium that protects the microorganisms. Different surface sampling techniques are described below.

1. *Surface slices*. Using sterile scalpels or forceps, a very thin slice of the food layers is removed. For example, the skin of the poultry and fish provides suitable sampling material. The slices are homogenized in a diluent (diluting solution) to obtain an initial 1:10 dilution.
2. *Washing and rinsing*. The food can be rinsed or washed in sterile diluent. The rinsing or washing is applicable to some types of foods (e.g., dried fruits, vegetables, poultry, and sausages). In rinsing or washing, the count represents bacteria from the surface area only. The microbial flora of some foods (e.g., poultry) is largely confined to the skin surface. The most appropriate sampling for the analysis of this type of food is surface-sampling. In some cases, it is essential to use one (or up to two) parts of rinse solution by weight of food or the whole or part of the food (e.g., poultry) may be rinsed in a known volume of diluent, for example, phosphate buffer (PB) solution. Poultry is placed into a sterile plastic bag in a container and enough PB solution is added into the plastic bag to cover the poultry. If poultry is not covered by the PB solution, add amounts up to two times the weight of food. Loosely tie the open-end of the plastic bag (not air-tight). Let it stand undisturbed for 60 min at room temperature and lightly mix by gently swirling.
3. *Swabs*. Cotton swabs are prepared from nonabsorbent cotton with the head firmly twisted to be 0.5 cm in diameter by 1–2 cm long on wooden sticks or stiff stainless-steel wire 12–15 cm in length. Swabs should be placed into individual tubes with the swab heads away from the closure. Commercially sterile swabs or sterilized swabs are used in the sampling. The swab head is moistening in a tube containing 10 mL sterile rinse solution (e.g., nutrient broth, 0.1% peptone water, PB solution, or quarter-strength ringer solution), and then the excess solution from the swab is removed by pressing against the interior wall of the tube with a rotating motion. If the sampled surface contains fatty materials, use 0.5%–1.0% Tween-80 in rinse solution. Sampling from a large surface (e.g., carcass and equipment), an area of 2 cm width and 25 cm length or 10 cm^2 area is marked. From a small area, 2 cm^2 can be sampled. During swabbing, the swab is held to make a 30-degrees angle contact with the surface. The swab is firmly rubbed over the approximately indicated area three times, then continue to swab in the reverse direction using parallel rubbing to the surface with a slow rotation. After sampling from the surface, the swab is immersed in 10 mL of transport broth (e.g., Cary-Blair medium or other transport medium) in a tube. The swab should be shaken up and down 10 times in the tube to allow bacteria to pass from the surface of the swab into liquid medium (Fig. 1.1). Finally, insert the swab about 5 cm within the neck of the tube, the swab is rinsed briefly in liquid and the excess fluid in the swab is removed by a short press on the inner surface of the tube over the liquid. The swab is carefully removed from the tube and thrown into the waste bin. Alternatively, break the swab tip off below where the analyzer has touched the swab stick, thereby dropping the cotton tip into solution. The swab washed medium is used in the microbiological analysis. A repeat sample can be removed from same area. In general, swabs are not suitable for culturing *Mycobacteria*.

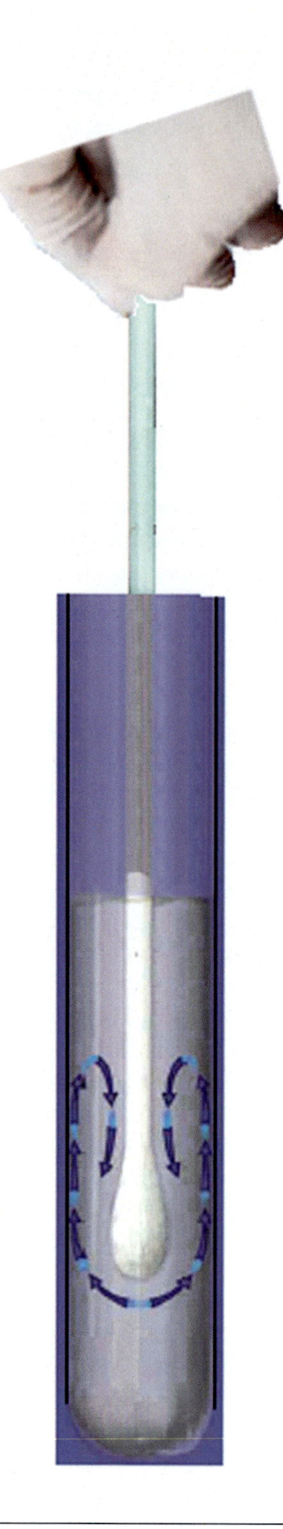

FIGURE 1.1

Shake the swab up and down 10 times in the transport broth in the tube to allow bacteria to pass from the swab to transport broth.

4. *Contact slide*. Press a disinfected microscope glass slide onto a surface area of the food to be analyzed. Then examine the slide microscopically after fixing and staining (e.g., simple staining). Alternatively, the slides pressed onto the surface area may be pressed onto the surface of the agar medium in a petri plate. The plates including slides are incubated. This plating does not allow quantitative estimation of microorganisms; it is useful for a rapid determination of hygienic conditions of surface area and the dominant microflora; especially on such foods as raw meat, poultry, soft cheese, etc.
5. *Agar-contact technique*. Pour selected medium (at 47°C) into each sterile Petri dish until the medium rises above the rim of the plate. Let the agar solidify. Aseptically, remove the cover of the plate and carefully press the surface of the agar onto the test area (contact area of plate is about 26 cm^2). After pressing the agar onto surface area, replace the cover of the plate and incubate the inverted plate at 32°C for up to 48 h.

1.3.8 Sampling for anaerobic bacteria

Samples should contain little free oxygen in the anaerobic media, and should not be exposed to normal atmospheric conditions of oxygen as would occur if small samples were taken. When small samples are unavoidable or swabs are used, they must be placed into a reduced anaerobic transport medium (e.g., Stuart transport medium). Swabs can be moistened in transport medium prior to use. Before using solid transport medium in the analysis, it should be exposed to 80°C to remove and "dissolve" the oxygen in the medium.

1.3.9 Air samples

Different techniques are used for air sampling. Some of these techniques are direct plating, air filtration, impingement, and others. In the direct plating technique, nonselective agar media (e.g., tryptose soy agar, plate count agar, and nutrient agar) are exposed to air in a variety of locations in the room for 15 min. In this exposure, the number of colonies forming on one plate represents the number of microorganisms settling on approximately 0.1 m^2 min^{-1}.

An air sample can be obtained from the room by filtration of air through a membrane filter (Technique 7.3.2.2). The air filtration apparatus includes a membrane filter (pore size 0.22 or 0.45 μm) placed on an appropriate holder and connected to a vacuum source through a flow rate controller. After filtration of an amount of air for a constant time from the processing room, membrane filter is placed on an agar surface.

An air sample can also be obtained from the room by impingement, as explained in Technique 7.3.2.2. Some parts of the air sampling device are sterilized in the autoclave by placing distilled water into glass parts of the impinger and all openings are closed with aluminum foil. The rubber parts are disinfected. In the air sampling process, add 20 mL of sterile 0.1% peptone water into the impinger, add 0.1 mL of sterile antifoam and label the liquid level on the impinger. Open the vacuum for a time to pass air through the liquid in the impinger. After the exposure time, remove 1 mL of fluid from the impinger using a 1 mL sterile pipette. Remember, the water volume in the impinger decreases during impingement because of evaporation and sample removal after each exposure. Add sterile water up to the labeled part of impinger during process.

1.4 Sample transport and storage

Samples for analysis should be transported to the laboratory as rapidly as possible to prevent microbiological and chemical changes. Samples should be examined within 36 h after sampling. Perishable samples that cannot be analyzed within 36 h should be frozen or retained under refrigeration. Nonfrozen samples of shellfish should be examined within 6 h after collection; they cannot be frozen. The transport conditions, time and data of arrival at the laboratory should be recorded. Samples collected from sources should be transported in the original unopened containers. The sample's original state should be maintained until the tests are carried out in the laboratory. Frozen foods should be transported frozen (at $-20°C$) by cold packs or using solid CO_2 in insulated containers, followed by storage in a deep freezer until being tested. When dry ice is used, the containers should have tight closures to prevent pH changes in the sample that will be caused by the adsorption of CO_2.

Perishable samples collected in a nonfrozen state must be refrigerated (at 3°C) from the time of collection until analysis in the laboratory. Refrigerated products must be transported in insulated containers at 3°C until arrival at the laboratory. If storage for a short period is unavoidable, the samples should be cooled and stored in a refrigerator at 3°C until being analyzed. Unfrozen food samples should not be frozen before testing. Refrigerated products should not be frozen to prevent destruction of certain microorganisms. The refrigeration of samples for 3 days or more will result in the multiplication of psychrotrophic microorganisms and may also cause the death of some mesophilic or thermoduric microorganisms.

1.5 Sample preparation and dilutions

The working area of a laboratory must be disinfected just before starting sample analysis. Samples for analysis should be taken under aseptic conditions. If the sample removal is necessary from the original container, it should be done aseptically within 3 min. The sample container exterior area should be cleaned with 70% ethyl alcohol and flaming, if possible, prior to opening to prevent contamination.

The fluids used in the dilution of the sample should not damage the microorganisms to be detected (e.g., osmotic shock). The common diluents used in microbiological laboratory are 0.1% peptone water, quarter-strength Ringer's solution, phosphate buffer solution, and 0.85% NaCl solution. Peptone water (0.1%) has higher protective effects on microorganisms than the other diluents. The preparation of dilution and inoculation into the medium should be completed within 15 min.

When specific microorganisms are analyzed, special diluents may be used. For such diluent and sample preparation, see the respective practice covering the relevant food types and microorganisms. For example, fatty foods require a wetting agent (e.g., tergitol anionic-7 solution) and it is added into the diluent to promote emulsification. When a sample is analyzed for anaerobic bacteria, a reducing agent is used in the diluent. In the anaerobic medium, dispersion of the sample in the diluent should be achieved by a technique that introduces the least oxygen into the medium.

The frozen sample should be thawed in its original container in a thermostatic shaking water bath. The water bath is agitated and sample is frequently shaken. It can also be thawed in the refrigerator (at 2°C–5°C within 8 h) or under running water, but water should not touch the sample. It should not be transferred to a second container for thawing the sample.

1.5.1 Sample preparation

1.5.1.1 Liquid samples

Liquid samples in flasks that are 2/3 to 3/4 full should be gently shaken 25 times for 7 s. Samples are measured volumetrically using a sterile pipette or graduated cylinder. Twenty-five mL of sample is added into 225 mL of homogenized solution (or 10 mL into 90 mL, 11 mL into 99 mL, 50 mL into 450 mL) in the Erlenmeyer flask. The sample is homogenized in the homogenizing solution (diluent). The mixture is shaken by a mechanical shaker for 15 s or gently shaken 25 times in 7 s over a 30 cm arc by hand. This homogenization provides an initial 1:10 dilution. If a pipette is used in removing a liquid sample, the pipette is not inserted more than 2.5 cm below the surface of the sample. Pipetin dışından fazla sıvıyı çıkarmak için pipet ucu tüpe veya şişe boynuna dokundurulur. The sample in the pipette should be added into the diluent by touching the tip of the pipette to the inside column of the tube. This will drain the liquid from the graduation mark to the liquid rest point of the pipette within 2 to 4 s. Do not rinse the pipette in the dilution fluid.

1.5.1.2 Solid and semisolid samples

Dry samples should be aseptically stirred with a sterile spoon, spatula, or other utensil to obtain a homogenous sample. Aseptically weigh 25 g (50 g) of solid or semisolid sample (using sterile forceps or spatulas) into 225 mL (450 mL) of homogenizing solution in a sterile blender (or stomacher flexible polyethylene bag). High-fat foods (e.g., butter) may require the use of warm (40°C) diluent or wetting agent to facilitate mixing. Foods containing bones, pits, or other hard objects may break or puncture the plastic bags used in the stomacher. Using a double bag can reduce the risk at plastic damage. When whole incubation of homogenized sample is necessary, carefully transfer the homogenate from the blender cup into a sterile Erlenmeyer flask.

If package contents are not homogeneous (e.g., a frozen dinner), its contents should be macerated first and then an adequate sample is removed or each food portion from the package can be used separately in the analysis depending on the specific analysis to be performed. Weigh the sample accurately (± 0.1 g) under aseptic conditions into a sterile high-speed blender or stomacher.

Homogenizing in blender. Samples in the diluent should be blended for up to 2 min at low speed (approximately 8000 rpm). Some blenders may operate at speeds lower than 8000 rpm. The blending time may vary depending on the type of food. The liquid in the blender must cover the blade completely. Caution should be taken to prevent excessive heating. Chilling (e.g., tempering in an ice water bath) may be employed to decrease the excessive heating of the blender.

Homogenizing in stomacher. The sample is placed into a sterile, disposable, and flexible polyethylene bag containing diluent. The bag is placed into the stomacher with a few cm of the bag projecting above the stomacher door; the door of the stomacher is firmly closed to seal the bag during blending. Operation for up to 2 min is sufficient for the dispersion of most samples. The advantages of the stomacher are that the samples do not come into contact with the apparatus as the sample is present in disposable bags; in particular no temperature rise occurs in the stomacher; and good dispersal is obtained even with deep-frozen samples.

1.5.1.3 Fine particle solid samples

The initial dilution of fine particle solid samples, for example, flour and milk powder, is accomplished easily. Weigh the solid food sample accurately ($\pm$ 0.1 g) under aseptic conditions. In aseptic conditions, 10 g of sample is weighed into a sterile Erlenmeyer flask and the content is filled up to 100 mL with sterile diluent to achieve a 1:10 dilution.

1.5.2 Sample dilutions

A sterile 1 mL pipette is held vertically, the tip of the pipette is immersed up to 2.5 cm under the surface of the homogenized sample, and the liquid is drawn up and down three times from the sample to the 1 mL mark. One mL of the homogenized or mixed sample is removed by touching the tip of the pipette to the inner surface (2.5 cm above the diluent level). The tip of the pipette should be touched to the upper side of the tube. The sample is transferred into 9 mL diluent. The tip of the pipette should not come into contact with the dilution fluid while adding the sample. The pipette contents are transferred within 3 s by blowing gently. The pipette should not be rinsed in dilution fluid. The used pipette is discarded. The discarded pipette is placed into pipette discard equipment (e.g., a graduated cylinder) containing a disinfectant and a piece of cloth at the bottom. The liquid transferred in the diluent should be mixed by vortexing the tube or turning it back and forth between the hands. Subsequent dilutions are made with a new sterile pipette. Depending on the possible bacterial number in the sample, dilution is made up to the suitable dilution range to obtain the most accurate colony count in the Petri plate of between 25 and 250 colonies.

Caution. (1) Some solid samples (e.g., poultry) can be placed into a suitable sterile container (plastic bag or sealed bottle) and a volume of sterile diluent added equal to the weight of the sample. Then the container is gently shaken to mix the sample. Each mL of homogenize represents 1 g of sample. (2) A sterile pipette is used for each transfer from each dilution. The temperature of sample and diluent should be 15–25°C. (3) Dilutions or pour plates should not be exposed to sunlight. (4) Pipettes must not be wiped or dragged across the lips and necks of tubes. (5) During sample transfer, the pipette is removed from the liquid and the pipette tip is maintained above the liquid level in the tube to remove excess liquid from external surface area of pipette. (6) Sterile pipettes should not be subjected to flame. (7) During inoculation of sample into the Petri plate or sterile Petri dish, the plate cover is removed high enough to insert the pipette and the pipette should be held at a 45–degree angle when touching the tip onto the inside of the agar in the Petri plate or Petri dish. Allow the pipette about 4 s for the draining of 1 mL of sample from the graduation mark to the rest point in the tip of the pipette; then holding the pipette in a vertical position, touch the tip of the pipette once against a dry spot on the plate. Do not blow out the liquid sample during inoculation. (8) When 0.1 mL of sample is measured using a pipette, use a 0.5 mL pipette, hold the pipette as directed above and allow the sample to drain from the graduation point down to the 0.4 mark from the 0.5 mark. Do not retouch the pipette to the plate. (9) Duplicate sampling and analysis (e.g., plating) should be prepared for each sample. (10) Do not remove less than 10% of the total volume of the pipette. For example, do not use a 10 mL pipette to remove 1 mL volume. (11) Dilution preparation and sample inoculation into media should be performed within 15 min.

PRACTICE

Plate count techniques

2

2.1 Introduction

Many quality characteristics of foods depend on the viable microbial numbers. The number of viable microorganisms in foods can provide information about certain conditions related to production, processing, and distribution. A high number of viable microorganisms in heat-treated foods (e.g., pasteurized milk) indicates inadequate heating, contamination after heat treatment from improperly cleaned and sanitized equipment, and growth of surviving bacteria during inadequate cooling. Aerobic plate count (APC) is the commonly used plate count technique in the estimation of viable microbial populations from foods. Basically, microorganisms can multiply and form visible colonies on or in the agar medium. The number of colonies are counted and the number of APC is calculated after considering the dilution rate and inoculation amount of sample. The count provides a numerical estimation of the number of living microorganisms in the sample. Bacterial cells can occur in pairs, chains, clusters, or clumps and form colonies, but they are included in plate counts as a single colony. Therefore, the number of colonies which develop on the plates would then be lower than the number of individual cells in the sample. Additionally, all microbial types will not grow on/in one type of medium and can grow at different incubation conditions. This can be due to deficiencies in medium nutrient composition, unfavorable oxygen, presence of inhibitory substance in the growth medium, unfavorable incubation temperature, injury on the microbial cell, length of incubation time, remaining inhibitory substances on equipment, improper sterilization of medium and glassware, improper inoculation of the sample, and others. Due to these facts, the counts obtained from the APC technique are reported as colony forming units (cfu) per unit (e.g., g or mL). Some other errors from the APC technique are (1) measuring quantities, (2) improper computing and results recording, (3) recontamination of medium and glassware, (4) possibility of presence of nutrients or growth inhibitor in water used in media preparation, (5) variation in the nutrient content, and (6) variation in pH of water used in the diluent and medium preparation.

These errors lead to the differences in the results of plate count from one lab to other. For this reason, a uniform testing procedure is of great importance to reach a reasonable degree of agreement between different laboratories. This approach then enables laboratories to compare results. Plate count is an agar plate count from foods and products carried out on a specific growth medium (e.g., plate count agar) with the use of standardized conditions of equipment, handling, incubation time, temperature, and procedure.

Minor modifications of the APC technique can be used to count special groups of microorganisms, for example, thermoduric, mesophilic, or psychrotrophic. Incubation conditions indicate the type of plate count. For example, if a count is performed from plates incubated at aerobic conditions and 35°C, the count is called an aerobic mesophilic count. There are two types of APC techniques: pour plate count and spread plate count. In this chapter, these counts are explained with basic steps and details on recording the results.

Microbiological Analysis of Foods and Food Processing Environments. DOI: https://doi.org/10.1016/B978-0-323-91651-6.00040-9

2.2 Plate count techniques

Sampling, sample preparation, and all experimental techniques should be performed under aseptic conditions and in duplicate, unless otherwise specified. All microbiological media to be used in applications must be sterilized. Again, all solutions, reagents, and equipment that may cause contamination of microorganisms must be sterilized or disinfected unless otherwise specified.

2.2.1 Equipment, reagents, and media

Equipment. Blender (or stomacher), Erlenmeyer flask, glass spreader (made from glass rods 20 cm long and 3.5 mm in diameter, and bent at right angles approximately 3 cm straight from the end), graduated cylinder, incubator, loop (4 mm internal diameter), Petri dishes, pipette discard equipment (e.g., a graduated cylinder) containing 10% bleach disinfectant solution and a piece of cloth at bottom, pipettes (1 and 10 mL) in pipette box, scissor, vortex, and water bath.

Reagents. Alcohol in a 50 mL beaker, ethanol in bottle, sterile distilled water, Erlenmeyer flasks and 0.1% peptone water,

Media. Plate count (PC) agar.

2.2.2 Sample preparation and dilutions

Refer to Technique 1.5 for detailed information about sample preparation and dilutions. For liquid samples, mix the sample well by hand shaking, and remove 25 mL (or 50 mL) of the sample by sterile graduated cylinder and add into a sterile Erlenmeyer flask containing 225 mL (or 450 mL) of 0.1% peptone water. Homogenize the sample by gently shaking the flask. For a solid sample, aseptically weigh 25 g (or 50 g) of solid food sample into a sterile blender (or sterile stomacher bag) containing 225 mL (or 450 mL) of 0.1% peptone water. Food products (e.g., vegetables and meat) may be cut into small pieces with scissors and then weighed into the blender. A solid sample is homogenized by blending for 1 min (or stomaching for 2 min). This homogenization results in a 1:10 dilution and is prepared further necessary serial dilution. In the preparation of further dilution, add 1 mL of homogenate using a sterile 1 mL pipette into 9 mL of 0.1% peptone water. The diluent is mixed by vortexing or shaking through hands (Fig. 2.1). High-fat foods (e.g., butter) may require the use of warm diluent (below 40°C or containing wetting agent) to mix the homogenate. Place the used pipette into the pipette discard equipment. From the time that the sample is homogenized, the dilution and inoculation to medium must be completed within 15 min. For better accuracy, use a 1 mL or 5 mL pipette for inoculation. Do not use a pipette to deliver $<10\%$ of its total volume; for example, a 10 mL pipette to deliver 0.5 mL.

2.2.3 Pour plate count technique

Labeling. Label all sterile Petri dishes and tubes with the sample number, dilution rate, date, name of medium, your initials, and any other desired information prior to sample dilutions and inoculation.

Melting medium. Melt the required amount of PC agar quickly in a boiling water bath or heater and avoid prolonged exposure to high temperatures (not greater than 75°C). Cool the melted

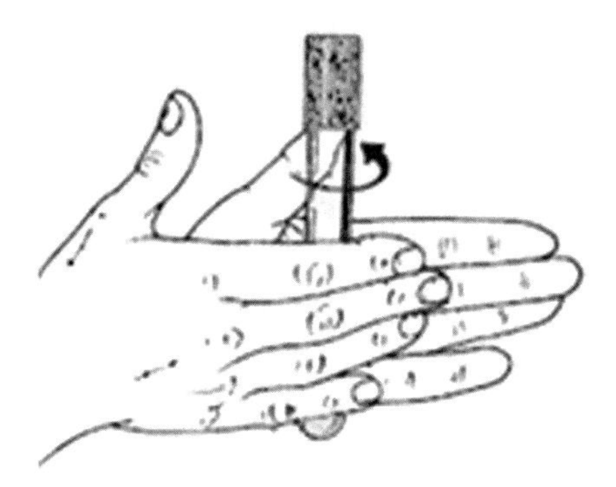

FIGURE 2.1

Sample and diluent are mixed by rolling the tube through hands.

medium to approximately 47°C and hold in a water bath at that temperature until use. Set a thermometer into water in a flask for temperature control. The temperature of the agar medium should not be controlled by touching the medium before pouring it into Petri dishes.

Inoculation and pouring. Place 1 mL of sample (nondiluted or diluted) into sterile Petri dishes in duplicate by using a sterile pipette. Place the used pipette into the pipette discard equipment. Pour 17–19 mL of melted agar medium (at 47°C) into each Petri dish by lifting the cover just high enough. Spilling the medium outside of the Petri dish or on the plate lid inside must be carefully avoided during pouring of the agar medium. After pouring the agar medium into Petri dish, thoroughly mix the medium with the sample inoculum by rotating the plates while drawing a figure 8 and while the plate is lying on the tabletop. During laying the plate on tabletop, take care not to splash the mixture over the edge of Petri dish and no bubbles should be formed on the surface of the agar medium. Allow plates to solidify (usually within 10 min) on a flat surface of the tabletop. After solidification of the medium, the plates should be inverted to prevent colony spreaders due to droplets of water condensation.

Incubation. Incubate inverted Petri plates depending on the nature of the food and the type of microorganisms to be counted. In general, the inverted plates are incubated at 35°C for 24–48 h for mesophilic aerobic count (MAC) and for dairy products, incubate the inverted plates at 32°C for 24–48 h. Excessive humidity in the incubator should be avoided to prevent water drops on the agar medium. These practices (inverting the plates and avoiding excessive humidity) reduce the tendency of the formation of colony spreaders. Excessive drying of the medium during incubation must be controlled by allowing suitable ventilation and circulation of air. During 48 h of incubation, agar in Petri plates should not lose weight by more than 15%. Count colonies as indicated in Technique 2.2.5 and report the result as indicated in Technique 2.2.6.

2.2.4 Spread plate count technique

In the spread plate count technique, colonies are formed on the surface of agar in the plates. This technique has certain advantages over the pour plate technique: (1) the use of translucent media is not essential with a spread plate, but it is necessary with the pour plate to see the location of colonies through the medium. (2) The colonial morphology can be observed better. (3) Different colony types can be distinguished easily. (4) Heating over 44°C will inactivate or injure some vegetative cells in a spread plate count. (5) Microorganisms are not exposed to the heat of the melted agar medium in the pour plate count. (6) Colonies are easily removed from surface area of agar medium for further study. (7) The spread plate technique may result in higher counts than the pour plate technique. Pour 18–20 mL of melted PC agar (47°C) into sterile Petri dishes. To facilitate uniform spreading, hold the plates at room temperature on a flat surface to solidify the agar. Inverted plates are incubated (at 30°C–35°C) for 18 h to control contamination. Transfer 1 mL of diluted sample onto the agar surface by using a sterile pipette. Use duplicate plates for each diluted (or undiluted) sample. Spread the inoculated sample as quickly and carefully as possible on the surface of the agar medium using a glass spreader. A separate sterile glass spreader or alcohol-flamed glass spreader must be used for each plate. Allow the plates to air dry (within 10 min) prior to incubation. Incubate the inverted plates as indicated in Technique 2.2.3. Count the colonies as indicated in Technique 2.2.5 and report the result as indicated in Technique 2.2.6.

2.2.5 Selecting and counting colonies from Petri plates

The colonies on the Petri plates are counted from the bottom of plates after incubation. If Petri plates are not counted at once, they must be stored at 0°C–4°C for a period of time of no more than 24 h in the inverted case. The following guidelines must be used for selecting and counting colonies from Petri plates.

1. *One Petri plate containing 25 to 250 colonies.* Only one Petri plate containing 25 to 250 colonies is counted. Calculate the count as the number of MAC colony forming unit (cfu) per g or mL of sample.
2. *Duplicate Petri plates containing 25 to 250 colonies.* Two Petri plates from the same dilution containing individual 25 to 250 colonies are counted and average the colony counts from two plates. Calculate the count as the number of AMC cfu per g or mL of sample.
3. *Petri plates from consecutive dilutions containing 25 to 250 colonies.* If Petri plates from consecutive decimal dilutions contain 25 to 250 colonies, count colonies from all plates separately. After counts, average the colony counts from same dilution and multiply by the dilution rate and average the counts from all dilutions. Record the result as the number of MAC cfu per g or mL of sample.
4. *No Petri plate between 25 to 250 colonies.* If there are no Petri plates with between 25 and 250 colonies and one or more Petri plates have nearest 25 colonies, count colonies from these plate(s). If necessary, average the colony counts for the same dilution and multiply by dilution rate. Calculate the average count from different dilutions, and report the result as the number of estimated (E) MAC cfu per g or mL of sample.
5. *All Petri plates containing fewer than 25 colonies.* If all Petri plates contain fewer than 25 colonies, count colonies from these plate(s). If necessary, average the colony counts from the same dilution, calculate the average count from each dilution, and report the result as EMAC cfu per g or mL of sample.
6. *No colonies on the Petri plates.* If petri plates have no colonies after incubation without inhibitory substances from all dilutions (or nondiluted), the count is reported as less than (<)

the corresponding lowest dilution. For example, if the colonies cannot appear on the 1:10 dilution (as lowest inoculated dilution), the counts are reported as "less than 10" (or "<10") EPAC cfu per g or mL of sample.

7. *Crowded (noncountable) colonies on the Petri plates (much more than 250 colonies)*. If the number of colonies on Petri plates is much more than 250 colonies (noncountable), the colonies are counted from portions of the plate that are representative of colony distribution. Place the Petri plates onto a 1 cm^2 ruler. the colonies are counted from portion (1 cm^2) of plate containing colonies fewer than 10, select six consecutive representative squares horizontally across the Petri plate and six consecutive squares at the reverse side. (1) When there are more than 10 colonies per cm^2, the colonies are counted in four such representative squares. In both cases, multiply the average number of counts per cm^2 by the number of areas of the Petri plate (appropriate factor) to determine the EMAC per plate. The area of standard 15×100 mm plate is approx. 56 cm^2; the appropriate factor is 56. (2) Where bacterial counts on crowded plates are greater than 100 colonies per cm^2, the counts are reported as greater than (>) for highest dilution. The result is reported as highest dilution EPAC cfu per g or mL of sample.
8. *Counting spread colonies*. Three types of colony spreaders can appear on the Petri plates: (1) a chain of colonies that is not significantly separated, which may result from the breakdown of a bacterial cluster that will occur when dispersed to or out of the proper location. If one or more such chains exists in separate locations, each is counted as a single colony. (2) Colony spreaders may develop in a film of water between the agar medium and the bottom of the Petri dish. They are counted as a single colony. (3) Count each spreader as one colony when the coverage is less than 25% of the plate. These spreaders develop largely due to the accumulation of moisture at the point from which the spreader forms.

2.2.6 Calculation and reporting counts

The numbers of counted colonies or estimated colonies from each plate for the dilutions are recorded. Calculate the average results from the same dilution rate and multiply by the dilution rate. For example, if two counts from the same dilution rate are 49 and 65 for 10^{-3} dilution, the average count for this dilution rate and multiplying by 10^3 will be 57×10^3. Average the counts for different dilutions. For example, if total counts from different dilutions are 57.0×10^3 and 1.96372×10^4, average the count and divide by inoculation amount. Report results as the number of MAC or EMAC cfu per g or mL of sample. Round the calculation to two significant figures. If the third digit after the calculation number is 6 or more, the number is rounded up (e.g., 58.6372 = 58.64); If it's 4 or less than 4, it is rounded to the preceding number (e.g., 38.6342 = 53.63). If the third digit of the calculation number is 5 and the second digit is even, round the number to the preceding number (e.g., 38.6352 = 38.63); If the second digit is odd, it is rounded up to the above number (e.g., 38.6252 = 38.63). Use the following formula in the calculation of microorganisms in the sample.

$$\text{MAC cfu per g or mL} = [(\text{AC1} \times \text{DR1})/\text{IA}] + [(\text{AC2} \times \text{DR2})/\text{IA}]$$

where AC1 is average count for first dilution; AC is average count for second dilution; IA1 is inoculation amount for first dilution rate; IA2 is inoculation amount for second dilution rate; DR1 is first dilution rate; DR2 is second dilution rate.

2.3 Interpretation of results

Total number of living (TVC) can only be determined by plate counting technique and quantify live bacteria in a sample and are widely used across the industry for microbiological monitoring of food and water samples. In many cases, the total number of bacteria, regardless of the type, is also an indicator of the quality and suitability for human consumption. TVC is widely used in the food industry in order to indiscriminately quantify the population of bacteria in a food sample. Each bacterium in the sample forms a colony and is counted through visual inspection. Several dilution steps of the original sample are often necessary in order to have isolated colonies that can be easily counted. In its basic form, TVC provides a rough estimate of the number of the microorganisms but it does not differentiate their types. It is however possible to selectively count or isolate specific types of bacteria by adding inhibitors to the growth medium. The required incubation time is a major shortcoming of the plate count technique as a late positive detection can be quite costly.

The specifications for many food ingredients and finished products contain microbiological criteria, often including a value for TVC. These criteria may be mandatory and set out in legislation, determined by a customer or may have been developed in-house. Building up a database of TVC values for the same food product over an extended time period can be a very useful means of monitoring the production process. A gradual increase in the number of TVC over time may indicate a problem with cleaning or a buildup of contamination within process conditions. Rapid rises in TVC are often the first warning of operating failures. Action taken on the basis of rises in TVC is important to prevent the development of food safety problems. TVC values are a useful means of following overall microbial growth in a stored product over time and estimating the interval before microbial spoilage occurs (microbial shelf life).

PRACTICE

Direct microscopic count techniques 3

3.1 Introduction

The direct microscopic count (DMC) consists of smear preparation from a sample or culture on a disinfected microscope glass slide, staining the smear with a suitable stain, and counting microbial cells using a light microscope. DMC techniques give only an estimated actual number of microorganisms present in the sample. When a high microbial number is expected from food products (e.g., processed fruits and vegetables), DMC is suitable to count the microorganisms (e.g., bacteria, yeasts, molds, and hyphae).

The advantages of DMC are (1) DMC is a rapid technique for counting the microorganisms from sample; (2) film on the slide prepared from sample may be stained and read later; (3) it requires the minimum of equipment; (4) it provides information about the microbial nature of microbial cells (e.g., cells morphology and Gram stain type); and (5) the prepared slides from a sample may be kept for further research. The disadvantages of DMC are (1) only a small amount of a sample (e.g., 0.01 mL) is examined; (2) it is useful only for foods containing higher numbers of microorganisms. A microbial population of about 10^5 mL^{-1} is minimum for statistical significance; (3) food particles prepared on a slide are not always distinguishable from microbial cells; (4) microbial cells on the prepared slide cannot be uniformly distributed to single cells; (5) viable and nonviable cells can possibly be distinguished from each other with staining; (6) in staining, some viable microbial cells do not take the stain well and they may not be counted; and (7) DMCs can give a higher number of count than plate counts or other counting techniques due to the counting both of viable and nonviable microbial cells, and nonmicrobial constituents (e.g., food components).

DMC is used extensively for the analysis of milk and milk products, it is also used for other foods, for example, egg products, frozen fruits and vegetables, fruit juices and concentrates. The detection by DMC of the number of leukocytes in raw milk is an important screening technique to indicate abnormal milk. Leukocyte counts of over 500,000 mL^{-1} of milk are considered high. When bacterial infections are responsible for mastitis, *Staphylococcus* or *Streptococcus* are also seen frequently in the film.

Essentially, there are three distinct steps in the DMC technique:

1. determination of the microscope factor (mf);
2. preparation of stained film; and
3. examination of stained film under microscope.

Once the MF for an objective of a microscope is determined, the same MF can be used in future DMC or microscopic analysis. There are different DMC techniques: Breed count, membrane filter count, Neubauer count, and Howard mold count (HMC).

Microbiological Analysis of Foods and Food Processing Environments. DOI: https://doi.org/10.1016/B978-0-323-91651-6.00016-1

Sampling, sample preparation, and all experimental techniques should be performed under aseptic conditions, and in duplicate, unless otherwise specified. Again, all solutions and equipment that may cause contamination of microorganisms must be sterilized or disinfected.

3.2 Breed count technique

In the breed count technique (BCT), bacteria or yeasts are counted from the sample with the smear. Any suitable staining technique (e.g., simple) may be used to count the number of microorganisms. BCT can be used in the indication of microbiological quality of raw milk, dairy products, dried and frozen foods, and other food products. BCT is also used for counting the animal cells, predominantly leukocytes, that can be found in mastitis milk.

Before the count, MF for an objective of a microscope must be determined. Each laboratory will determine the MF for the objective of microscope using a stage micrometer. Microscope objectives used for this procedure should have a MF between 300,000 (0.206 mm field diameter) and 600,000 (0.146 mm field diameter). The average number of microorganisms obtained from microscope counts for a sample is multiplied by the objective MF. Dilution of the sample is considered in the calculation of the microbial number for the original sample, for example, food product. The MF for the objective may be specified at the time the microscope is purchased.

3.2.1 Equipment and reagents

Equipment. Cover slip, pipette (0.5 and 1.0 mL) in pipette box, incubator, microscope, microscope glass slides in a bottle containing alcohol, paper tissue, pipette discard equipment, stage micrometer, vortex, waterproof pen and wire loop (0.01 mL; 4 mm internal diameter).

Reagents. Peptone water (0.01%), alcohol (95%), Gram stains, immersion oil (with a refractive index of 1.51 to 1.52 at 20°C), Levowitz–Weber stain, LiOH solution (0.1 N), Liefson's methylene blue stain, North's aniline oil-methylene blue stain, safranine, and xylol.

3.2.2 Count from microbial culture

3.2.2.1 Determination of microscope factor

Place a stage micrometer ruled in 0.01 mm divisions (1 mm divided into 100 units of 10 μm each) on the stage of microscope. Begin by observing the stage micrometer first with the low-power objective. Add an immersion oil drop on the stage micrometer, place a cover slip on to the immersion oil and magnify the stage micrometer with the oil-immersion objective. The microscopic field diameter, the oil-immersion objective field diameter, is measured to the third decimal place in mm, for example, 0.146 mm (diameter, D). To determine the area of the field in mm using the formula (radius, $r = D/2 = 0.073$ mm) for oil-immersion objectivee:

$$\text{Area of field} = 3.1416 \times r^2 = 3.1416(0.073\ \text{mm})^2 = 0.01674\ \text{mm}^2$$

Convert mm^2 area of one field to cm^2; divide by 100 as: 0.01674/100

To calculate the number of fields in 1 cm^2 as MF:

$$MF = (1)/(0.01674/100) = 5973.72$$

or the MF can be calculated using the following formula (r in cm):

$$\text{MF in cm}^2 = (100)/(3.1416 \times r^2)$$

Count the number of microorganisms from at least 10 fields of oil-immersion objective and then find the average number (n):

$$\text{Average number of microorganisms per field } (n) = (\text{Average count})/\text{Number of counted fields})$$

When the average number of microorganisms (n) per field is multiplied by MF of oil-immersion objective, it will represent the total number of microorganisms present in 1 cm^2 of film.

$$\text{Total number of microorganisms in 1 cm}^2 = \text{MF} \times n$$

Since only 0.01 mL of sample is spread over the 1 cm^2 area of slide.

$$\text{Number of microorganisms in 1 mL of sample} = \text{MF} \times n \times 100$$

Then the number of microorganisms per mL is reported as the number of microorganisms from DMC per mL of sample.

3.2.2.2 Smear preparation and staining

Refer to Technique 1.5 for details of sample preparation and dilutions. The sample is thoroughly mixed and a necessary dilution must be prepared using 0.1% peptone water to produce smears no more than 20 microbial cells per field (10^6 cells per mL). Mix diluents on a vortex mixer or through hands. Do not exceed 15 min from the time of homogenization, preparation of all dilutions, and preparation of stained film.

Take a microscope glass slide from slide bottle, flame slide through Bunsen burner flame and put on a dust free area. Draw a circular 1 cm^2 film area (1 cm dimension) on the back side of slide using a waterproof pen. With a clean sterile pipette (0.5 mL), remove sample and wipe the exterior of the pipette with a clean paper tissue. Adjust the length of the column to the 0.5 mark. Place the tip near the center of the film area on the slide, and spread the sample over the circular 1 cm^2 area and expel the 0.01 mL of sample (drain sample from the 0.05 graduation mark down to the 0.04 mark. Do not retouch the pipette to the slide). Alternatively, use a 0.01 mL calibrated loop (4 mm internal diameter) in transferring the sample. Withdraw the loop vertically from the sample and spread over the circular 1 cm^2 area.

The smear is immediately dried in an incubator (at 40°C–45°C). Smear staining should be completed within 5 min to prevent possible bacterial multiplication, but too rapid drying can cause cracking or peeling of the slide. Perform a simple staining technique on the smear. Heat fix the slide through a Bunsen burner flame before staining. Stain with an appropriate simple stain (e.g., Liefson's methylene blue or safranine) for 1–2 min. The preparation is called Breed's smear.

3.2.2.3 Microscopic count

Preliminary analysis of the Breed's smear should be performed using the high-power objective of microscope. After finding microbial cells under this objective, rotate the oil-immersion objective halfway and place one drop of immersion oil on to the smear, place oil-immersion objective onto film and count the number of microorganisms present in the field of oil-immersion objective (Fig. 3.1). This can be accomplished by the selection of successive rows at right angles to each

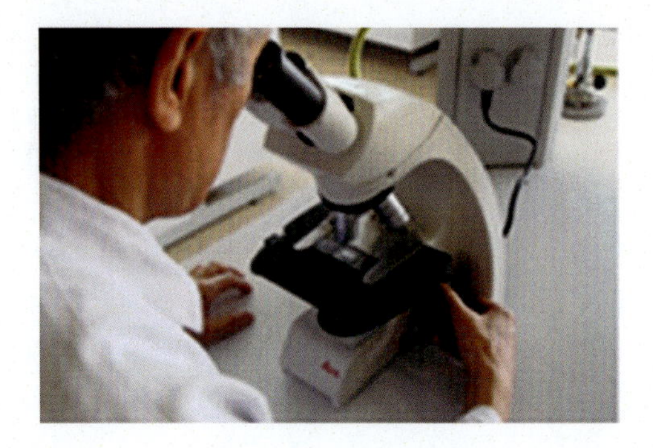

FIGURE 3.1

Microorganism count under microscope.

other across the slide, beginning midway on any side with two to three fields from the edge. Count separate fields in a series across the film. Then start midway at the top or bottom of the film and count a series of separate fields in a line perpendicular to the first series. If more fields are to be examined, repeat the selection of fields about 2 mm away from the first series. Depending on the cell density in the film, count from 10 to 100 fields (microscope fields are counted until reaching 100 or more bacterial cells). Each colony clump or chain on the slide would probably give rise to one colony only, therefore they are counted as one. Clumps or cell groups are separated by a distance equal to or greater than twice the smallest diameter of the two cells nearest to each other. A diplococcal cell form is counted as single. Observe and note the morphology of the microorganisms.

Calculate the number of microorganisms per g or mL of sample using MF of the oil-immersion objective, average number of microorganisms (n) per field, amount of sample (0.01 mL), and the reciprocal of the dilution rate (Dr).

$$\text{Direct microscopic count (DMC) per g or mL} = (\text{MF} \times n \times \text{DR})/0.01$$

Round off the count to two significant figures and report the result as the number of microorganisms from DMC per g or mL of sample.

3.2.3 Counting technique from foods

A general microscopic microbial counting technique is described here for some types of foods. Microscopic counting of microorganisms from food may involve a special preparation technique, in this case relevant food practices should be considered. Take a microscope glass slide immersed in alcohol from a bottle and pass it over a Bunsen burner flame. Draw a 1 cm^2 area under the slide. Shake the milk sample vigorously to obtain a homogeneous suspension. Using 0.5 mL pipette or a calibrated loop (4 mm internal diameter), remove 0.01 mL of the homogenized (or no homogenized liquid food) sample and place it in the center of a 1 cm^2 area on slide. In this transfer, withdraw a sample slightly above the graduation mark. Wipe the exterior of the pipette tip with clean dry paper tissue. Then, touch dry paper to the tip of pipette and absorb liquid at the tip to give exactly

0.01 mL. Place the tip of pipette to the center of the slide and expel the test portion (0.01 mL). Promptly spread the sample over the marked 1 cm^2 area using a loop or needle. Allow the film to dry in an incubator at 40°C within 5 min and protect the surface from contamination with dust. Heat fix the slide by passing it three times through a Bunsen burner flame. Submerge the dried and fixed slide into Liefson's methylene blue stain or Levowitz–Weber (LW) stain for 2 min (other stains can also be used depending on the type of microorganism or food). The slide is rinsed in water (at 40°C) by submerging into water three times, then drain the water. Excess water-stain from the slide is drained by resting the slide edge on absorbent paper, and then allow the slide to air dry by forced air if available.

After drying the film, place the slide into the microscope. Preliminary analysis of the Breed's smear should be performed using the low-power objective and then high-power objective of microscope. After finding microbial cells under objectives, rotate the oil-immersion objective halfway and place one drop of immersion-oil on the film, and then place oil-immersion objective onto film. Count microbial cells from 10 to 100 fields of microscope as indicated in Technique 3.2.2.3. Calculate the average number of cells per field and report result as DMC per g or mL of sample. Observe and note the morphology of microorganisms.

Fixing. If the whole fatty food product is being examined and defatting is impossible with direct staining, defat the smear by flooding with xylol for 1 min or less and draining. Fix the smear by immersing in alcohol (95%) for 1 min or less and dry before staining.

Precautions: Immerse the slide in stain solution in the covered container to prevent evaporation during staining. Repeated use of stain in an uncovered container may result in the precipitation of stain. The evaporation may result in precipitation of the stain on the slide, and add fresh stain to prevent drying stain during staining.

3.3 Membrane filter technique

3.3.1 Equipment and reagents

Equipment. Cover glass slip, forceps, incubator (at 35°C), membrane filter (pore size 0.22 μm), membrane filter apparatus, microscope, microscope glass slides in bottle containing alcohol, Neubauer improved counting slide, nutrient pads (47 mm diameter), Pasteur pipette, Petri dishes, pipette (10 mL) in pipette box, pipette discard equipment, vacuum source, and wrap paper.

Reagents. Basic fuchsin, 0.4% erythrosine stain solution (or 0.4% tryptone blue solution), Immersion oil, Liefson's methylene blue stain, and sterile distilled water.

3.3.2 Filtration and counting

The sample is filtered as indicated in Technique 5.2.2. After filtration of 100 mL sample, the vacuum pump should be turned off and the membrane filter isn't removed from the filtration device. Sufficient staining solution (e.g., Liefson's methylene blue stain) is poured onto the membrane filter up to complete covering. Leave the stain for 5 min and then the vacuum pump should be started once more to remove the stain residues from the filter. Vacuuming is continued by adding distilled water until a colorless membrane filter is obtained. The membrane filter is then removed and left to air dry. After the membrane filter is completely dry, the filter is lubricated by placing the membrane filter in a Petri dish containing a small amount of oil. This procedure makes the membrane filter transparent. The membrane filter is removed from the Petri dish and placed on a microscope glass slide (if the filter is large, the filter is first divided into 5 cm pieces and placed on the slide) and examined under the oil-immersion objective of the microscope. Randomly select 10 fields under the microscope and count all cells within each field. Calculate the number of microorganisms in the sample considering the membrane filter area (in cm^2) and the amount of sample passed through the filter. Report the results as the number of microorganisms per g or mL of sample.

Calculation of number of fields on membrane filter area: The diameter of the circle membrane filter will be 18.00 mm. The oil-immersion objective field diameter is measured to the third decimal place, for example, 0.146 mm. The following calculations will be performed:

$$\text{Membrane filter area} = 3.14(9.00)^2 = 254.47 \text{ mm}^2$$

$$\text{Microscope field area} = 3.1416 \times r^2 = 3.1416 \times (0.073 \text{ mm})^2 = 0.01674 \text{ mm}^2$$

$$\text{Number of fields area on membrane filter: } 254.47 \text{ mm}^2/0.01674 \text{ mm}^2 = 15,201$$

$$\text{Microbial number per mL of sample} = (\text{Average count per field}) \ (15201 \text{ fields})/100 \text{ mL sample}$$

3.4 Microbial count using counting slide

The device used for cell counting under microscope is called a counting chamber. Various counting chambers are available for microbial counts. Each consists of a grid of squares of a given area. Only the Neubauer improved counting slide is included here.

3.4.1 Microbial count with Neubauer improved counting slide

3.4.1.1 Equipment and reagents

Equipment. Loop (4 mm in diameter), Neubauer improved counting slide, cover glass slip, Pasteur pipette, 0.4% trypan blue (or erythrosine) stain, physiological saline solution (PSS), and test tubes (10 × 75 mm).

Reagents. 0.1% peptone water (containing 0.1% lauryl sulfate), phosphate saline (PS) solution, Liefson's methylene blue stain, basic fuchsin (in ethanol solution), trypan blue (0.4% aq.) Erythrosine stain (0.4% aq.), and alcohol.

3.4.1.2 Neubauer improved counting slide

This slide not only counts microbial cells in the sample, but also the percentage of living cells can be determined by staining. It has two polished counting areas (Fig. 3.2A,B); each area displays a precisely ruled and subdivided grid (Fig. 3.2C). The grid on the chamber consists of 25 primary small squares, each of small square measuring 1 mm on each side (1 mm^2 area) and is limited by three closely spaced lines (2.5 μm apart). These three lines are used to differentiate the cells lying within or outside the grid. Each of the small squares is further divided to 16 smallest squares of width 0.2 mm. At both sides of each polished slide surface, there are beveled edges (V-shaped

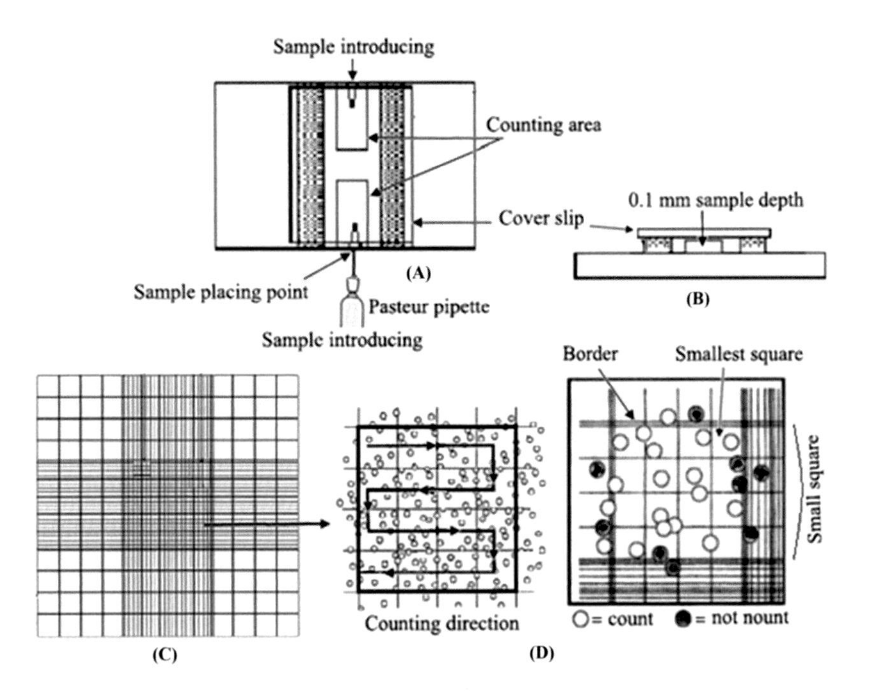

FIGURE 3.2

Neubauer improved counting slide: (A) introducing the sample; (B) front view; (C) Neubauer improved counting slide area; and (D) counting a small square.

wells) for sample placing, where the sample is introduced to be drawn across the grid by capillary action.

1. *Sample preparation*

   Refer to Technique 1.5 for details of sample preparation and dilution. Dilute the samples if necessary. Disinfect the Neubauer improved counting glass slide and cover slip by immersing into alcohol and flaming. The cover glass slip is placed over the counting surface of chamber prior to placing sample (Fig. 3.2A,B). The sample is placed into the V-shaped wells slide on both sides with a sterile Pasteur or other pipette (Fig. 3.2A). The area under the cover slip fills with the sample by capillary action. Enough liquid sample should be introduced into the wells to cover the surface area under the cover slip. Carefully touch the well surface of the slide with pipette tip and do not overfill or underfill the well of the slide. Allow the sample to remain for 5 min to fill the area under the cover slip with the test suspension by capillary action and settle the cells. The counting slide is then placed on the microscope stage and a grid on the counting slide is brought into focus at a low-power objective. Each small square on the counting slide has a surface area of 1 mm^2 and the depth of the counting slide is 0.1 mm.

2. *Counting*

   Count enough squares up to about 600 total cells in the counting direction (Fig. 3.2D). The cell motion should be minimal during count.

Count the cells from the 16 smallest squares (in one small square) using the following steps:

1. Cells touching the upper or left borders should be included in the count, cells touching the lower or right borders should be excluded from the count (Fig. 3.2D).
2. If more than 10% of the particles contain cell aggregation, the sample suspension should be made homogeneous and prepared again.
3. If less than 200 or more than 500 microbial cells are observed in 10 squares, an appropriate dilution can be made from the original sample and the appropriate count can be repeated. A dilution buffer or dye may be used in the dilution of sample. The diluent to be used must be isotonic (e.g., 0.85% NaCl solution).
4. For accuracy and repeatability of the count, counts should be done in the same way every time.
5. Perform the same count from second counting grid.
6. The Neubauer improved counting slide volume (25 small squares) is 0.1 mm^3 ($01 \times 1 \times 0.1$ mm) or 10^{-1} mm^3 or 10^{-4} mL (cm^3). Find the average microbial count per small square (n). Total microbial number per mL in the sample is then calculated as:

$$\text{Number of microorganisms per mL of sample} = (n \times \text{dilution rate})/10^{-4}\ \text{mL}$$

3. *Counting viable and nonviable cells*

Living (viable) microbial cells are stained and can decolorize the stain, whereas dead (nonviable) cells are stained but cannot decolorize. Trypan blue or erythrosine stain (0.4%) solution is used in the differentiation of viable and nonviable microbial cells. The cell suspension is prepared in the PS solution. In this technique, 0.5 mL of erythrosine stain (or trypan blue stain solution), 0.3 mL of PS solution, and 0.2 mL of the cell suspension (dilution factor = 5) are added in that order into a sterile test tube and mixed thoroughly. The mixture is allowed to stand for 5–10 min.

Note. If microbial cells are exposed to stain longer than 15 min or extended periods, viable cells may begin to take up stain.

After that time, a disinfected cover slip is replaced over the counting slide. A small amount of stain cell mixture is introduced into the counting slide well as indicated above. Count the stained (nonviable) and nonstained (viable) cells separately. Nonviable cells will stain blue and viable cells will not stain (decolorize).

The count is repeated to ensure accuracy from a second sample. Calculate the average viable and nonviable cell numbers.

$$\text{DMC of viable (or nonviable) cells per mL of sample} = (n \times \text{dilution factor})/10^{-4}\ \text{mL}$$

$$\text{Total cells per mL} = \text{viable cells} + \text{nonviable cells.}$$

$$\text{Cell viability } \% = (\text{total viable cells}/\text{total cells}) \times 100.$$

3.5 Howard mold count technique

The HMC technique determines the percentage of microscopic fields containing hyphae of mold. HMC magnification shows the presence or absence of mold filaments, but this technique cannot be used to differentiate mold genera or species. The HMC technique is applicable to indicate the presence or absence of mold in different fruit, vegetable, and tomato products (canned, pizza, juice, catsup, sauce, paste, sauce, soup, etc.), frozen berries, cranberry, pineapple juices, fruit nectars and purees, etc. HMC analysis (1) can give an indication of the extent of mold spoilage of food and (2) indicates the presence of hyphae in raw material where the raw material has not been properly used in the food production.

The HMC slide is a 76 × 35 mm (diameter × depth) glass slide with a central circular part (Fig. 3.3A) and is used for counting mold hyphae and mold spores from different types of foods, especially from tomato products. The HMC slide has 25 calibrated fields of 1.382 mm diameter through which to view the mold filaments or spores. The cover slip is supported on the shoulders and leaves a depth of 0.1 mm between the underside of the cover slip and the plane surface.

An appropriate stain (e.g., erythrosine, Liefson's methylene blue, or basic fuchsine) may be used to magnify the hyphae.

$$\text{Area of circle on HMC slide} = pr^2 = \left(3.141 \times r^2\right) = (3.1.41)\ (0.691)^2 = 1.5\ \text{mm}^2$$

$$\text{Volume of sample in circle} = \text{Area} \times \text{Height} = 1.5\ \text{mm}^2 \times 0.1\ \text{mm} = 0.15\ \text{mm}^3$$

$$\text{Volume of sample in circle} = 1.5 \times 10^{-4}\ \text{cm}^3\ \text{or mL}$$

$$\text{Volume of sample per field} = 1.5 \times 10^{-4}\ \text{mL}/25\ \text{fields} = 6.0 \times 10^{-6}\ \text{mL}$$

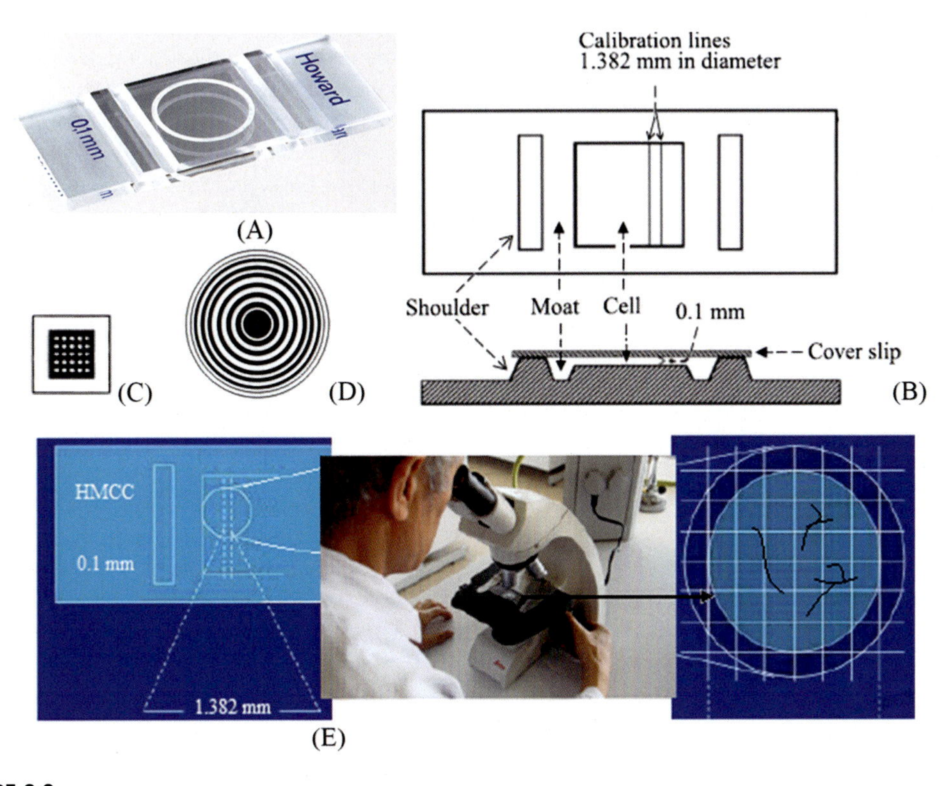

FIGURE 3.3

Howard mold count (HMC) slide (A), (B), cover glass (C), Newton's rings (D); and microscopic magnification of hyphae from HMC slide (E).

3.5.1 Equipment and reagents

Equipment. Erlenmeyer flask, high-speed blender (rotating from 10,000 to 12,000 rpm), a Warring blender (rotating from 3000 to 3500 rpm), HMC slide, cover glass slip, Microscope, refractometer, scalpel, and loop.

Reagents and water. Capryl alcohol, 0.5% sodium carboxymethylcellulose (stabilizer solution), sterile distilled water, NaOH (0.1N), and HCl.

3.5.2 Sample preparation for Howard mold count

1. *Apple pomace.* Weigh 200 g of apple pomace in a 1500 mL sterile beaker and add 1.0 L of distilled water. Heat to boil and allow to remain for 2 h. A sufficient amount of sterile distilled water is added up to the original volume and mixed thoroughly. Add 50 mL of 3% pectin or another stabilizer solution to a 250 mL beaker and add 50 g of homogenized pomace. Mix thoroughly and undertake a hyphae count.
2. *Ground capsicums (red pepper, etc.) and garlic powder.* Weigh 10 g of a thoroughly mixed sample of ground capsicum or garlic powder and transfer to a high-speed sterile blender. Add 200 mL of 1.0% NaOH solution in three or four successive portions, after each addition, blend for 1 min and wash down any material sticking to the blender wall. Add two or three drops of capryl alcohol to break foam and mix. Mix 100 g of this mixture with 50 mL of stabilizer solution and undertake a hyphae count.
3. *Citrus juice.* Pour the can contents into a sterile Erlenmeyer flask, and mix thoroughly by gently shaking. After mixing, transfer 50 mL of juice into sterile centrifuge tubes. Centrifuge for 10 min at 2200 rpm. Remove the tube and decant the supernatant without disturbing the sediment. Add water to the centrifuge tube, bring the volumes to the 10 mL mark, and add 5 mL stabilizer solution. Thoroughly mix the sediment in water and stabilizer solution, and pour into a small sterile beaker. Mix thoroughly by pouring back and forth between two beakers for six or more times, and undertake a hyphae count.
4. *Citrus juice concentrates.* Make a hyphae count on a well-mixed sample. If a high number of hyphae are present or juice concentration does not allow the examination, dilute the sample using sterile distilled water.
5. *Cranberry.* (1) Can of sauce. Sauce in the unopened can is placed into a boiling water bath for 30–40 min to easily break the gel. Then the can is removed from the water bath, the opening side of can is disinfected and aseptically opened carefully to avoid the loss of sauce through a sudden release of pressure. Transfer the content into a beaker. Stir the sauce to break the gel (in a slow speed Warring blender at 350–450 rpm). Mix 50 g of the stirred sauce with 50 mL stabilizer solution and undertake a hyphae count. (2) Whole cranberry (including seeds and skins). Remove skin and seeds, and prepare homogenous pulp. Mix 50 g of this pulp with 50 mL stabilizer solution and undertake a hyphae count.
6. *Grape pulp, heat processed.* Weigh 100 g of well-mixed pulp in a sterile blender. Add 100 mL of 0.5% NaOH solution and mix for 3 min. If necessary, break the foam by adding two or three drops of capryl alcohol and stir. Add 20 g of blended pulp to 20 mL of a stabilizer solution. Mix thoroughly and undertake a hyphae count.

7. *Grape, steamed and crushed (not heat processed).* Mix the pulp sample thoroughly and make the hyphae count.
8. *Infant food, pureed.* Mix 6 g of sample in 100 mL of NaOH solution (0.5%) and mix thoroughly by gently shaking. Make the hyphae count.
9. *Jams and preserves.* Pulp the entire sample. (1) Weigh 50 g of 3% pectin or another stabilizer solution in a sterile beaker. Add 50 g of the pulped jam or preserves to a beaker and shake gently until most of the air bubbles are removed (within 10 min). Make the hyphae count. (2) Add approximately 100 g of pulped sample into a sterile 250 mL suction flask. Heat the flask gently under vacuum until most of the air bubbles are removed from the sample. Mix thoroughly and make the hyphae count.
10. *Pineapple.* (1) Juice. Proceed as for citrus juice, but after decanting the supernatant in the centrifuge tube, add 0.5 mL of concentrated HCl (37%) to dissolve the oxalate crystals. (2) Concentrate. Make the hyphae count on a well-mixed sample without dilution if possible. (3) Crushed. Drain the pineapple and make the hyphae count.
11. *Tomato.* (1) Catsup. Place 50 mL of stabilizer solution into a 100 mL beaker, add 50 mL of well-mixed sample and mix thoroughly. Make the mold count. (2) Canned. Drain juice from contents of can and pass the drained tomatoes through a laboratory cyclone. Make hyphae counts on both drained juice and pulped tomatoes. (3) Juice. Mix the juice thoroughly and make the hyphae count on the undiluted sample. (4) Powder. Weigh 11 g of thoroughly mixed sample into a high-speed blender containing 150 mL of sterile distilled water to produce a mixture equivalent to tomato juice, or use 17 g of tomato powder and 150 mL of sterile distilled water to obtain a mixture equivalent to a tomato puree. Blend mixture for 30 sec and rub down any material adhering to walls. Rinse walls with 50 mL of water and blend again for 1 min. Add two drops of capryl alcohol to break foam and make the hyphae count. (5) Puree and paste. Add water to make a mixture having a tomato soluble solid to obtain a refractive index of the filtered liquid portion (1.3448–1.3454°B at 20°C). Make the hyphae count. (6) Sauce. Mix thoroughly and make the hyphae count on the undiluted sample. (7) Soup. Place an unopened can in hot water and heat until the contents are thoroughly warmed, then open. Transfer 10 mL of thoroughly mixed soup to a 50 mL centrifuge tube and add 3 mL of NaOH solution, if starch is absent, omit the NaOH. Stir until the starch dissolves and tissues clear. Add enough water to fill the tube and centrifuge. Pour supernatant and add enough sterile water to the residue in the tube to bring it to the original volume of soap, mix and make the hyphae count.

3.5.3 Counting

Clean the HMC slide and cover slip by washing with dipping in a solution of soapy water or alcohol. Place cover slip over the circular part of the HMC slide to observe Newton's rings between the HMC slide and cover slip (Fig. 3.2D). Remove the cover slip with a scalpel. If necessary, add water into the sample to give a total solid content in the refractometer reading of 45.0–48.7 at 20°C (or a refractive index of 1.3448–1.3454°B). Place a portion (4–5 loopful) of well-mixed sample over the Howard cell with loop, and spread the sample over the calibrated fields of slide using the loop. Carefully place the cover slip onto the slide, lowering from one side to exclude air bubbles. The cover slip is placed at an approximately 45 degree angle with one edge resting on the chamber, then lower the glass in place on the HMC slide. Lowering the cover slip too fast will splash the sample onto the slide; and lowering the cover slip too slowly may lead to uneven distribution of the sample on the slide. Place the HMC slide on the microscope.

Analyze at least 25 fields from each of two or more preparations using a low-power objective (Fig. 3.2E). While the presence (positive) of mold hyphae is observed, record the results as positive field when the aggregate length of not more than three hyphae exceeds 1/6 of the diameter of the field. If there is enough mold hyphae in the field, it is recorded as positive. If there are not enough mold hyphae, then all parts of the field should be carefully examined. Careful examination includes varying the intensity of the light occasionally and continuous use of the fine adjustment on the microscope. In some cases, a 20 × magnification (20 × objective) may be necessary to positively identify mold hyphae. Then return to 10 × magnification and magnify. Calculate the proportion of positive fields from the results of the analysis of all observed fields (F) using the total hyphae count (T) and report the result as the percent of positive fields containing mold hyphae.

$$\text{Number hyphae per mL of sample} = (\text{T}/\text{F}) \times 100$$

Calculate the number of mold hyphae per mL of sample by multiplying average hyphae count (n) per field by volume of sample per field of HMC slide (6.0×10^{-6} mL):

$$\text{Number of mold hyphae per mL} = n \times 6.0 \times 10^{-6}\ \text{mL}$$

3.6 Interpretation of results

There is a sufficiently close relation between the mold count and the amount of decay tissue in a fruit or vegetable product that the hyphae count may be used as an index for determining whether good manufacturing technique has been followed or not. The US Food and Drug Administration has set a maximum limit of 40% "positive" fields for tomato paste, and 20% "positive" fields for tomato juice. There is not any good correlation between the amounts of spoilage of the product and organoleptic changes, since the amount of spoilage will depend on the mold species, its metabolic activity and physiological ability. DMC is a rapid, simple, and easy technique requiring the minimum of equipment. The morphology of the bacteria can be observed as they are counted through staining. Very dense suspensions can be counted if they are diluted appropriately.

PRACTICE

Most probable number technique

4

4.1 Introduction

The most probable number (MPN) technique is used to detect the number of microorganisms from foods, especially when the number of microorganisms is less than 10 g or mL of sample. Foods of different quantities (e.g., 100, 10 or 1 g or mL) can be used in the MPN technique. This is statistical and the results usually yield a higher number of microorganisms than plate count techniques. In this technique, the liquid medium in multiple tubes (usually three or five) is used for each of the diluted (or nondiluted) samples. After the sample is inoculated, tubes are incubated at a suitable temperature. After incubation, the number of positive tubes is determined for each dilution by looking at the growth (turbidity) and gas formation in the tubes. The MPN value is obtained from the MPN table and the number of microorganisms is calculated from the formula to be used for this technique. Selective media can be used for specific microorganisms. This technique can be used to count the total number of microorganisms, *Salmonella*, *Staphylococcus*, *Enterococcus*, *Vibrio parahaemolyticus*, *Escherichia coli*, coliforms, fecal coliforms, and others. Only living microorganisms are counted from the sample in the MPN technique. The MPN technique predicts growth units (GUs) instead of individual bacteria. The sample to be analyzed in the MPN technique is always diluted to obtain a tube without microbial growth. The growth of microorganisms in dilution provides an estimate of the number of microorganisms in the undiluted sample.

MPN results obtained in a laboratory: (1) are more likely to be the same as those in another laboratory than those obtained by a standard plate count; (2) total microbial number is indicated; (3) specific microbial groups can be indicated using the appropriate selective and differential media. The disadvantages of the MPN technique are (1) the use of a large number of glassware or materials; (2) the lack of opportunity for morphological observation of microorganisms; (3) the lack of sensitivity in determining the exact number of microorganisms; and (4) the inhibitory agent(s) that may be present in the medium prevent the growth of microorganisms.

Confidence intervals. The MPN tables have a 95% confidence interval which means that before the tubes are inoculated, the probability of the result-related confidence interval to cover the actual concentration is at least 95%.

4.2 Most probable number count

Sampling, sample preparation, and all experimental techniques should be performed under aseptic conditions, and in duplicate, unless otherwise specified. All microbiological media to be used in applications must be sterilized. Again, all solutions and equipment that may cause contamination of microorganisms must be sterilized or disinfected.

Microbiological Analysis of Foods and Food Processing Environments. DOI: https://doi.org/10.1016/B978-0-323-91651-6.00042-2

4.2.1 Equipment, reagents, and media

Equipment. Test tube racks, Loop (4 mm), microscope, microscope glass slide, pipettes (1 and 10 mL) in pipette box, incubator, blender (or homogenizer) and pipette discard equipment (e.g., a graduated cylinder) containing a disinfectant and a piece of cloth at bottom.

Reagents. 0.1% Peptone water.

Media. Lauryl sulfatetryptose (LST) broth containing inverted Durham tube, Brilliant green lactose bile (BGLB) broth containing inverted Durham tube and EC broth containing inverted Durham tube, Eosin methylene blue (EMB) agar, Brain heard infusion (BHI) agar slants, and BHI broth.

4.2.2 Presumptive test

Refer to Technique 1.4 for details of sample preparation and dilution. The ratio of inoculating sample to medium should be one part of the sample to 10parts of the medium (e.g., 1 g or mL of sample into 10 mL of medium or 10 g or mL of sample into 100 mL of medium).

In this part, an outline procedure for a coliforms test is performed as an example of the MPN technique. Aseptically, pipette 1 mL of sample (diluted or nondiluted) into each of three or five MPN tubes of LST broth. Incubate tubes at 35°C, and examine tubes after 24 and 48 h to predict positive tubes. LST broth tubes showing gas production in Durham tube (and acid with yellow color) with growth (turbidity) indicate the positive tubes for a presumptive test. If very low numbers of microorganisms are expected, 10 mL or 100 mL of sample may be added into equal volumes of double-strength LST broth.

Detection of positive tubes (Fig.13.2)

1. *Turbidity.* The development of turbidity after incubation indicates the growth of microorganisms.
2. *Gas production.* A positive result is recorded when gas bubbles appear as space in the inverted Durham tube at the end of the incubation.
3. *Acid production.* Acid or base production can be detected by a medium containing a pH indicator dye that gives a color change with changes in pH. Electron acceptors or a pH indicator (e.g., resazurin, methylene blue or triphenyl-tetrazolium chloride) can be incorporated into the medium. Reduction of the pH indicator can cause a color change with the production of acids in the medium (yellow color).

Record all presumptive positive tubes for each dilution to indicate coliform numbers. Perform confirmed tests on all presumptive positive tubes.

4.2.3 Confirmed test

All positive tubes showing gas within 48 h at 35°C should be used in the confirmed test. From each presumptive positive tube, inoculate a loopful of culture into a tube of BGLB broth. Incubate the inverted BGLB broth tubes and plates at 35°C for 24 and 48 h. After incubation, the number of BGLB broth tubes showing positive results (gas and/or acid production) is recorded. All confirmed positive tubes must be followed with a completed test. Positive BGLB broth tubes indicate positive correlation for coliforms of the respective presumptive positive tubes.

4.2.4 Completed test

From each gassing BGLB broth tube, inoculate a loopful of culture to PC agar slant and incubate at 35°C for 24 h. After incubation, prepare a Gram stain as explained in Appendix B under the topic "Gram Staining Reagents" from PC agar slant culture. Examine for the presence of Gram-negative, nonspore-forming rods.

The formation of gas in LST broth and BGLB broth, and the demonstration of Gram-negative, nonspore-forming rods constitute a positive test for coliforms for the sample. Predict the number of positive LST broth tube(s) and refer to the MPN Table 4.1 for three tubes (or MPN Table 4.2 for five tubes) to read the MPN number. Calculate the number of MPN using the formula (Technique 4.2.5). Report the results as coliforms MPN per g or mL of sample. The absence of gas and the presence or absence of rods constitute a negative completed test and also negative presumptive test.

Table 4.1 Most probable number estimates for three dilutions in three-tube MPN analysis (in g or mL sample) and 95% confidence intervals.

Positive tube[a]			MPN estimate[b]	Conf. limit		Positive tube[a]			MPN estimate[b]	Conf. limit	
0.10	0.01	0.001		Low	High	0.10	0.01	0.001		Low	High
0	0	0	*<3.0*	–	9.5	2	2	0	*21*	4.5	42
0	0	1	*3.0*	0.15	9.6	2	2	1	*28*	8.7	94
0	1	0	*3.0*	0.15	11	2	2	2	*35*	8.7	94
0	1	1	*6.1*	1.2	18	2	3	0	*29*	8.7	94
0	2	0	*6.2*	1.2	18	2	3	1	*36*	8.7	94
0	3	0	*9.4*	3.6	38	3	0	0	*23*	4.6	94
1	0	0	*3.6*	0.17	18	3	0	1	*38*	8.7	110
1	0	1	*7.2*	1.3	18	3	0	2	*64*	17	180
1	0	2	*11*	3.6	38	3	1	0	*43*	9	180
1	1	0	*7.4*	1.3	20	3	1	1	*75*	17	200
1	1	1	*11*	3.6	38	3	1	2	*120*	37	420
1	2	0	*11*	3.6	42	3	1	3	*160*	40	420
1	2	1	*15*	4.5	42	3	2	0	*93*	18	420
1	3	0	*16*	4.5	42	3	2	1	*150*	37	420
2	0	0	*9.2*	1.4	38	3	2	2	*210*	40	430
2	0	1	*14*	3.6	42	3	2	3	*290*	90	1,000
2	0	2	*20*	4.5	42	3	3	0	*240*	42	1,000
2	1	0	*15*	3.7	42	3	3	1	*460*	90	2,000
2	1	1	*20*	4.5	42	3	3	2	*1100*	180	4100
2	1	2	*27*	8.7	94	3	3	3	*>1100*	420	>4100

[a]*Number of positive tube(s) per three tubes.*
[b]*MPN per g or mL of sample.*

Table 4.2 Most probable number (MPN) estimates for three dilutions in five-tube MPN analysis (in g or mL sample).

Positive tube[a]			Most probable number (MPN) estimate[b]	Positive tube			MPN estimate	Positive tube			MPN estimate
0.1	0.01	0.001		0.1	0.01	0.001		0.1	0.01	0.001	
0	0	0	0	1	0	0	2	2	0	0	4.5
0	0	1	1.8	1	0	1	4	2	0	1	6.8
0	0	2	3.6	1	0	2	6	2	0	2	9.1
0	0	3	5.4	1	0	3	8	2	0	3	12
0	0	4	7.2	1	0	4	10	2	0	4	14
0	1	5	9	1	0	5	12	2	0	5	16
0	1	0	1.8	1	1	0	4	2	1	0	6.8
0	1	1	3.6	1	1	1	6.1	2	1	1	9.2
0	1	2	5.5	1	1	2	8.1	2	1	2	12
0	1	3	7.3	1	1	3	10	2	1	3	14
0	1	4	9.1	1	1	4	12	2	1	4	17
0	2	5	11	1	1	5	14	2	1	5	19
0	2	0	3.7	1	2	0	6.1	2	2	0	9.3
0	2	1	5.5	1	2	1	8.2	2	2	1	12
0	2	2	7.4	1	2	2	10	2	2	2	14
0	2	3	9.2	1	2	3	12	2	2	3	17
0	2	4	11	1	2	4	15	2	2	4	19
0	3	5	13	1	2	5	17	2	2	5	22
0	3	0	5.6	1	3	0	8.3	2	3	0	12
0	3	1	7.4	1	3	1	10	2	3	1	14
0	3	2	9.3	1	3	2	13	2	3	2	17
0	3	3	11	1	3	3	15	2	3	3	20
0	3	4	13	1	3	4	17	2	3	4	22
0	4	5	15	1	3	5	19	2	3	5	25
0	4	0	7.5	1	4	0	11	2	4	0	15
0	4	1	9.4	1	4	1	13	2	4	1	17
0	4	2	11	1	4	2	15	2	4	2	20
0	4	3	13	1	4	3	17	2	4	3	23
0	4	4	15	1	4	4	19	2	4	4	25
0	5	5	17	1	4	5	22	2	4	5	28
0	5	0	9.4	1	5	0	13	2	5	0	17
0	5	1	11	1	5	1	15	2	5	1	20
0	5	2	13	1	5	2	17	2	5	2	23
0	5	3	15	1	5	3	19	2	5	3	26
0	5	4	17	1	5	4	22	2	5	4	29
0	5	5	19	1	5	5	24	2	5	5	32

Table 4.2 Most probable number (MPN) estimates for three dilutions in five-tube MPN analysis (in g or mL sample). ***Continued***

Positive tube[a]			Most probable number (MPN) estimate[b]	Positive tube			MPN estimate	Positive tube			MPN estimate
0.1	0.01	0.001		0.1	0.01	0.001		0.1	0.01	0.001	
3	0	0	*7.8*	4	0	0	*13*	5	0	0	*23*
3	0	1	*11*	4	0	1	*17*	5	0	1	*31*
3	0	2	*13*	4	0	2	*21*	5	0	2	*43*
3	0	3	*16*	4	0	3	*25*	5	0	3	*58*
3	0	4	*20*	4	0	4	*30*	5	0	4	*76*
3	1	5	*23*	4	0	5	*36*	5	0	5	*95*
3	1	0	*11*	4	1	0	*17*	5	1	0	*33*
3	1	1	*14*	4	1	1	*21*	5	1	1	*46*
3	1	2	*17*	4	1	2	*26*	5	1	2	*64*
3	1	3	*20*	4	1	3	*31*	5	1	3	*84*
3	1	4	*23*	4	1	4	*36*	5	1	4	*110*
3	2	5	*27*	4	1	5	*42*	5	1	5	*130*
3	2	0	*14*	4	2	0	*22*	5	2	0	*49*
3	2	1	*17*	4	2	1	*26*	5	2	1	*70*
3	2	2	*20*	4	2	2	*32*	5	2	2	*95*
3	2	3	*24*	4	2	3	*38*	5	2	3	*120*
3	2	4	*27*	4	2	4	*44*	5	2	4	*150*
3	3	5	*31*	4	2	5	*50*	5	2	5	*180*
3	3	0	*17*	4	3	0	*27*	5	3	0	*79*
3	3	1	*21*	4	3	1	*33*	5	3	1	*110*
3	3	2	*24*	4	3	2	*39*	5	3	2	*140*
3	3	3	*28*	4	3	3	*45*	5	3	3	*180*
3	3	4	*31*	4	3	4	*52*	5	3	4	*210*
3	4	5	*35*	4	3	5	*59*	5	3	5	*250*
3	4	0	*21*	4	4	0	*34*	5	4	0	*130*
3	4	1	*24*	4	4	1	*40*	5	4	1	*170*
3	4	2	*28*	4	4	2	*47*	5	4	2	*220*
3	4	3	*32*	4	4	3	*54*	5	4	3	*280*
3	4	4	*36*	4	4	4	*62*	5	4	4	*350*
3	5	5	*40*	4	4	5	*69*	5	4	5	*430*
3	5	0	*25*	4	5	0	*41*	5	5	0	*240*
3	5	1	*29*	4	5	1	*48*	5	5	1	*350*
3	5	2	*32*	4	5	2	*56*	5	5	2	*540*
3	5	3	*37*	4	5	3	*64*	5	5	3	*920*
3	5	4	*41*	4	5	4	*72*	5	5	4	*1600*
3	5	5	*45*	4	5	5	*81*	5	5	5	*>2400*

[a]*Number of positive tubes per five tubes.*
[b]*MPN per g or mL of sample.*

4.2.5 Most probable number result

MPN Tables 4.1 and 4.2 show the MPN of microorganisms for three and five tubes, respectively, corresponding to the frequency of positive tubes. The positive tubes should be determined from each of the three selected subsequent dilutions. To calculate the MPN per g or mL of the sample use the following formula:

$$\text{MPN per g or mL of sample} = (\text{MPN from Table}/100) \times \text{middle tube dilution rate}$$

Example: (1) Results of a three-tube MPN determination showing three positive 0.1 g or mL sample (10^{-1}), two positive 0.01 g or mL sample (10^{-2}), and one positive 0.001 g or mL sample (10^{-3}) are read (3–2–1) from Table 4.1 as 150. It is multiplied by the middle tube dilution rate (here 10^2 for 10^{-2}dilution) and divided by 100 to find 150 as coliforms MPN per g or mL of sample. (2) If these results will be obtained for other dilutions: determination showing three positive 0.01 g or mL sample (10^{-2}), two positive 0.001 g or mL sample (10^{-3}), and one positive 0.0001 g or mL sample (10^{-4}) are read (3–2–1) from Table 4.1 as 150. It is multiplied by the middle tube dilution rate (here 10^3 for 10^{-3}dilution) and divided by 100 to find 1500 as coliforms MPN per g or mL of sample. Other examples in the calculation of MPN number are given in MPN Tables 4.3 or 4.4 for three or five tubes respectively.

4.3 Interpretation of results

The US Food and Drug Administration (FDA) and AOAC present their preferred MPN technique due to the statistical interpretation of serial dilution tests for microorganisms. This technique includes the adjustments chosen to create flexible techniques for calculating MPNs, confidence

Table 4.3 Examples to determine most probable number estimation from three tubes.

Examples	Sample dilution rate					Number of positive tubes	Most probable number (MPN) estimate (per g or mL)
	0.1	0.01	0.001	0.0001	0.00001		
a	3[a]	2	1	1	0	2-1-1	*280*
b	3	3	2	2	1	2-2-1	*2100*
c	3	3	2	0	0	3-2-0	*930*
d	3	3	3	2	0	3-2-0	*9300*
e	2	3	1	0	0	3-1-0	*430*
f	3	2	2	1	0	2-2-1	*280*
g	3	3	2	0	1	2-2-1	*1500*
h	3	3	2	1	1	3-2-2	*2100*
i	3	3	3	3	3	3-3-3	*> 110,000*
j	3	3	3	3	1	3-3-1	*46,000*
k	0	0	1	0	0	0-0-1	*3*

[a] *No. of positive tubes.*

Table 4.4 Examples to determine most probable number estimation from five tubes.

Examples	Sample dilution rate					Number of positive tubes	MPN estimate (per g or mL)
	0.1	0.01	0.001	0.0001	0.00001		
a	5[a]	5	2	0	0	5-2-0	**490**
b	5	5	5	2	0	5-2-0	**4900**
c	5	4	5	2	0	5-2-0	**4900**
d	5	4	4	1	0	4-4-1	**400**
e	5	5	3	1	1	5-3-2	**1400**
f	5	4	4	0	1	4-4-1	**400**
g	5	5	5	5	5	5-5-5	**>240,000**
h	5	5	5	5	2	5-5-2	**54,000**
i	0	0	1	0	0	0-0-1	**1.8**
j	4	4	1	1	0	4-4-2	**47**

[a]*No. of positive tubes.*

intervals, and measures of improbability for dilutions. The technique is easy to use, and is offered free in the interest of promoting the microbial safety of food and water. MPN is used to estimate the number of viable microorganisms in a sample by means of replicated liquid broth growth in 10-fold dilutions. It is commonly used in estimating microbial populations in soils, waters, agricultural products, and is particularly useful with samples that contain particulate material that interferes with plate count. MPN is most commonly applied for quality testing of water. The summary of MPN application is(1) ease of interpretation, either by observation or gas emission; (2) toxins in samples are diluted; (3) effective technique of analyzing highly turbid samples such as sediments, sludge, mud, etc.; (4) MPN analysis is a statistical technique based on the random dispersion of microorganisms per volume in a sample; (5) a volume of water is added to a series of tube containing a liquid indicator substance in growth medium; (6) the media receiving one or more indicator bacteria show growth and a characteristic color change; and (7) from the number and distribution of positive and negative reactions, the MPN of indicator microorganisms in the sample may be estimated by reference to statistical tables.

PRACTICE

Membrane filter techniques

5

5.1 Introduction

Membrane filter (MF) disks consist of cellulose nitrate, cellulose acetate, or esters of mixed cellulose. MFs are available with a pore size ranging from 10 nm to 8 μm or more. Bacteriological MFs with a pore size of 0.22–0.47 μm will retain microbial cells. A few bacteria with a very small cell diameter require a filter with a pore size of 0.22 μm. In membrane filtration, liquid foods, water, or solutions are filtered by passing through MF under vacuum. Microbial cells retained on the MF surface can be analyzed for microorganisms by staining, culturing on selective/differential media, or culturing on an absorbent pad saturated with liquid medium. The membrane capillary pores transfer the nutrients for microorganisms. In the case of selective agar media, their suitability for use with MFs should be determined for different diffusion rates of the selective agent. MFs can be used in the microbiological analyses of sugar solutions, air, water, beverages, milk, and other dairy products. But fat globules in the milk should be first broken down by a wetting agent, for example, iso-octylphenoxypolyethoxyethanol (Triton X-100). MFs are also used for a variety of other purposes, for example, the sterilization of fluids and gases; the sterility testing of sterile liquids; and the separation of viruses from bacteria.

Microorganisms will form colonies on MF placed on agar medium or absorbent pad after incubation. The standard media for the MF technique contains different concentrations of nutrients. The MF has the advantage that small numbers of microorganisms can be detected from large amounts of a sample. A microscopic examination can also be carried out after staining of MF retaining microbial cells. Suitable stains for MFs are Liefson's methylene blue and crystal violet.

5.2 Microbial count

Sampling, sample preparation, and all experimental techniques should be performed under aseptic conditions and in duplicate, unless otherwise specified. All microbiological media to be used in applications must be sterilized. Again, all solutions and equipment that may cause contamination of microorganisms must be sterilized or disinfected.

5.2.1 Equipment, reagents, and media

Equipment. Membrane filter system, forceps, pipettes (1 and 10 mL) in pipette box, Petri dishes, membrane filter paper, wrap paper, incubator, and pipette discard equipment (e.g., a graduated cylinder) containing a disinfectant and a piece of cloth at bottom.

Microbiological Analysis of Foods and Food Processing Environments. DOI: https://doi.org/10.1016/B978-0-323-91651-6.00039-2

Reagents. 0.1% peptone water, quarter-strength ringer's (GSR) solution, Malachite green stain (0.01%, aq.), and Liefson's methylene blue stain.

Media. Nutrient agar plates.

Sterilization of MF equipment. The MF is placed between the sheet of the filter paper and wrapped, the filter holder should be loosely assembled, wrapped with aluminum foil, and they are autoclaved at 121°C for 15 min. The funnel top is sterilized by dipping into alcohol and flaming.

5.2.2 Filtration

Sample filtration and incubation of MF. Aseptically assemble the MF apparatus. Remove the cover of a sterile MF (Fig. 5.1A) and place MF using sterile forceps, grid side up, on the flask of the filter unit (Fig. 5.1B). Lock the funnel into place and connect the vacuum system (Fig. 5.1C). The liquid sample should be shaken vigorously. Pour the liquid sample into the funnel, and draw the sample through the MF by applying vacuum to the filtering apparatus (Fig. 5.1D). Allow the liquid to pass completely through the filter into the flask. Do not turn off the vacuum used in filtration. After all samples have been passed through the MF, the inside of the funnel is rinsed with sterile QSR solution. Decrease the vacuum of the filtration device, remove the filtration funnel from the top under aseptic conditions, transfer the MF with sterile forceps and place the MF onto nutrient agar in the plate (Fig. 5.1E). The MF is placed on the agar medium with the rolling action to avoid trapping air bubbles between the medium and the MF. Replace the lid of the Petri plate (Fig. 5.1E) and incubate plates with the lid uppermost in the incubator at respective microorganism optimum growth temperature. The incubation time for the MF will be less than plate techniques.

Colony count without staining. After incubation of the MF, observe the colony formation on the MF (Fig. 5.1F) and count the number of colonies on the MF. An acceptable range of colony formation for count per MF should be 20 to 200. Calculate the MF viable microbial cells in colony forming unit (cfu) per mL or g of sample based on the amount of sample filtered.

Colony counting with staining. There may sometimes be insufficient contrast in visualizing between colony color and MF; in this case, colonies on the MF can be stained and counted.

MF staining with malachite green. After the incubation of the MF, the malachite green stain is added over the MF, it is allowed to flow for 3–10 s and the excess stain is poured into a jar containing the disinfectant. In this staining technique, the MF is stained without staining the microbial colonies. After staining, colonies on the MF are counted against stained background.

Bacterial colonies staining with Liefson's methylene blue stain. After the incubation of the MF, a fresh pad is saturated with methylene blue stain and the MF carrying microbial colonies is placed on the stained pad and left for 5 min. Then, the MF to be taken over the pad is placed on a pad saturated with distilled water. Methylene blue stain first passes through the MF and stains colonies in the pad quickly. As soon as there is a sufficient contrast between the colonies and the membrane, count the stained colonies.

5.3 Interpretation of results

Membrane filtration technique is an effective, accepted technique for testing fluid samples for microbiological contamination. It involves less equipment and media preparation than many

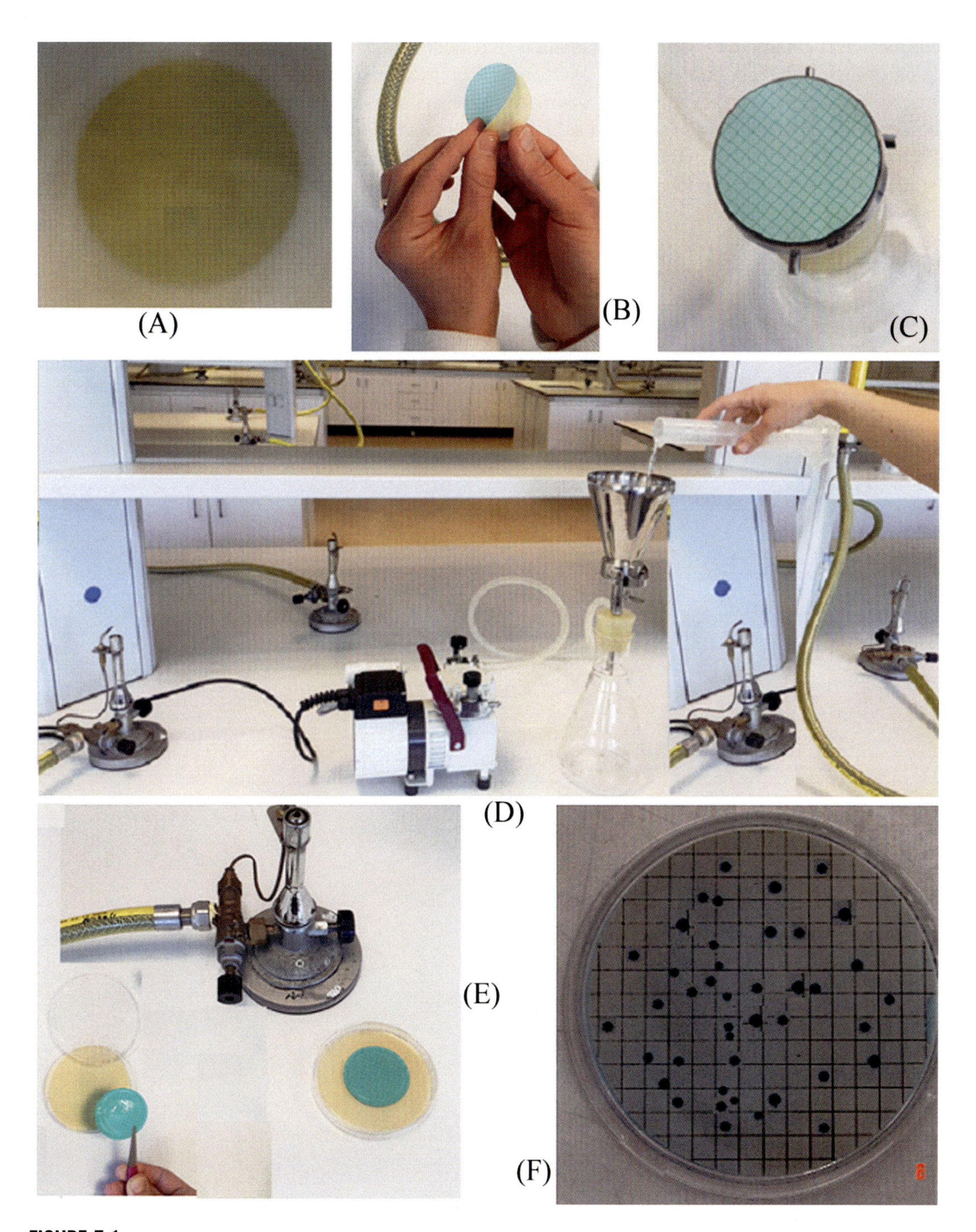

FIGURE 5.1

Membrane filtration: (A) a sterile membrane filter (MF); (B) remove MF cover and place MF on a filtration flask; (C) replace the funnel into place and connect the vacuum; (D) attach the sample vessel and filter the sample through MF into liquid collection flask; (E) placed MF on a Petri plate; and (F) formation of microbial colonies on MF after incubation.

traditional techniques, will allow the isolation and counting of microorganisms, and permits the testing of large sample volumes. The membrane can be transferred from one medium to another for purposes of selection or differentiation of microorganisms, thus allowing isolation and counting of discrete colonies of microorganisms. Results can be obtained more rapidly. Membrane filtration is used extensively in the laboratory and in the industry to sterilize heat-labile fluids. It is an effective and acceptable technique to monitor drinking water for microorganisms in different industrial areas (e.g., pharmaceutical, cosmetics, electronics, food and beverage industries). It allows for the removal of bacteriostatic or bactericidal agents that would not be removed in pour plate, spread plate, or MPN (most probable number) techniques.

PRACTICE

Yeasts and molds counting techniques

6

6.1 Introduction

The molds are obligate aerobic microorganisms (they need free oxygen). Most yeasts are aerobic and others are facultative anaerobic. Fungi (yeasts and molds) can grow from pH 2 to 9. They can grow in temperatures from 7°C to 35°C; a few species of molds can grow below or above this temperature range. Moisture requirements of most foodborne fungi are relatively low; they are able to grow at a water activity (a_w) of 0.85 or less. They are widely distributed in the environment and they may be associated with the normal flora of foods, and inadequately sanitized equipment and air. They grow in foods with high sugar content, high salt, low moisture and pH, and low storage temperatures. Additionally, they can cause problems through (1) formation of toxic metabolites (mycotoxins); (2) resistance to different processes, e.g., heat, freezing, antibiotics, or irradiation; (3) causing off-odors and flavors; and (4) discoloration of food surface.

Fungi (sing. fungus) absorb nutrients from the environment (external source). Three groups of fungi important for food are: yeasts, molds, and mushrooms. Fungi have important activities in natural ecosystems; they decompose organic compounds in soil, wood, plants, animals, and others into smaller units, e.g., inorganics. They play an important role in the minerals and carbon recycling in nature. Some types of fungi, mainly molds, produce enzymes, organic acids, antibiotics, and hormones. Some fungi damage can cause different spoilages and decompositions on agricultural crops, cause disease in animals and humans, and produce mycotoxins in foods. Most mycotoxins are retained in processed foods since processing conditions cannot destroy them. Food processing inactivates molds but their preformed mycotoxins may still be present in the foods. Some fungi may also cause allergic reactions or infections, for example, on skin. Some of the mycotoxins are retained in foods after heat processing or cooking. Some mold species are opportunistic pathogens and can cause diseases, especially in virus-infected people, immunocompromised peoples, and persons receiving chemotherapy or antibiotic treatment.

They can easily contaminate, survive, and grow on different types of foods; they can contaminate crops (e.g., grains, nuts, beans, and fruits) during harvesting and storage. They can grow on different foods depending on intrinsic and extrinsic factors. Their isolation and counts from foods depend on their types and numbers. The foods contaminated with molds and yeasts may be slightly spoiled, severely spoiled, or completely spoiled. The spoilage can appear with the formation of rot spots in various sizes and changes in colors, slimy, white cottony mycelium, or highly colored sporulation. They can also produce abnormal flavors and odors. Sometimes, a food can appear mold-free when examined with the naked eye but the mycotoxins will remain in the food. Fungal contamination of foods can result in economic losses.

Many molds produce organic acids (e.g., citric and gluconic acids), vitamins (such as riboflavin and vitamin B12), and plant growth regulators (e.g., gibberellic acid). Some molds can produce

Microbiological Analysis of Foods and Food Processing Environments. DOI: https://doi.org/10.1016/B978-0-323-91651-6.00050-1

commercially important enzymes (such as cellulase, glucoamylase, and rennet). Some of the medically important antibiotics (e.g., penicillin, griseofulvin, cyclosporine, and cephalosporin) are produced by molds. Molds are important tools in the decomposition in the environment of chemicals (e.g., pesticides and others) that destroy many ecosystems and also in bioremediation. Some molds can be used to control insects, pathogens, parasites, and other microorganisms causing damage and disease on agricultural crops. Some molds are used in cheese ripening (e.g., *Penicillium roquefortii* in roquefort cheese and *Penicillium camembertii* in camembert cheese). Some yeasts, e.g., *Saccharomyces*, produce vitamins, minerals, and other nutrients in the fermented food product.

Yeasts can be oxidative, fermentative, or both. The oxidative yeasts can grow as a film, or pellicle, on a liquid surface and are called film yeasts. Fermentative yeasts grow throughout the liquid. Some yeasts are facultative anaerobic. When oxygen is not available, yeasts carry out fermentation, i.e., anaerobic respiration. Yeasts are microscopic, unicellular, usually larger than bacterial cells; their size ranging from 1 to 5 μm in width and 5 to 30 μm or more in length. Yeast cells are mostly oval shaped, but sometimes spherical, elongated, and other shapes.

Molds grow in the form of a mass (mycelium) that spreads rapidly and may cover an area in 2 or 3 days. Molds are multicellular and look like filaments under low-power magnification. The mass of hyphae is called mycelium. The body or thallus (plural, thalli) of a mold consists of the mycelium (plural, mycelia) and the dormant spores (Fig. 6.1). Each hypha includes many cells joined together. The rigid walls of hyphae are made of cellulose, chitin, and glucan. Hyphae may be septated (presence of cross walls between cells) or nonseptated (no cross walls between cells). Some hyphae extend into solid media or foods and are called nutritional hyphae (rhizoids). Reproductive hyphae may grow in air and produce spores. Spores are reproductive structures in molds and yeasts. Yeasts also reproduce with budding.

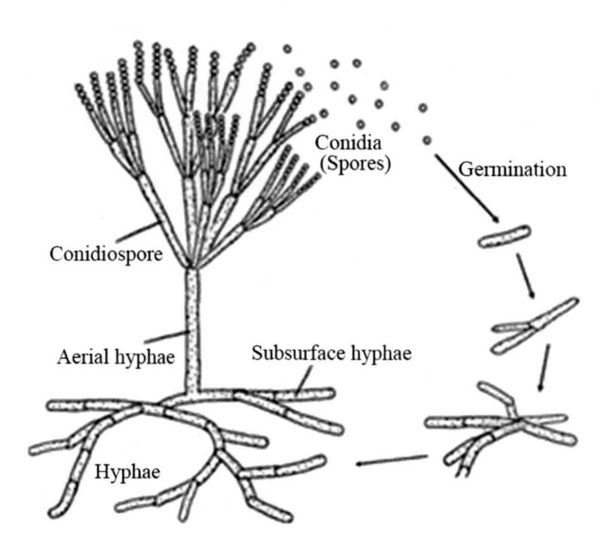

FIGURE 6.1

Mold life cycle.

6.2 Fungal classification

Depending on their reproduction with spores, fungi are classified as perfect and imperfect fungi. Perfect fungi reproduce by asexual and sexual spores. Imperfect fungi reproduce only by asexual spores. Fungi are classified in the Kingdom Fungi, also known as the Kingdom Mycetae. The Fungi kingdom can be divided into five main groups: Ascomycota, Basidiomycota, Chytridiomycota, Deuteromycota, and Zygomycota.

Kingdom: Fungi (Myceteae)	
Division: Ascomycota Class: Ascomycetes *Division:* Basidiomycota • Class: Basidiomycetes (Mushrooms) • Class: Uredinomycetes (Rusts) • Class: Ustomycetes (Smuts)	*Division:* Chytridiomycota Class: Chytridiomycetes *Division:* Deuteromycota Class: Deuteromycetes *Division:* Zygomycota • Class: Zygomycetes • Class: Trichomycetes

Chytridiomycota. Chytridiomycota have zoospores that are motile cells with a flagellum and lack a true mycelium. They usually associate with freshwater ecosystems and wet soils. Most of them are parasitic fungi for algae and animals. They live on organic debris (as saprophytes) in the soil. A few species can cause disease on plants (e.g., Olpidium). They reproduce asexually with the zoospores and can also reproduce sexually. Chytridiomycota are unicellular and anaerobic. They associate with the gut of herbivores and survive synergistically with the degradation of carbohydrates in the diet of ruminants.

Ascomycetes. The ascomycetes, known as sac fungi, include mold and yeast species. They reproduce with both sexual and asexual spores. They produce powdery asexual ascospores called conidia. Certain ascomycetes produce fruiting bodies with sexual spores. Some of the genera in the ascomycetes are *Byssachlamyse*, *Eurotium*, and *Eupenicillium* (molds), and *Debaryomyces* and *Saccharomyces* (yeasts).

Basidiomycetes. The basidiomycetes, known as club fungi, include mushrooms, bird's nest, jelly, rusts, smuts, and bracket fungi. They reproduce with sexual and asexual spores. Their reproductive cells are called basidia sexual spores (basidiospores). Basidiomycetes include plant-decaying parasites and they cause brown rot and white rot on wood. Basidia forms line gills or tubes on the fruiting bodies which are the main components of most mushrooms. Asexual reproduction occurs with conidia and conidia spores. Basidiomycetes produce incomplete septate hyphae. There are edible (e.g., *Agaricus bisporus*) and poisonous (such as *Amanita phalloides*, *Psilocybe semilanceata* and *Amanita muscaria*) mushrooms. One of the yeasts that is human pathogen infecting the skin, brain, and lungs is *Cryptococcus neoformans*. Smuts (e.g., *Ustilago* and *Tilletia*) invade flowering plants, e.g., cereal grasses, and cause serious economic loss. Rusts (e.g., *Puccinia*) invade agricultural crops (e.g., wheat) and forest trees, and cause economic loss.

Zygomycetes. The zygomycetes include parasites of spiders and insects. Hyphae are usually nonseptated. They produce sexual and asexual spores. One of the most common black bread molds is *Rhizopus*. It has a fuzzy growth with tiny black dots that are sporangia growth at the end of hyphae. Other zygomycetes are *Mucor* and *Thamnidium*.

Deuteromycetes. The deutoromyctes reproduce asexually by spores called conidia on septated hyphae. The deuteromycetes include mold genera (e.g., *Alternaria*, *Aspergillus*, *Botrytis*, *Blastomyces*, *Fusarium*, *Coccidioides*, and *Microsporum*) and yeast genera (e.g., *Candida*, *Brettanomyces*, and *Rhodotorula*).

Most fungi can grow with an optimum pH lower than that of most bacteria. Fungi mostly grow on media containing inorganic salts and carbohydrate, some fungi require B-group vitamins or other growth factors. These requirements can be provided by adding 0.1% yeast extract into growth medium. All molds are saprophytes, and grow and decay organic matter. There are many media used for culturing the fungi; the following media are more useful in culturing the fungi: malt extract agar, potato dextrose agar, orange serum agar, osmophilic agar, and rose Bengal agar. Phosphate buffer solution or 0.1% peptone water is a suitable diluent in the dilution of sample for culturing yeasts and molds. An appropriate solution, e.g., 20% sucrose, should be used as the diluent when counting osmotrophs from high-sugar-containing foods (e.g., syrups and fruit juice concentrates). The most probable number technique may be appropriate for some foods when counting yeasts and molds. The direct plating techniques may be used to count and detect yeasts and molds from foods. Membrane filter techniques are also applicable for beverages, clear liquids, and certain solid foods after homogenization. Sometimes, entrapment of mycelia and food particles in membrane pores limits the applicability of a membrane filter.

Sampling, sample preparation, and all experimental techniques should be performed under aseptic conditions, and in duplicate unless otherwise specified. All fungi media to be used in applications must be sterilized. Again, all solutions and equipment that may cause contamination of microorganisms must be sterilized or disinfected.

6.3 Plate count techniques

Refer to Technique 1.5 for details of sample preparation and dilutions. Sample dilutions and inoculation into the medium must be completed within 15 min. Spread plating gives more appropriate and better results than the pour plating for yeasts and molds count, since mold colonies on the surface grow faster than in pour plating and colonies are often obscured within or under the surface in pour plating. Some of the fungi can be inhibited by heat in the pouring agar medium. This can result in less accurate counting than the spread plate technique. The spread plate technique gives a more identifiable growth and allows easy colony isolation. DRBC agar can be used in spread plate to prevent mold spread on agar medium.

6.3.1 Equipment, materials and media

Equipment. Cover glass slip, filter paper, incubator, microscope, microscope glass slide, Pasteur pipette, Petri dishes, pipettes (1 mL) in pipette box, U-shaped glass rod, vortex, and pipetted discarded equipment.

Reagents. Glycerol solution (20%), Gram stains, lactophenol cotton blue, Liefson's methylene blue stain, physiological solution (0.85% NaCl), and tartaric acid solution (10%).

Media. Davis yeast salt (DYS) broth, dicloran rose Bengal agar containing 100 μg per mL chloramphenicol (DRBC), malt extract (ME) agar, malt yeast extract agar (MYE) agar containing 0.2% sodium propionate for yeast, nutrient agar containing 0.04% glucose and 0.14% anhydrous sodium acetate, nutrient agar containing 5% glucose and 0.5% tartaric acid, plate count (PC) agar containing 100 μg per mL chloramphenicol, potato dextrose (PD) agar (pH 3.5), and potato dextrose (PD) agar containing tartaric acid for mold.

6.3.2 Antibiotic technique

Mycological media is prepared with the addition of antibiotics to inhibit bacteria. Chloramphenicol can be used for this aim, because it is stable during the sterilization of media in the autoclave conditions. The recommended concentration of this antibiotic is 100 mg per liter of medium. Filter-sterilized 50 mg per liter chlortetracycline can also be used in the preparation of a medium. Other antibacterial antibiotics can also be used in the preparation of media.

Inoculate 1 mL of diluted (or nondiluted) sample to DRBC agar plate in duplicate and spread over the entire surface using a disinfected glass spreader. Incubate plates in the incubator in dark at 25°C for up to 5 days; do not stack plates higher than three and do not invert the plates. Do not disturb plates until colonies are to be counted. After incubation, examine plates for characteristic yeasts and molds colonies. Count the plates containing 15 to 150 yeast and mold colonies. If excessive mold growth develops on the plates, count first after 2 or 3 days and then again after 5 days. Calculate the number of yeasts and molds by multiplying the average count by dilution rate and dividing by inoculation amount. Overgrowth of molds should not a problem on the DRBC agar. Report results as yeasts and molds colony forming unit (cfu) per g or mL of sample. Select at least three colonies per plate and carry out streak plate (Technique 13.4.1) to obtain a pure culture. Use pure cultures in the identification of yeasts and molds (Technique 6.4).

6.3.3 Acidified technique

The technique is identical to those described in Technique 6.3.2 for the antibiotic technique except the medium is made selective for fungi by acidification instead of antibiotic in the preparation of medium, usually to a pH of 3.5 (in some cases pH 5.5). PD agar acidified with sterile 10% tartaric acid is the medium of choice. Inoculate 1 mL of diluted (or nondiluted) sample to PD agar plate in duplicate and spread over the entire surface using a disinfected glass spreader. Incubate plates in the incubator in the dark at 25°C for up to 5 days; do not stack plates higher than three and do not invert the plates. Do not disturb plates until colonies are to be counted. After incubation, examine plates for characteristic yeast and mold colonies. Count the plates containing 15 to 150 yeast and mold colonies. If excessive mold growth develops on the plates, count first after 2 or 3 days and then again after 5 days. Report results as the number of yeasts and molds cfu per g or mL of sample.

Count colonies (from antibiotic containing or acidified medium) at the end of the incubation of plates because extended incubation could result in unwanted growth of molds from dislodged spores, this can result in invalid final fungi counts. If yeasts are counted from plates, the plates with 150 colonies are usually countable. However, if substantial numbers of molds grow on the plates, the plates should be discarded. Calculate the number of yeasts and molds by multiplying the average count by dilution rate and dividing by inoculation amount. When plates from all dilutions have no colony formations, report mold and yeast counts (MYC) as less than 1 times the lowest dilution used. Purify individual colonies from Petri plates, if further analysis and identification are necessary. Select at least three colonies per plate and carry out streak plate (Technique 13.4.1) to obtain pure culture. Use pure cultures in the identification of yeasts and molds (Technique 6.4).

6.3.4 Analysis of nonsurface-disinfected (NSD) foods

Remove the intact NSD food item (particle) by 95% ethanol-flamed forceps, place NSD food items on the surface of PD agar plate; place 5–10 items per plate (depending on size of NSD food item). Overheating of forceps can be avoided by using several forceps in removing the food sample. When the food sample has visible mold growth or includes blemished food items, they should not be used. During incubation, align three Petri plates in stacks and label the plate with the sample number and date. Petri plates are incubated of up to 5 days at 25°C in the incubator. If there is no microbial growth on plates after 5 days of incubation, they should be incubated for another 2 days to allow sufficient time for growth of heat or chemically injured fungi and their spores.

Analyze the plates for growth of molds. Indicate the percentage of mold on the plate. If mold growth appears from all 50 NSD food items, record the moldiness as 100%; if mold growth appears from 32 items, record the moldiness as 64%. Analyze for mold genera and indicate the percent of individual mold genera and species if possible. For this, analyze growth directly from plates with microscopic magnification (10–40×) to identify *Aspergillus*, *Penicillium*, or most other foodborne mold genera.

6.3.5 Most probable number technique

Use DYS broth in counting of yeasts by MPN technique. Low numbers of yeasts may be detected using this medium. Perform MPN technique for counting yeasts as indicated in Technique 4.2.

6.4 Isolation and identification techniques for yeasts and molds

6.4.1 Slide cultures

Prepare slide cultures from yeasts and molds on PD agar (or ME agar). Place a sterile filter paper into a sterile Petri dish. Immerse a U-shaped glass rod into alcohol and then flame the glass rod. Place disinfected glass rod onto filter paper in the Petri dish. Remove microscope glass slide from bottle containing alcohol and flame the slide. Place disinfected glass slide onto U-shaped glass rod in the Petri dish. Aseptically add a sterile 20% glycerol solution on to the filter paper for wetting. This will prevent drying of the slide culture during incubation. Pour PD agar (at 45°C) to the surface area of the slides to form a very thin layer of agar medium. In this way, prepare two Petri dishes. After solidifying the agar medium on slides, inoculate the surface area of one slide with yeast and the other slide with mold in a series of streaks with a loop. Place a sterile disinfected cover slip (with immersion into alcohol and flaming) over the inoculated area of medium. Place the cover of the Petri dishes and incubate the Petri dishes at 25°C for up to 5 days. During incubation, analyze slides for yeast and mold growth. After incubation, examine the slide cultures with the eye first and then with the aid of a microscope. Indicate and compare the types of growth under the cover slips. Gently remove the cover slip from the slide culture and mount the culture on the cover slips with a drop of lactophenol cotton blue. Place the cultured and stained surface of the cover slips onto the slide with paraffin wax at the corners of the cover slip. Perform the following yeasts and molds analyzing technique.

6.4.1.1 Analysis of yeasts

Identify the yeasts from slide cultures (from Technique 6.4.1) on the basis of morphology and cultural characteristics, and then by biochemical tests from pure cultures.

6.4.1.1.1 Cultural characteristics

After incubation of the slide medium at 25°C for up to 5 days, young yeasts produce characteristically dormant spherical colonies with the entire edge with a moist and butyrous consistency, with a shiny, semimat or mat smooth surface. The color of the yeast colony will be white, cream, pink, or salmon. On further incubation, the colonies tend to become wrinkled and dry. The yeast colonies resemble some bacterial colonies, but most bacterial colonies tend to grow sparsely on the low pH media or selective media used for yeasts and molds. Yeast subsurface colonies in pour plates often have a characteristic satellite appearance.

6.4.1.1.2 Morphological characteristics

1. Place a drop of water onto a clean and disinfected microscope glass slide. Transfer one loopful of culture near the water on slide and homogenize in the water. Flood one drop of Gram iodine (or Liefson's methylene blue stain) onto the homogenized culture and place a cover glass slip onto the preparation. Magnify with the high-power objective of microscope.
2. Prepare a heat fixed smear in the usual way from yeast culture and stain by simple staining (Appendix B under the topic "Crystal Violet Stain") using Liefson's methylene blue stain.
3. Microscopic analysis reveals unicellular spherical, ellipsoidal, ovoid, or cylindrical yeasts. The formation of a pseudomycelium is often revealed only by slide cultures. The reproductive structures form ascospores and blastopores, ferment carbohydrate, utilize ethanol as a carbon source, and are resistant to actidione (cycloheximidine). Yeasts can reproduce by dividing into two by budding. Ascospore-forming yeasts are difficult to identify because yeasts rarely produce ascospores in ordinary media. These yeasts are allowed to form ascospores by subculturing the yeasts twice on a solid media containing 5% glucose and 0.5% tartaric acid and then culturing in solid media containing only 0.04% glucose and 0.14% anhydrous sodium acetate.

6.4.1.2 Analysis of molds

Record the macroscopic mold colonial characteristics from slide culture (from Technique 6.4.1). Examine the mold colonies under low-power and high-power objective of the light microscope. Take a portion of the mold culture with a needle and place it into a drop of lactophenol picric acid or lactophenol cotton blue on a microscope glass slide. Cover the preparation with a disinfected cover slip; exclude air bubbles through preparation. Examine the prepared culture under the light microscope, first using the low-power and then with the high-power objective. Identify morphological characteristics of molds on the bases of the following:

1. The mold colony characteristics are identified according to the surface appearance, size, texture, and color of the colonies.
2. The mold hyphae are identified by the presence or absence of cross-walls (septation or nonseptation) and their diameter.
3. The asexual and sexual reproductive structures of molds are identified by the formation of conidial heads, sporangia, arthrospores, and zygospores.

6.4.2 Counting and isolation of yeasts and molds separately by plate technique

In the isolation of molds and yeasts from sample, add 10 g of the sample to sterile 90 mL of physiological solution (0.85% NaCl) in the Erlenmeyer flask. Prepare further sterile dilutions using physiological solution. Spread plate 1 mL of diluted (or nondiluted) sample onto each MYE agar containing 0.2% sodium propionate for yeasts and PD agar containing tartaric acid for molds. Incubate plates at 25°C up to 48 h. After incubation, count the colonies of yeasts and molds separately. Calculate the number of yeasts and molds by multiplying the average count by dilution rate and dividing by inoculation amount. Record results as the number of yeasts and molds cfu per g or mL of sample separately.

Pick different types of colonies from plates (least three colonies per plate) and purify colonies by streak plating on MYE agar for yeasts, and PD agar for molds (Technique 13.4.1). Purification is done by streaking on agar plate and repeat two or three times or until pure cultures are obtained, as confirmed by colony and microscopic examinations. Purified yeast and mold cultures are identified through morphological, cultural, and biochemical tests.

6.4.3 Identification of yeasts and molds species by foodomics techniques

6.4.3.1 MALDI-TOF Ms identification of yeasts and molds

Yeasts and molds can be rapidly identified with the matrix-mediated laser desorption ionization-time-of-flight-mass spectrometry (MALDI-TOF Ms) technique, which is based on the principle of ionizing specific protein profile of microbial cells. Yeasts and molds can be easily identified by comparing these profiles with the reference spectrum. MALDI-TOF Ms identification is explained in Technique 13.5.1

6.4.3.2 Molecular identification of yeasts and molds

Species of yeasts and molds can also be identified by molecular techniques. One of the genotypic identification techniques for the identification of yeasts and molds is the polymerase chain reaction (PCR). PCR is a molecular biology technique used in the detection of various yeasts and molds species according to the genes encoding toxins, metabolic or other characteristics of yeasts and molds. Some yeast- and mold-specific primer sets that can be used in their identifications are given in "Appendix A: Gene Primers in Table A-1." Using yeast- and mold-specific primers, techniques, and materials, their species can be identified by PCR analysis, as described in Technique 15.5.2.

6.4.4 Determining toxin-producing molds

Equipment. Blender, boiling chips (silicon carbide), cotton, Erlenmeyer flasks, filter paper (18 cm in diameter), fume hood equipped with steam bath (air-flow rate, 250 $cm^2 min^{-1}$), funnels (90–100 mm diameter), high-performance liquid chromatograph, and incubator.

Materials. Chloroform, long- or short-grain polished rice, methanol, mycotoxin standards, and NaOCl solution (5%).

6.4.5 Toxin-producing molds

Place 50 g of rice and 50 mL of distilled water into Erlenmeyer flasks in duplicate. Plug Erlenmeyer flasks with cotton and sterilize flask in the autoclave at 121°C for 20 min. After sterilization, cool the flask and inoculate spores onto the rice from individual mold cultures. Incubate inoculated rice in the flasks at 25°C in the incubator until the surface area of rice is covered with mold mycelium, and mycelia penetrate through the rice within 15–20 days. After mold growth, add 150 mL chloroform (150 mL methanol if the toxin is deoxynivalenol). Place the flask into a water bath and heat contents until boiling. During boiling, break up moldy rice cake with a spatula for 10 min. After boiling, transfer the contents of the flask into a sterile blender and blend contents at high speed for 1 min. After blending, filter blender contents through filter paper inserted into a glass funnel. Pour filtrate into a sterile Erlenmeyer flask. Return rice cakes on the filter paper into the blender, add 100 mL unheated solvent and blend at high speed for 1 min. Filter as above and combine filtrates in flask. Add boiling chips into a flask containing filtrates and heat the flask in the heater. Evaporate water from filtrates until 20–25 mL remaining in flask and use in the toxin analysis. If analysis is not to follow immediately, evaporate to dryness and store flask in the dark. Add filtration from filter paper into the glassware containing water into NaOCl (5%) solution in glassware. Allow rice cake to stand in NaOCl solution for 72 h before autoclaving.

Toxin analysis. Use mycotoxin standards in both qualitative and quantitative analysis of toxin. Use high-performance liquid chromatography to determine mycotoxins in rice cake produced by mold culture.

PRACTICE

7 Sanitation detection techniques in food processing plants

7.1 Introduction

Information on the sanitation conditions of the food processing plant and equipment can be obtained from microbiological analysis. The samples obtained from processing plant are examined for aerobic plate count, coliform count, and other microorganisms. Surfaces of equipment, for example, containers and utensils, which come into contact with food must be cleaned and sanitized properly. Aseptic conditions should be provided during production to prevent the contamination of equipment and product. Microbial contamination can occur during food production from equipment, air, employees, and others. They should not increase the microflora of a product significantly. For example, an animal slaughtered under sanitary conditions will yield a carcass with a low bacterial microflora. However, inadequate cleaning and sanitation in the cutting room may increase the microbial population in meat products. Another example is the effect of postpasteurization contamination of fluid milk products from air or equipment. Poorly cleaned and sanitized pipelines and pumps may contribute psychrotrophic bacteria to products, for example, fluid milk.

Basically, there are four techniques to check the sanitary (hygienic) conditions of equipment and utensils: (1) the rinse technique, (2) the swab contact technique, (3) a direct agar-contact technique, and (4) hygiene monitor. When the rinse technique is used, the equipment or utensil is rinsed with an amount of sterile liquid. The microbial content of the rinse solution is then determined by plate count technique. The swab technique is conducted by moistening a sterile swab in a sterile buffer and then a predescribed area of the equipment is swabbed. Following swabbing, the swab is rinsed thoroughly in the sterile buffer. This is followed by counting of the microorganisms by plate count technique. In the direct-agar contact technique, the medium is pressed lightly against the surface area of equipment to be checked. Hygiene monitoring is based on swabbing a special strip onto the surface of equipment, and the chemical and residues give a reaction to produce different color changes. These sanitation indication techniques can be recommended to routinely check the effectiveness of a cleaning and sanitizing program.

An effective monitoring program to reserve food sanitation and hygiene practices should be applied in the food production plant. Food can be contaminated at any point from farm to table during slaughter, harvest, processing, storage, distribution, transportation, and preparation. “Food sanitation” refers to the cleaning of equipment and facilities, and the safe disposal of waste. “Food hygiene” refers specifically to practices that prevent microbial contamination of food at all points along the food chain from farm to table. Inadequate food sanitation and hygiene can lead to foodborne illness and consumer death. Some examples of illnesses that can spread easily due to poor hygiene and sanitation are diarrhea, flu, hepatitis A, and even COVID-19. The goals of food hygiene and sanitation are:

- Preventing food spoilage due to microbial or chemical contamination.
- Educating people about simple and practical ways to keep food safe to avoid foodborne illnesses.

Microbiological Analysis of Foods and Food Processing Environments. DOI: https://doi.org/10.1016/B978-0-323-91651-6.00017-3

- Protecting food from adulteration.
- Providing appropriate practices in the food trade to prevent the sale of unsuitable food in terms of quality.
- Preventing spread of pathogens in environment.

7.2 Monitoring microorganisms in food plant

An effective sanitation program should remove food debris, organic material, and soil from areas that may support the survival or growth of microorganisms. The master cleaning schedule and sanitary standard operating procedures are key programs to guide cleaning activities. In general, food contact surfaces (tools and equipment used in production), water, air, and employees in a food production factory are the primary sources of microbial contamination. The way to prevent contamination from these sources can be achieved by including an environmental monitoring program, called the "regional" approach. Sanitation programs in a food plant can be typically separated into different zones, for example, three or four zones. An effective microbiological regional monitoring program in a food production plant will provide early warning of potential microbiological hazards, identify microbiological problems early, provide scientific information for resource exploration, guarantee regional overall microbiological control, and prevent product and economic losses.

Successful sanitation and hygiene can be measured by visual inspection and microbiological analysis for indicator microorganism or pathogens. Such a program can also determine whether the food production environment is under microbiological control and can indicate potential problems before they affect the finished product. There are several ways to monitor the microbiological quality of a food production environment. These are to determine the microbiological potential of the tools and equipment used in the food production, the air in the production area, the water used in production and the employees. The way to determine the microbiological potential or the efficiency of sanitation efforts is to take samples at a certain frequency from designated points in the food factory and perform microbiological analyses in these samples. Microbiological analyses to be made after sanitation do not fully reflect the microbiology during production. A general microbiological sampling program should include sampling during manufacture. In-process samples will be a direct indicator of the overall microbiological control of the finished product.

An effective microbiology monitoring program should include analysis for indicators and pathogens. Results of analyses should be interpreted and recorded by qualified quality assurance and food safety personnel. Data from microbiology monitoring program should reveal problems in sanitation, air quality, personnel practices, standard operation, and sanitary design that affect the microbiological quality of the finished product. A microbiological monitoring program should include following:

1. *Air monitoring*. It is important to monitor the microbiological load of the air, since everything from the air (microorganisms, dust, etc.) can fall on surfaces that come into contact with food. For this, certain air monitoring programs should be applied according to the situation of the plant. Total microbial count or yeasts and molds count can be recommended in the program to be monitored. For example, in a weekly air monitoring sample, the total microbial number for 15 min sampling, <250 colony forming units (cfu) can be considered an acceptable limit.

2. *Preproduction*. Such microbiological monitoring can provide verification that equipment used in food production is properly sanitized. Total microbial counts for samples (such as swab, wash solution) can be considered to have an acceptable limit of <100 cfu per unit.
3. *In-Process*. This allows determining whether hygienic conditions are met during food production. A program should be implemented to verify microbial conditions in food production. In this program, surfaces in contact with food and air condition must be microbiologically monitored. For this purpose, total bacteria or Enterobacteriaceae from samples (swab, air, etc.) can be recommended as indicators of appropriate microbiological monitoring. In this monitoring, <100 or <1000 cfu per unit can be used as the acceptable limit. Both food contact and non-food contact surfaces should be included as sampling areas.
4. *Pathogenic microorganism*. In a food-producing plant, a pathogen monitoring program should be applied. This program should include specific pathogens with respect to food products or ingredients used in the food production. For example, in the case of chicken and poultry products, *Salmonella*; in freshly consumed dairy products or foods with high meat ingredients, *Listeria*; or in dairy products produced from raw milk, *Brucella* can be added to the pathogen monitoring program.
5. *Employees*. In food production, pathogenic microorganisms can easily be transmitted to food, air, and equipment from employees. Programs for tracking pathogenic microorganisms should be implemented for employees. The health status of employees should be constantly monitored, and those who are sick or carriers of pathogens at risk levels should not be employed in the food production. *Staphylococcus aureus* can be added to the pathogen monitoring program to be followed by employees.

7.3 Determination of sanitation and hygienic conditions in food processing plant

7.3.1 Equipment, reagents, and media

Equipment. Cotton wool swabs, incubator, and ice-water bath.

Reagents. Ringer solution containing 0.05% sodium thiosulfate and quarter-strength Ringer's (QSR) solution containing 0.5% sodium thiosulfate.

Media. Davis's yeast salt (DYS) agar, orange juice (OJ) agar, lauryl sulfate tryptose (LST) broth, plate count (PC) agar, potato dextrose (PD) agar, and yeast extract milk (YEM) agar.

7.3.2 Sampling and sample preparation

Microbiological monitoring practices in a food production facility include sampling from the surface, air, water, employees, and others. There are several sampling and analytical techniques and devices for verifying the effectiveness of sanitation and hygiene efforts. When selecting any sampling, screening, or testing technique to determine the effectiveness of sanitation and hygienic conditions, one must ensure that the application fits the operation.

7.3.2.1 Sampling from surfaces

The most common ways of sampling surfaces include swabs, wipes, sponges, contact plates/strips, and rinses. Typically, swabs are best used when sampling a small defined surface area, while wipes and sponges provide good sampling from larger, undefined areas. The contact plate and strips are useful for flat, defined surface areas. A contact plate is turned upside down and held onto a flat, even surface, such as stainless steel or plastic. Rinse water is a common sampling technique in cleaned-in-place equipment; once the cleaning and sanitation has been completed, a sample of the final rinse may be taken.

Swabbing. Refer to Technique 1.4.7 for details of surface sampling by swabbing. In testing any pieces of equipment, more than one area should be swabbed. The most common sampling area size is a 10 cm × 10 cm (100 cm^2) or a very small or large area of equipment using a cotton wool swab. Before swabbing, the swab is moistened in the QSR solution in the tube and the swab is pressed against the side of tube to express excess liquid. Rub the cotton wool swab firmly over the sampling surface with a slow rotation of the swab, and swab a second time using parallel strokes at right angles to the first set. Insert cotton wool swabs into the QSR solution and mix by swirling the swab vigorously 10 times.

Swabbing. Rinsing. Refer to Technique 1.4.7 for details of surface sampling by rinsing. Food containers can be examined by rinsing with 500 mL of QSR solution. Replace the lid of the container, lay the container on its sides, and roll it through 12 complete revolutions. The container is allowed to stand for 5 min and rolling is repeated. When cleaned-in-place pipeline systems are examined by a rinse technique, the rinse liquid is moved from the system. After rinsing, it should be used in the microbiological tests within 6 h.

For a bottle, it is selected for analysis after washing. The QSR solution is added (from 2 to 100 mL) depending on the size of bottle and the bung of bottle is replaced. Hold the bottle horizontally in the hands and rotate gently 12 times in one direction thoroughly wetting the internal surface. After thoroughly mixing, prepare necessary dilutions using QSR solution.

7.3.2.2 Sampling from air

The quantitative determination of airborne microorganisms is possible by sedimentation, impaction on solid surfaces, impingement in liquids, filtration, and others. Sampling stress may injure the microorganisms. The air sample collection time may affect a representative bioaerosol sampling. Too long sampling time may cause sampling stress on microorganisms or collection of an overload of particles which may prevent microbial colony formation. Air sample collection time should be moderately short. Least duplicate air samples can be used and their analyses are expressed as an average of the replicated data observations. Refer to Technique 1.4.9 for details of air sampling.

Sedimentation (plate exposure to air). Settle or exposure plates are used to sample sedimentation in the air. The Petri plates are exposed to the air for a certain time period, after which they are covered and taken to the laboratory. Sedimentation sampling is a static technique where the gravity force and air currents cause airborne microorganisms to settle onto agar media, for example, plate count (PC) agar for APC and potato dextrose (PD) agar for yeasts and molds count. Selective agents can be incorporated into media to inhibit unwanted microorganisms. Open the Petri plate lid and expose the plate to air in a variety of locations of the processing room for 15 min. After exposure, the inverted PC agar plates are incubated at 30°C for up to 48 h for APC and inverted PD agar plates at 25°C for up to 5 days for yeast and mold counts. After incubation, colonies are counted and the results are expressed as respective microorganism cfu per min of exposure. This technique is inexpensive, easy to perform, and collects bioaerosols in their original state. The disadvantage of this technique is its inability to count all viable particles per volume of air, it involves a long sampling time, has a great reliance on air current, shows a greater susceptibility toward detecting large particles, and it has a low correlation with counts obtained using other techniques.

Impaction. Most of the food industry obtains air samples (bioaerosols) with the bioaerosol sampler (Fig. 7.1A). Different types of bioaerosol samplers (impactors) are commercially available. In bioaerosol sampling, the inertia of particles is separated from the air currents. The bioaerosol sampler collects airborne microorganisms onto an agar surface with the use of a vacuum by rotating the plate on a turntable to create an even distribution of particles on an agar medium. In this technique, a sampler works at a vacuum to draw a constant air flow rate of about 28.3 L min^{-1}. After exposure of Petri plates to air for 60 min, incubate the inverted plates at 30°C for up to 48 h for APC and invert PD agar plates at 25°C for up to 5 days for yeasts and molds count. After incubation, colonies are counted (Fig. 7.1B) and the results are expressed as respective microorganism cfu or particles per min of exposure or volume of air. Other microorganisms can also be counted using selective/differential agar plates exposed to air. This technique obtains higher air recovery than other air sampling techniques. It is used when bioaerosol levels are expected to be low in air. This technique results in a low sampling stress on microorganisms.

Impingement. The impingement technique uses a liquid medium for the airborne microorganism collection. Air particles are entrapped in the liquid during the dispersal of air through the liquid. Liquid impingers can collect air by low velocity or high velocity. Low -impingers do not efficiently collect small particles. The liquid preserving the viability of the microorganisms from air should not permit growth. Some of the common collection liquids are phosphate buffer (PB) solution, buffered gelatin solution, and peptone water (0.1%). After the collection of microorganisms in liquid from air, the sampled liquid is serially diluted and inoculated onto the medium. The total volume of air sampled and the volume of collection fluid must be measured to determine the number of cells per unit. This technique allows the collection of air by 12.5 L min^{-1}. The airstream can cause sonic speed and destruction or injury effect on vegetative cells. Another limitation of the impingement technique is its failure to collect particles smaller than 1 μm. There can be an overestimation

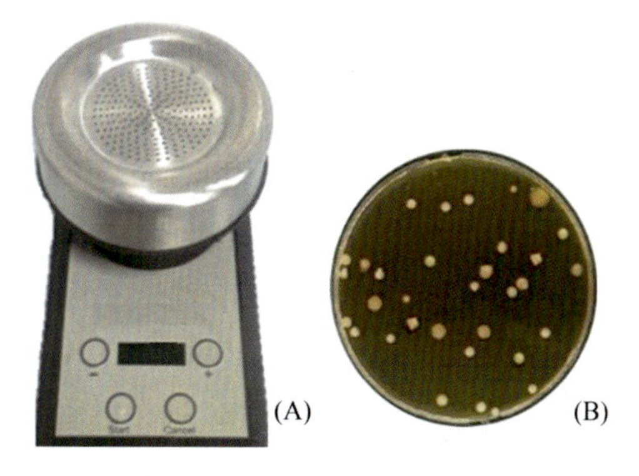

FIGURE 7.1

(A) Bioaerosol sampler and (B) colonies on petri plate exposed to sampler for 60 min.

of microbial count since high air sampling velocity breaks up clumps of bacteria. The air sample can be obtained from food-producing or research rooms by impingement. All parts of impingement devices are sterilized in the autoclave by placing distilled water into the glass and closing all openings with aluminum foil. Rubber and glass parts are disinfected using hypochlorite solution (0.2% aq.) or H_2O_2 solution (6.0% aq.). After disinfection, sterile distilled water must be passed on disinfected equipment to remove the remaining disinfectant. In the air sampling process, add 20 mL of sterile PB solution into the impinger and add 0.1 mL of sterile antifoam into the PB solution. Label the liquid level on the impinger. Set up air sampling impingement as shown in Fig. 7.2. Open the vacuum to pass air through the liquid in the impinger. After each exposure time, remove 1 mL of fluid using a sterile pipette from the impinger and then continue to collect the air sample. Remember, the water volume in the impinger decreases during impingement because of the evaporation of water and sample removal after each exposure. Add sterile distilled water up to the labeled part of the impinger after each removal of the sample. Perform a second sampling operation and remove the sample again. Follow a further sampling operation, if necessary.

Filtration. The filtration technique collects airborne microorganisms onto a membrane filter which is mounted on a holder and connected to a vacuum source with a flow rate controller. The membrane filter may consist of sodium alginate, cellulose fiber, glass fiber (pore size 3 μm), or a synthetic membrane filter (pore size 0.45 μm or 0.22 μm). Filter collection devices are used for sampling of air. They are less effective on vegetative cells because of the stress associated with desiccation; further shorter sampling times may reduce stress. Filtration devices are low in cost and simple to operate. The air sample is removed from the room by filtration of air through a membrane filter. The membrane filter is placed in an appropriate holder and connected to a vacuum source through a flow rate controller. The amount of air passing (in liter) through a membrane filter per filtration time (in min) should be recorded. After filtration of an amount of air in a constant time from a processing room, the membrane filter is placed directly on the agar surface. The inverted PC agar plates are incubated at 30°C for up to 48 h for APC and inverted PD agar at 25°C for up to 5 days for yeast and mold counts. After incubation, colonies are counted from Petri plates. In another method, the membrane filter may also be agitated in a liquid in the flask to disperse the particles from filter paper and mixed by swirling the swab vigorously 10 times. Spread plate the liquid sample to Petri plates (Technique 2.2.4) and incubate as per direct plating of membrane filter. Calculate the number of APC, and yeasts and molds by multiplying the average count by dilution rate, and dividing by the inoculation amount and the volume of filtered air. Report the results as respective number of microorganisms cfu per L of air.

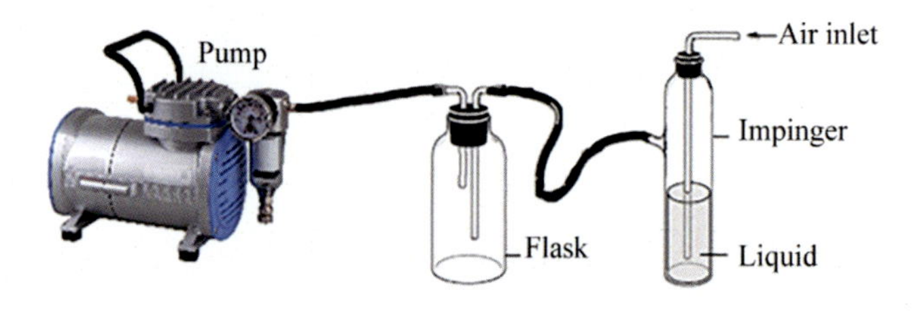

FIGURE 7.2

Biological aerosols collection from air in impinger.

7.3.2.3 Sampling from water

Water sampling is fairly routine in food manufacturing facilities and typically the processor is looking for general coliforms as indicators of cleanliness. In the food production plant water is used as an ingredient and in cleaning and washing facilities. All food businesses should frequently take water samples from different outlets. Water samples should be collected from the line in the facility and from the water storage tank.

7.3.2.4 Transport of sample

It is important to have a transport fluid that is buffered or that will maintain the bacteria in whatever state they exist, that is, dead, alive, or injured. The transport fluids should be isotonic and produce minimal stress, so that if injured cells exist in the sample, carryover sanitizer or salt in the fluid will not kill any bacteria in the sample. There are sanitizer neutralizing solutions, and many of these contain a surfactant that helps to suspend the bacteria and prevent them from clumping onto residual particles. The neutralization of the sanitizers involves some common broths, most notably Dey-Engley buffering (DEB) broth and lecithin broth.

7.3.3 Analysis of samples

Sampling, sample preparation, and all experimental techniques should be performed under aseptic conditions, and in duplicate, unless otherwise specified. All microbiological media to be used in applications must be sterilized. Again, all solutions and equipment that may cause contamination of microorganisms must be sterilized or disinfected.

7.3.3.1 Sampling from water counting techniques

Plate count. Sample should be tested as soon as possible after being obtained (within 6 h). When the sample is not analyzed in place, transport temperatures and conditions should be regulated and recorded. After thorough mixing, prepare necessary dilutions using QSR solution. Analyze the samples for aerobic plate count (Technique 2.2.4) using YEM agar, coliforms count (Technique 15.3.2), and yeasts and molds count using PD agar (Technique 6.3.2), and pathogens using appropriate selective count medium, as indicated in the respective Practice of this book. Incubate at 30°C for 24 h for aerobic plate count (APC) and at 25°C for Technique 2.2.4 for yeasts and molds count. For the air sample, incubate the inverted PC agar plates at 35°C for 24 h for APC and inverted PD agar plates at 25°C for up to 5 days for yeasts and molds count.

After incubation, examine colonies from the plates. Count colonies from agar plates. Calculate the number of APC, and the yeast and mold count. Record results as the number of microbial types cfu per cm^2 of utensil or mL or L of sample, etc.

In an alternative technique of analyzing washed clear glass bottles for the presence of contaminants, add PC agar (at 47°C) into the bottle in sufficient volume to form a thin (5 mm) layer over the inside surface. The bottle is rolled slowly horizontally until the agar sets on the inside surface (this can be provided by rolling the bottle in a shallow dish of iced water). The bottle should be incubated bung down. After incubation, count colonies and record the result as the number of APC cfu per bottle.

Agar-contact technique (RODAC plate count). Pour PC count agar (at 47°C) into a sterile Petri dish, filling until the agar rises above the rim of the plate. Let the agar solidify. Remove the cover and carefully press the surface of the agar against the test surface contact area which is about 26 cm^2. Replace the cover and incubate inverted plates at 32°C up to 48 h. After incubation, count the colonies and record the results as the number of APC cfu per cm^2.

Yeasts and molds count. Refer to Technique 6.3.2 for the yeast and mold count. This should be performed using DYS agar.

Coliforms and fecal coliform counts. Perform coliforms and fecal coliform counts by MPN, as explained in Technique 15.3.2.

Calculation for air sample. The air passing through liquid may be 12.5 L min^{-1}. Calculate the numbers of microorganisms per liter of air as follows:

- Volume of air (L) = Air sampling time (min) × air flow rate L/min.
- Microorganisms cfu per L of liquid = count × liquid remaining in impinger.
- Microorganisms cfu per L of air = count/Volume of air (L).

7.3.3.2 Hygienic monitor

Using hygiene monitoring to evaluate the effectiveness of cleaning procedures is a critical aspect of an environmental monitoring program, which can help to ensure the sanitary condition of the food and beverage manufacturing environment. Hygiene monitoring can be used without laboratory facilities in the food industry from raw material to finished product. Hygiene monitoring vials are commercially available, the analysis procedure is specific to the type of hygiene monitoring and is noted on the device package's insert. Hygiene monitoring is based on a special dip paddle in the hermetically sealed vial. Each paddle contains a specialized growth medium. Analysis can be carried out by any person in the production area or on the processing line as well as in kitchens, tool shops, dairy farms, fish farms, clean rooms, air fall-out in homes, offices, etc. No laboratory is needed; the only equipment required is a hygiene monitoring vial and a standard incubator. Hygiene monitoring is particularly useful for identifying the types and numbers of microorganisms commonly present: aerobic bacteria, lactic acid bacteria, coliforms, yeasts and molds, *S. aureus*, *Salmonella*, and others. Examples of hygiene monitoring systems are ATP swab read in luminometer, protein swab colorimetric assay (presence/absence), and colorimetric food residues assay using test strips and reagents.

ATP bioluminescence. The ATP technique consists of a nonmicrobiological rapid test for checking the sanitary condition of equipment and utensils. The system includes reagents necessary for the light-producing reaction. The ATP necessary to activate this reaction is obtained from the surface area of the equipment. The sample from the surface is obtained by swabbing. The more light is produced in the analysis, the more ATP is collected and the more microorganisms or food residues are present on the equipment surface. The amount of light is read in a luminometer, which gives a reading in light units. With the sampling kit, swab a surface of approximately 25 cm^2 (5 × 5 cm) from the test area. Then place the swab back in the sampling kit and press the bulb at the top of the kit to release the washing solution. Shake the kit so that the solution will wash the swab. With pliers, break the glass tube in the sampling kit to allow the wash suspension to dissolve the bioluminescence reagents. Place the kit in the luminometer, set to read mode, and wait for 11 s before reading. Any results indicating more than 3.5 light units are considered unacceptable equipment use in the food production.

ATP bioluminescence. Test strip technique. Different test strips are available for the indication of proteins (protein test strip), hygienic control (hygiene test strip), etc. Protein test strip techniques offer results in less than 5 minutes, and typically are used to sample surfaces for the presence of protein, which is an indicator of inadequate sanitation. The test strip is placed onto the surface, it is brought into contact with certain chemicals, and the reaction gives different color changes depending on the protein concentration. The hygiene test strip provides a technique for assessing the general cleanliness of surfaces and rinse water. Hygienic evaluation of food and hand contact surfaces are performed after cleaning. The reaction on the hygiene strip gives different color changes depending on the detectable levels of residue on the surface.

7.4 Interpretation of results

Total microbial count will be variable among food producing plants due to different sanitation and hygiene programs. The results will depend on the analyzing techniques, type of process used in food producing, sampling instrument, the amount of samples taken, the type of food production, and the quality of the established baseline. In terms of plant/line-specific standards, 100 cfu/100 cm^2 usually may be an acceptable standard. The accepted standard for potable (sanitary) water is no coliforms in 100 mL. In a clean room or aseptic area, in-house bacterial standards are usually applied by the company. As data are recorded, a trend analysis can be obtained. The company will have a historical microbiological record of the result. Such recording allows the company to narrow the variance or reduce the average count to lower levels in order to foster continuous and progressive improvement. If pathogens are found in the samples, the use of bacterial subtyping techniques can aid in better interpretation of the results and finding the root sources of the contamination. Plate count limits for equipment used in food plant are given in Table 7.1.

Table 7.1 Plate count limits for equipment used in food plant.

(i) Aerobic plate count per cm^2	**Conclusion**
Not more than 5 Between 5–25 More than 25	Satisfactory Requires further investigation Unsatisfactory; take immediate action.
(ii) Aerobic plate count	
(i) Bottles and small container	
Not more than 200 200–1000 More than 1000	Satisfactory. Improve cleaning. Unsatisfactory; take immediate action.
(ii) Churn and can	
Not more than 10,000 10,000–100,000 More than 100,000	Satisfactory. Improve cleaning. Unsatisfactory; take immediate action.
(iii) Coliform counts. The equipment used for carrying, dispersing or holding heat-treated foods should contain less than 10 coliforms per 100 cm^2 of equipment. No coliform bacteria found on 100 cm^2 of equipment can be regarded as satisfactory.	

SECTION

Counting of important microbial groups from food products

Microbiological assessment of the quality and safety of foods relies upon the counting of microbial groups. Microbial isolation and counting techniques involve the use of growth media. Quality management and control systems (e.g., good manufacturing technique and hazard analysis at critical control points) involve the counting of contaminated survival microbial groups to monitor the safety and quality of products, and efficiency of processes. The microbial quality of foods depends on the result of microbial contamination and their growth. The process efficiency and food quality will be guaranteed when the overall process sufficiently inactivates microorganisms or reduces microorganisms to safe levels. Therefore, the microbiological quality assurance of foods is not only a matter of control, but also of a careful design of the total process chain. There is a need to predict the efficiency of a process in the food industry, and adapt to the technology and logistics of specific production processes. This section gives basic microbiological techniques with regard to counting microbial groups that are important in food processing and storage. Counting of microbial groups indicates the extent of microbial survival in foods. Counting has been included for the following microbial groups: injured microorganisms; cold-tolerant microorganisms; mesophilic and thermophilic microorganisms; halophilic, osmophilic, and xerophilic microorganisms; and thermoduric microorganisms.

Practice 8. Injured Microorganisms and Viable but Nonculturable Cells

Practice 9. Counting of Cold-Tolerant Microorganisms

Practice 10. Counting of Mesophilic and Thermophilic Sporeformers

Practice 11. Counting of Halophilic, Osmophilic, and Xerophilic Microorganisms

Practice 12. Counting of Thermoduric Microorganisms

Injured microorganisms and viable but nonculturable cells

8

8.1 Introduction

Food processing techniques can have three antimicrobial effects on microorganisms. Foods may contain physiologically and biochemically deficient viable cells. Microorganisms can be killed, survive, or be injured in foods, depending on the degree of process application on foods. The injury may result from different food processing techniques: heat, refrigeration, frozen, drying, irradiation, osmotic pressure, low water activity, and physical treatments (e.g., pulsed electrical field, pulsed light, high hydrostatic pressure, and CO_2 pressure). Injured cells can repair injuries themselves under favorable conditions which can result in potential safety hazards in the processed and stored foods. Injured microorganisms are important with respect to public health and food spoilage. They can start to grow as a normal cell after repairing the injury. Selective agents in media inhibit injured microbial cells. Injured microorganisms can be recovered from foods after repairing cellular injury in an appropriate repair (preenrich) medium at optimum growth conditions. The growth on the nonselective media represents both the injured and noninjured microbial cells, while only the noninjured cells grow on the selective media. The difference between the numbers of colonies on these two media gives the number of injured cells in the culture or food. In this practice, some injured microbial cells recovery techniques are explained. A significantly higher number of injured cells is recovered when the surface plate technique is used instead of the pour plate technique. Liquid media can also be favorable in the recovery of injured microorganisms.

Many microbial species, including foodborne pathogens, develop stress resistance mechanisms that enable them to enter into a temporary state of low metabolic activity. Under these conditions, cells can survive for extended periods without growth, called a viable but nonculturable (VBNC) state. VBNC is a survival strategy, where bacteria cannot grow on routine microbiological media but are alive and are able to renew metabolic activity. VBNC cells have different changes, including morphological and compositional variations, which allow them to have a higher resistance to chemical and physical stresses compared with vegetative cells. VBNC cells maintain their intact cell membrane and are metabolically active, continuing gene expression, having the ability to become culturable once resuscitated. The occurrence of VBNC bacterial pathogens in food creates a public health risk. The routine microbiological analysis cannot differentiate dead and VBNC cells. Although VBNC bacteria lose culturability on normal microbiological medium, this does not mean that these cells are equivalent to dead cells. For example, the membrane of dead cells is damaged so that the genetic material in the cell cannot be preserved and expressed, while VBNC cells have a complete membrane structure so that the genetic material is retained. Moreover, while VBNC cells retain absorptive capacity and are metabolically active, dead cells lose these characteristics. A series of physiological changes occur during the transition from the normal state to the VBNC state, including slowing down the nutrient absorption, the level of macromolecular synthesis and metabolism reduction, the concentration of the cytoplasm, and total proteins

Microbiological Analysis of Foods and Food Processing Environments. DOI: https://doi.org/10.1016/B978-0-323-91651-6.00014-8

reduction. At present, the technique used to detect VBNC cells is mainly based on two key characteristics of VBNC cells: viability and nonculturability. Generally, if bacteria lose culturability but are still viable, they can be considered to have entered the VBNC state. Thus using the conventional plate counting or isolation techniques, culturing is impossible. The estimation of VBNC cells can be possible by using other techniques: Fluorescent microscopic techniques, immunological techniques, genetic-based techniques, ATP bioluminescence, and cytometry.

8.2 Repairing and counting techniques

Sampling, sample preparation, and all experimental techniques should be performed under aseptic conditions, and in duplicate, unless otherwise specified. All microbiological media to be used in applications must be sterilized. Again, all solutions and equipment that may be contaminated with microorganisms must be sterilized or disinfected.

8.2.1 Equipment, reagents and media

Equipment. Blender (or stomacher), conical tube (50 mL), membrane filter system, membrane filter (0.22 μm pore size), Petri dishes, thermometer, vortex, pipettes (1 and 10 mL) in pipette box, pipette discarding equipment, incubator, and Erlenmeyer flask.

Reagents. 0.1% Peptone water, 0.2% peptone water, and ice-water.

Media. Baird-Parker (BP) agar, bismuth sulfite (BS) agar, brain heart infusion (BHI) agar, lactose broth, MacConkey agar, trypticase soy (TS) agar, TS broth, violet red bile (VRB) agar, VRB-2 agar (containing 200 μg per mL 4-methylumbelliferone glucuronide, MUG), and xylose lysine deoxycholate (XLD) agar.

8.2.2 Counting of injured microorganisms from processed foods

Sample preparation. Refer to Technique 1.5 for further sample preparation and dilutions. Aseptically weigh 50 g of processed solid food sample into a sterile blender (or stomacher plastic bag) and add 450 mL of room temperature 0.1% peptone water or others depending on the type of food or microorganisms. Homogenize sample by blending at low speed (8000 rpm) for 1 min (or stomaching for 2 min). For a liquid sample, add 50 mL of sample into a sterile Erlenmeyer flask containing 450 mL room temperature 0.1% peptone water or others depending on the type of food and microorganism. Homogenize sample by gently shaking the flask. Prepare further dilutions using 0.1% peptone water or others depending on the type of microorganism.

Counting. Pour plate 1 mL of diluted (or nondiluted) sample to each melted (at 47°C) nonselective agar (e.g., BHI agar or TS agar) in duplicate (Technique 2.2.3). The high temperature (47°C) of the medium will inactivate injured microbial cells. Spread plate 1 mL of diluted (or nondiluted) sample to each nonselective agar (e.g., BHI agar or TS agar) in duplicate (Technique 2.2.4). Incubate inverted plates at 35°C or at the microbial group's optimum growth temperature for up to 48 h. After incubation, count the colonies from spread plate agar medium (total number of microorganisms) and pour plate agar medium (number of noninjured microorganisms) from plates containing 25 to 250 colonies.

Calculate the number of microbial cells separately by multiplying average counts by dilution rate and dividing by inoculation amount. Finally, calculate the number of injured microbial cells by subtracting the pour plate counts from the spread plate counts. Record the results as injured and total number of microorganisms colony forming unit (cfu) per g or mL of sample.

8.2.3 Counting of injured particular microorganisms

Prepare sample as indicated in Technique 8.2.2. Spread plate 1 mL of diluted (or nondiluted) sample to each nonselective (e.g., BHI agar or TS agar) and selective agar medium (e.g., BS agar for *Salmonella*) in duplicate (Technique 2.2.4). Incubate inverted plates at a particular microorganism's optimum growth temperature for up to 48 h. After incubation, count the colonies from both nonselective medium (total count) and selective medium (noninjured count) from plates containing 25 to 250 colonies. Calculate the number of noninjured microbial cells and total microbial cells separately by multiplying the average counts by dilution rate and dividing by inoculation amount. Finally, calculate the number of injured microbial cells by subtracting the selective count from the nonselective count. Record the results as injured and total microorganisms cfu per g or mL of sample.

8.2.4 Counting of injured microorganisms by membrane filtration technique

Membrane filtration has the advantage that it can repair injured microbial cells without the danger of overgrowth by the others. This technique is applicable to any microorganisms by altering the repair medium and the incubation conditions. It allows a high amount of sample use in the analysis. So a low number of injured microbial cells are easily recovered by membrane filtration technique.

Prepare the sample as indicated in Technique 8.2.2. Filter 100 mL samples through the membrane filter (0.22 μm pore size) as explained in Technique 5.2. After filtration, place the membrane filter on a selective agar medium (e.g., bismuth sulfide agar for *Salmonella*) to count noninjured microorganism. Add 50 mL of TS broth into 50 mL of sample in the Erlenmeyer flask and mix by gently shaking the flask. Incubate the flask at 35°C for 2 h (or at 25°C for frozen foods) to repair injured cells. Filter preenriched mixture through membrane filter. After filtration, place the membrane filter on another selective agar medium (e.g., BS agar for *Salmonella*) for total count. Incubate inverted plates at 35°C or at particular microorganism optimum growth temperature up to 48 h. After incubation, count the colonies from both enriched plates (total count) and nonenriched plates (noninjured count) from plates containing 25 to 250 colonies. Calculate the number of noninjured microbial cells and total microbial cells separately by multiplying average counts by dilution rate and dividing by inoculation amount. Finally, calculate the number of injured microbial cells by subtracting the nonenriched count (noninjured count) from the enriched count (total count). Record the results as injured and the total number of microorganisms cfu per g or mL of sample.

8.2.5 Counting of coliforms from processed foods

Prepare sample as indicated in Technique 8.2.2, but use lactose broth in the homogenization of sample. Count injured and total coliforms with or without repair. Spread plate 1 mL of diluted sample to VRB agar in duplicate (Technique 2.2.4) for noninjured coliform count. Incubate the remaining homogenized sample at 35°C for 2 h (or at 25°C for frozen foods) to repair injured cells (recover) in lactose broth. After preenrichment, make additional decimal dilutions from preenriched sample using sterile 0.1% peptone water as required. Spread plate 1 mL of preenriched sample on VRB agar in duplicate for total coliform count. After drying the surface area of plate, incubate inverted plates at 35°C for up to 48 h. After incubation, examine the formation of colonies from plates. Count red to pink colonies from plates separately containing 25 to 250 colonies. Calculate the number of coliforms separately by multiplying the average count by the dilution rate and dividing by the inoculation amount. Finally, calculate the number of injured coliforms by subtracting the noninjured coliform count from the total coliform count. Report the results as injured coliforms and coliforms cfu per g or mL of sample. Confirm representative colonies as indicated in Technique 13.3.2.2.

8.2.6 Counting of fluorogenic *Escherichia coli* from processed foods

Prepare the sample as indicated in Technique 8.2.2, but use lactose broth in the homogenization of the sample. Spread plate 1 mL of diluted sample to Petri plates containing 15 mL of VRB agar (selective medium for noninjured count). Spread plate 1 mL of sample to Petri plates containing 15 mL of TS (nonselective medium for total count) and allow plates to dry surface area of both agar medium. Incubate inverted plates at 35°C for 24 h. After incubation of plates, overlay the surface area of the agar medium in plates with 5 mL of VRB-2 agar containing MUG. Incubate inverted plates at 35°C for 12–20 h. After incubation, examine the layered plates for the formation of fluorescent halos; *E. coli* produces fluorescent halos on this medium. Count the number of *E. coli* from plates containing 25 to 250 colonies. Calculate the number of noninjured and total coliforms separately by multiplying the average count by the dilution rate and dividing by the inoculation amount. Finally, calculate the number of injured coliforms by subtracting the selective coliform count (selective count) from the nonselective coliform count. Some non-*E. coli* and other species of Enterobacteriaceae can also produce colonies with fluorescent halos. Thus confirmation tests are required (Technique 13.4). Record the results as the number of injured and total number of *E. coli* cfu per g or mL of sample.

8.2.7 Counting of *Staphylococcus aureus* from processed foods

Prepare the sample as indicated in Technique 8.2.2, but use TS broth in the homogenization of the sample. Prepare further dilutions using 0.1% peptone water as required. Spread plate 1 mL of diluted sample to BP agar (Technique 2.2.4) to count noninjured *S. aureus*. Incubate the remaining homogenized sample at 35°C for 2 h to repair injured cells (preenriching). After incubation, make additional decimal dilutions using 0.1% peptone water as required. Spread plate 1 mL of sample to BP agar to count total number of *S. aureus*. Incubate inverted plates at 35°C for up to 48 h. After incubation, count the shiny black colonies surrounded by a clear zone from the hydrolysis of egg yolk from plates containing 25 to 250 colonies. Calculate the number of noninjured and total number of *S. aureus* separately by multiplying the average count by the dilution rate and dividing by the inoculation amount. Finally, calculate the number of injured *S. aureus* by subtracting the noninjured counts from the total counts. Record the results as the number of injured and total *S. aureus* cfu per g or mL of sample.

8.2.8 *Salmonella* count from processed foods

Prepare the sample as indicated in Technique 8.2.2, but use lactose broth in the homogenization of the sample. Prepare further dilutions using 0.1% peptone water as required. Spread plate 1 mL of diluted sample to BS agar (Technique 2.2.4) to count noninjured *Salmonella*. Incubate the remaining homogenized sample at 35°C for 2 h to repair injured cells (preenriching). After incubation, make additional decimal dilutions using 0.1% peptone water as required. Spread plate 1 mL of sample to each BS agar to count total number of *Salmonella*. Incubate inverted plates at 35°C for up to 48 h. After incubation, examine colonies from the plates. Count the brown, gray, or black colonies; sometimes they have a metallic sheen from plates containing 25 to 250 colonies. Calculate the number of noninjured and total *Salmonella* separately by multiplying the average count by dilution rate and dividing by the inoculation amount. Finally, calculate the number of injured *Salmonella* by subtracting the noninjured counts from total counts. Record the results as the number of injured and total number of *Salmonella* cfu per g or mL of sample.

8.2.9 Other microbial counts from processed foods

Perform sample preparation, selective counting, and repairing (preenriching) techniques for microbial cell as indicated in the respective practices for particular microorganisms. Use the preenrichment medium, repair with incubation from 2 to 6 h, and then perform selective counting of microorganisms. Some examples are given here.

For *Vibrio parahaemolyticus*, add extra NaCl into the preenrichment medium to obtain 3% concentration in the medium (Technique 23.4.2), spread plate preenriched culture to the selective medium after 2 h preenrichment at 35°C (Technique 23.4.3).

For *Yersinia enterocolitica*, incubate the preenrichment medium at 25°C for 4 h (Technique 18.2.2) and spread plate the preenriched culture to a selective medium (Technique 18.2.3) and incubate at 25°C.

For *Campylobacter jejuni*, incubate the preenrichment medium at 37°C for 6 h under microaerophilic conditions (Technique 17.2.2), spread plate the preenriched culture to a selective medium (Technique 17.2.3), and incubate at 42°C for 24 h under microaerophilic conditions.

For *Listeria monocytogenes*, incubate the preenrichment medium at 30°C for 2 h (Technique 16.2.3), spread plate the preenriched culture to a selective medium (Techniques 16.2.4), and incubate at 35°C.

For *E. coli*, incubate the preenrichment medium at 35°C for 2 h (Technique 13.5.2.1), spread plate the preenriched culture to a selective medium (Technique 13.5.2.2), and incubate at 35°C.

8.2.10 Repairing and counting by MPN technique from processed foods

This technique is suitable for MPN determination of coliforms, fecal coliforms, *V. parahaemolyticus*, *Enterococcus*, and *S. aureus*. Refer to Technique 4.2 for the complete details of the MPN count. Prepare the sample as indicated in Technique 8.2.2, but homogenize the sample in diluent (liquid) according to the particular microorganisms. Again, use diluent in the preparation of required dilution according to the particular microorganisms. Add 1 mL of diluted sample into three-MPN tubes of selective broth for the particular microorganism. Incubate the remaining homogenate at the optimum growth temperature of the microorganism for 2–6 h. After incubation, dilute the sample using 0.1% peptone water. Add 1 mL of diluted sample into another three-MPN tubes of selective broth. Incubate the tubes at the particular microorganism's optimum growth temperature for up to 48 h. After incubation, examine the tubes for growth (turbidity and possible color change). Indicate the numbers of positive tubes for each dilution. Refer to MPN Table 4.1 and calculate the MPN of injured and noninjured particular microorganisms. Record the result as the MPN of injured and noninjured microorganisms per g or mL.

8.3 Heat injuring of microorganisms and their count

Inoculate TS broth with one loopful of the particular microbial culture and incubate at the optimum growth temperature for 24 h. After incubations, add 1 mL of incubated TS broth culture into each of five 50 mL conical tubes containing 12 mL of preheated (at 55°C) 0.2% peptone water (PW) in duplicate. Place tubes into a thermostatic shaking water bath (at 55°C). Immerse the PW in tubes completely in the water bath. Place another conical tube containing 12 mL of PW inserted with a thermometer for temperature control. Perform heat treatment at 55°C for 5, 15, 25, and 35 min. One heated tube in duplicate is removed after each heating interval and the tube is immediately cooled using ice-water. Then immediately prepare further 10-fold dilutions with 0.2% PW at room temperature. Spread plate 1 mL of diluted sample to nonselective TS agar and on selective agar medium according to the particular microorganism (e.g., xylose lysine deoxycholate agar for *Salmonella* and sorbitol MacConkey agar for *E. coli*). Incubate the inverted plates at the optimum growth temperature of the particular microorganisms for up to 48 h. After incubation, count the number of colonies from selective and nonselective agar media separately containing 25 to 250 colonies. Calculate the number of the particular microbial species from nonselective and selective agar media separately by multiplying the average count by the dilution rate and dividing by the inoculation amount. Finally, calculate the count differences between TS agar and selective agar to determine the number of sublethally injured cells. Record the results as the number of injured and total number of microorganisms cfu per g or mL of sample.

8.4 Fluorescent microscopic techniques to detect viable but nonculturable bacteria

Various fluorescent staining procedures can be used to determine VBNC organisms. Frequently used stains are acridine orange, 4,6-diamino-2- phenyl indole, fluorescein isothiocyanate, indophenyl-nitrophenyl-phenyltetrazolium chloride, and 5-cyano-2,3-ditolyl tetrazolium chloride. (1) Nalidixic acid (20–40 mg per L) is used to stop cell division. After exposure to nalidixic acid, viable cells continue to grow and will appear elongated, whereas the VBNC will retain their original shape and size. The cells are then observed under a microscope. Viable cells will be seen as elongated, whereas VBNC cells will be seen as oval and large. (2) In another technique, VBNC can be detected using the acridine orange direct count (AODC). In this count, acridine orange (AO) is used to differentiate between viable and nonviable cells. VBNC produces green fluorescence, whilst viable cells produce red/orange fluorescence. This differential fluorescence is used to differentiate between viable and nonviable forms of microorganisms.

8.5 Interpretation of results

Foodborne bacterial pathogens cause public health risks and economical losses. Pathogenic microorganisms are exposed to many environmental stresses during food production, processing, storage, and distribution. Injury of microorganisms can be induced by sublethal heat, freezing, freeze-drying, drying, irradiation, high hydrostatic pressure, aerosolization, dyes, sodium azide, salts, heavy metals, antibiotics, essential oils, sanitizing compounds, and other chemicals or natural antimicrobial compounds. Exposure to stress can cause varying degrees of cellular damage (injury) depending on the stress intensity and the physiological state of the individual cells in the population. Damaged cells can reproduce actively again by repairing the damaged parts under suitable conditions. The potential for hazard is still a concern because injured foodborne microorganisms are capable of repair and toxin production. Damaged cells cannot multiply on selective media used for counting the microorganisms and therefore may not be detected during routine analysis. These damaged cells can cause a significant potential risk for foodborne illnesses. Therefore it is necessary to ensure that damaged cells are repaired by using nonselective media for routine monitoring or counts. Moreover, the analysts will ultimately lead to more efficient design and sequencing of microbiological analyzing stages, thus maintaining the nutritional value of the food while providing maximum safety to the consumer. VBNC cells have higher physical and chemical resistance than culturable cells, which might be due to their reduced metabolic rate and a cell wall strengthened by increased peptidoglycan cross-linking. The ability to enter the VBNC state poses a risk to human health. İndication of VBNC will increase food safety.

PRACTICE

Counting of cold-tolerant microorganisms

9

9.1 Introduction

The microorganisms growing at low temperature can be categorized into two groups: *psychrotrophic* (also known as facultative or psychrotolerant psychrophile) and *psychrophilic*. *Psychrotrophic* bacteria are a subgroup of the mesophilic bacteria adapted to cold temperature. *Psychrotrophic* microorganisms have a maximum temperature for growth of 35°C. Psychrotrophic microorganisms are more ubiquitous than psychrophilic microorganisms; they are widely distributed in natural environment and foods. They have optimum growth temperature from 20°C to 25°C and are able to grow at 4°C. *Psychrophilic* microorganisms have an optimum temperature for growth between 12 and 15°C, a maximum temperature for growth is 20°C. Some psychrophilic species can grow slowly at temperatures as low as −5°C to −12°C. Psychrophilic microorganisms are restricted to cold habitats, e.g., oceans and lakes (below 5°C), alpine soils (in mountainous regions), and ice and snow-fields. They mostly are associated with environments where the temperature is mostly below 15°C. They can successfully compete with psychrotrophs for nutrients at low temperatures. Both microbial groups grow at low and high substrate concentrations, but psychrophiles grow faster than psychrotrophs at 10°C or below.

The reason why psychrotrophic microorganisms can proliferate at low temperatures is that they increase the unsaturated fatty acids in their lipid membrane structures. This makes more fluidity and carrier proteins are functional at low temperature. The growth of these microorganisms at low temperature is also related with the thermolability of one or more essential cellular components, particularly enzymes, exhaustion of cell energy, and intracellular substances leakage. Psychrotrophic microorganisms have higher degradative activities on foods than psychrophiles due to being widespread in nature or foods and favorably grow at room temperature. Some cold-tolerant microorganisms are pathogenic or toxinogenic for humans, animals, or plants. In natural microbial ecosystems, they have an important role in the biodegradation of organic matter during cold seasons. Psychrophilic species are important in fish harvested from cold water. Refrigerated foods are mainly spoiled by *psychrotrophic* microorganisms since psychrophilic microorganisms are rarely present in foods. Some psychrotrophic microorganisms present in raw milk are also capable of producing heat-stable enzymes (e.g., proteinases and lipases). When foods are heated (e.g., pasteurized or commercially sterilized), these enzymes are not inactivated, they can cause quality defects in this type of food products due to the decomposition of fat and/or proteins during storage. Psychrotrophic bacteria that adapt to the cold environment from their mesophilic habitats can more widely contaminate foods than psychrotrophilic microorganisms. Within the cold-tolerant bacteria, psychrotrophic microorganisms are primarily responsible for the spoilage of chilled food of animal origin, while molds are mostly responsible with the spoilage of fruit and vegetables. The major *psychrotrophic* species of bacteria associating with poultry, meats, dairy products, and seafood include the genera *Aeromonas*, *Acinetobacter*, *Achromobacter*, *Alcaligenes*, *Arthrobacter*, *Brochothrix*, *Clostridium*, *Enterobacter*, *Escherichia*, *Flavobacterium*, *Hafnia*, *Leuconostoc*, *Klebsiella*, *Lactobacillus*, *Microbacterium*, *Moraxella*,

Microbiological Analysis of Foods and Food Processing Environments. DOI: https://doi.org/10.1016/B978-0-323-91651-6.00038-0

Campylobacter, Yersinia, Pediococcus, Proteus, Pseudomonas, Serratia, and *Streptococcus.* In addition, certain psychrotrophic human pathogens can grow at refrigeration temperatures, e.g., *Aeromonas hydrophilia,* proteolytic *Clostridium botulinum* type E, nonproteolytic *C. botulinum* types B and F, *Bacillus cereus, Campylobacter jejuni, Escherichia coli, Listeria monocytogenes, Yersinia enterocolitica,* and *Salmonella.* The major psychrotrophic bacterial species in vacuum or modified atmospheric packaged foods are *Brochothrix, Lactobacillus, Leuconostoc,* and *Pediococcus.* Some psychrophilic species include the genera *Alcaligenes, Aeromonas, Flavobacterium* (e.g., *Flavobacterium bomense*), *Pseudomonas, Serratia, Vibrio, Bacillus, Clostridium,* and *Micrococcus.*

Psychrotrophic fungi predominate in the spoilage of refrigerated or chilled foods with low water activity and high acidity. They commonly associate with refrigerated fresh fruits and vegetables, and ready-to-eat foods. Some psychrotrophic yeast species include the genera *Candida, Hansenula, Pichia, Debaryomyces, Torulapsis, Saccharomyces, Cryptococcus, Leucosporidium,* and *Trichosporan.* Psychrotrophic molds include some species in the genera *Alternaria, Aspergillus, Botrytis, Cladosporium, Penicillium, Fusarium, Mucor, Geotrichum, Rhizopus,* and *Trichothecium.*

Large numbers of psychrotrophic bacteria can associate with refrigerated foods and cause off-flavor and physical defects. The presence of psychrotrophic bacteria in cold stored foods (particularly in large numbers) indicates a high potential for spoilage of foods within a shorter time. Most cold-tolerant microorganisms are destroyed by a mild heat treatment, e.g., pasteurization. In most cases, the presence of psychrotrophic bacteria in heat-processed foods implies postprocess contamination. They cannot grow in frozen foods, but they can grow and cause spoilage during thawing and in partial thawing at refrigeration temperature, e.g., poultry. The counting of psychrotrophic and psychrophilic microorganisms involves the ability to grow and form colonies on a solid agar medium after incubation at low temperature. The various incubation conditions are used in the determination of the cold-tolerant microorganisms. A standard reference technique involves the incubation of spread plated media at 7°C for 10 days. In some techniques, bacteria on selective and nonselective media are incubated at higher temperatures for shorter times. The count at low temperature mostly includes psychrotrophic bacteria since psychrophilic bacteria mostly associate with extreme conditions (e.g., cold water seafoods, cold environment) and foods obtained from these sources.

9.2 Counting of cold-tolerant bacteria

Sampling, sample preparation, and all experimental techniques should be performed under aseptic conditions, and in duplicate unless otherwise specified. All microbiological media to be used in applications must be sterilized. Again, all solutions and equipment that may contaminate microorganisms must be sterilized or disinfected.

9.2.1 Equipment, reagents, and media

Equipment. Incubator, pipette discarding equipment, Pipettes (1 and 10 mL) in pipette box, polypropylene container (nonrigid plastic), thermostatic shaking water bath, and vortex.

Solution. Sterile distilled water and 0.1% peptone water.

Media. Plate count (PC) agar.

9.2.2 Sampling

Refer to Technique 2.2.2 for sample preparation and dilutions. When a low number of microorganisms are present in a sample, analyze cold-tolerant bacteria with a minimum of 100 g or mL of sample. Add 100 g or mL of sample into a sterile polypropylene (nonrigid plastic) container. Add 200 mL of room temperature 0.1% peptone water into an Erlenmeyer flask and homogenize by gently shaking the flask. Prepare the required decimal dilutions using 0.1% peptone water. Refrigeration should be minimized because psychrotrophic bacteria will grow. Refrigerated samples should not be frozen because many psychrotrophs are sensitive to freezing and can be injured or killed. If samples need to be frozen for shipment, the possibility of some microbial death should be considered. When a high number of microorganisms is present in the sample, add 25 g or mL of sample into an Erlenmeyer flask containing 0.1% peptone water and homogenize by gently shaking the flask for 5 min. Prepare further decimal dilutions using 0.1% peptone water. Homogenization and inoculation should be completed within 10 min.

9.2.3 Counting of cold-tolerant bacteria

Spread plate 1 mL of diluted sample (or nondiluted) to nonselective medium (e.g., PC agar) in duplicate (Technique 2.2.4). Cold-tolerant bacteria are easily injured or killed when medium is added by pour plating from 44°C to 50°C. Incubate inverted plates at 7°C for 7–10 days. Alternatively, incubation for 16 h at 17°C can be followed by incubation for 3 days at 7°C. Visible colonies do not develop during the first 16 h but growth at 7°C is much more rapid. Incubation can be included at 25°C for 24 h or at 21°C for 25 h. After incubation, examine colonies from plates. Count colonies from plates containing 25 to 250 colonies. Calculate the number of psychrotrophic (or psychrophilic or cold-tolerant bacteria multiplying the average counts by dilution rate and dividing by inoculation amount. Report results as the number of psychrotrophic (or psychrophilic or cold-tolerant) bacteria cfu per g or mL of sample.

In the counting psychrophilic bacteria, incubation at 12°C–15°C is more suitable and for psychrotrophic bacteria from 20°C–25°C. But low temperature incubation is mostly dominated by psychrotrophs, since samples mostly contain psychrotrophic bacteria and psychrophilic bacteria are mostly restricted to foods in extreme conditions.

Cold-tolerant pathogenic bacteria can be counted using selective media. Spread plate (Technique 2.2.4) the sample on a selective medium for pathogenic bacteria, as given in the respective practice of this book.

The counting of psychrotrophic aerobic bacteria by membrane filter technique with incubation at 7°C for 10 days is also possible (Technique 5.2).

9.3 Interpretation of results

The counting of cold-tolerant bacteria in refrigerated foods gives an indication of the potential spoilage, keeping quality, or safety of the food. If the food has been refrigerated for some time, the numbers can represent a normal increase in these bacteria. Processing can kill or injure cold-tolerant bacteria, and analyzing such foods may need time to recover injured cells. If processed foods are stored in the refrigerator for extended periods, even a few cells can grow to large enough numbers to cause spoilage in a few days or weeks. Also, cold-tolerant pathogens can grow at low temperature and cause foodborne disease.

Counting of mesophilic and thermophilic sporeformers

10

10.1 Introduction

There is a clear association between soilborne endospore-forming bacteria and food contamination. The endospore-forming anaerobic (*Clostridium*) and aerobic (*Bacillus* and other genera) microorganisms are phylogenetic groups. Endospores are formed at the end of the stationary phase within the vegetative cells acting as sporangium (endospore-formers) and released as survival structures. One bacterial cell produces one endospore. Endospores (1) commonly associate with soil, (2) are resistant to common industrial processes (e.g., pasteurization), (3) have adhesive characteristics facilitating their attachment to surface (e.g., processing equipment), (4) are able to germinate to vegetative cells in favorable conditions, and (5) tolerate specific conditions and are activated by subprocess (e.g., pasteurization). Adaptation of soilborne endospores to unfavorable environmental conditions might result in increasing their tolerance and resistance to food processing conditions (e.g., ultrahigh heat treatment or commercial sterilization) and survive the food processing. Therefore endospores can cause spoilage and health problems in the food industry. They can cause foodborne disease (e.g., *Bacillus cereus* and *Clostridium botulinum*), food spoilage, and reduction of shelf life of foods. Microbial spoilage of food is usually indicated by changes in texture and development of off-flavors. In this practice, the isolation and counting techniques of important endospore-forming bacteria are explained with specific techniques.

Caution: Sampling, sample preparation, and all experimental techniques should be performed under aseptic conditions, and in duplicate, unless otherwise specified. All microbiological media to be used in applications must be sterilized. Again, all solutions and equipment that may contaminate microorganisms must be sterilized or disinfected.

10.2 Mesophilic aerobic sporeformers

Members of the mesophilic aerobic sporeformers (MAS) in the spoilage of food belong to species in the genera *Bacillus* and *Sporolactobacillus*. The genus *Sporolactobacillus* contains a single species, *Sporolactobacillus inulinus*. It is a microaerophilic, motile, catalase negative, homofermentative rod and grows at 35°C. This bacterium has little importance in food spoilage. *Bacillus* species are aerobic to facultative anaerobic rods, catalase positive or negative, and grow better at 35°C. They have a great importance in the food spoilage and foodborne diseases. *Bacillus macerans* and *Bacillus polymyxa* are strong gas producers. *Bacillus licheniformis* produces small amount of gas during spoilage of thermally processed foods. *Bacillus subtilis* endospores cause a ropy bread spoilage (production of extracellular slime polysaccharides). Mucoid variant strains of *B. subtilis* have been classified as *Bacillus mesentericus*. They can cause rope, most commonly on bread with water

Microbiological Analysis of Foods and Food Processing Environments. DOI: https://doi.org/10.1016/B978-0-323-91651-6.00033-1

activity of about 0.95. Any ingredient used in bread production (e.g., flour, yeast, and others) may contribute rope-forming endospores. Rope formation on bread is the main spoilage problem. Since the higher the rope-forming endospore content of ingredients, the greater the potential of rope spoilage. Enzymes produced by these bacteria hydrolyze the starch in the bread loaf to a gummy substance. In addition to ropiness, the spoiled bread will have an off-aroma sometimes characterized as fruity or pineapple-like. These bacterium spores can survive at ordinary baking temperatures in the center of the loaf, where the internal temperature rises only to about 100°C. *B. cereus* is the most important pathogenic or spoilage bacterium in the dairy and cereal industry. In the ready-to-eat food industry, *B. cereus* and other aerobic endospore-forming bacteria are introduced via vegetables, fruits, herbs, and spices,

MAS cause spoilage in the low acid foods (pH > 5.0), for example, fruit juices hot filled or pasteurized. Different techniques are used to count MAS; first vegetative cells are heat inactivated and then endospores are heat activated, and then endospores are counted using suitable medium.

10.2.1 Equipment, reagents, and media

Equipment. Blender (or Stomacher), Erlenmeyer flask, graduated cylinder, incubator, petri dishes, pipette discarding equipment, pipettes (10 and 1 mL) in the pipette box, thermostatic shaking water bath, and vortex.

Solution. 0.1% Peptone water.

Media. Dextrose tryptone bromocresol purple (DTBP) agar, Jensen's pork sucrose nitrite (JPSN) medium, and tryptone glucose yeast extract (TGYE) agar.

10.2.2 Counting of mesophilic aerobic sporeformers

Refer to Technique 2.2.2 for details of sample preparation and dilutions. Weigh 50 g of ingredient or food sample into sterile blender (or stomacher plastic bag) and add 450 mL of sterile 0.1% peptone water. Homogenize by blending (at 8000 rpm) for 1 min (or stomaching for 2 min). In the case of liquid sample, add 50 mL of liquid sample into a sterile Erlenmeyer flask containing 450 mL of 0.1% peptone water and homogenize by gently shaking the flask.

Add 10 mL of homogenate into 90 mL, and 1.0 and 0.1 mL of homogenate into separate 100 mL of melted TGYE agar (at 47°C) in sterile Erlenmeyer flasks in duplicate while holding at 47°C. Place flasks into the thermostatic shaking water bath at 80°C. Agitate flasks gently to disperse the sample through the medium. Place an additional uninoculated control Erlenmeyer-flask containing 100 mL of same medium and thermometer. The water level should be above the sample level in the flasks. Agitate the flasks with shaking of water bath during heating. After temperature reach 80°C, hold the tubes for 15 min at that temperature. After heating, adjust the temperature of the water bath to 47°C. Remove the flasks from the bath and rapidly cool the contents of the flasks in tap water (or in cold water) to 47°C.

Pour 100 mL heated TGYE agar medium (at 47°C) into sets of five sterile Petri dishes in approximately equal volumes. Allow the plates to solidify on a flat surface (within 10 min), invert the plates to prevent the spreaders, and promptly place them into the incubator. Incubate inverted plates at 35°C for 48 h. After incubation examine the plates for the formation of yellow colonies. Count the surface and subsurface colonies from each set of five plates. Calculate the average MAS counts from three sets of counts. Multiply the average by dilution rate and divide by inoculation amount. Report results as the number of

MAS colony forming unit (cfu) per g or mL of sample. Also count the yellow (acid-producing) colonies separately and report results as the number of acid-producing MAS cfu per g or mL of sample.

10.2.3 Counting of mesophilic aerobic rope sporeformers

Refer to Technique 10.2.2 for details of sample preparation and dilutions. Inoculate 10 mL of homogenate into 90 mL, and 1.0 and 0.1 mL of homogenate into a separate 100 mL of melted DTBP agar in Erlenmeyer flasks in duplicate while holding at about 47°C. Agitate the flasks gently to disperse the sample through the medium. Place flasks into the thermostatic shaking water bath. Place an additional uninoculated temperature control Erlenmeyer flask containing 100 mL of the same medium and a thermometer. The water level in the bath should be above the sample level in the flasks. Gently shake the flasks by shaking the water bath during heating. After the internal temperature of the medium reaches 94°C (within 5 min heating), hold the tubes for 15 min at that temperature. After heating, adjust the temperature of the water bath to 47°C and remove the flasks from the bath and rapidly cool the contents of the flasks in tap water (or in cold water) to about 47°C. Place the flasks in the 47°C water bath and hold there until pouring into plates.

Pour each 100 mL of the heated medium into five sterile Petri dishes in approximately equal volumes. Allow the plates to solidify on a flat surface (within 10 min) and incubate inverted plates at 35°C for 48 h. After incubation, count gray–white, vesicle-like, becoming at first dry, and finally wrinkled colonies on DTBP agar. Calculate the average counts for each set of five plates and then for three sets. Calculate the number of mesophilic aerobic rope sporeformers by multiplying the average counts by the dilution rate and dividing by the inoculation amount. Report the results as the number of mesophilic aerobic rope sporeformers cfu per g or mL of sample.

Rope formation mostly is caused by *B. mesentericus* in cereal products. *B. mesentericus* should be identified by morphological and biochemical tests (Technique 19.3). The isolate should be examined microscopically for morphology; ellipsoidal or cylindrical spores do not produce a distinct swelling of the sporangium (spore-forming cell). Perform the following biochemical tests: catalase; acetoin production; nitrate reduction; utilization of citrate; growth in 7% NaCl; acid production from glucose, arabinose, xylose and mannitol; growth at pH 5.7; and hydrolysis of starch, casein, and gelatin. *B. mesentericus* gives positive results for each of these tests.

10.2.4 Counting of mesophilic aerobic gas-producing sporeformers

Prepare the sample and perform heating as indicated in Technique 10.2.2 but place flasks containing homogenate into a water bath at 80°C. After the temperature reaches 80°C, hold the tubes for 15 min at that temperature. After heating, adjust the temperature of the water bath to 47°C. Remove the flasks from the bath and rapidly cool the contents of the flasks in tap water (or in cold water) to 47°C. Prepare necessary dilutions from heated homogenate using 0.1% peptone water. For counting mesophilic aerobic gas-producing sporeformers, inoculate 1 mL of heated and diluted sample into 5-MPN tubes of JPSN broth and mix the contents by vortexing the tubes. Incubate the tubes at 35°C for 3 days. After incubation, look at for gas production in Durham tube (empty space). Indicate the number of positive tube(s) containing gas formation for each dilution. Refer to MPN Table 4.2 and calculate the number of mesophilic aerobic gas-producing sporeformers. Report the results as the number of mesophilic aerobic gas-producing sporeformers MPN per g or mL of sample.

10.3 Mesophilic anaerobic sporeformers

The mesophilic anaerobic sporeformers (MAAS) are Gram-positive rods, distributed in nature, and commonly associated with soil, intestinal tracts, dried-milk products, milk, meats, vegetables spices, cereals, cereal products, dried eggs, dried milk, milk products, dried vegetables (e.g., onions and garlic), and aquatic environments. They belong to the genus *Clostridium* and those of greatest interest in foods fall into two main groups: (1) Proteolytic (putrefactive): *Clostridium sporogenes*, *Clostridium bifermentans*, *Clostridium putrefaciens*, *Clostridium histolyticum*, and *C. botulinum* types A and B; and (2) Nonproteolytic: *Clostridium perfringens*, some strains of *C. botulinum*, and the butyric *Clostridium butyricum* and *Clostridium pasteuranum*.

Proteolytic *Clostridium* decompose proteins, peptides, or amino acids anaerobically with the production of foul-smelling sulfur-containing products (e.g., hydrogen sulfide, methyl and ethyl sulfide, and mercaptans). Ammonia, amines, putrescine, and cadaverine are usually produced along with indole, carbon dioxide, and hydrogen gases. They will grow at 10°C–50°C, with the exception of *C. putrefaciens*, which will grow at 0°C–30°C. Proteolytic *Clostridium* spores have a high heat resistance and are associated with unprocessed low-acid canned food spoilage. They do not grow in foods with a pH of less than 4.6. Nonproteolytic MAAS do not digest protein and they grow at a pH as low as 4.3.

Endospores of *Clostridium* can germinate and grow in low-acid foods. They can cause spoilage in canned food with a pH of 4.8 or above. *Clostridium perfringens* can cause spoilage in cooked meat dishes. *C. botulinum* is the most important anaerobe to control in food processing due to the public health significance of their toxins. The spores of many nontoxic strains of *C. botulinum* are much more heat resistant than the toxic strains; these nontoxic strains can cause severe economic losses through spoilage if it is not controlled by adequate thermal processing.

10.3.1 Equipment, reagents and media

Equipment. Alcohol in beaker (50 mL), Erlenmeyer flask, glass spreader, Petri dishes, pipette discarding equipment, pipettes (1 and 10 mL) in pipette box, Plastic bags, refrigerator, swab, thermostatic shaking water bath, and vortex.

Water. Cold water, sterile distilled water, and sodium thiosulfate (3%) solution.

Media. Liver agar, liver broth, PE-2 medium, fluid thioglycolate (FT) broth, and FT agar.

10.3.2 Sample preparation and isolation of mesophilic anaerobic sporeformers

Place tubes containing 7 mL of double strength liver broth (or double strength PE-2 broth) in tubes in a thermostatic shaking water bath at 75°C. After the internal temperature of the liver broth (or PE-2 broth) reaches 75°C, hold the tubes for 15 min at that temperature to exhaust tubes. Gently swirl the flasks containing the homogenized sample from time to time. After heating cool the medium to 47°C.

Prepare following sample. (1) Transfer 25 g or mL of samples into a sterile Erlenmeyer flask containing 225 mL FT broth and gently shake the flask until mixing the sample; (2) add swab samples into 100 mL of FT broth in the Erlenmeyer flask and gently shake the flask until mixing the sample; (3) add 0.74 mL of sterile 3% sodium thiosulfate into 100 mL of cooling water sample in the Erlenmeyer flask. Gently shake the flask until mixing the sample; (4) add 10 g of sugar or 30 mL of liquid sugar (approximately 67 Brix) and complete to 100 mL with sterile distilled water. Shake the flasks until mixing the sugar in water; (5) add 10 g of starch, flour, or other cereal products and dried eggs, and complete to 100 mL with sterile distilled water. Gently shake the flasks until mixing the sample in water; (6) add 10 g of cutting vegetable into 90 mL of sterile distilled water and shake the flask to mix. Rehydrate for 30 min at refrigerator temperature. Then shake vigorously for 2 to 3 min; and (7) add 10 g of whole spice, 2 g of bulk spice, or 1 g of ground spice and complete to 100 mL with sterile distilled water.

Place homogenate into a thermostatic water bath at 95°C, also place another Erlenmeyer flask containing 100 mL of water, and place a thermometer to read the temperature. The water level should be above the sample level in the flasks. Agitate the flasks by shaking the water bath during heating. After the temperature reaches 95°C, heat the homogenate at that temperature for 20 min. After heating, quickly cool the medium in a flask to 45°C in the cold water. After cooling, allow the particles to settle out for 5 min before culturing. Remove 20 mL of heated sample and distribute equally among six freshly exhausted tubes containing 6 mL of double strength liver broth (or double strength PE-2 broth). Mix the contents and incubate the tubes anaerobically at 35°C for 7 days. After incubation, examine the tubes for the growth (turbidity) of MAAS and indicate the presence or absence of putrefaction of meat in the broth. Decomposition of meat in the broth indicates putrefactive activities of bacteria.

10.3.3 MPN counting of mesophilic anaerobic sporeformers

Refer to Technique 10.3.2 for sample homogenization using 25 g of sample and 225 mL of liver broth. Heat the homogenate as indicated in Technique 10.3.2 and after heating, cool to 45°C. Prepare necessary dilution using FT broth. Inoculate 1 mL of diluted sample into each 3-MPN tube containing 10 mL of liver broth (Technique 4.2). Incubate the tubes at 35°C for 3 days in anaerobic atmosphere by evacuation and refilling of the incubator with carbon dioxide (or in the gas pack system). After incubation, record the number of positive tube(s) for each dilution showing growth (turbidity). Mesophilic anaerobic putrefactive sporeformers (MAAPS) can also be indicated for each dilution by analyzing the tubes for decomposition of meat in the medium. Refer to the MPN Table 4.1 for the calculation of MAAS and MAAPS. Record results as the number of MAAS and MAAPS MPN per g or mL of sample. If tubes are negative, continue to incubate and the final reading should be made within 3 weeks because of the slow germination of spores.

10.3.4 Plate counting of mesophilic anaerobic sporeformers

Pour plate. Refer to Technique 10.3.3 for details of sample preparation and heating of the homogenized sample. Dilute the heated homogenate in FT broth. Add 1 mL of diluted sample into a sterile Petri dish in duplicate and pour 13–14 mL of FT agar (Technique 2.2.3). As each plate is poured, thoroughly mix the medium with the sample in the Petri dish by rotating the plates in the route, drawing a figure 8 while it is lying on the table top, taking care not to splash the mixture over the edge. Allow the plates to solidify (within 10 min) on a flat surface. After solidification, overlay the plates with a heavy layer of FT agar (3–5 mL). Allow the medium to solidify within 10 min.

Invert the plates to prevent spreaders, and promptly place them into an incubator and incubate at 35°C up to 6 days in an anaerobic atmosphere using an evacuated incubator refilled with carbon dioxide (or in the gas pack system). After incubation, count all the colonies from plates containing 25–250 colonies. Calculate the number of MAAS by multiplying the average count by the dilution rate and dividing by the inoculation amount. Record the result as the number of MAAS cfu per g or mL of sample.

Spread plate. Add 1 mL of the diluted sample into the Petri plate containing 13–15 mL of the liver agar in duplicate and spread plate over agar medium using a sterile bent glass rod (Technique 2.2.4). Allow the surface area of the plates to dry (within 10 min). Overlay the plates with a layer of melted liver agar (at 47°C) by adding 4–5 mL and allow it to solidify (within 10 min). Incubate the inverted Petri plates at 35°C for 2 days in an anaerobic atmosphere by evacuating the incubator and refilling with carbon dioxide (or in the gas pack system). After incubation, count all the colonies from plates containing 25–250 colonies. Calculate the number of MAAS by multiplying the average count by dilution rate and dividing by inoculation amount. Record the result as the number of MAAS cfu per g or mL of sample.

10.4 Thermophilic aerobic flat sour sporeformers

Thermophilic aerobic flat sour spoilage in medium acid foods (pH 4.3–5.3), for example, tomato products (pH 4.2–4.8), is caused by *Bacillus coagulans* (formerly: *Bacillus thermoacidurans*). It is a nonpathogenic, motile, Gram-positive, catalase and VP-positive, spore-forming and aerobic bacterium; it grows at 50°C; cannot grow in 7% NaCl; and has as many as 10 flagella per cell. It grows at 21°C–55°C. It is a common soil bacterium, present in canned tomato products (particularly tomato juice, tomato puree, tomato soup, and tomato–vegetable juice mixes), cream, evaporated milk, cheese, and silage. It causes spoilage and curd formation on evaporated milk and spoilage in foods at a temperature as low as 27°C.

Thermophilic flat sour spoilage in low-acid (pH > 5.3) canned foods is caused by *Geobacillus stearothermophilus*. It is a thermophilic facultative aerobic bacterium and grows at a temperature as low as 35°C, unless the product temperature exceeds 43°C and the pH falls due to acid production. Proper cooling after thermal processing and avoiding high temperatures during storage are essential for the prevention of bacterium growth. *G. stearothermophilus* spores have high thermal resistance. The D_{121} value ranges between 4.0 and 5.0 min, and the Z-value between 7.8°C and 10.2°C. Sources of its spores are soil, raw foods, and ingredients (e.g., sugar, spices, starch, and flour). It grows within the thermophilic temperature range from 35°C to 75°C. Insufficient cooling of canned foods after heat treatment and keeping canned foods at a high temperature for a long time allows the growth of thermophilic bacteria.

10.4.1 Equipment, reagents and media

Equipment. Blender (or stomacher), Erlenmeyer flask, incubator, Petri dish, pipettes (1 and 10 mL) in pipette box, screw capped tube, thermometer, thermostatic shaking water bath, and vortex.

Reagents. Gum Arabic, 0.1% peptone water, NaOH (0.02 N), sterile distilled water, and tragacanth.

Media. DTBP agar, litmus milk, and thermoacidurans (TA) agar (pH = 5.0).

10.4.2 Sample preparation

General sample preparation. Refer to Technique 2.2.2 for details of sample preparation and dilutions. Prepare two sets from the following in duplicate. Add 25 g of sample into 225 mL of 0.1% peptone water in the blender (or stomacher). Homogenize by blending for 1 min (or stomaching for 2 min). If the sample is liquid, add 25 mL of sample into 225 mL of 0.1% peptone water in the Erlenmeyer flask and homogenize by gently shaking the flask. Transfer 100 mL of the homogenized sample ino a sterile Erlenmeyer flask. Completely immerse the flask in a thermostatic shaking water bath adjusted to 88°C. Use two extra flasks (containing same medium); one for the control medium inoculated with *B. coagulans* and the other inserted with a thermometer for checking the medium temperature. After reaching a temperature of 88°C (within 5 min), hold the tubes for 10 min at that temperature. After heating, immediately cool the flasks in cold water. The flasks should be gently agitated while cooling as rapidly as possible to 47°C. If necessary, prepare further dilutions by adding 1 mL of homogenized sample into a sterile 99 mL of 0.1% peptone water.

Nonfat dry milk. About 10 g of the nonfat dry milk sample is weighed into a sterile Erlenmeyer flask and completed to 100 mL by NaOH solution (0.02 N) and the bottle should be gently shaken to dissolve the sample completely in hot water. Heat for 10 min at 108.4°C (5 psi) in an autoclave. After heating, the flask should be gently agitated while cooling in cold water as rapidly as possible to 47°C. Bring the volume back to mark with sterile NaOH (0.02 N) solution. If necessary, prepare further dilutions by adding 1 mL of heated homogenate into 99 mL of sterile 0.1% peptone water in the Erlenmeyer flask.

Cream. Add 2 g of tragacanth and 1 g of gum Arabic into a sterile Erlenmeyer flask containing 100 mL of water and homogenize by gently shaking. The mixture is sterilized at 121°C for 20 min. Add 10 mL of cream sample into a sterile Erlenmeyer flask and add the sterilized gum mixture up to the 100 mL mark of the flask, and carefully shake the flask to homogenize the sample in hot water. Heat the mixture in the flask at 108.4°C for 5 min in an autoclave. After heating, the flask should be gently agitated while cooling in cold water as rapidly as possible to 47°C. Bring the volume back to the mark with sterile gum mixture. If necessary, prepare further dilutions by adding 1 mL of heated homogenate into 99 mL of sterile 0.1% peptone water in the Erlenmeyer flask.

Dry sugar. Add 10 g of dry sugar into a sterile Erlenmeyer flask in duplicate and complete to 100 mL with sterile distilled water. Homogenize by gently shaking. Place the flask into a thermostatic shaking water bath at 88°C. Use a second flask containing distilled water and a thermometer for temperature control. After the temperature of the flasks reach 88°C, continue to heat for 5 min. After heating, the flask should be gently agitated while cooling in cold water as rapidly as possible to 47°C. Bring the volume back to the mark with sterile distilled water. If necessary, prepare further dilutions by adding 1 mL of heated homogenate into 99 mL of sterile 0.1% peptone water in the Erlenmeyer flask.

Dry starch. Place 10 g of dry starch in a sterile Erlenmeyer flask and complete to 100 mL with sterile cold water with shaking. Transfer flask into thermostatic shaking water bath at 55°C, and shake the flask in water bath for 3 min to thicken the starch. Then heat at 108.4°C for 10 min in an autoclave. After heating, the flask should be gently agitated while cooling in cold water as rapidly as possible to 47°C. Bring the volume back to the mark with sterile distilled water. If necessary, prepare further dilutions by adding 1 mL of heated homogenate into 99 mL of sterile 0.1% peptone water in the Erlenmeyer flask.

10.4.3 Counting of thermophilic aerobic flat sour sporeformers

Distribute each set of heated 100 mL of sample equally into five sterile Petri dishes in duplicate. Prepare two sets of Petri dishes. Pour 18–19 mL of DTBP agar (at 47°C) into the first set of Petri dishes and pour 18–19 mL of TA agar (at 47°C) into the second set of Petri dishes. Thoroughly mix the medium with the sample in the Petri dish by rotating the plates by drawing a figure of eight while it is lying on the tabletop, taking care not to splash the mixture over the edge. Allow the plates to solidify (usually 10 min) on a flat surface area. After solidification, incubate the inverted Petri plates at 55°C for 3 days. After incubation, examine the plate for colony formation. *B. coagulans* will produce slightly moist, convex, and pale-yellow colonies on the surface of DTBP agar. *B. coagulans* subsurface colonies will be compact with fluffy edges, and slightly yellow. A yellow zone will surround both the surface and subsurface colonies due to acid production. After 48 h incubation, the medium color may turn completely to yellow. *G. stearothermophilus* will produce pinhead-size brown colonies on DTBP agar. Each characteristic colony (three colonies per plate) should be inoculated into litmus milk, *B. coagulans* will cause coagulation in litmus milk. *G. stearothermophilus* will not grow on the TA agar at pH 5.0, and therefore the *B. coagulans* count on the acid medium may have more significance. Typical colonies of *B. coagulans* on the TA agar are large and white to cream in color. Count the typical colonies from each Petri plate. Combine the counts from five plates for each dilution and multiply the average count by dilution rate and divide by inoculation amount. Calculate the number of *G. stearothermophilus* by subtracting the TA count from the DTBP agar or calculate the number *G. stearothermophilus* and *B. coagulans* from the DTBP agar. Record the results as *B. coagulans* (thermophilic aerobic aciduric flat sour sporeformers) and *G. stearothermophilus* (thermophilic aerobic flat sour sporeformers) cfu per g or mL of sample.

10.5 Thermophilic aerobic *Alicyclobacillus acidoterrestris*

A. acidoterrestris is a Gram-positive, obligately aerobic, and rod-shaped bacterium. The heat resistance of the endospores is extremely high ($D_{95°C}$ 1.0–14.5 min, z-value = 6.4°C–11.3°C). *A. acidoterrestris* survives in fruit juices with high acidity (pH < 4.0) for the combination of pasteurization (85°C–96°C for 2 min) with hot-filling and causes spoilage. Endospores have high heat resistance in orange juice (D-values of 65.6 min at 85°C and 11.9 min at 91°C). Endospores can germinate following heat treatment and vegetative cells start to grow in the product up to a level of 10^5–10^6 cfu per mL. It can also have unique physiological characteristics: it grows from pH 2.2 to 6.0 with optimum pH from 3.5 to 5.0 and grows from 20°C to 60°C with an optimum at 42°C–53°C. Pasteurized fruit juices storage below 20°C would be sufficient to prevent the germination of endospores and growth of bacterium, but these juices are usually distributed at ambient conditions (over 20°C). Other species of this genus are *Alicyclobacillus acidiphilus*, *Alicyclobacillus herbarius*, and *Alicyclobacillus pomorum*, and they associate with spoiled herbal tea. Spoilages by *A. acidoterrestris* associate with a slight increasing turbidity, white sediment at the bottom of packages, and faulty taint of strong medicinal or antiseptic flavor due to production of guaiacol, 2,6-dibromophenol, and 2,6-dichlorophenol. *Alicyclobacillus* can cause spoilage on pasteurized fruit juices (mainly apple and orange juices), fruit juice blends, carbonated fruit drinks, berry juice containing iced tea, and diced canned tomatoes. Spoilage requires a combination of adequate conditions, for example, available oxygen, sufficient nutrients, taint precursors, and high temperature storage (over 20°C) for a long time. All fruit juice manufacturers should have a quality assurance programs to monitor and control levels of *Alicyclobacillus* species in raw materials and fruit concentrates.

10.5.1 Equipment, reagents and media

Equipment. Thermostatic shaking water bath, incubator, pH meter, blender (or stomacher), Petri dishes, graduated cylinder, vortex, pipettes (1 and 10 mL) in pipette box, and Erlenmeyer flask.

Solution. Serum physiological (SP) solution (0.85% NaCl)

Media. Orange serum (OS) agar (pH 5.5) and BAM broth.

10.5.2 Isolation and counting techniques of *Alicyclobacillus*

10.5.2.1 Isolation of Alicyclobacillus

Enrichment isolation of Alicyclobacillus. The following samples should be homogenized and used. (1) About 100 mL or g of sample (e.g., soil, fruits, and fruit juices) is obtained randomly and placed into a sterile Erlenmeyer flask in duplicate. Pour 100 mL of sterile SP solution into this flask containing sample. (2) For fruit juice concentrate by ~70 °Brix, add 10 mL of concentrate into 90 mL of BAM broth in the Erlenmeyer flask in duplicate. Then place flasks in a thermostatic shaking water bath for 10 min at 120 rpm. The same amount of medium should be included in another Erlenmeyer flask for temperature control with an inserted thermometer. After that, place the flask in the water bath and incubate at 50°C by 120 rpm stirring for 5 days. After incubation, heat flasks at 80°C for 10 min. This treatment activates the germination of spores and inactivation of vegetative bacteria, yeasts, and molds. After temperature treatment, streak plate one loopful of heated sample to OS agar or spread plate 1 mL of heated culture sample to OS agar (Technique 2.2.4). Incubate plates at 45°C for 48 h. After incubation, examine the plates for *Alicyclobacillus* typical colonies. Identify colonies by colonial morphology. Indicate the presence or absence of bacterium in sample (Technique 10.5.2.1).

Counting of Alicyclobacillus spores. Aseptically weigh 25 g of sample into a sterile blender (or stomacher plastic bag) containing 225 mL of BAM broth. Homogenize by blending for 1 min (or stomach for 2 min) and then pour into Erlenmeyer flasks. In the case of liquid sample, add 25 mL of sample into a sterile Erlenmeyer flask containing 225 mL of BAM broth. Then place the flask containing the homogenized sample in a thermostatic shaking water bath and tighten the cap of the flask securely. Completely immerse the liquid in the tubes in the water bath and adjust the temperature of the water bath to 80°C with 120 rpm shaking. The same amount of medium should be included in another Erlenmeyer flask for temperature control with inserted thermometer. After reaching temperature to 80°C (within 5 min), hold the tubes for 10 min at that temperature. This treatment activates the germination of spores and the inactivation of vegetative bacteria, yeasts, and molds. After heating, the flask should be gently agitated while cooling as rapidly as possible to 35°C in cold water. Prepare further required dilutions using 0.1% peptone water, add 1 mL of heated (diluted) sample into a sterile Petri dish, pour 18–19 mL of melted OS agar (at 47°C), and mix by rotating the plates by drawing a figure of eight while it is lying on the table top, taking care not to splash the mixture over the edge and that there are no bubbles on the surface. Allow the plates to solidify (usually 10 min) on a flat surface. After solidification, incubate the plates at 45°C for 3 days. After incubation, examine the plates for *Alicyclobacillus*' characteristic colony formation. Count the number of typical colonies on the Petri plate containing 25–250 colonies. Calculate the number of *Alicyclobacillus* spores by multiplying the average count by the dilution factor and dividing by the inoculation amount. Record the result as the number of *Alicyclobacillus* spores cfu per g or mL of sample.

10.5.2.2 Identification of Alicyclobacillus

Confirm the presumptive isolates with morphological (Technique 13.4.2), biochemical, and molecular (Technique 15.3.4) identification tests. The molecular technique shows promise for the identification of *Alicyclobacillus* species by PCR assays of the genes coding for 16S rRNA and 23S rRNA (Appendix B, Table A.1).

Growth and taint production. A. acidoterrestris isolate is inoculated into four sets of orange juice, apple juice, and a noncarbonated fruit juice-containing drink at low numbers (20–2000 cells) by using a syringe. Incubate one set of inoculated products at each of the following temperatures; 4°C, 25°C, 35°C, and 44°C for 21 days. Incubated products are removed and assessed by odor for taint formation. Samples showing taint by aroma can be analyzed by GC-Ms.

10.6 Thermophilic anaerobic sporeformers

The type of bacterial species in this group is *Thermoanaerobacterium thermosaccharolyticum*. It is an obligate anaerobic; Gram-positive rod; strongly saccharolytic; and produces acid and gas from glucose, lactose, salicin, and starch. This bacterium cannot hydrolyze proteins and cannot reduce nitrate to nitrite. Spores are terminal and swollen. The Z-value is about 6°C–7°C. Its optimum growth temperature ranges from 55°C to 65°C, it cannot grow below 32°C, and it can cause spoilage on canned foods within 14 days at 37°C after germination of spores at a high temperature. It has an optimum growth pH from 6.2 to 7.2 but grows in food products having a pH of 4.7 or higher, and spoils tomato products at pH 4.7. Soil, mushrooms, onion, spaghetti with tomato sauce, green beans, sweet potatoes, asparagus, and ingredients (e.g., cereals, dehydrated milk, flour, sugar and starch) are the main sources of thermophilic anaerobic sporeformers.

10.6.1 Equipment, reagents, and media

Equipment. Blender (or stomacher), Erlenmeyer flask, thermostatic shaking water bath, pipettes (1 and 10 mL) in pipette box, vortex, and tubes.

Reagents and water. Vaspar, cold water, and sterile distilled water.

Media. Liver broth and PE-2 broth.

10.6.2 Sample preparation

(1) Weigh 10 g of starch, flour, or powdered milk into a sterile Erlenmeyer flask and add sterile distilled water up to a final volume of 100 mL. Homogenize the sample by gently shaking. (2) Weigh 50 g of well-mixed cereal into a sterile blender and add sterile distilled water up to a final

volume of 450 mL. Homogenize by blending for 1 min. (3) Cut mushrooms to small pieces. Weigh 100 g of mushrooms into a sterile blender containing 900 mL of 0.1% peptone water. Blend the sample until the pieces are homogenized. Shake the blender to ensure proper blending.

Prepare further necessary dilutions by transferring 1 mL homogenate into 99 mL of 0.1% peptone water in Erlenmeyer flask.

10.6.3 Counting thermophilic anaerobic sporeformers

Heat freshly prepared sterile 10 mL PE-2 broth (or liver broth) at 75°C in a thermostatic shaking water bath and hold for 20 min at that temperature to remove the oxygen and then cool to 55°C before use. Add 1 mL of each homogenized (or diluted) sample into 5-MPN tubes of PE-2 broth. Place the tubes into the water bath at 88°C. Use a second tube containing distilled water and a thermometer for temperature control. After the temperature reaches 88°C in the flasks, continue to heat the broth in the tubes for 15 min. Then the tubes are removed from the water bath and they are quickly cooled in cold water while shaking. After the cooling of the broth, add 3 mL of sterile 2% agar (at 45°C) into the tube to cover (layer) the surface area of broth in the tubes, allow the tubes to solidify the agar on surface area of broth, heat the tubes again to 55°C, and incubate the tubes at 55°C for 3 days. After incubation, examine tubes for growth (turbidity) and predict the number of positive tubes for each dilution. Refer to MPN Table 4.2 and calculate the number of thermophilic anaerobic sporeformers. Record the result as the number of thermophilic anaerobic sporeformers MPN per g or mL of sample.

10.7 Thermophilic anaerobic sulfide spoilage sporeformers

The sulfide spoilage *Desulfotomaculum nigrificans* (formerly: *Clostridium nigrificans*) grows at 43°C but not at 37°C and grows between 43°C and 70°C with the optimum at 55°C under anaerobic conditions. It is thermophilic and may present in canned foods, for example, peas, mushroom products, sweet corn, infant foods, and other nonacid foods.

10.7.1 Equipment, reagents and media

Equipment. Erlenmeyer flasks closed with a rubber stopper, incubator, Petri dishes, screw-cap tubes (20 × 150 mm) with rubber stoppers, thermometer, vortex, and thermostatic shaking water bath.

Reagents and water. 0.1% Peptone water, sterile distilled water, cold water, ferric citrate (5%), gum Arabic, sulfide iron (SI) agar, tragacanth, HCl, and NaOH solution (0.02 N).

Media. Sulfite agar.

10.7.2 Sample preparation

If necessary, dilute the homogenate by adding 1 mL of homogenate into 99 mL of 0.1% peptone water in a sterile Erlenmeyer flask and follow the respective heating.

(1) Add 20 g of food sample (e.g., dry sugar, flour or starch) into the sterile Erlenmeyer flask and add sterile distilled water to the flask to complete the volume to 100 mL. Shake the flask to dissolve the sample in water. In the case of starch and flour, the tubes should be swirled manually several times before heating and during the 15 min heating period in the water bath to ensure even dispersion of the starch and flour. Place the flask into the thermostatic shaking water bath at 88°C. Use a second flask containing distilled water and a thermometer for temperature control. After the sample reaches a temperature of 88°C, continue to heat for 10 min. After heating, cool the mixture immediately in cold water to 55°C. (2) Add 10 g of the skim milk into a sterile Erlenmeyer flask, add NaOH (0.02 N) and complete to 100 mL. Shake the flask to dissolve the sample completely. Heat at 108°C in an autoclave for 10 min, and then cool the mixture immediately in cold water to 55°C. (3) Add 2 g of tragacanth and 1 g of gum Arabic into a sterile Erlenmeyer flask containing sterile 100 mL of distilled water and mix. Sterilize at 121°C for 20 min. Add 10 mL of sample into a sterile Erlenmeyer flask, add the sterilized gum mixture and complete to 100 mL. Carefully shake the flask to mix. Heat at 108°C for 5 min in an autoclave. After heating, cool the mixture immediately in cold water to 55°C. (4) Prepare a 10% soy protein solution in sterile 0.1% peptone water in the Erlenmeyer flask. Adjust pH to 7.0. Heat in an autoclave at 108°C for 20 min. After heating, cool the mixture immediately in cold water to 55°C.

10.7.3 Counting thermophilic anaerobic sulfide spoilage sporeformers

Thermophilic anaerobic sulfide spoilage sporeformers are extremely sensitive to oxygen; the inoculum should be added below the surface of the medium in the tube to obtain the maximum count. Equally distribute 20 mL of heated sample among six tubes of 10 mL double strength SI agar (at 55°C) from each dilution in duplicate. Heat SI agar in tubes to 55°C in thermostatic shaking water bath while gently shaking. Incubate the tubes at 55°C for 14 days. After incubation, *D. nigrificans* will appear throughout the agar medium in the tube as jet-black spherical areas due to the iron sulfide formation. No gas (bubble) production through medium. Preliminary counts should be made from the six tubes after 2 days, 7 days, and 14 days incubation. Combine counts from six tubes for each dilution. Calculate the number of thermophilic anaerobic sulfide spoilage sporeformers by multiplying the average count by the dilution rate and dividing by 20. Report the results as the number of thermophilic anaerobic sulfide spoilage sporeformers per g or mL of sample.

Thermophilic anaerobic nonsulfide spoilage sporeformers can also grow without black color in this medium. Count the thermophilic anaerobic nonsulfide spoilage sporeformers and calculate their numbers. Record the result as the number of thermophilic anaerobic nonsulfide spoilage sporeformers cfu per g or mL of sample.

10.8 Interpretation of results

The data obtained from the plant survey will provide the information about sanitation program in the canning plant. The aciduric flat sour spores, *B. coagulans*, can survive in canned products from pH range from 4.1 to 5.0. *G. stearothermophilus* will not grow in this pH range. Thermophilic flat sour spores in sugar or starch (as ingredients) shall be not more than 50 spores per 10 g of sugar (or starch). The total thermophilic spore counts for canners shall be not more than 125 spores per 10 g of sugar (or starch). Thermophilic spores do not grow in commercially sterile canned foods during storage and distribution where the temperature will not exceed about 32°C.

Nonhydrogen sulfide-producing thermophilic anaerobic sporeformers should not be present in more than 60% of the samples tested or in more than 66% of the tubes for any single sample. Canned foods with a pH below 4.0 are not spoiled by thermophilic bacteria. The presence of these bacteria on equipment and food processing systems suggests that equipment is in need of thorough cleaning and sanitation. Sulfide spoilage spores should be present in not more than two of the five samples, and in any samples to the extent of not more than five spores per 10 g. The presence of rope-forming bacteria in bread indicates the absence of good sanitation, modern baking practices, and preservatives combined to keep rope spoilage under control conditions.

PRACTICE

Counting of halophilic, osmophilic, and xerophilic microorganisms 11

11.1 Introduction

Three major types of extreme environment provide habitats for some groups of microorganisms: foods preserved by dehydration, enhancing amounts of sugars, and increasing hypersaline environments. The availability of water in these environments is limited by a high concentration of salts, sugars, or other solids. The microbial habitats in low water activity (dehydrated) foods are dominated by xerophilic molds (dry-loving). The microbial habitats in high-sugar foods are dominated by osmophilic yeasts. The mcrobial habitats in high-salt foods or high salt environment are dominated by halophilic bacteria, yeasts, and molds. Generally, xerophilic microorganisms grow at water activity (a_w) values below 0.85. Xerophilic molds (e.g., *Aspergillus patulum*, *Aspergillus glaucus*, *Aspergillus conicus*, and *Aspergillus flavus*) grow up to a_w 0.75 and another xerophilic mold *Xeromyces bisporus* grows up to a_w 0.70. *Zygosaccharomyces* spp. (osmophilic yeast) grow up to a_w 0.60. Microorganisms that require minimum concentrations of salt (sodium chloride, other cations, or anions) are called halophilic. Halophilic bacteria (e.g., *Halobacterium salinarum*) can spoil brined foods, such as fish. Halophilic molds (e.g., *Wallemia sebi*) will also grow on fish. Halophilic lactococci (e.g., *Tetragenococcus halophilus* and *Tetragenococcus muriatianus*) can be used as starter cultures in the manufacturing of soy sauce and fermented liver sauce.

Caution: Sampling, sample preparation, and all experimental techniques should be performed under aseptic conditions, and in duplicate, unless otherwise specified. All microbiological media to be used in applications must be sterilized. Again, all solutions and equipment that may contaminate microorganisms must be sterilized or disinfected.

11.2 Halophilic microorganisms

Halophilic microorganisms require a minimum level of salt for growth. Halophilic bacteria cannot grow in an environment containing less than 0.5% NaCl. They can be classified depending on the level of salt requirement for growth. Slightly halophilic microorganisms grow optimally in media containing 0.5%–3.0% salt. Human pathogens, for example, *Staphylococcus aureus*, *Clostridium perfringens*, and some strains of *Clostridium botulinum*, are slightly halophilic and responsible for food poisoning outbreaks involving low-salted foods. Slightly halophilic spoilage microorganisms are *Pseudomonas*, *Moraxella*, *Acinetobacter*, and *Flavobacterium*. Moderate halophilic microorganisms grow optimally in media containing 3.1%–15% salt. They are *Micrococcus halodinitrificans*, *Paracoccus halodenitrificans*, *Planococcus halophilus*, *Vibrio*, and *Achromobacter*. Extremely halophilic microorganisms grow optimally in salted foods with 15.1%–30% salt. Extremely halophilic bacteria *Halococcus* and *Halobacterium* produce bright red or pink pigments on the salted foods.

Microbiological Analysis of Foods and Food Processing Environments. DOI: https://doi.org/10.1016/B978-0-323-91651-6.00043-4

They present in aquatic environments containing higher amount of salt and in solar evaporated sea salts. Additionally, microorganisms that can grow in NaCl concentrations over 5% as well as grow without salt are called halotolerant. Halotolerant bacterial species are present in the genera *Micrococcus*, *Bacillus*, and *Corynobacteria*. Some molds are halophilic, being well adapted to salty environments, for example, salted fish. Mold species *Basipetospora halophila* and *Polypaecilum pisce* grow more rapidly in media containing NaCl. Several species of yeasts grow in salted foods. Salt-tolerant yeast species are present in the genus *Debaryomyces* (*Debaryomyces hansenii*), *Hansenula* (*Hansenula anomala*), and *Candida* (*Candida pseudotropicalis*). They grow well on cured meats and pickles at NaCl concentrations up to 11% (a_w, 0.93).

Low-salt foods (containing 1%–7% salt by weight) are more susceptible to microbiological spoilage and are more likely to contain human pathogens. High-pickled foods (more than 8% NaCl) will not deteriorate easily unless stored at high temperatures. As a result of the dilution of seafood with distilled water, many halophilic microorganisms are lysed. Avoid washing the food sample with distilled water since a hypertonic environment inactivates halophilic microorganisms, and therefore diluents containing less than 3% NaCl cannot be used in the homogenization and dilution of samples for a halophilic count. Halophilic bacteria are counted by plate count techniques.

11.2.1 Counting halophilic microorganisms

11.2.1.1 Equipment, reagents, and media

Equipment. Pipettes (1 and 10 mL) in pipette box, incubator, blender (or stomacher), Erlenmeyer flask, and vortex.

Reagents. Phosphate buffer (PB) solution supplemented with NaCl equivalent to the salt concentration of sample, PB-3% NaCl solution, and PB containing 40% glucose (PB40G) solution.

Media. Halophilic agar, halophilic broth, dichloran glycerol (G18) agar, polypectate gel (PG) agar, seawater (SW) agar, and trypticase soy (TS) agar supplemented with NaCl equivalent to salt concentration of sample (TSA-NaCl agar).

11.2.1.2 Sampling

Refer to Technique 2.2.2 for details of sample preparation and dilutions. Low-salted foods should be tested within 24 h for halophilic spoilage microorganisms, otherwise growth will occur. Samples should be stored at 5°C until tested for a short time. Prepare the following homogenate: (1) General sample. Add 50 g of sample (e.g., slightly salted meat, vegetable, shellfish, and fish) into a sterile blender (or stomacher plastic bag) containing 450 mL of sterile PB solution (containing salt equivalent to the salt concentration of sample). Homogenize sample by blending for 2 min (or stomaching for 3 min). (2) Fish skin sample (1.6 cm diameter). Weigh 50 g of skin to a sterile blender (or stomacher plastic bag) containing 450 mL of sterile PB-3% NaCl solution or NaCl equivalent to salt concentration of food sample. Homogenize sample by blending for 2 min (or stomaching for 3 min). (3) Fish flesh, fillets, or small whole fish (less than 6 cm). Remove slices (2.5 cm^2) and weigh 50 g of slices into a sterile blender (or stomacher plastic bag) containing 450 mL of sterile PB-3% NaCl solution, and homogenize sample by blending for 2 min (or stomaching for 3 min). (4) Shellfish. Weigh 50 g of slices into a sterile blender (or stomacher plastic bag) containing

450 mL of sterile PB-3% NaCl solution and homogenize sample by blending for 2 min (or stomaching for 3 min). (5) Lightly salted meats and vegetables (from 1% to 7% NaCl), weigh 50 g of sample into a sterile blender (or stomacher plastic bag) containing 450 mL of sterile PB-3% NaCl solution or NaCl equivalent to salt concentration of food sample. Homogenize sample by blending for 2 min (or stomaching for 3 min). Prepare further required dilutions from homogenized sample using diluent as used in the homogenization of sample.

11.2.1.3 Counting of slightly halophilic bacteria

Spread plate 1 mL of diluted sample to each of two sets of TS agar-3% NaCl (or SW agar) plates (Technique 2.2.4). Incubate one set of plates at 7°C for 7 days (or at 25°C for 5 days) for the counting of psychrotrophic slightly halophilic bacteria; and the other set at 32°C for 3 days for the counting of mesophilic slightly halophilic bacteria. After incubation, examine colonies from Petri plates. Count colonies from plates containing 25 to 250 colonies. Calculate the number of each slightly halophilic bacteria separately by multiplying the average count by dilution rate and dividing by the inoculation amount. Report the results as the number of psychrotrophic slightly halophilic bacteria and mesophilic slightly halophilic bacteria colony forming unit (cfu) per g or mL of sample.

11.2.1.4 Counting of moderate halophilic bacteria

Spread plate 1 mL of sample (diluted or nondiluted) to TS agar containing the salt equivalent of the salt concentration of sample (Technique 2.2.4). Incubate inverted plates at 25°C for 3 days. After incubation, examine the colonies from Petri plates. Count the colonies from plates containing 25 to 250 colonies. Calculate the number of moderate halophilic bacteria by multiplying the average count by the dilution rate and dividing by the inoculation amount. Report the result as the number of moderate halophilic bacteria cfu per g or mL of sample.

11.2.1.5 Counting of extremely halophilic bacteria

Extreme halophilic bacteria normally are present in aquatic environments (usually high salt concentrations) and in solar evaporated sea salts. The microorganisms require a minimum of 24% NaCl for growth. They can grow optimally in media and foods containing 15%–30% NaCl. Spread plate 1 mL of sample (diluted or nondiluted) to TS agar containing the salt equivalent of the salt concentration of the sample. Incubate inverted plates at 25°C for 3 days. After incubation, examine the colonies from the Petri plates. Count colonies from plates containing 25 to 250 colonies. Calculate the number of extremely halophilic bacteria by multiplying the average count by the dilution rate and dividing by the inoculation amount. Report the results as the number of extremely halophilic bacteria cfu per g or mL of sample.

11.2.1.6 Counting of halotolerant yeasts

Follow the plate count for yeasts as described in Technique 6.2.3 and use NaCl in agar medium (PG agar) equivalent to the concentration of the sample. Incubate inverted plates at 25°C for 3 days. After incubation, count colonies from plates containing 25 to 250 colonies. Calculate the number of halotolerant yeasts by multiplying the average count by the dilution rate and dividing by the inoculation amount. Report the results as the number of halotolerant yeasts cfu per g or mL of sample.

11.2.2 Enrichment isolation of halophilic bacteria

When a low number of halophilic bacteria is present in the sample, add 50 g or mL of sample into 450 mL halophilic broth in the Erlenmeyer flask in duplicate and homogenize by gently shaking. Incubate flasks at 32°C for 12 days. During incubation, examine the flasks for growth (turbidity). Then streak plate from incubated halophilic broth onto halophilic agar plates (Technique 13.4.1.1) and incubate inverted plates at 32°C for 12 days. After incubation, the presence of pink or red colonies on plates is indicative of halophiles in the sample. Report the results as the presence or absence of halophilic bacteria in the sample.

11.2.3 Interpretation of results

In general, fish contaminated with more than 10 psychrotrophic bacteria per cm^2 of skin or per g is considered to be spoiled fish. The most common seafood spoilage bacteria are *Pseudomonas* species, which are psychrotrophic and actively proteolytic. Organoleptic tests (odor and visual evidence of spoilage, such as slime or gas formation) can be used together with bacterial counts to determine the extent of spoilage. For heavily brined foods, only an organoleptic observation (presence of red or pink slime and putrefaction) can be used to check spoilage. *S. aureus* food poisoning may be associated with salted foods. Bacterial activity can be indicated in salted foods with the evolution of high total volatile nitrogen and trimethylamine values.

11.3 Osmophilic microorganisms

Microorganisms growing in high concentrations of organic matter (especially sugar) are called osmophilic. The osmophilic microorganisms most commonly associating with foods are yeasts. They can grow in highly concentrated sugar solutions. They can cause spoilage on chocolate candy, honey, jams, molasses, flavored syrups, corn syrup, concentrated fruit juices, and other similar products. Osmophilic yeasts are not important for public health, but they are of economic importance for the food industry due to their spoilage characteristics. Osmophilic yeast species are present in the genus *Zygosaccharomyces* (e.g., *Zygosaccharomyces mellis*, *Zygosaccharomyces bailii*, and *Zygosaccharomyces rouxii*; formerly: *Saccharomyces mellis*, *Saccharomyces bailii*, and *Saccharomyces rouxii*).

Sterilization of high-sugar media may result in the formation of toxic components for osmophilic microorganisms. By heating the medium containing high-sugar concentration at 100°C for 30 min, all microorganisms except bacterial spores are destroyed. The pour plate technique is used in the counting of osmophilic microorganisms from food products. The membrane filtration technique can be used for products containing a low number of microorganisms using a higher amount of sample. The membrane filtration technique in combination with fluorescent staining for counting the number of osmophilic yeasts is a recent technique. Different media can be used for the counting of osmophilic microorganisms.

11.3.1 Counting of osmophilic yeasts

11.3.1.1 Equipment, reagents, and media

Equipment. Blender (or Stomacher), Erlenmeyer flask, filter paper pad, filtration system, forceps, incubator, membrane filter (0.45 μm pore size), Petri dishes, pipettes (1 and 10 mL) in pipette box, vacuum source, and vortex.

Reagents and water. Glycerol formaldehyde solution, Liefler's methylene blue (LMB) stain, phosphate-buffer water containing 40% glucose (PBW40G) solution, and sterile distilled water.

Media. Malt extract-yeast extract 40% glucose agar (MEYE40G) agar, MEYE40G broth, and potato dextrose (PD) containing 40% glucose (PD40G, pH 5.3).

11.3.1.2 Plate count technique

Diluents with reduced a_w by sugar are needed to detect and count osmophilic yeasts from food samples. When a hypotonic solution is used in the preparation and dilution of sample, osmophilic yeasts may be destroyed. Use PBW40G solution (hypertonic solution) in the homogenization and dilutions of sample. Aseptically, weigh 50 g of processed solid food sample into a sterile blender (or stomacher bag) and add 450 mL of room temperature PBW40G solution. Blend solid foods at low speed (8000 rpm) for 1 min (or stomach for 2 min). For a liquid sample, add 50 mL of the sample into a sterile Erlenmeyer flask containing 450 mL of room temperature PBW40G solution and homogenize by gently shaking. Prepare further dilutions using PBW40G solution.

Spread plate 1 mL of sample onto MEYE40G agar (or acidified PD40G agar) (Technique 2.2.4). Incubate inverted plates at 25°C for 7 days. After incubation, count colonies from plates with the help of a colony counter that will facilitate counting because many types of yeast grow slowly and thus their colonies are difficult to see with the unaided eye. Report results as the number of osmophilic yeasts cfu per g or mL of sample.

11.3.1.3 Membrane filter technique

Weigh 75 g of the 67–70 Brix liquid sugar sample aseptically into a sterile Erlenmeyer flask containing 175 mL of sterile distilled water and homogenize by gently shaking. This results in a solution of approximately 20 Brix. Refer to Technique 5.2 for the complete details of sample filtration. Connect the filter flask to a vacuum source, apply suction, and filter the homogenized sample under aseptic conditions through a membrane filter (0.45 mm). Filtration requires less than 5 min. Rinse walls of sample flask with about 100 mL of sterile distilled water. Likewise, flush sides of funnel with two portions of 100 mL of sterile distilled water. Allow each sample to be pulled through the filter before the next is added.

Follow one of the culture techniques:

Petri plates. Remove the membrane filter with sterile forceps and place grid side up on an absorbent pad saturated with broth. The membrane filter should be carefully rolled on the surface of the pad to avoid the entrapment of air bubbles. Place filter paper on MEYE40G agar. Incubate the plates at 30°C for 5 days. After incubation, count colonies from membrane filter containing 15 to 150 colonies. Calculate the number of osmophilic yeasts by dividing by 75 g and record the results as the number of osmophilic yeasts cfu per g or mL of the sample.

Pad in sterile Petri dish. Place a sterile absorbent pad in bottom of a sterile Petri dish. Pipette 1.8–2.0 mL of MEYE40G broth onto the pad. The pad should be saturated uniformly wet.

Incubate Petri dish at 30°C for 5 days. After incubation, examine colonies from the membrane filter. Count colonies from the membrane filter containing 15 to 150 colonies. Calculate the number of osmophilic yeasts by dividing by 75 g and record the results as the number of osmophilic yeasts cfu per g or mL of the sample.

Staining. Differentiation between colonies and membrane filter may not be possible. This can be provided by staining the membrane filter. After incubation, transfer the membrane filter (colony side up) to an absorbent pad in a Petri dish and saturate with LMB stain. Allow for 2 min to stain. Fix colonies for 4 min by transferring the disk onto another absorbent pad soaked in glycerol formaldehyde solution. Carefully transfer the disk containing fixed colonies to a dry Petri dish. Place into an oven at 80°C for 10 min or until drying (pad changes from dark to pale blue when dry). Count the colonies under microscope and record result as above.

11.4 Xerophilic molds

Xerophilic is defined as microorganisms that grow at low water activity (a_w). Specific solutes that affect microorganisms are important in the isolation and counting of xerophilic molds. Most xerophilic microorganisms (e.g., *Eurotium* spp.), extreme xerophiles (e.g., *X. bisporus* and *Chrysosporium* species), and osmophilic yeasts (e.g., *Zygosaccharomyces*) generally grow better when a_w is controlled by glucose or sucrose, rather than an ionic solute, for example, NaCl.

Some "osmophilic" yeast species (such as *Z. rouxii, Z. bailii*, and *Zygosaccharomyces bisporus)* grow in high-sugar containing foods (such as jams, honey, and syrups) with low a_w level up to 0.60. Molds grow at much lower a_w than those of other foodborne microbial groups. The minimum a_w value for the growth of most molds is 0.85. Most mold spore germination fermentations may be inhibited by a_w requirements below 0.93. The minimum a_w for xerophilic molds (e.g., *A. glaucus*) ranges from 0.71 to 0.77, while the optimum a_w is 0.96. True xerophilic (dry-loving) molds (e.g., *Monascus bisporus*) do not grow at a_w levels greater than 0.97 and lower than 0.65.

11.4.1 Equipment, reagents, and media

Equipment. Blender (or Stomacher), Petri dishes, pipettes (1 mL), incubator, and vortex.

Reagents and water. Distilled water, glycerol formaldehyde solution, and phosphate buffer water (PBW) containing 40%–50% glucose.

Media. Dichloran glycerol 18% (G18) agar, malt extract yeast extract 40% glucose agar (MY40GA), and potato dextrose (PD) agar (acidified).

11.4.2 Counting of xerophilic molds

Dissolve 10 g of food sample (e.g., honey) in 90 mL of 0.1% peptone water in a sterile Erlenmeyer flask. Prepare further required dilutions using 0.1% peptone water. Spread plate 1 mL of homogenized sample onto each of two G18 agars (without dichloran) (Technique 2.2.4) in duplicate. Once the medium absorbs the water, incubate the inverted plates, one in the dark at 15°C and the other at 25°C for up to 2 months. After incubation, examine the colonies on the plates. Count the colonies from plates containing colonies from 15 to 150 colonies. Calculate the number of xerophilic microorganisms by multiplying the average count by the dilution rate and dividing by the inoculation amount. Record the result as the number of xerophilic microorganisms cfu per g or mL of sample.

11.5 Interpretation of results

Acceptable levels of osmophilic microorganisms are based on the type of product. The presence of osmophilic microorganisms in high-sugar products may result in spoilage. High counts of yeast (>10 cfu/mL or g) may be indicative of an inadequate process that did not destroy yeasts originating from raw materials or be indicative of postprocessing contamination from equipment, air, or packaging materials. The presence of a particular osmophilic microorganism at any level is unacceptable. Counting 10 colonies or fewer per g or mL of products may be considered as highly significant if the yeast is *Zygosaccharomyces*.

PRACTICE

Counting of thermoduric microorganisms

12

12.1 Introduction

Thermally processed fruits and fruit products can be spoiled by heat-resistant thermoduric microorganisms. Thermoduric microorganisms survive heat treatment. Molds include thermoduric species. Thermoduric mold genera are *Byssochlamys* (e.g., *Byssochlamys fulva* and *Byssochlamys nivea*), *Neosartorya* (e.g., *Neosartorya fischeri*), *Talaromyces* (e.g., *Talaromyces flavus*), and *Eupenicillium* (e.g., *Eupenicillium brefeldianum*). They are characterized by the formation of ascospores that frequently show heat resistance, in some cases comparable to bacterial spores. Ascospores survive at hot-fill temperatures (below 100°C) of processed acidic foods (e.g., fruit products). As a result of the germination of ascospores, visible mycelium may occur in fruit products. *Byssochlamys* spp. are important spoilage thermoduric molds compared with others, producing pectic enzymes that can cause deterioration in the tissues of the fruits with the production of bad taste. Some *Byssochlamys* species produce mycotoxins (e.g., patulin, byssotoxin A, and byssochlamic acid). *N. fischeri* produces fumitremorgin A, B, and C, and terrein. Toxin-producing heat-resistant molds may cause public health hazards as well as spoilage problems. Heat-resistant mold ascospores require heat activation before growth in foods. *B. fulva* and *N. fischeri* ascospores' maximum activation occurs at 70°C within 30 min in fruit juices (e.g., grape juice). *Byssochlamys* ascospores survive from 60°C to 90°C thermal processing. The minimum, maximum, and optimum growth temperatures of *Byssochlamys* are 9°C, 52°C, and 32°C. *Byssochlamys* produces ascospores at 32°C within 4–7 days. The minimum water activity (a_w) level is 0.88. *B. nivea* can produce patulin at a_w 0.92°C and 21°C. *Byssochlamys* ascospores have a D-value at 90°C between 1 and 12 min.

Thermoduric bacterial species are present in the genera *Arthrobacter*, *Bacillus*, *Clostridium*, *Enterococcus*, *Lactobacillus*, *Microbacterium*, *Micrococcus*, *Pediococcus*, and *Streptococcus*. Thermoduric bacteria are "heat resistant" and able to survive at pasteurization temperature, and have the potential to cause loss of quality and shelf life in foods (e.g., dairy products). The thermoduric bacteria in raw milk are able to grow under refrigeration (e.g., *Bacillus cereus* and *Bacillus circulans*). Packaging of milk extends shelf life against *Bacillus* spp. Thermoduric bacteria have an optimum growth temperature between 50°C and 55°C, and grow from 20°C to 65°C.

Many media support the germination of ascospores and the vegetative growth of thermoduric molds. When molds are counted from foods where bacterial spores will be abundant, acidification or the addition of chloramphenicol to the medium can inhibit bacteria. But antibiotic use is not suitable for an acidified medium for spore germination. *Byssochlamys* spores germinate and grow very well in media at pH 3.0 and lower. Thermoduric bacteria can be counted selectively from heat-treated samples by plate count techniques.

Caution: Sampling, sample preparation, and all experimental techniques should be performed under aseptic conditions, and in duplicate, unless otherwise specified. All microbiological media to be used in applications must be sterilized. Again, all solutions and equipment that may contaminate microorganisms must be sterilized or disinfected.

Microbiological Analysis of Foods and Food Processing Environments. DOI: https://doi.org/10.1016/B978-0-323-91651-6.00032-X

12.2 Heat resistant molds

12.2.1 Equipment, reagents, and media

Equipment. Blender (or stomacher), incubator, and thermostatic shaking water bath.

Media. Malt extract (ME) agar, orange serum agar, plate count (PC) agar, Czapek yeast autolysate (CYA) agar, fruit juice, potato dextrose (PD) agar containing 8.0 μg rose Bengal dye per mL of medium (pH 5.6), and PD agar at pH 5.6.

12.2.2 Counting of heat resistant ascospores

There are low incidences of *Byssochlamys* ascospores on many foods, so 100 g or more of grapes, apples, and other fruits samples are cultured for the counting of spores. Centrifugation may be used to concentrate the spores from fruit juices and other liquid foods. Food samples can be frozen prior to culturing for *Byssochlamys* spores. The isolation and counting of *Byssochlamys* may also depend on heating the sample to activate dormant spores and to destroy aciduric vegetative microorganisms.

Weigh approximately 100 g of fruit into a sterile screw cap blender containing 100 mL of sterile distilled water. Blend until the mixture appears homogeneous (within 2 min) and pour homogenized mixture into a sterile Erlenmeyer flask. For a liquid sample, add 100 mL of fruit juice into an Erlenmeyer flask containing 100 mL of distilled water. Homogenize by gently shaking. Place the flask into a thermostatic shaking water bath at 70°C. The water level in the water bath should be 2 cm higher than the liquid level in the flask. Heat the homogenized sample in the water bath at 70°C for 1 h. After heating, cool the mixture immediately to 45°C in cold water and prepare further dilutions using 0.1% peptone water (add 1 mL of heated sample into 99 mL of diluent). Distribute 100 mL of the diluted sample into sterile Petri dishes (10 plates) with about 10 mL per plate. Add equal volumes of acidified double strength PD agar (at 47°C) into the plates. Mix by rotating the plates by drawing a figure of eight while lying on the table top, taking care not to splash the mixture over the edge and that there are no bubbles on the surface. Allow the plates to solidify (within 10 min) on a flat surface. After solidification, incubate inverted plates at 32°C for up to 5 days. After incubation, count colonies from plates containing 15–150 colonies. Combine counts from 10 plates for each dilution. Calculate the number of thermoduric mold ascospores by dividing the average count by the dilution rate and the inoculation amount (100 mL). Report results as the number of thermoduric molds ascospores colony forming units (cfu) per g or mL of sample.

Most probable number (MPN) technique using a respective broth has the advantage that relatively large samples can be analyzed. This technique can be applicable after heat treatment of the sample and culturing in acidified PD broth in MPN tubes.

12.2.3 Identification of isolates

Confirm isolated colonies by microscopic techniques. The acidified medium can also permit the growth of other molds, e.g., *Penicillium* and *Aspergillus*. *Byssochlamys* are recognized by the structure of their conidiophores and nonenclosed asci. *Byssochlamys* form abundant conidiophores and ascospores after culturing for 1 week. Some species of the genus *Byssochlamys* can reproduce asexually by the production of conidiospores and ascospores (Fig. 12.1). When observed under the

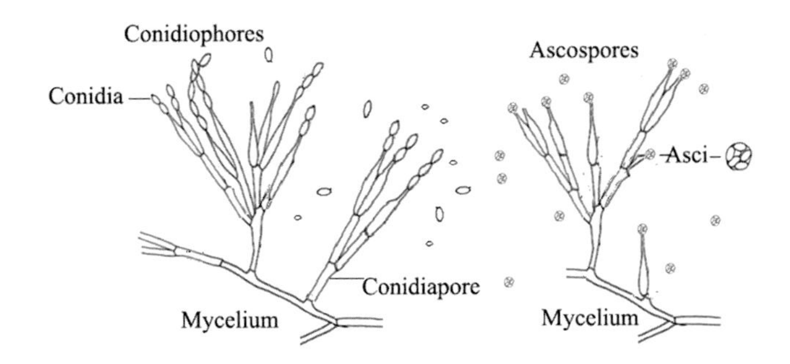

FIGURE 12.1

Appearance of *Byssochlamys* hyphae and spores.

microscope, asci (9–12 μm in diameter) are seen as single, naked spherical clusters, each of which contains eight ascospores (each ascospore is 5–7 μm long); they rarely release free ascospores. Conidial colors are buff to brown or pale brown to yellow. Mature cultures of *B. nivea* release free ascospores from asci. The heat-resistant *Byssochlamys* species can be distinguished by the color of their colonies on acidified agar medium after 14 days of incubation: *B. nivea* predominantly produces white colonies, and *B. fulva* produces dull yellow color colonies. The colonies on agar start as small mycelia zones and spread rapidly to cover the entire plate. The spreading of colonies prevents accurate counts when the colony number is over 10–15 per plate. The addition of rose Bengal dye into PD agar (pH 5.6) restricts colony spread and permits as many as 150 colonies per plate to be readily counted. This concentration of dye cannot inhibit the viable recovery of ascospores subjected to thermal stress.

Select at least three characteristic colonies per plate and use in identification tests. Inoculate isolated mold colonies onto two sets of CYA agar and ME agar by streak plating (Technique 13.4.1.1). Incubate inverted one set of CYA agar and ME agar plates at 25°C, and the others at 32°C for up to 7 days. After incubation, examine the plates, measure colony diameter with a ruler, and make a wet mount (Technique 13.4.2) to magnify mold morphology under a microscope. The following key will assist in the identification of heat-resistant ascospore forming molds.

1. *B. fulva* colonies on CYA agar and ME agar are predominantly buff or brown color. *B. nivea* colonies on these media are persistently white or cream color.
2. *N. fischeri* has discrete bodies of asci with totally enclosed walls.
3. *T. flavus* has asci in enclosed hyphae, and open asci on hyphae and fine hyphae may be present.

12.3 Counting of thermoduric bacteria

Weigh 50 g of ingredient or food sample into a sterile blender (or stomacher plastic bag) containing 450 mL of sterile 0.1% peptone water. Homogenize by blending for 1 min (or stomaching for 2 min). A liquid food sample can also be directly used without homogenization. Add 5 mL of the sample into a sterile test tube in duplicate. Use caution to avoid contamination of sample from the above level of tube filled by the sample. Place tubes into a thermostatic shaking water bath at

62.8°C ± 0.5°C. The tubes may be immersed with the water line approximately 4 cm above the sample level. Include a control tube containing 5 mL of 0.1% peptone water and inserted with a thermometer. Hold the tubes at that temperature for 30 min. After heating, cool the sample immediately below 35°C by immersion in ice water during shaking. Prepare further necessary dilutions using 0.1% peptone water. Determine the thermoduric bacteria by the pour plate technique using PC agar (Technique 2.2.3). Incubate inverted plates at 42°C for up to 3 days. After incubation, count colonies from plates containing 25–250 colonies. Calculate the number of thermoduric bacteria by multiplying the average count by dilution rate and dividing by the inoculation amount. Record the result as the number of thermoduric bacteria cfu per g or mL of sample.

12.4 Interpretation of results

The presence of *Byssochlamys* ascospores is not uncommon on fruit when received from the processing plant. The average *Byssochlamys* ascospores count is generally low at 10 ascospores per 100 g or mL of foods. This level of contamination on raw fruit usually does not cause a spoilage problem to the processor because various processing steps (e.g., washing and filtering) remove the majority of the ascospores. The heat is high enough the ascospores are destroyed. *Byssochlamys* should be absent in 100 g or mL of processed foods. Thermoduric bacterial count is widely used in the milk industry for utensil sanitation and detecting sources of thermoduric microorganisms in pasteurized products. Thermoduric standard plate count of milk should not exceed 2×10^4 cfu mL^{-1}. The levels of thermoduric microorganisms tolerated in processed milk varies in different situation, but counts of greater than 10^3 cfu mL^{-1} would require correction to be performed.

SECTION

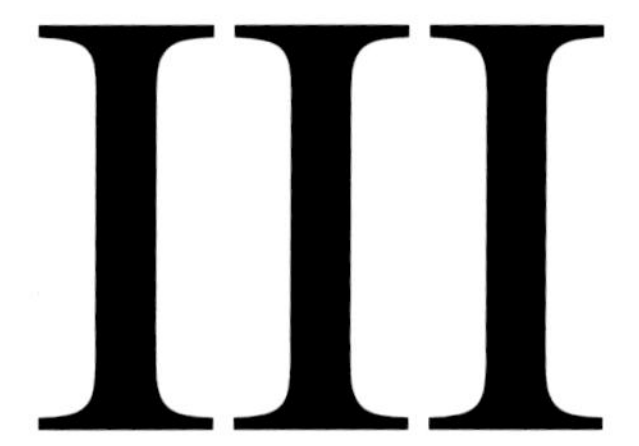

Isolation and counting of indicator and pathogenic microorganisms

The quality and safety of water and food are always important aspects for healthy living. Most of the infections in humans result from improper handling and processing conditions. The contamination of drinking water and foods with pathogenic microorganisms are associated with the spread of illnesses that can cause health problems and mortality. According to the WHO, about 5 million people die each year due to untreated water and poor hygiene practices during food handling, production, and storage. Isolation and counting of all microorganisms from foods and water are impossible. Some microbial groups can be used to indicate the possible presence of pathogenic and spoilage microorganisms. These microbial groups are called indicator microorganisms. Indicator microorganisms are preferred to be nonpathogens and reliably detectable at low numbers with simple techniques. The indicator microorganisms should be present in higher numbers than the pathogens and have similar or higher survival rates. Different indicator microorganisms are used in food and water quality monitoring. The efficiency of the indication of pathogens by indicators depends on their detection limit, their resistance to environmental stresses, and other microbial contamination levels. In this section, the isolation and counting techniques are explained for the following pathogenic microorganisms and microbial groups.

Practice 13. Isolation and Counting of Coliforms and Escherichia coli

Practice 14. Isolation and Counting of Enterococcus

Practice 15. Isolation and Counting of Salmonella

Practice 16. Isolation and Counting of Listeria monocytogenes

Practice 17. Isolation and Counting of Campylobacter jejuni

Practice 18. Isolation and Counting of Yersinia enterocolitica

Practice 19. Isolation and Counting of Bacillus cereus

Practice 20. Isolation of Clostridium perfringens

Practice 21. Isolation and Counting of Staphylococcus aureus

Practice 22. Isolation and Counting of Clostridium botulinum

Practice 23. Isolation and Counting of Vibrio

Practice 24. Isolation and Counting of Shigella dysenteriae

Practice 25. Isolation and Counting of Brucella

Practice 26. Isolation and Counting of Aeromonas hydrophila

Practice 27. Isolation and Counting of Plesiomonas shigelloides

PRACTICE

Isolation and counting of coliforms and *Escherichia coli*

13

13.1 Introduction

The basic feature of the food is its hygienic quality, which depends on its production and processing being under hygienic conditions. Lack of good hygiene practices during production, transportation, storage, and consumption may result in the loss of food quality and the spoilage or presence of health hazards in foods. Foods must be free from hazardous microorganisms (either pathogenic or spoilage) or those present should be at a safe low level. The sanitary quality of foods can be determined by the content of certain indicator microorganisms. It would not be feasible to examine each food for all hazardous microorganisms. The routine microbiological analysis of foods to determine the sanitary quality is not based on the isolation and identification of all pathogenic or spoilage microorganisms due to the following reasons:

1. Each microorganism requires complex and specific media.
2. Their isolation or counting requires specific procedures.
3. Generally pathogenic microorganisms present in low numbers and they may not survive for long time periods in foods or water.
4. If very small numbers of pathogenic or spoilage microorganisms are present in the foods, they may not be easily counted or isolated.
5. Isolation of all the microorganisms is a time-consuming job.
6. Their isolation and identification take 72 h or longer to obtain results from a routine laboratory analysis. By the time the pathogens are found, many people would have consumed the food and would be exposed to pathogenic microorganisms before action could be taken to discard or reduce pathogens from foods and water.

For these reasons, microbiologists have developed food analysis techniques that are based on counting or the isolation of indicator microorganisms. Indicator microorganisms serve as an "alarm" system about hazardous microorganisms and usually associate with the intestinal tract of human and animals. There are several microorganisms that can be used as sanitary quality indicators of water and foods. The most common indicator microorganisms are coliforms, *Escherichia coli*, *Enterococcus*, and total plate count. Other individual microorganisms can also be used as indicators depending on the type of food and process conditions. There should be some criteria for the indicator to meet for use as a sanitary index microorganism. Some of the important characteristics of an indicator microorganism are the following:

1. It should require simple, accurate, rapid, and standard techniques to count.
2. It should be easy to detect in very low numbers.
3. It should be generally present in the intestinal tract.

Microbiological Analysis of Foods and Food Processing Environments. DOI: https://doi.org/10.1016/B978-0-323-91651-6.00051-3

4. It should be present in food when pathogens are present.
5. The quantity of the indicator microorganism correlates with the amount of pollution.
6. It should survive better and longer than the pathogens.
7. It should be uniform and has stable properties.
8. It should be harmless to humans and animals.
9. It should be present in higher numbers than pathogens.
10. It should be detected by standard laboratory techniques.
11. It can be used to detect microbial conditions of foods:
    a. fecal contamination;
    b. the presence of pathogens;
    c. potential spoilage microorganisms;
    d. sanitary conditions of food processing, production, and storage; and
    e. adequate for consumption.

13.2 Indicator microorganisms

Sampling, sample preparation, and all experiments should be performed under aseptic conditions. Sampling and all analysis should be performed in duplicate, unless otherwise specified. All microbiological media to be used in this application must be sterilized. Again, all solutions and equipment that may cause contamination must be sterile or disinfected.

The number of indicator microorganisms (such as coliforms) is used to predict water and food sanitary quality, and to indicate sanitary conditions during food production. In general, the presence of indicator microorganisms in the raw and processed foods and water is indicative of one or more of the followings:

1. The initial number of microorganisms is so high that the treatment is inadequate to reduce microbial number to safe level.
2. Postprocess conditions may allow multiplication of survivors until their numbers become detectable.
3. Microorganisms can be contaminated during and after processing.
4. The presence in higher numbers may indicate a lack of good sanitary practice or conditions during handling and processing.
5. Presence of pathogenic and spoilage microorganisms in foods or water would be due to fecal contamination.
6. Foods and products can be produced under inadequate handling conditions and improper processing.

This practice of the book considers the indicator microbial sources, and isolation and counting techniques. The most common indicator microorganisms are coliforms, fecal coliform, *E. coli*, *Enterococcus*, and total plate count.

13.2.1 Enterobacteriaceae

The Enterobacteriaceae family can be used as an indicator. The members of the family are Gram-negative, rod-shaped, motile or nonmotile, and counted by adding 1% glucose into a medium. They ferment glucose rapidly with or without gas production, reduce nitrate to nitrite (except *Erwinia* and *Yersinia*), are oxidase negative and catalase positive. They can be associated with the spoilage of different types of foods and drinks. They can be a useful indicator for heat-processed food's sanitation, postprocess contamination, and in the indication of food safety and quality. This count is an effective technique to indicate the sanitation conditions of environments, e.g., postprocess food contact surfaces and potential contamination sources. It may be a good indicator of sanitation due to resistance to the unfavorable environmental conditions for the coliforms that naturally inhabit the gastrointestinal tract as a glucose and lactose utilizer and have the ability to survive in the natural environments. Enterobacteriaceae includes species from different bacterial genera: *Escherichia*, *Klebsiella*, *Citrobacter*, *Enterobacter*, *Proteus*, *Hafnia*, *Serratia*, *Providencia*, *Erwinia*, *Edwardsiella*, and some of the important enteric pathogens, e.g., *Salmonella*, *Shigella*, *Yersinia enterocolitica*, and pathogenic *E. coli*. All species in the Enterobacteriaceae family are able to ferment glucose with the production of acid and reduce nitrates. This family may also include some physiological groups, e.g., psychrotrophic species. The psychrotrophic species in the genera *Enterobacter*, *Hafnia*, and *Serratia* can grow at temperatures as low as found in refrigeration.

13.2.2 Coliforms

The name Coliform is not a taxonomic classification but is used to describe a microbial group that is Gram-negative, rod-shaped, facultative anaerobic, and ferments lactose with acid and gas production within 48 h at 35°C. In the counting of coliforms, the incubation is performed at 35°C to detect the potential pathogenic risk in foods and at 30–32°C to get an idea about process conditions. Coliform count is a more suitable standard technique compared to other indicators in terms of its accuracy and acceptability. Coliforms include four species: *E. coli*, *Enterobacter aerogenes*, *Klebsiella pneumoniae*, and *Citrobacter freundii*. Other minor coliform species are present in the genera *Proteus*, *Serratia*, and *Hafnia*. *E. coli* is a common intestinal bacterium and also presents in soil, water, and on plants; *E. aerogenes* is a common plant bacterium and occasionally present in the intestine; *K. pneumoniae* is a less common intestinal inhabitant and is present in industrial wastes or nature, and *C. freundii* (slow fermenting lactose) cannot associate in the intestine and generally associates with fresh vegetables, usually on market vegetables. Coliform species can be differentiated from each other depending on their cultural, morphological, and biochemical characteristics. Coliaerogenes bacteria in the coliform group (*E. coli* and *E. aerogenes*) can grow in the presence of bile salts or other equivalent selective agents.

The significance of coliform bacteria in water and foods must be interpreted with great caution. The higher numbers of coliforms in water and foods indicate the lack of good sanitary practices. A small number of coliforms will spoil raw milk under normal conditions of production and handling. A high number in a coliform count in raw milk indicates the degree of contamination in production through unsanitary practices or conditions. High numbers of coliforms in pasteurized milk indicate postpasteurization contamination or improper heat-processing. Washing and other practices will remove soil and coliforms from surface area of foods, but some may remain.

13.2.3 Fecal coliform

Fecal coliform is a group of bacteria that can inhabit the gastrointestinal tract and spread with the fecal excrement of humans, animals, livestock, and wildlife. They aid in the digestion of food. The most common member is *E coli*. Fecal coliform in the coliform group is differentiated based on the Eijkman characteristics that fecal coliform grows with the fermentation of lactose at high incubation temperature compared with the other coliforms, and they can also be referred to as thermotolerant coliforms. The incubation temperature differentiates the fecal coliform from coliforms. Fecal coliform analyses are performed at 44.5 ± 0.2°C. Fecal coliform can grow and produce acid and gas from lactose at an elevated temperature. The incubation temperature is a critical factor, and a water bath should always be used in the incubation. High quantities of fecal coliform in water and foods suggest the presence of pathogenic microorganisms and indicate the potential health risk of foods and water. The presence of fecal coliform in water indicates the contamination of water with the feces of humans or other animals. Fecal coliform in feces can be transmitted to river waters from mammals and bird feces, from agriculture and storm flows, and through the direct draining of untreated sewage. Swimming in waters containing high numbers of fecal coliform and pathogens can result in the enteric pathogens entering of through the mouth, nose, ears, or skin with the cuts during swimming. This can increase the likelihood of illness (fever, nausea, or stomach cramps). The main enteric diseases from such water are typhoid fever, hepatitis, gastroenteritis, dysentery, and ear infections. Fecal coliform and enteric bacteria are sensitive to chlorine; therefore, water should be disinfected by chlorination.

In the examination of foods and water for fecal coliform, there are two approaches: (1) Inoculation of presumptive positive LST broth culture to EC broth and incubation at 44.5 ± 0.2°C for 24 h or inoculation of presumptive positive LST broth culture to EC broth and incubation at 35°C for 3 h followed by incubation at 44.5 ± 0.2°C for an additional 21 h. (2) In a membrane filter technique, inoculate presumptive positive LST broth culture to EC broth and incubate at 44.5 ± 0.2°C for 24 h.

13.2.4 Escherichia coli

E. coli is a Gram-negative, rod-shaped, nonspore-forming, oxidase negative, catalase positive, and fermentative bacterium. *E. coli* ferments lactose and only 10% of isolates are slow or nonlactose fermenting. It differs from other Enterobacteriaceae members by its biochemical characteristics and serological tests. The possibility of fecal contamination and unsanitary processing conditions can be indicated by the presence of *E. coli*. Its natural habitat is the lower part of the intestines of humans and animals. The strains of *E. coli* colonizing the human intestine are harmless commensals. *E. coli* strains are used as an indicator of fecal contamination of foods and water and the possible presence of enteric pathogens. There are pathogenic strains of *E. coli* that can cause distinct diarrheal diseases and are associated with foodborne illnesses.

13.3 Counting of coliforms, fecal coliform, and *E. coli*

They can be counted using different techniques: most probable number (MPN), plate count, membrane filter, and rapid techniques. The most commonly used technique is MPN.

13.3.1 Equipment, reagents, and media

Equipment. Blender (or Stomacher), cover slip, depression slide, Erlenmeyer flask, forceps, incubator, loop (3 mm in diameter), membrane filter (0.45 μm), membrane filtration system, microscope, microscope glass slide, needle, Petri dishes, pipette discard equipment (e.g., a graduated cylinder) containing 10% bleach solution and a piece of cloth at bottom, pipettes (1 mL) in the pipette box, test tubes (13 × 100 mm), and vortex.

Reagents. Barritt's reagent, creatin crystal, Gram stains, H_2O_2, Kovac's reagents, Liefson's flagella stains, methyl red solution (0.1% aq.), 0.1% peptone water, phosphate buffer (PB) solution, and 2,3,5-triphenyltetrazolium chlorite (TTC).

Media. Brain heart infusion (BHI) agar, BHI broth, brilliant green lactose bile (BGLB) broth with inverted Durham tube, buffered glucose (Mr-VP) broth, Dextrose Calcium Carbonate (DCC) medium, chromocult coliform (CC) agar, EC broth with inverted Durham tube, eosin methylene blue (EMB) agar, KCN broth, lauryl sulfate tryptose (LST) broth with inverted Durham tube, LST broth-MUG (4-methylumbelliferyl-beta-D-glucuronide), M-Endo agar, LES-Endo agar, motility test (MT) medium, orthonitrophenyl-β-galactopyranoside (ONPG) peptone broth, carbohydrate fermentation (CF) broth containing (1%) individual carbohydrate with inverted Durham tube, Simmon's citrate (SC) agar slant, triple sugar iron (TSI) agar slant, tryptic (tryptone) soy (TS) agar, TS broth, urea agar slant, violet red bile (VRB) agar containing 1% glucose, and VRB agar-MUG.

13.3.2 Most probable number technique

The coliform count flowchart by MPN technique is given in Fig. 13.1.

13.3.2.1 Presumptive test

Refer to Technique 2.2.2 for details of sample preparation and dilutions. Aseptically, 50 g of sample is added into a sterile blender (or stomacher plastic bag) containing 450 mL of sterile PB solution (or 0.1% peptone water) and the mixture is blended for 1 min (or stomached for 2 min). Frozen samples can be thawed with the storage for <18 h at 2–5°C. In the case of a liquid sample, 50 mL of sample is added into a sterile Erlenmeyer flask containing 450 mL of sterile PB solution (or 0.1% peptone water) and mixed by gently shaking the flask (25 times in 30 cm arc). If <50 g or mL of sample is available, take a portion of the sample and add a sufficient volume of sterile diluent to make a 1:10 dilution. Prepare decimal dilutions with sterile PB solution. The level of decimal dilutions is prepared depending on the expected number of coliforms. Not more than 15 min should elapse from the time the sample is blended until all dilutions are inoculated to the appropriate medium.

One milliliter of sample from each dilution is removed and added into each of 3-MPN tubes of LST broth containing inverted Durham tube. During transfer, the pipette is lowered at an angle in the tube. The pipette is allowed on the inside wall of the tube for 2–3 s to drain the sample. If very low numbers of coliform are expected from the sample, 10 or 100 mL of the lowest dilution or undiluted liquids may be added into double strength LST broth.

After inoculation, all tubes are incubated at 35°C for 24 and 48 h. After incubation, tubes are examined and the results are recorded after 24 h for growth with the production of gas and acid for each dilution (some strains are late acid producers). Tubes showing gas

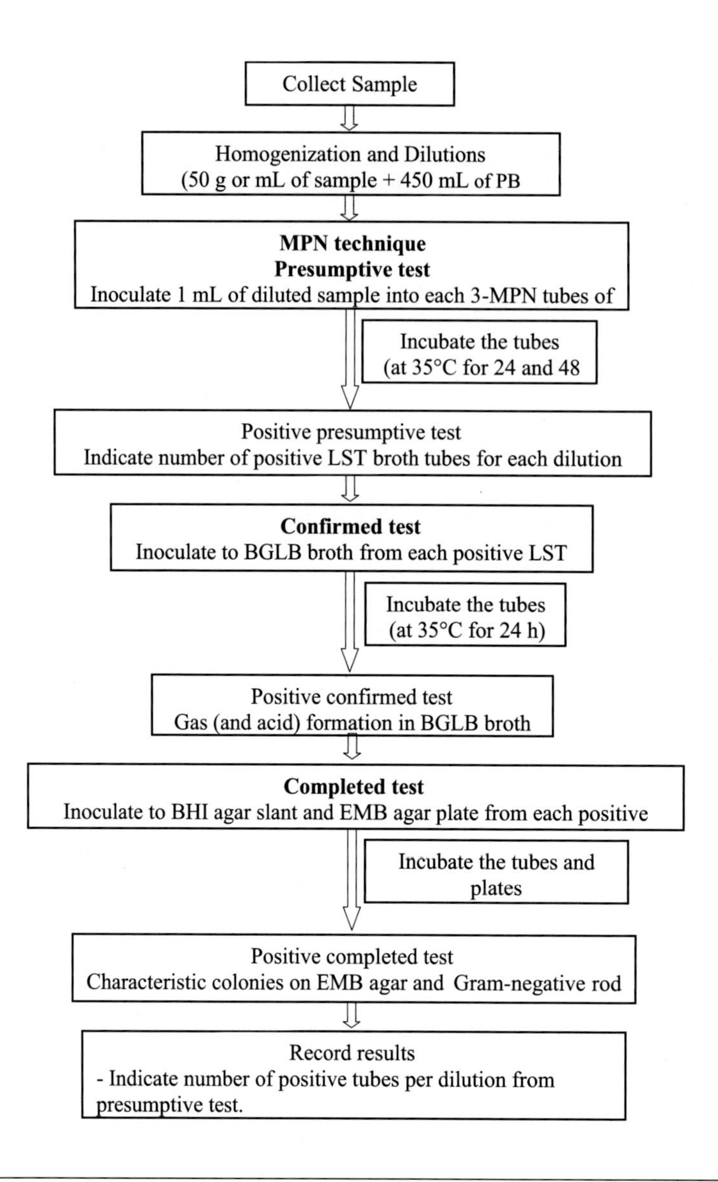

FIGURE 13.1

Coliform count flowchart by most probable number (MPN) technique.

production (space) in the inverted Durham tube (and yellow color with acid production) are recorded as positive (Fig. 13.2b). The tubes without gas production are reincubated for an additional 24 h. After incubation, tubes are examined and reported again. All presumptive positive tubes must be confirmed. When the LST broth tubes do not have gas formation, they are considered to be negative for coliforms.

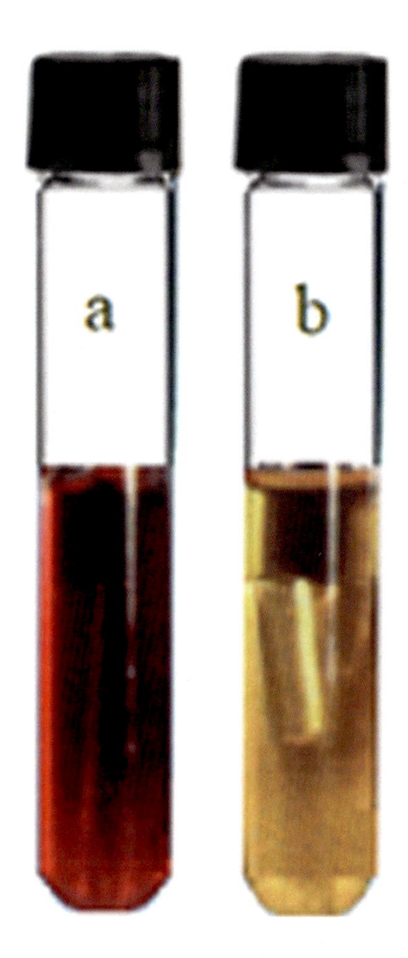

FIGURE 13.2

Coliform results in the lauryl sulfate tryptose (LST) broth. (A) Red color and no gas production indicate nonlactose fermentation. (B) Yellow color indicates acid production and space in the inverted Durham tube indicates gas production.

13.3.3 Confirmatory test

1. Confirmed test for coliforms

   A loopful of culture from each gas forming LST broth is inoculated into BGLB broth in the tube containing inverted Durham tube. Tubes and Petri plates are incubated at 35°C for 24 and 48 h. After incubation, the number of BGLB broth tubes showing positive results (gas and/or acid production) is recorded. All confirmed positive tubes must be followed with completed test.
2. Confirmed test for fecal coliform and *E. coli*

One loopful of culture from each positive LST broth tubes is inoculated into EC broth in the tube containing inverted Durham tube. The tubes are incubated at 44.5 ± 0.2°C for 24 h. After incubation, they are examined for acid production (yellow color) and gas formation (space in the Durham tube). Indicate the positive EC broth tube against the positive LST broth tube. If tubes are negative for gas production, they are reincubated for an additional 24 h and examined again after incubation. All confirmed positive results for fecal coliform and *E. coli* must be followed with a completed test.

13.3.3.1 Completed test

1. Completed test for coliforms and fecal coliform

   Perform the following tests for both confirmed positive coliforms and fecal coliform. From each gassing BGLB broth and EC broth tube, inoculate a loopful of culture to BHI broth, and incubate tubes at 35°C for 24 h. After incubation, prepare a Gram stain and motility test (Technique 13.4.2) and examine for the presence of Gram-negative rods and motile bacterium (if necessary, apply spore staining and examine for nonspore-forming bacterium).

   Correlate positive LST broth tubes with these results (confirmed and completed tests) and indicate the final number of positive LST broth tubes again for each dilution. Refer to MPN Table 4.1 and calculate coliforms (or fecal coliform) MPN per g or mL of sample. If 10 mL or 100 mL of inoculations is performed in the presumptive test, multiply MPN calculation by inoculation amount in the calculation of coliform (or fecal coliform) MPN per g or mL of sample. Coliform (or fecal coliform) species can be differentiated from each other with the tests indicated in Table 13.1.

2. Completed test for *E. coli*

Each gas (and acid) forming EC broth culture is gently agitated and one loopful of EC broth culture is streak plated onto an L-EMB agar. Incubate inverted plates at 35°C for 24 h. After incubation, examine plates for characteristic *E. coli* colony formation on L-EMB agar. Inoculate one loopful of culture from a single colony into BHI broth and incubate the tubes at 35°C for 24 h. Perform morphological analysis using the BHI broth culture. *E. coli* ferments lactose with the

Table 13.1 Characteristics of coliform species.

Gas from		Tests								Bacterium
Glucose	Lactose	ET	I	MR	VP	C	H	M	O	
+	+	+	+	+	-	-	-	v	v	Typical *E. coli* type 1
+	-	-	-	+	-	-	-	v	v	Atypical *E. coli* type 2
+	-	-	+	+	-	+	-	v	v	Typical intermediate
+	-	-	-	+	-	+	-	v	v	Atypical intermediate
-	v	-	+	+	-	-	-	-	v	Inactive *E. coli*
+	+	-	-	-	+	+	-	v	-	Typical *E. aerogenes*
+	+	-	+	-	+	+	-	+	-	Atypical *E. aerogenes*
+	+	-	-	v	+	+	-	-	v	*Klebsiella pneumoniae*
+	s	-	+	+	-	+	+	+	v	*Citrobacter freundii*

C, *Citrate test;* ET, *Eijkman test;* H, *hydrogen sulfide test;* I, *indole test;* M, *motility test;* Mr, *methyl red test;* O, *ornithine test;* s, *slow lactose fermenter;* v, *variable;* VP, *Voges–Proskauer test.*

formation of gas and acid in LST broth, BGLB broth, and EC broth, and is a Gram-negative, motile, and nonsporulating bacillus. *E. coli* forms a flat, deep purple or dark centered colonies with a metallic green sheen on L-EMB agar. These constitute a positive completed test for *E. coli*. Correlate positive LST broth tubes with these results and indicate the number of positive LST broth tubes again for each dilution. Refer to MPN Table 4.1 and calculate *E. coli* MPN per g or mL sample. If further identifications of *E. coli* are necessary, use BHI agar slant pure culture in the phenotypic identification of *E. coli* species, as given in Table 13.1, and Tables 13.3–13.6.

Characteristic colony formation on L-EMB agar: Some Gram-negative lactose-fermenting bacteria produce acid. Coliforms produce the following colonies on L-EMB agar: (1) strong acidity due to lactose fermentation by *E. coli* produces a flat, red colonies with a metallic green sheen (Fig. 13.3A); (2) less acidity of coliform species may produce a brown-pink coloration of colony (Fig. 13.3B); and (3) other lactose fermenter coliform species produce large, mucoid colonies, often purple in their center (Fig. 13.3B). On the other hand, a nonlactose fermenter produces colorless, translucent, or pink colonies. The eosin and methylene blue (aniline dye) inhibit Gram-positive and fastidious Gram-negative bacteria.

13.3.4 Counting of coliforms on solid medium

In the counting of coliforms and *E. coli* from drinking water and processed foods, CC agar can be used as a selective/differential medium. The Tergitol-7 in the CC agar can inhibit Gram-positive and some Gram-negative bacteria but does not have a negative effect on the growth of coliforms. The characteristic enzyme producing by coliforms, ß-D-galactosidase, hydrolyzes salmon-GAL substrate (chromogenic substrate mixture) in the CC agar and produces a medium color change from salmon to red color on the coliform colonies. Other Gram-negative bacteria produce colorless colonies.

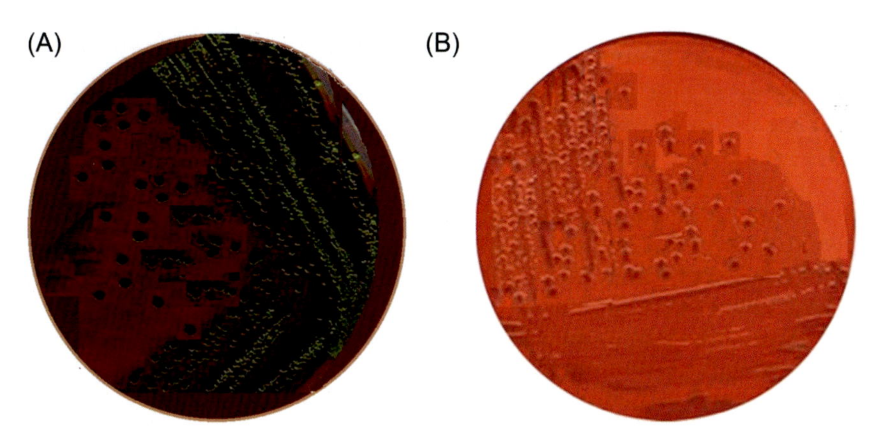

FIGURE 13.3

(A) Lactose fermenting coliforms producing a flat, red-centered colonies with a metallic green sheen and (B) nonlactose fermenting coliforms producing a brown-purple-centered colony with or without mucoid colonies.

Prepare the sample as indicated in Technique 13.3.2.1. Streak plate 1 mL of diluted sample to CC agar in duplicate and allow the plate medium at room temperature to absorb the liquid of the inoculum. Incubate inverted Petri plates at 35°C for 24 h. After incubation, analyze the characteristic colony formation on the CC agar medium. Coliforms produce red, dark-blue–violet colonies on CC agar. Count the number of characteristic colonies from the plates containing 25–250 colonies. Calculate the number of coliforms by multiplying the average count by dilution rate and dividing by inoculation amount. Report the results as the number of coliform colony forming units (cfu) per mL or g of sample.

For the confirmation of characteristic colonies, pick at least 10 representative colonies per plate and inoculate each colony to BHI broth separately to obtain pure culture (Technique 13.4.1). Use pure cultures in a confirmed test (Technique 13.3.2.2).

13.3.5 Counting of coliforms by membrane filter technique

Perform the membrane filter (MF) technique as explained in Technique 5.2. Water sanitary quality is commonly determined by this technique. The simultaneous detection of coliforms and *E. coli* can be performed using CC agar, depending on the formation of specific colony colors. The technique permits the use of large volumes of sample to count the coliforms. Place a sterile MF (0.25 μm pore size) on the filter base while grid side up and place the funnel to the base; MF is now held between the funnel and the base. The liquid sample (such as water) is homogenized by gently shaking about 25 times, and 100 mL of sample (diluted or nondiluted) is added into the sterile funnel. The sample is filtrated under vacuum and the funnel sides are rinsed at least twice by 20–30 mL of sterile 0.1% peptone water. After filtration, the vacuum is turned off and the funnel is removed from the filtration apparatus. The MF is removed from the filtration apparatus by flamed forceps and it is rolled onto the surface of the CC agar plate (or M-Endo agar or LES-Endo agar) while avoiding bubble formation between the filter and the surface of the medium. If bubbles occur, filtration should be repeated. The MF rolled plates are incubated at 35°C for 24 h. After incubation, the plates are examined for the formation of characteristic coliform colonies on CC agar. Coliforms produce red, dark-blue–violet colonies on CC agar (pink to dark red colonies with a green metallic sheen colony on M-Endo agar or LES-Endo agar). Count the number of characteristic colonies from MF containing 20–60 colonies. Calculate the number of coliforms using the formula:

$$\text{Coliforms cfu/100 mL} = (\text{Average colony count})(\text{DF})$$

where DF is the dilution rate of sample.

For the confirmation of characteristic colonies, pick least three representative colonies per CC agar plate and inoculate each colony into LST broth. Incubate tubes at 35°C for 24 h. After incubation, any gas positive LST tubes should be subcultured in BGLB broth and incubated at 35°C for 48 h. Gas production in BGLB within 48 h is a positive coliform test. Confirm the isolates by completing tests using L-EMB agar (Technique 13.3.2.3).

13.3.6 Most probable number coliforms test for bottled water

In the analysis of bottled water, add 10 mL of water sample (without dilution) into each of 10 tubes of 10 mL double strength LST broth. Incubate the tubes at 35°C for 24 h. After incubation, examine the tubes for growth (turbidity and color change) and gas formation (empty space) in the Durham tube. Indicate the number of positive tubes. If tubes are negative after 24 h, reincubate tubes for an additional 24 h and examine again for gas formation. Perform a confirmed test on all presumptive positive (gassing) tubes, as indicated in Technique 13.3.2.2. Calculate MPN for coliforms using MPN Table 42.1 and multiply the count by dilution rate (if diluted water sample is used). Report results as the number of coliforms MPN per mL of bottled water.

13.3.7 Counting *E. coli* with MUG test

The MUG test depends on the β-glucuronidase (GUD) enzyme activity. This enzyme hydrolyzes the substrate MUG (4-methylumbelliferyl-beta-D-glucuronide) to release the fluorescent 4-methylumbelliferone (MU). When the medium including this enzymatic reaction is exposed to long wave (365 nm) ultraviolet (UV) light in the dark, the product (MU) shows a bluish fluorescence around the colonies. The MUG test is used to determine GUD enzyme-producing bacterium. Over 95% of *E. coli* produce GUD. *Enterohemorrhagic E. coli* (EHEC) strains are GUD negative and cannot produce GUD enzyme. This test is also used to differentiate EHEC strains from other *E. coli* strains. Other strains of the family Enterobacteriaceae rarely produce the GUD enzyme, except for some *Shigella* (44%–58%) and *Salmonella* (20%–29%). The media (such as EMB agar) containing fluorescent components are not suitable for the MUG test. This test identifies *E. coli* within 24 h.

13.3.7.1 Lauryl sulfate tryptose broth containing MUG

Prepare samples as described in Technique 13.3.2.1. Three MPN tubes of LST broth-MUG are inoculated with each diluted sample. One set (three MPN) of LST broth-MUG tube is also inoculated with GUD-positive *E. coli* control and another set with *E. aerogenes* as a negative control. Incubate the tubes at 35°C for 24–48 h. After incubation, each tube is examined for growth (turbidity) and gas (empty space) in Durham tube. Then tubes are examined in the dark under long-wave UV lamp 6-watt or equivalent (at 365 nm) lamp. A bluish fluorescence (beside gas production) indicates a positive presumptive test for *E. coli*. A 24 h fluorescence reading can identify *E. coli* positive tubes by 83%–95%. All presumptive positive tubes (for fluorescence and gas formation) are indicated for each dilution, and streak plate one loopful culture from each positive tube to L-EMB agar to confirm the *E. coli* isolates. Incubate inverted L-EMB plates at 35°C for 24 h. The complete test is performed as indicated in Technique 13.3.2.3 with inoculation of at least three representative pure colonies per L-EMB agar plate into BHI broth. After identification of *E. coli*, record the numbers of positive tubes (LST broth-MUG) for each dilution based on test results, refer to MPN Table 4.1 and calculate *E. coli* MPN per g or mL of sample.

13.3.7.2 Violet red bile agar containing MUG

Homogenize and dilute the sample as described in Technique 13.3.2.1. Transfer 1 mL of diluted sample into sterile Petri dishes in duplicate. Pour 17–19 mL of melted VRB agar containing MUG (at 47°C) into Petri dishes. Mix agar and 1 mL sample by thoroughly rotating the plates by drawing a figure of eight while lying on the table tap, and allow the agar to solidify in the plates (within 10 min). Incubate inverted plates at 35°C for 24 h. After incubation, analyze plates for characteristic colony formations. Purple-red colonies surrounded by a reddish haze (precipitation zone) are indicators from the plates. The plates are examined under long-wave UV lamp 6-watt or equivalent (at 365 nm) lamp. Count all bright blue-white fluorescent *E. coli* colonies (Fig. 13.4). Follow

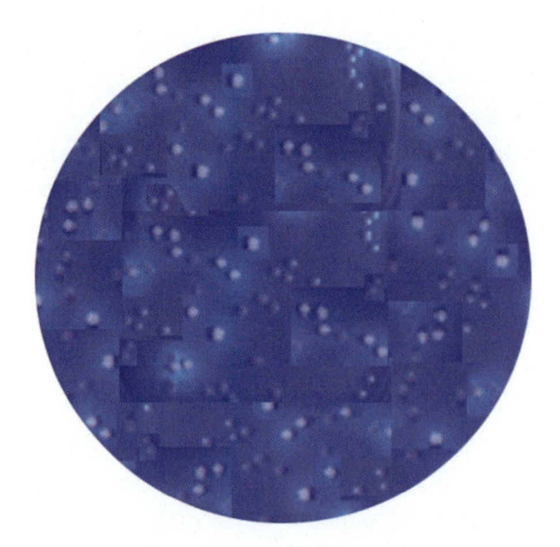

FIGURE 13.4

Appearance of *E. coli* colonies on violet red bile (VRB) agar-MUG.

the confirmed test as indicated in Technique 13.3.2.2. Calculate the number of *E. coli* and record the result as the number of *E. coli* cfu per g or mL of sample.

13.3.8 Enumeration of Enterobacteriaceae

Refer to Technique 13.3.2.1 for sample preparation and dilutions. Add 1 mL of diluted sample into sterile Petri dishes in duplicate and pour 13–14 mL of melted VRB agar (at 45°C) containing 1% glucose. Mix thoroughly by rotating the plates by drawing a figure of eight while lying on the table tap. Allow the Petri plates to solidify the medium and overlay plates with 4–5 mL of additional VRB agar containing 1% glucose. After solidification of the overlayered agar, incubate the inverted plates at 35°C for 24 h. After incubation, analyze Petri plates for characteristic colony formations. Count the number of colonies from plates by a dark reddish zone of precipitated bile from the plates containing 25–250 colonies. Calculate the number of Enterobacteriaceae by multiplying the average count by the dilution rate and dividing by the inoculation amount. Confirm the presumptive isolates with morphological and biochemical tests, as indicated in Technique 13.4. Calculate the number of Enterobacteriaceae and record the result as the number of Enterobacteriaceae cfu per g or mL of sample. Characteristic growths of Enterobacteriaceae members on different agar media are given in Table 13.2.

13.4 Identification techniques

13.4.1 Pure culture technique

For mixed culture. Select at least three typical colonies from isolated agar media. Touch the center of a selected colony with a sterile loop, inoculate to BHI broth, and incubate broth at 35°C for 18 h. Streak plate (Technique 13.4.1.1) one loopful of BHI broth culture onto EMB agar plates.

Table 13.2 Characteristic growth of Enterobacteriaceae members on different agar media.

Bacterium	MacC	EMB	HEC	XLD	SS	DCC	BS	BG
E. coli (lac$^+$)	R-P	Met	Y	Y	P	P	NG	(Y,G)
E. coli (lac$^-$)	C	C	Y or G	Y	C	C	NG	NG
K. pneumonia	P(Y),M	V	Y	Y	P	P	NG	Y,G
E. aerogenes	P(Y)	LV-C	Y	Y	P	P	NG	Y,G
Citrobacter	P(Y)	LV-C	C	Y	C	C	NG	G(A)
Serratia	P, Pic	LV-C	C	R	C	C	NG	R,Y
Hafnia	P	LV-C	C	R,Y,C	C	C	NG	NG
Providencia	P	LV-C	C	Y	C	C	NG	(Y,G)
Proteus	C	C	C	Y(.)	C	C	NG	R
Morganella	C	C	C	R,Y,C (.)	C	C	NG	NG
Edwardsiella	Pal	C	C	R(.)	C	C	NG	NG
Salmonella	Pal	C	G,B,G	R(.)	C (.)	C	G-Y	P,W,R
Shigella	C	C	G,B,G	C	C	C	NG	NG
Yersinia	Pal	C,V	B,P	Y	C	C	NG	(G)(A)

(.), black center; (), weak; A, *acid;* B, *black;* C, *colorless;* G, *green;* L, *light,* M, *mucoid;* MacC, *MacConkey;* Met, *metallic;* NG, *no growth;* P, *pink;* Pic, *pigment;* R, *red;* Y, *yellow;* V, *violet;* W, *white.*

Incubate inverted plates at 35°C for 18 h. After incubation, look for single pure characteristic colonies on plates. If colonies are not pure, repeat inoculation from single colonies into BHI broth and restreak onto EMB agar plates.

For pure culture. After obtaining pure culture, select at least three typical pure colonies from EMB agar per plate. Touch the center of a selected colony with a flamed loop, and inoculate to BHI broth and BHI agar slant. Incubate the tubes at 35°C for 18 h. Use pure cultures in identification tests.

13.4.1.1 Streak plate technique

The microbial pure culture is necessary in the many types of microbiological research as well as for microbial identification. The streak plate is commonly used in the pure culturing of microorganisms and in a dilution method for microbial cells in which results with a single colony form from a single cell. In streak plate culture, the confluent microbial growth as a colony will occur on the medium at the inoculation lines, the colony formation decreases during further streaking, and a single colony from a single cell will be appear over or near final streaking. The single cell is picked using an inoculating loop or needle by touching onto the center of colony, transferred to a liquid medium, and incubated.

In streak plating, place two or three loopfuls of broth culture onto one side of the agar medium (Fig. 13.5A). For the first streaking, streak over the agar medium in about one-quarter of the plate with a rapid, smooth, back-and-forth motion of the loop. During streaking hold the loop like a pencil at a 10–20° angle to the medium (Fig. 13.5B). After completing the first streaking, reflame the loop and turn the Petri plate 45° for a second streaking starting from the end of the first streaking lines area. Using the back-and forth pattern, cross over the last half of the streaks in the first

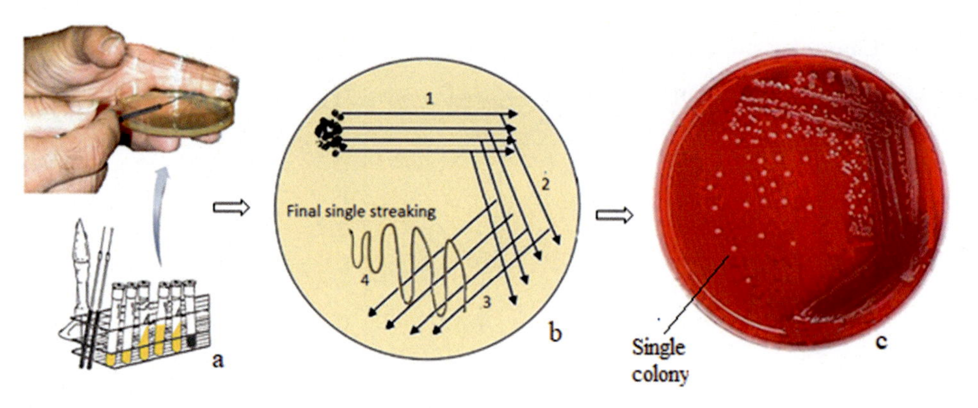

FIGURE 13.5

Streak plating and single colony formations on the plate: (A) Placing broth culture and streak plating, (B) results of streaking, and (C) single colony formation on agar medium.

streaking, then move the loop onto the empty second streaking area and reflame the loop. Repeat a third streaking as well as the second streaking starting from the end of the second streaking area. Perform a final single line streaking starting from the end of the third streak without flaming the loop (Fig. 13.5B). Incubate the inverted plate at 35°C for 24 h. After incubation examine the plate for the formation of a single colony (Fig. 13.5C).

13.4.2 Morphological identification

Gram staining. Prepare the Gram stain from 18 h BHI agar slant culture (or BHI broth culture), as given in Appendix B under the topic "Gram Staining Regents." The Gram stain of coliforms will show Gram-negative (red color) rod-shaped bacteria (Fig. 13.6).

Motility test. Direct microscopic analysis, flagella staining or MT medium is used to search motility of bacterium. (1) *Motility test medium*. Inoculated MT medium with one loopful of BHI broth culture by stabbing down the medium center and incubate the MT medium at 35°C for 18 h. After incubation, analyze the medium for growth. Motile bacterium will show diffuse growth (colony formation) away from the inoculation line (through medium) (Fig. 13.7B) and nonmotile bacterium will show growth only along the inoculation line (Fig. 13.7A). *E. coli* produces diffuse growth in a semisolid medium and this indicates the motility of bacterium by means of flagella.

If TTC is added into the MT medium, microorganisms reducing TTC in the medium will appear as a red coloring of the medium along the growth area. This substrate makes it easier to differentiate bacterial growth in the medium.

Motility (or nonmotility) results can be confirmed by the wet mount, hanging drop or flagella staining technique. Use fresh culture in these techniques. These techniques are performed by inoculating one loopful of BHI broth culture onto BHI agar slant. The slants are incubated at 35°C for 6–8 h (maximum 18 h). Add 1 mL of 0.1% sterile peptone water onto BHI agar slant to wash the microbial cells from slant. (2) *Wet mount*. Place a small drop of washed culture on the center of the microscope glass slide using a sterile 11 mL pipette or a flamed loop. Carefully place a clean cover

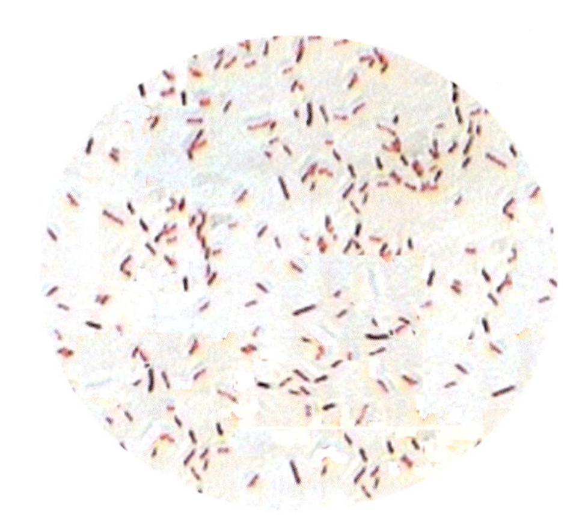

FIGURE 13.6

Staining appearance of *E. coli*.

slip over the drop. Place preparation on the microscope and use the low-power objective in magnification of the microbial cells. Find microbial cells with the help of fine and coarse knobs. Then switch to the high-power objective, magnify with the coarse knob, and adjust the light. Indicate cells as motile or nonmotile. Movement of bacterial cells in different directions indicates the motility of the bacteria. (3) *Hanging drop*. Place one drop of the washed microbial culture from slant on the center of a cover slip by a sterile pipette or flamed loop. Place a small amount of water or jelly on the cover slip corners. Place a depression slide over the cover slip (depressed area facing down onto the culture drop). Slides are rotated over the depression slide and lightly touched on the cover slip. The jelly holds the cover slip to the slide and also keeps the suspension from drying out. Place the preparation on the light microscope. Use the low-power objective to locate the microbial culture, and find microbial cells with the help of fine and coarse knobs. Then switch to the high-power objective, magnify with the coarse knob, and adjust the light. Indicate cells as motile or nonmotile bacteria. Movement of bacterial cells in different directions indicates the motility of bacteria. (4) *Flagella staining*. The results may also be confirmed by flagella staining, as indicated in Appendix B under the topic "Liefson Flagella Stain."

13.4.3 Biochemical identification

Phenotypic characteristic results for Enterobacteriaceae members are given in Tables 13.3–13.5.

13.4.3.1 Triple sugar iron agar slant reactions

Inoculate one loopful of BHI broth culture on a TSI agar slant (Fig. 13.8A) by stabbing the butt one time and streaking the slant. Incubate the TSI agar slant tubes at 35°C for 24 and 48 h. After incubation, examine the slants for growth and color changes in the medium. Record the results as follows: the red (alkaline) slant and butt indicates that the lactose and glucose, respectively, are not fermented by bacteria (Fig. 13.8B); yellow butt indicates that the glucose is fermented by bacterium and the red slant indicates that the lactose is not fermented (Fig. 13.8C); yellow slant and butt indicates that lactose and glucose, respectively, are fermented by bacterium; raising of the medium from

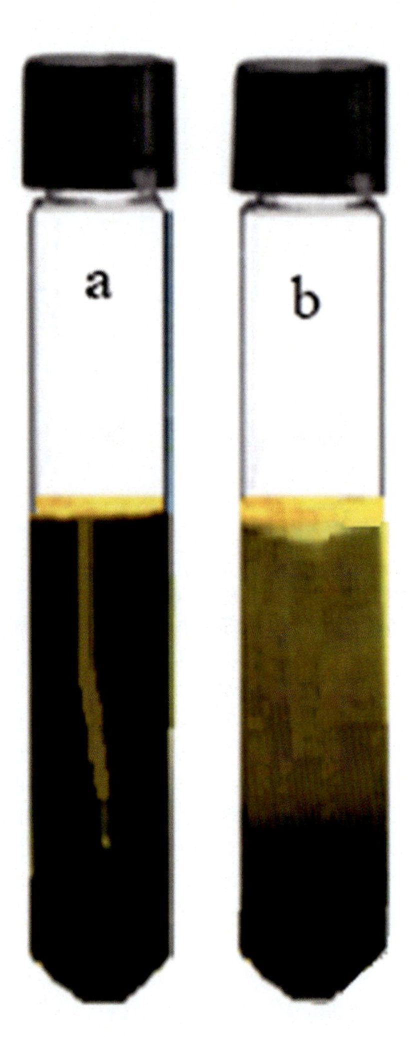

FIGURE 13.7

Growth of bacterium in motility test (MT) medium. (A) Nonmotile bacterium shows growth only along inoculation line and (B) motile bacterium shows diffuse growth away from the inoculation line.

the butt indicates gas production (Fig. 13.8D); and black color on the slant indicates H_2S production (Fig. 13.8E). *E. coli* produces gas and acid from glucose and lactose and cannot produce H_2S.

Characteristics of Enterobacteriaceae on TSI agar and LI agar slants, and IMViC tests are given Table 13.3.

13.4.3.2 Indole test

Some bacteria are able to metabolize tryptone and produce indole. Inoculate tryptone broth (5 mL) with one loopful of BHI broth culture and incubate tubes at 35°C for 48 h. After incubation, add

Table 13.3 Characteristics of Enterobacteriaceae on triple sugar iron (TSI) agar and LI agar slants, and IMViC tests.

Bacteria	TSI/LI agar slants				IMViC tests					
	G	H_2S	Butt	Slant	I	MR	VP	C	U	M
E. coli	+	-/-	Y/Y	Y/R	+	+	-	-	-	v
Salmonella	+	v/ +	Y/R	(R)/R	-	+	-	+	-	v
S. Paratyphi A	+	-/-	Y/Y	R/R	-	+	-	+	-	+
S. Arizona	-	V/ +	Y/R	R/R	-	+	-	+	-	+
S. Typhimurium	+	+/ +	Y/R	R/R	-	+	-	+	-	+
S. Enteritidis	+	+/ +	Y/R	R/R	-	+	-	+	-	+
S. dysenteriae	-	-/-	Y/Y	R/R	v	+	-	-	-	-
Citrobacter	v	+/-	Y/Y	R(Y)/R	v	+	-	v	v	+
Edwardsiella	v	+/ +	Y/R	R/R	+	+	-	-	-	+
K. pneumonia	+	-/-	Y/R	Y/R	-	(-)	+	+	+	-
Hafnia	+	-/	Y/R	R/R	-	v	(+)	-	-	(+)
E. aerogenes	+	-/-	Y/R	Y/R	-	-	+	+	v	+
E. cloacae	+	-/-	Y/Y	Y/R	-	-	+	+	v	+
E. agglomerans	+	-/-	Y/Y	Y/R	-	-	+	+	v	+
Serratia	-	+/-	Y/R	R/R	-	(-)	+	+	(-)	+
P. vulgaris	+	+/(+)	Y/Y	Y/R	+	+	-	(-)	+	+
P. mirabilis	+	+/(+)	Y/Y	(R)/R	-	+	(-)	v	+	+
M. morgani	v	-/-	Y/Y	R/R	+	+	-	-	+	+
Y. enterocolitica	-	-/-	Y/Y	R/R	v	+	-	-	+	-
Providence	v	-/-	Y/Y	R/R	+	+	-	v	v	+

+, *Positive;* -, *negative;* ( ), *sometime positive or negative;* C, *citrate;* G, *gas;* I, *indole;* M, *motility;* R, *red;* U, *urease;* v, *variable;* Y, *yellow.*

0.2–0.3 mL Kovacs' reagent into a tryptone broth culture and shake the tube vigorously. The formation of a red color on the upper layer of broth within 15 min indicates a positive indole test (Fig. 13.9). The test is negative if there is no red color formation in the tube. *E. coli* is an indole positive bacterium.

13.4.3.3 Methyl red and Voges–Proskauer (Mr-VP) tests

Bacteria produce acid and gas by either mixed acid fermentation or 2,3-butylene glycol fermentation. The methyl red test shows the presence of a high acid concentration, at a pH of 4.4 and below, by the mixed acid fermentation. VP reaction demonstrates the presence of neutral substances (e.g., acetylmethylcarbinol) formation by the 2–3-butylene glycol fermentation. Inoculate one loopful BHI broth culture to the Mr-VP broth (10 mL) in duplicate and incubate the tubes at 35°C for 24 h. After incubation, use in Mr-VP tests.

1. *VP test.* Aseptically transfer 2 mL of 24 h Mr-VP broth culture into a sterile test tube. Add 0.5 mL of Barritt's reagent-A (5% alcoholic α-naphthol solution) and shake vigorously from

Table 13.4 Carbohydrate fermentation characteristics of Enterobacteriaceae.

Carbohydrate	*Escherichia coli*	*Escherichia coli*[a]	*Shigella boydii*	*Shigella dysenteriae*	*Shigella sonnei*	*Enterobacteria aerogenes*	*Enterobacteria cloacae*	*Klebsiella pneumonia*	*Klebsiella ozaenae*	*Hafnia alvei*	*Serratia marcescens*	*Salmonella typhi*	*Salmonella gallinarum*	*Salmonella typhimurium*	*Salmonella salamae*	*S. arizona*	*Salmonella houtenae*	*Salmonella bongori*	*Citrobacter freundii*	*Citrobacter amalonaticus*
Glucose	+	+	+	+	+	+	+	+	+	+	+	+	+	+	+	+	+	+	+	+
Adonitol	-	-	-	-	-	+	(-)	+	+	-	v	-	-	-	-	-	-	-	-	-
Arabinose	+	(+)	+	v	v	+	+	+	+	+	-	-	+	+	+	+	+	+	+	+
Dulsitol	v	v	-	-	+	-	(-)	v	-	-	-	-	+	-	+	-	-	+	v	-
Lactose	+	(-)	-	-	-	+	v	+	v	-	-	-	-	-	-	v	-	-	v	v
Maltose	+	(+)	(-)	(-)	v	+	+	+	+	+	+	+	+	+	+	+	+	+	+	+
Mannitol	+	+	+	-	-	+	+	+	+	+	+	+	+	+	+	+	+	+	+	+
Mannose	+	+	+	+	+	+	+	+	+		+	+	+	+	+	+	+	+	+	+
Salicin	v	-	-	-	-	+	(+)	+	+	(-)	+	-	-	-	-	-	-	-	-	v
Sorbitol	+	(+)	v	v	v	+	+	(+)	+	-	+	+	-	+	+	+	+	+	+	+
Sucrose	+	(-)	-	-	-	+	+	(+)	v	-	+	-	-	-	-	-	-	-	v	-v
Raffinose	v	(-)	-	-	v	+	+	+	+	-	-	-	-	-	-	-	-	-	v	-
Rhamnose	(+)	v	-	v	-	+	+	+	v	+	-	-	-	-	+	+	+	+	+	+
Trehalose	+	+	(+)	(+)	v		+	+	+	+	+	+	v	v	+	+	+	+	+	+
Xylose	+	v	(-)	-	-	+	+	+	+	+	-	+	v	v	+	+	+	+	+	+

[a]*Inactive* E. coli *strains are nonmotile, nonlactose fermenter and included in enteroinvasive group, +, positive, -, negative, v, variable.*

Table 13.5 Some phenotypic characteristics of Enterobacteriaceae.

Characteristic	*E. coli*	*E. coli*[a]	*S. boydii*	*S. dysenteriae*	*S. sonnei*	*E. aerogenes*	*E. cloacae*	*K. pneumonia*	*K. ozaenae*	*H. alvei*	*S. marcescens*	*S. Typhi*	*S. Gallinarum*	*S. Typhimurium*	*S. salamae*	*S. arizona*	*S. houtenae*	*S. bongori*	*C. freundii*	*C. amalonaticus*
Phen. deam.	-	-	-	-	-	-	-	-	-	-	-	-	-	-	-	-	-		-	-
Lysin dec.	(+)	(+)	(-)	-	-	+	-	+	v	+	+	+	+	+	+	+	+	+	+	+
Arginine dec.	(-)	(-)	(-)	-	-	-	+	-	-	-	-	+	+		+	+	v		-	-
Ornithin dec.	v	v	(-)	v	v	+	+	-	-	+	-	-	-	+	+	+	+	+	-v	+
Lipase	-	-	-	-	-		-	-	-	-	-	-	-		-	-	-		-	-
DNAse	-	-	-	-	-		-	-	-	-	+	-	-		-	-	-		-	-
NO_3 red.	+	+	+	+	+		+	+	(+)	+	+	+	+		+	+	+		+	+
Oxidase	-	-	-	-	-		-	-	-	-	-	-	-		v	+	-		-	-
β-galactosidase	+	+	v	v	-		+	+	(+)	+	+	-	-		+	+	-	+	+	+
Malonate	-	-	-	-	-		(+)	+	-	v	-	-	-		+	+	-	-	-v	-v
Gelatin hydr.	-	-	-	-	-	-	-	-	-	-	+	-	-		+	+	-	-	-	-
KCN/ONPG	-/+	-/+	-/v	-/v	-/v	+/+	+/+	+/+	+/+	+/+	+/+	-/+	-/+	-/+	-/v	-/+	+/-	+/-	+/-	-/+
D-Tartarate											+	-	-	-	-	-	-	-	-	+
Inositol	-					+	(-)		+	-	(+)				(-)	-	v			

[a]*Inactive* E. coli *strains are nonmotile, nonlactose fermenter and included in enteroinvasive group, +, positive; -, negative; v, variable; phen. deam., phenylalanine deaminase; dec., decarboxylase; hydr., hydrolysis; red., reduction.*

Table 13.6 Some pathogenic characteristics of diarrheagenic *E. coli* subgroups.

Adhesion	ETEC	EPEC	EHEC	EIEC	EAEC	DAEC
	Fimbria	**Nonfimbria**	**Fimbria**	**Membrane protein**	**Fimbria**	**Fimbria**
Plasmid	+	+	+	+	+	+
Toxin production	HST/HLT	-	Stx or VT	-	HST/CY	ET/CY
Inflammation	-	Low	+	+	-	-
Fever	Low	+	-	+	-	-
Serological type	V	O26, O111, others	O157:H7, O26, O111, others	V	O126:H2, others	EC7372, C1845, others
Infective dose	Least 10^8 cells	Least 10^6 cells	Low (10–100 cells)	Least 10^6 cells	Least 10^8 cells	Least 10^8 cells
Invasive	-	-	+	+	+	+
Intestine side	Small intestine adherence.	Small intestine adherence, invasion on epithelial cells.	Large intestine adherence.	Mucosal invasion, inflammation on small and large intestine.	Small and large intestine adherence.	Small intestine, produce diffuse adherence to invade epithelial cells.
I/D	10–72 h/ 1–5 days	9–12 h/12 days	3–4 days/5–7 days	10–18 h/4–7 days	8–48 h/ 3–14 days	-
Syndrome	Acute watery diarrhea	Watery and bloody diarrhea.	Watery and very bloody diarrhea.	Mucoid and bloody diarrhea.	Watery-mucoid diarrhea.	Watery diarrhea.
Frequent association	Water, soft cheese, raw vegetables.	Water, some meat products.	Water, undercooked ground beef, raw milk, sandwiches, raw apple juice, vegetables.	Facal-oral route, water, flies, hamburger, raw milk.	Infant foods, water, sprouts and other vegetables.	Undercooked ground beef, water, livestock.
Motility	Motile	Motile	Motile	Nonmotile	Motile	Motile
Lactose fermentation	+	+	+	-	+	+

CT, *Cytotoxin;* ET, *enterotoxin;* HST, *heat stable toxin;* HLT, *heat labile toxin;* Stx, *shiga toxin;* VT, *vero toxin;* V, *variable;* I/D, *incubation/duration.*

side to side and then add 0.5 mL of Barritt's reagent-B (40% KOH solution) and shake vigorously from side to side again. To intensify and speed reactions, add a few creatine crystals to the test culture. Shake and leave the tubes to stand, and read results within 4 h of adding reagents. The appearance of a deep red color (eosin pink) indicates the VP test is positive for the tested culture or no color change indicates the VP test is negative (Fig. 13.9). *E. coli* is a VP negative bacterium.

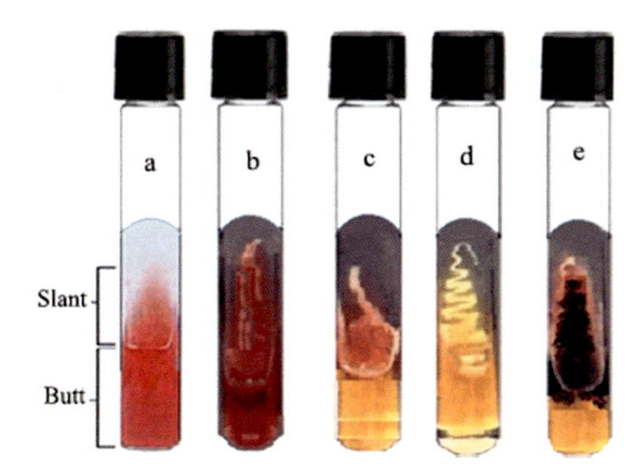

FIGURE 13.8

Triple sugar iron (TSI) agar slant and reactions. (A) TSI agar slant. (B) Red slant and butt. (C) Yellow butt and red slant. (D) Yellow slant and butt and raising of medium from butt. (E) Blackening.

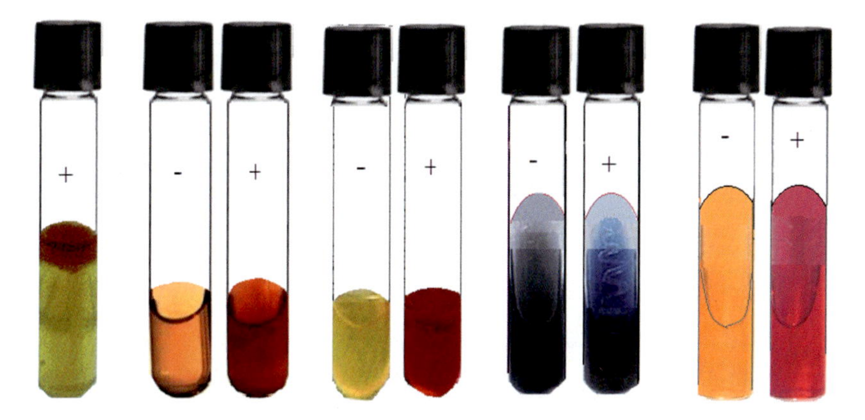

FIGURE 13.9

Some biochemical tests results for *E. coli* (+, positive and –, negative).

2. *Mr test.* After removing 2 mL of culture for VP test, incubate the remainder of the Mr-VP culture for an additional 24 or 48 h. Do not perform tests on cultures incubated less than 48 h. Transfer 2 mL of Mr-VP culture into a sterile test tube and add five to six drops of methyl red solution. The color change should appear immediately; if the pH of the culture is acidic, the color of the reagent will be red and the test is positive. If the pH of the culture is neutral or basic, the color of the test will be yellow and the test is negative (Fig. 13.9). *E. coli* is a Mr positive bacterium.

13.4.3.4 Citrate utilization test

The citrate utilization test shows the ability of a microorganism to utilize citrate as a carbon source. Inoculate one loopful of BHI broth culture onto SC agar slants. Incubate the slants at 35°C for

96 h. After incubation, examine the slants for growth (colony formation) and color change. A color change from green to blue with the formation of colonies indicates a positive test for the utilization of citrate as a carbon source. If there is no growth or color change, the test is negative (Fig. 13.9). *E. coli* is a citrate negative bacterium.

13.4.3.5 Urease test

Inoculate urea agar slant with one loopful of BHI broth culture and incubate the tubes at 35°C for 48 h. After incubation, examine the slants for growth (colony formation) and color changes. Production of ammonia is indicated by the hydrolysis of urea and this increases the pH of the medium which is shown by a medium color change from yellow to red and this indicates a positive test. If pH of the culture is acidic without ammonia production and the medium retains its original yellow color, this is a negative test (Fig. 13.9). *E. coli* cannot utilize urea and the test is negative.

13.4.3.6 Orthonitrophenyl-β-galactopyranoside test

Inoculate one loopful BHI broth culture to ONPG peptone broth (3 mL) in the tube and incubate the tubes at 35°C for not more than 18 h. After incubation, examine broth for growth (turbidity) and color changes in the medium. Beta-galactosidase production by bacterium and its activity on the substrate (ONPG) causes the formation of a yellow color in the medium due to the release of orthonitrophenol (Fig. 13.10). The reaction will usually be detectable within 3 h of incubation. This indicates that the ONPG test is positive. This test is positive for *E. coli*.

13.4.3.7 Potassium cyanide test

Inoculate one loopful of BHI broth culture into KCN broth and incubate the tubes at 35°C for 48 h. After incubation, examine the broth for growth (turbidity) and color changes. When the bacteria cannot grow in KCN broth, this indicates a negative KCN test. The growth of bacteria in the KCN broth indicates a positive KCN test. *E. coli* cannot grow in KCN broth.

13.4.3.8 Cellobiose and donitol utilization tests

Refer to Technique 15.3.2.2 for the carbohydrate fermentation test. Use CF broth in tubes containing cellobiose and donitol separately. Inoculate O-F broth with one loopful of BHI broth culture and incubate the tubes at 35°C for 2–7 days. After incubation, examine the broth for growth (turbidity) and color changes. Growth of bacteria and the formation of a yellow color indicates acid production from cellobiose and donitol. An empty space in the inverted Durham tube indicates gas production. *E. coli* cannot utilize cellobiose or donitol.

13.4.3.9 Eijkman test

Some species in the coliforms (such as fecal coliform) produce acid and gas from lactose after incubation at 44.5 ± 0.2°C. This growth of bacteria with lactose fermentation at this temperature is called the Eijkman test. *E. coli* type 1 and a few coliform species produce acid and gas at that temperature. EC broth is heated to 44.5 ± 0.2°C and inoculated with one loopful of BHI broth coliform or *E. coli* culture. Immediately incubate at a constant temperature of 44.5 ± 0.2°C in a water bath for 24 h. After incubation, examine broth for growth (turbidity) and color change. A tube showing growth with yellow color (acid) and gas (space in the Durham tube) indicates a positive test. For it to work successfully, a fluctuation of temperature of less than ±0.2°C is essential. If the

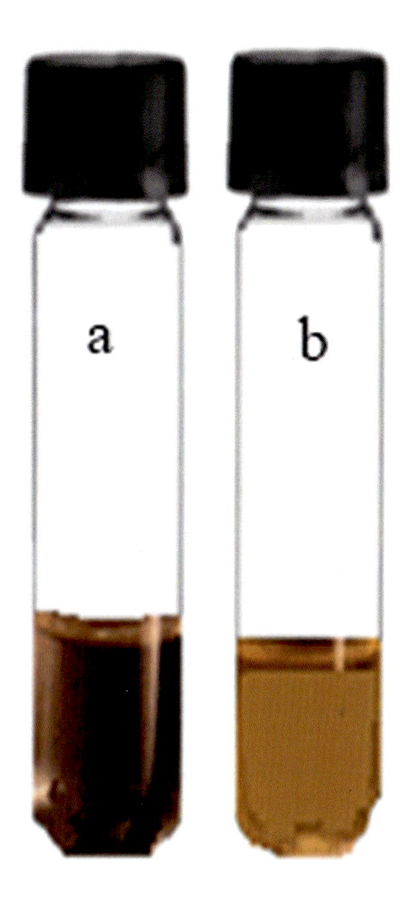

FIGURE 13.10

Orthonitrophenyl-β-galactopyranoside (ONPG) test: (A) a negative result (red color) and (B) positive result (yellow color).

temperature rises over this, some *E. coli* type 1 strains cannot produce gas. If the temperature falls to 42°C, other coliforms produce gas.

13.4.3.10 Catalase test

For this test, hydrogen peroxide (H_2O_2, 10%) should be freshly prepared each day, and stored in the refrigerator between tests. A catalase test can be performed from solid or liquid culture. (1) From TS agar plate culture. One milliliter of H_2O_2 is poured over the surface of the colony on the agar plate. Alternatively, place a loopful of the colony from the plate or slants onto a microscope glass slide and pour one drop of H_2O_2 onto the colony. (2) From TS broth culture. Aseptically, transfer 1 mL of TS culture into a small clean test tube and add 1 mL of H_2O_2. After adding H_2O_2 onto the culture, free oxygen released as gas bubbles indicates the presence of catalase enzyme in the culture and the catalase test is positive. Coliform species produce catalase enzyme and are catalase positive.

Other carbohydrate and phenotypic characteristics of Enterobacteriaceae are given in Table 13.4 and Table 13.5, respectively.

13.4.4 Interpretation of results

Total coliform bacteria and *E. coli* count can be used as indicators of unfavorable hygienic conditions and fecal contamination in foods. Recovery of coliforms, fecal coliform, and *E. coli* from foods and water indicate the following public health concerns.

1. *Coliforms*. The recovery of coliforms from foods and water implies that other bacteria of fecal origin may be present, pathogenic or spoilage bacteria could be present, and there are poor processing conditions and sanitation in the production. The recovery of coliforms from processed foods is a useful indicator of postprocess and postsanitation contamination. The failure to recover coliforms from a sample does not indicate wholly the absence of pathogens.
2. *Fecal coliform*. The fecal coliform gives greater fecal specificity because of the high possibility of *E. coli* incidence within the group and in intestinal sources. The recovery of fecal coliform from foods implies that other bacteria of enteric origin may be present and high numbers of spoilage microorganisms could be present. The recovery of fecal coliform from processed food is a useful indicator of postsanitation and postprocess contaminations.
3. *E. coli*. The recovery of *E. coli* from foods implies that other bacteria of fecal origin may be present and pathogens may be present. The failure to recover *E. coli* from foods does not assure wholly the absence of enteric pathogens, because *E. coli* is not a perfect indicator, like coliforms and fecal coliform. However, it is the best-known fecal indicator at present together with fecal coliform. Bottled water is not permitted to contain *E. coli*.

13.5 Diarrheagenic *E. coli*

Nonpathogenic *E. coli* strains associate with the human gastrointestinal system and are harmless to the human. They play an essential role in keeping the digestive system healthy, helping to digest food, and producing vitamin K. However, some *E. coli* strains are pathogenic and can cause foodborne disease. The pathogenic strains can be transmitted through contaminated food, water, and contact with animals. Pathogenic *E. coli* strains can cause diarrheal disease in humans and are called diarrheagenic *E. coli* (DEC). Analysis of samples for DEC requires isolation and identification tests before testing for virulence factors. The pathogenic *E. coli* strains are classified into six major enterotoxigenic groups based on their virulence factors, clinical syndromes, interactions with the intestinal mucous, differences in epidemiology, and distinct O:H serogroups. These groups are Enterohemorrhagic *E. coli* (EHEC), Enterotoxigenic *E. coli* (ETEC), Enteropathogenic *E. coli* (EPEC), Enteroinvasive *E. coli* (EIEC), diffusely adherent *E. coli* (DAEC), and Enteroadhesive *E. coli* (EAEC). The identification of pathogenic strains requires phenotypic, serological, pathological, physiological, and molecular analysis. Some pathogenic characteristics of these subgroups are given in Table 13.6.

ETEC can cause traveler watery diarrhea with little and without fever. ETEC foodborne disease most commonly occurs in underdeveloped countries. Its illness appears with the enterotoxin production in the intestines. It produces heat-labile enterotoxins (HLT) and heat-stable enterotoxins (HST). The infective dose of ETEC for health people is least 10^8 viable cells per g or mL of food or water; but young and elderly people are susceptible to lower numbers as infective doses. In the foodborne illness, HLT binds to the epithelial cells of the intestinal tract and activates adenylate cyclase. This increases cyclic adenosine monophosphate (cAMP) level in the intestinal cells. The subsequent increasing amount of cAMP inhibits Na^+ and Cl^- absorption by the villus cells from intestinal lumen while stimulating the secretion of Cl^-, HCO^-, and Na^+ ions by cells into the intestinal lumen. The electrolyte transfer occurs with flow of water into the intestinal lumen from the cell to maintain osmotic balance. This results in profuse watery diarrhea. After contaminated food ingestion, the symptoms appear within 12–36 h and last about 2–3 days. The most common vehicles of outbreaks are cheese (especially soft cheese), mayonnaise, prepared food (e.g., in restaurants, cafeterias, and cruise ships), raw vegetables, and water. The main reservoirs are man, food handlers, and sewage.

EIEC causes invasive dysenteric diarrhea in human and its pathogenicity resembles *Shigella*. The primary source for EIEC is infected human, food handlers, and sewage. The infective dose of EIEC is at least 10^6 viable cells per g or mL of food or water in order to cause illness in healthy adults. EIEC is nonmotile, does not decarboxylate lysine, and does not ferment lactose which is different from the other *E. coli* strains. EIEC pathogenicity is primarily due to invading and destroying colonic tissue. EIEC invades the epithelial cells of the colon and grows, and produces lesions (ulcerations) and symptoms of bacillary dysentery. The most common vehicles of EIEC outbreak are water, soft cheeses, potato salad, and raw vegetables.

EPEC causes watery diarrhea and infantile diarrhea in developing countries. The main reservoirs are man, food handlers, and sewage. Outbreaks of EPEC most commonly result from the consumption of meat products, contaminated drinking water, and coffee substitute. The EPEC infectious dose for healthy adults can range from 10^6–10^9 viable cells per g or mL of food or water. EPEC adheres to the intestinal mucous epithelial cells and colonizes. It destroys microvilli and results in lesions. The ingestion of contaminated food can cause symptoms within 12–36 h and last about 2–3 days. EPEC pathogenesis is caused by plasmid-encoded protein that causes the attachment of EPEC cells to epithelial tissue. This protein is an adherence factor.

EHEC primarily causes hemorrhagic colitis, bloody and watery diarrhea, and hemolytic uremic syndrome. EHEC produces verotoxin that is known as Shiga toxin (Stx). Although Stx1 and Stx2 toxins are most commonly associated with human illnesses, there are many EHEC serotypes. *E. coli* O157:H7 is the most common pathogenic strain often associated with worldwide foodborne illness. The EHEC infectious dose for foodborne diseases is as low as 10–100 viable cells per g or mL of food or water. The symptoms in people appear within 3–8 days after ingestion of contaminated food. EHEC infections are most commonly associated with undercooked ground beef, cold sandwiches, hamburger, poultry, raw milk, raw apple juice, water, sprouts, and vegetables. This strain is phenotypically different from other *E. coli* strains; it can slowly or not ferment sorbitol and does not has glucuronidase activity. The known reservoirs are cattle, cattle feces, dairies, and meat handling facilities.

EAEC adheres to Hep-2 cells of intestinal mucous and causes mild diarrhea in children and infants. It forms a stack brick appearance due to the production of aggregative adherence on HEp-2 cells. EAEC has a plasmid that is responsible for the synthesis of fimbriae and specific outer membrane protein, and the production of heat-labile toxin and cytotoxin. EAEC foodborne disease is commonly associated with infant foods, water, sprouts, and other vegetables. Cytotoxin can cause damage to intestinal mucosa with the formation of lesions. This results in very mucoid diarrhea without vomiting. EAEC causes bloody diarrhea in children. The transmission of EAEC may occur through food or water contaminated with human or animal feces. Person-to-person transmission may also occur.

DAEC. It mostly causes diarrhea in young children (1–5 years old). A typical symptom associated with DAEC is mild watery diarrhea without leukocytes or blood. DAEC produces a characteristic diffuse-adherent attachment agent to epithelial HEp-2 or HeLa cells. DAEC cannot produce toxin. It produces diffuse adherence factors and invades intestinal epithelial cells. DAEC forms biofilm on equipment and foods such as EPEC and EAEC. The most common vehicles are undercooked beef, water, and livestock.

13.5.1 Equipment, materials, and media

Equipment. Blender (or Stomacher), Erlenmeyer flask, microscope glass slide, Petri dishes, pipette discard equipment (e.g., a graduated cylinder) containing 10% bleach solution and a piece of cloth at bottom, pipettes (1 mL) in the pipette box, tubes (13 × 100 mm), and thermostatic shaking water bath.

Reagents. Formalized physiological saline (FPS) solution, glycerol (20%), McFarland standard (No. 2), $NaHCO_3$ (10%), phosphate buffer saline (PBS) solution, and physiological saline (PS) solution (0.5%).

Media. Brain heart infusion (BHI) agar slant, BHI agar, BHI broth, EC broth, EHEC enrichment broth, lauryl sulfate tryptose (LST) broth, L-eosin methylene blue (L-EMB) agar, Luria Bertani (LB) broth, MacConkey agar, Sulfide indole motility medium, mBPWp broth with and without selective supplement, Lysine iron (LIA) agar, tellurite cefixime sorbitol MacConkey (TCSMAC) agar, and tryptic (tryptone) soy (TS) broth.

13.5.2 Diarrheagenic E. coli isolation techniques

13.5.2.1 Counting by plate count technique

If higher numbers of DEC are present in the sample, perform the direct plating technique. Aseptically, weigh 50 g of sample into a blender containing 450 mL of PBS solution and blend for 1 min to homogenize the sample. In the case of a liquid sample, add 50 mL of sample into a sterile Erlenmeyer flask containing 450 mL of PBS solution and gently shake to homogenize the sample. Dilute the homogenized sample using PBS solution, spread plate the diluted sample to MacConkey agar (Technique 2.2.4), and incubate inverted plates at 35°C for 18 h. DEC produces typical brick red colonies on MacConkey agar. Lactose nonfermenting *E. coli* biotypes produce slightly pink colonies on both media. Count the number of DEC colonies and calculate the number of DEC cfu per mL or g of sample. The count from the plate is most effective if the numbers of *E. coli* present at a level of >1000 cfu per g or mL sample. Select as many as three typical and three atypical colonies for further identification tests.

13.5.2.2 Isolation by enrichment technique

DEC strains are identified depending on their virulence characteristics. Before testing for the specific virulence factor, DEC should be isolated and identified by enriching, plate counting, phenotyping, and genotyping. A DEC enrichment isolation flowchart is given in Fig. 13.11. The enriching of the sample permits the qualitative isolation of DEC. Add 50 g of sample into a sterile blender (or stomacher plastic bag) containing 450 mL of mBPWp broth (without selective supplement) in duplicate and homogenize by blending for 20 s (stomaching for 1 min). In the case of a liquid sample, add 50 mL of sample into a sterile Erlenmeyer flask containing 450 mL of mBPWp broth (without selective supplement) in duplicate and gently shake to homogenize the sample. In both cases, allow the homogenate to stay at room temperature for 10 min with periodic shaking and then allow it to stand for a further 10 min without shaking to settle large particles by gravity. Incubate the flasks containing the liquid homogenate at 35°C for 3 h to repair injured DEC cells. Then the supernatant is poured into a sterile Erlenmeyer flask containing selective supplement (with the concentration using for mBPWp broth) under aseptic conditions. Incubate the flasks at 44.5 ± 0.2°C for 20 h in a water bath without shaking. After incubation, examine the flask for growth (turbidity) and color changes. Streak plate the incubated homogenate to L-EMB agar and TCSMAC agar (Technique 13.4.1). Incubate the inverted plates at 35°C for 18 h. After incubation, examine the colonies on the agar medium in Petri plates. DEC produces brick red, dark centered and flat colonies with or without a metallic green sheen due to the typical lactose-fermentation on

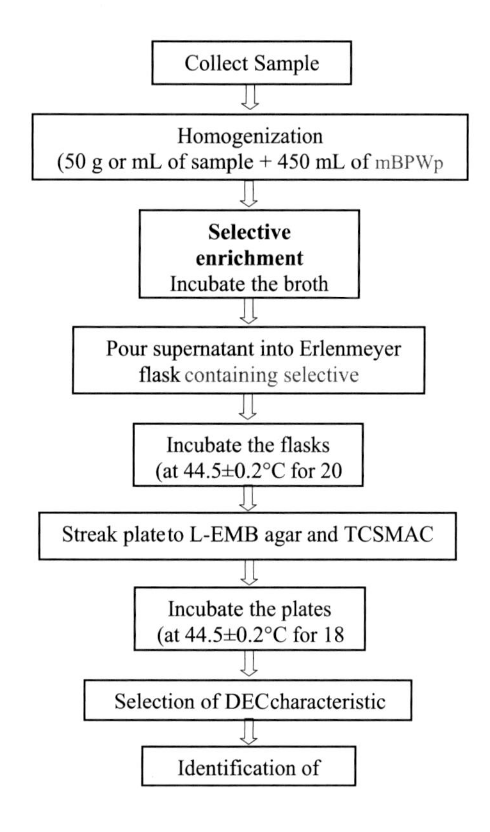

FIGURE 13.11

Diarrheagenic *E. coli* (DEC) selective enrichment isolation flowchart.

L-EMB agar. DEC produces brick red colonies on TCSMAC agar. Lactose nonfermenting *E. coli* biotypes produce slightly pink colonies on both media. Select least three colonies per plate for identification tests.

13.5.3 Interpretation of DEC Results

There is no vaccine for *E. coli* infection and there are no medications recommended to prevent DEC infections. Taking antibiotics can adversely affect the intestinal microbiota and increase susceptibility to gut infections. Food and water are the primary sources of *E. coli* infection, so travelers should be reminded of the importance of adhering to food and water precautions. People working with livestock should be instructed about the importance of handwashing in preventing and spreading infection. People should use hand sanitizer containing 70% alcohol. During outbreaks of *E. coli*, travelers to areas should be warned of the disease and be aware of possible infections among returning travelers.

13.6 Identification of *E. coli*

A DEC identification flowchart is given in Fig. 13.12.

13.6.1 Phenotypic identification *E. coli*

13.6.1.1 Morphological and biochemical identification techniques

Identify *E. coli* before foodomic identifications by the phenotypic techniques (morphological and biochemical tests) and perform these tests on *E. coli* pure culture, as described in Technique 13.4. Some of these identification tests are given here.

1. *Primary tests*. BHI broth pure cultures are cultured on TSI agar slant, used in the identification of *E. coli* (Technique 13.4.3.5): SIM medium (Technique 13.4.2), Citrate agar slant (Technique 13.4.3.4), ONPG broth (Technique 13.4.3.6), Mr-VP broth (Technique 13.4.3.3), and CF broth (Technique 15.3.2.2) containing carbohydrate (e.g., arabinose, lactose, and glucose). Characteristic results from TSI agar slant, indole positive, Mr positive, VP negative, citrate negative, and fermenting arabinose, lactose, and glucose are identified as *E. coli*.
2. *Secondary tests*. Incubate BHI broth tubes for 24 h at 35°C after inoculation with one loopful of BHI broth culture. After incubation, further morphological (Technique 13.4.2) and biochemical (Technique 13.4.3) and serological (Technique 13.5.4) tests should be performed. In the differentiation of *E. coli* from *Shigella*, motility, slow lactose fermentation, indole test, lysine decarboxylating (Technique 15.3.2.1), and acetate reactions (Technique 17.4.2.4) can be used. *Shigella sonnei* may also grow in the same enrichment of *E. coli*, but it cannot decarboxylate lysine and nonmotile, gives negative indole test and slow or nonlactose fermenter.
3. Alternatively, further biochemical tests as indicated in Tables 13.3–13.5 can be used to identify *E. coli*.

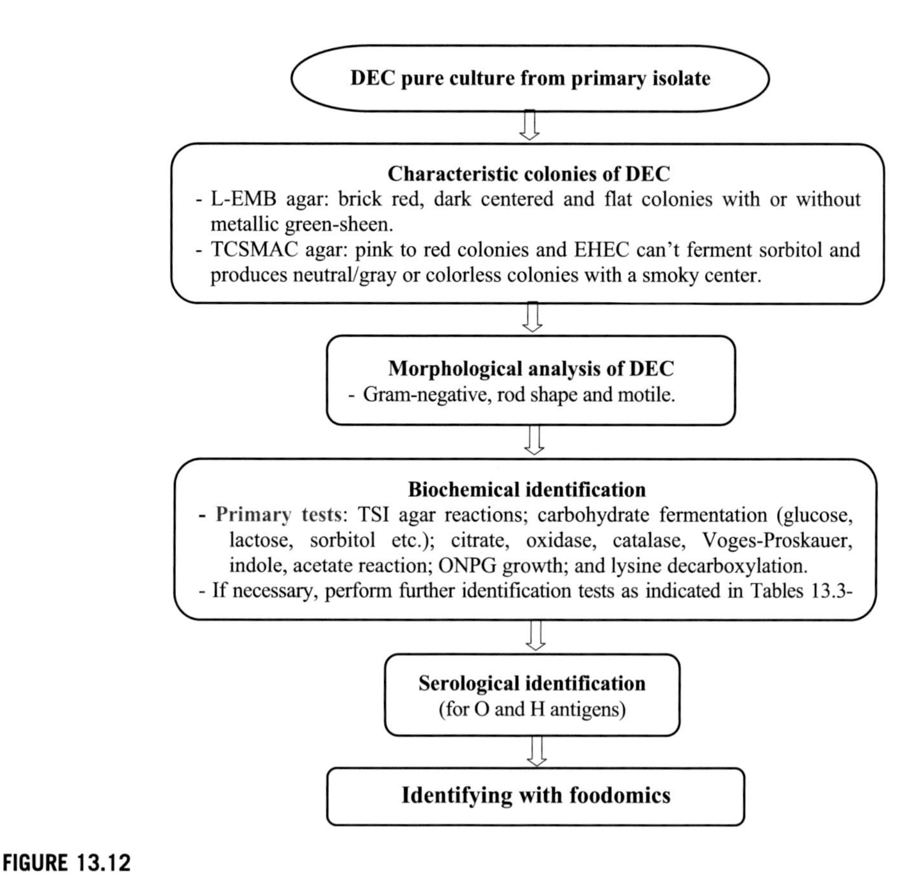

FIGURE 13.12

Diarrheagenic *E. coli* (DEC) identification flowchart.

4. Once phenotypically and serologically identified, foodomic identification techniques for DEC isolates, as described in Technique 13.6.3.1 and Technique 13.6.3.2, should be performed, but DEC-specific procedures and materials should be used.

13.6.2 Serological identification of *E. coli*

1. Tests for Somatic (O) Antigens
   a. *Slide agglutination test*

      Test each *E. coli* isolate by using BHI agar slant culture with *E. coli* polyvalent O antiserum. Homogenize colonies in 5 mL of PS solution to a density corresponding to McFarland standard 4. Discard rough cultures, i.e., those failing to yield homogenous, stable suspensions. Using a waterproof pen, divide slide into two sections on the bottom of a microscope glass slide. Place one drop of homogenized culture onto the upper part of each marked Technique. Add one drop of *E. coli* polyvalent O antiserum to one section and place one drop of PS solution onto the other section. Emulsify the culture in PS solution with the same transfer loop. Mix drops and gently tilt the mixtures back and forth for about 3 min. Examine for agglutination against a dark background with overhead illumination. Classify with any degree of agglutination of somatic (O) test: (1) *positive*; agglutination reaction in the mixture, but no agglutination reaction in the saline control; (2) *negative*; no agglutination reaction in test mixture; no agglutination reaction in the saline control; (3) *nonspecific*; agglutination reaction in the test and control mixtures.

If negative in all sera, heat bacterial suspension at 100°C for 15 min to hydrolyze the interfering surface factors (e.g., capsule). Reexamine in antiserum. If positive in polyvalent antiserum, reexamine with *E. coli* monovalent antiserum.

b. *Tube agglutination test*

Test each *E. coli* isolate for O antigen by using colonies from BHI agar slants. Homogenize colony from BHI agar slant in 0.5% saline up to MacFarland standard 3. Heat homogenate at 100°C for 60 min. Dilute 0.1 mL monovalent O antiserum in 0.9 mL of sterile PS solution. Prepare up to of 1:20 to 1:1280 and OB antiserum up to 1:40 to 1:1280 twofold dilutions by successive transfers of 0.5 mL suspension to 0.5 mL of sterile PS solution in 13 × 100 mm tubes. Add 0.5 mL of heated culture, and unheated culture to each dilution of antiserum separately. For a control, add 0.5 mL of heated suspension and 0.5 mL of PS solution to control tubes as given in Table 13.7. Gently agitate the tubes and cover with aluminum foil. Incubate tubes at 49°C for 16 h and then chill for 1 h at 4°C. Examine tubes carefully for agglutination of cells at the bottom of the tube that do not resuspend after gentle agitation. If the culture agglutinates to the titer, the test is positive for tested culture.

2. Test for Flagellar (H) Antigens

Test each confirmed *E. coli* isolate by using BHI agar slant culture and *E. coli* polyvalent H antiserum. Transfer a small amount of colony from BHI agar slant culture into 5 mL of BHI broth and incubate at 35°C for 4–6 h until visible growth appears (to test on same day); or TS broth at 35°C for 24 h (to test on the following day). Hold broth culture at room temperature for 1 h prior to use. Add 2.5 mL FPS solution to 5 mL of broth culture (or adjust turbidity with sterile FPS solution to MacFarland standard No-2). Test formalized broth culture with *E. coli* polyvalent H antiserum. Place 0.5 mL diluted antiserum into a 13 × 100 mm serological test tube and add 0.5 mL of formalize broth culture (antigen). Prepare saline control in the tube by mixing 0.5 mL of FPS solution with 0.5 mL of formalized broth culture. Incubate mixtures in a water bath at 49°C. Examine at 15 min intervals for agglutination and record final results after 1 h. Classify with any degree of agglutination with polyvalent flagellar (H) test: (1) *positive*; agglutination in test tube with

Table 13.7 Semiquantitative determination of O and B antigens.

Tube no	O serum 0.5 mL (1:40)	OB serum 0.5 mL (1:20)	O antigen 0.5 mL	B antigen 0.5 mL	PS solution 0.5 mL
1	-	-	+	-	+
2	+	-	+	-	-
3	+	-	-	+	-
4	-	+	+	-	-

antigen-antiserum mixture and no agglutination in FPS solution control; (2) *negative*; no agglutination in test tube with antigen-antiserum mixture and in FPS solution control; (3) *nonspecific*; agglutination in both test and control mixtures.

3. Serological Analysis of Enriched Culture

For a liquid sample, aseptically add 25 mL of sample into an Erlenmeyer flask containing 225 mL of BHI broth. Mix well by gently shaking. For a solid sample, aseptically weigh 25 g of sample into a sterile blender (or stomacher plastic bag) containing 225 mL of BHI broth. Homogenize the sample by blending for 1 min (or stomaching for 2 min) and pour the homogenate into a sterile Erlenmeyer flask. Incubate the homogenate in the flask at 35°C for 2 h. After incubation, inoculate 1 mL of homogenate into LST broth and incubate the tubes at 35°C for 18 h. After incubation, inoculate one loopful of LST broth culture into EC broth and incubate tubes at 35°C for 2 h. After incubation, characteristic growth is detected. Neutralize EC broth cultures with 10% $NaHCO_3$ that can be observed by color change. Inoculate one loopful of EC broth culture into BHI broth and BHI agar slant and incubate tubes at 35°C for 2 h. Refer to Technique 13.5.4 for serological identification of pathogenic *E. coli*. Reject all enrichments giving (1) no agglutination in any of the polyvalent antiserum within 3 min or (2) agglutination in polyvalent antiserum and saline control.

13.6.3 Identification of E. coli by foodomics techniques

The phenotype of a microorganisms is determined by the DNA–RNA–protein and other characteristics, and these characteristics are known as "omics." Foodomics is used in the fields of food and nutrition studies as an omics technology to improve the consumer's well-being, health, and confidence in food products with standards. Foodomics involves four main omics areas:

- Genomics involves the investigation of the chromosomal genome and its pattern by molecular techniques (e.g., PCR, genome sequence).
- Transcriptomics explores a set of gene (e.g., mRNA and cDNA) and identifies the difference depending on conditions, microorganisms, and circumstance by several techniques (e.g., microarray analysis).
- Proteomics depends on an indication of proteins and peptides by several techniques (e.g., mass spectrometry as an analytic omics).
- Metabolomics explores chemical processes involving metabolites.

The foodomics application can help to solve some of the problems in food safety and quality, e.g., the indication of chemical and microbial contaminants, the indication of food origin, discovery of biomarkers, and production of novel functional foods. Foodomics is a powerful tool in the indication of microbial food safety. Whole-genome sequence (WGS) and metagenomics are used for the subtyping and detection of foodborne pathogens and viruses from foods, and to characterize the microbial ecology during the production of fermented foods (e.g., cheese, kefir, and kimchi). Omics tools can also be used to indicate the physiological state of pathogens after stresses.

13.6.3.1 MALDI-TOF mass spectrometry identification of E. coli

Mass spectrometry (Ms) is the important analytical technology on which the "omics" approaches are based. The identification of various pathogens with mass spectrometer-based techniques offers an option for microbiology laboratory. Pathogenic microorganisms can be rapidly identified with the matrix-mediated laser desorption ionization-time-of-flight-mass spectrometry (MALDI-TOF Ms) technique, which is based on the principle of ionizing specific protein profile of microbial cells. Microorganisms can be easily identified by comparing these profiles with a reference spectrum. In this technique, protein fingerprints of microorganisms are compared with the references in the system's database. In MALDI-TOF Ms, the sample is dropped on a metal plate, a matrix solution is placed on it and dried in air. After placing it in the Ms device and pulsing with laser beams, the matrix absorbs the light and converts it into the molecule of interest (sample, DNA, or proteins). Usually, ions with multiple charges are formed. The separation of ions is made by the mass-to-charge ratio in mass analysis. For those with a single load, the simple spectrum is created for a MALDI-TOF Ms device. Therefore, MALDI-TOF Ms is more commonly used in food microbiology for its ease of obtaining the spectrum. In this technique, only peptides with a low mass and a large amount of easy ionizability in the mass range of 2000–10,000 Da can be detected. Computer programs that recognize patterns are often used. Unfortunately, there may be differences from study to study or from sample to sample, which makes it difficult to evaluate the results.

A MALDI-TOF Ms *E. coli* identification flowchart is given in Fig. 13.13. Use pure cultures of isolates in MALDI-TOF Ms analysis. A pure culture is analyzed from a single colony or broth culture. One loopful of pure colony or one drop of pellet (from 6000 g centrifugation of liquid culture for 5 min) is suspended in sterile distilled water (or in 70% alcohol) in a small serological tube. The cell suspension is mixed with acetonitrile/formic acid/water (50:35:15, v/v) solution in the extraction process. α-Cyano-4-hydroxycinnamic acid (CHCA; 50% acetonitrile and 2.5% trifluoroacetic acid in pure water) is chosen as the matrix solution. For analysis, the extracted cell suspension in the matrix (1:1) is applied dropwise with an automatic dispenser onto the MALDI target plate and then the mixture is dried at room temperature. After the matrix and sample mixture is dried on the target plate, the plate is inserted in the MALDI-TOF Ms device and the proteomic spectra are generated. The mass spectrometry is calibrated with the manufacturer bacterial test standard *E. coli* extracts. Ions mass spectra in the linear mode are executed on a Microflex LT mass spectrometer. The spectra are initially recorded at 60 Hz and the frequency is reduced to the minimum. The range of recorded masses is 3.6–17 kDa in the positive ion mode. The spectrum of the sample is then matched with reference spectrum in the MALDI Biotyper software and the resulting spectra are analyzed automatically with software. The integrated software generates an outcome list of identified microorganisms. Results obtained from the spectrum are given as score values depending on the type of devices. Score values >2.0 (green color) and from 1.7 to 2 (yellow color) are recognized as reliable at the species and genus level, respectively. When the score is <1.7, the fingerprint is not reported as "reliable identification," meaning that it cannot match sufficiently with any reference microbial species or genus in the database.

13.6.3.2 Molecular identification of E. coli

The genotypic identification techniques are whole-genome analysis and polymerase chain reaction (PCR) assays in the DEC strains identification. Genes encoding virulence factors in DEC strains can be detected by PCR assays. DEC demonstrates a pattern of diffuse adherence. A fimbriae adhesin is responsible for the diffuse cell adherence (polypeptide) of a DEC. The polypeptide is encoded by the *daad* gene. DEC can be identified molecularly by a PCR assay.

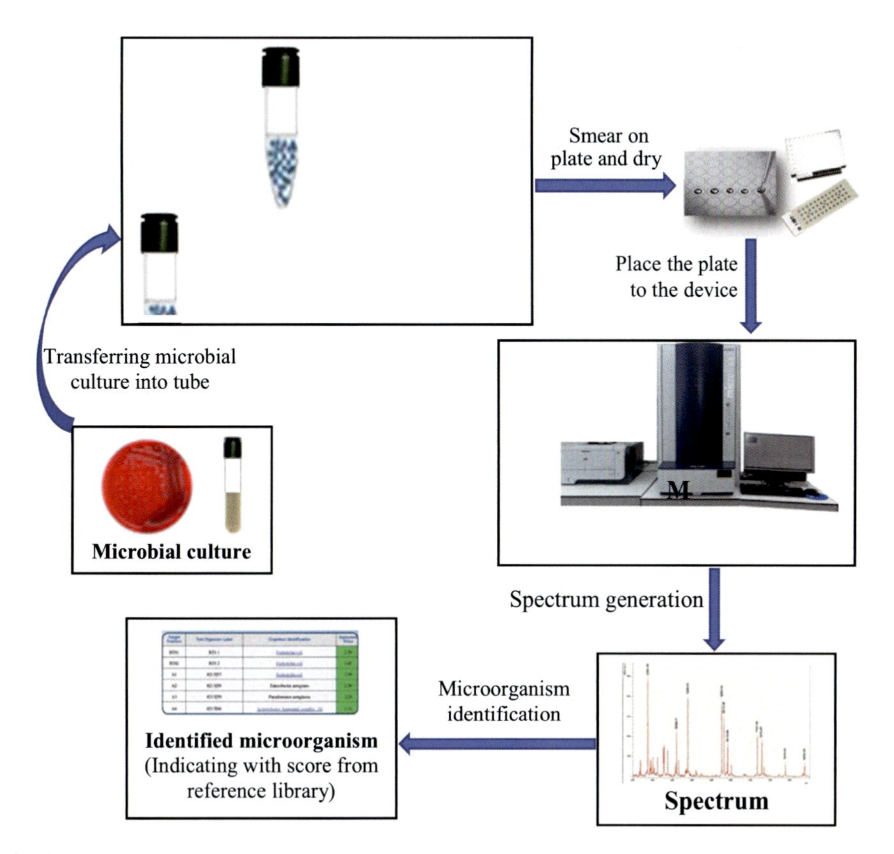

FIGURE 13.13

Matrix-mediated laser desorption ionization-time-of-flight-mass spectrometry (MALDI-TOF Ms) *E. coli* identification flowchart.

1. *PCR assay in the identification of E. coli*

Culture preparation. DEC isolates are cultured in LB broth (or tryptic soy broth). The culture is stored in LB broth (or tryptic soy broth) containing 20% glycerol at −20°C for long-term storage. If it needs to be used in a molecular assay technique, one loopful of LB broth culture is inoculated into BHI broth in tubes and BHI broth tubes are incubated at 35°C for 24 h. This should be repeated once more to activate the culture. Finally, the BHI broth culture is streak plated on BHI agar plate to obtain single colonies (Technique 13.4.1) and inverted plates are incubated at 35°C for 24 h. A single colony is inoculated into BHI broth and incubated at 35°C for 24 h, and the BHI broth culture is used in molecular assay techniques.

DNA extraction with thermal treatment. DNA extraction is performed by the boiled-BHI broth culture. One milliliter of fresh BHI broth culture of *E. coli* is centrifuged at 10,000 rpm for 4 min. The supernatant is decanted and the pellet is resuspended in 1 mL of 10 mM Tris−1 mM EDTA (TE pH 8) solution, the mixture is vortexed for 10 s in order to resuspend the pellets. The mixture is again centrifuged twice and the supernatant is decanted. In each case, the pellet is resuspended in 500 μL of TE and vortexed for 10 s. The final resuspended solution in the tube is placed on a heater at 95°C for 10 min. After heating, the mixture is centrifuged at 12,000 g for 1 min and the supernatant is decanted into a sterile tube and stored at −20°C until use. This is used as a template (extracted microbial DNA) in the PCR assay (amplification).

Oligonucleotide primers. Oligonucleotide primers for DEC strains are obtained from commercial suppliers. The forward (F) and reverse (R) PCR primers for DEC strains are given in Appendix A: Gene Primers in Table A-1.

Polymerase chain reaction. For PCR amplification, the nucleotide sequences of the primers specifically amplify *invA* gene regions. Primer sets for each isolate are used in the separate amplifications. The amplification is performed in 25 μL of reaction mixture containing 5 μL of 5 × buffer (10 mM Tris−HCl pH 9.0, 50 mM KCl, and 0.1% Triton X-100), 3 μL of 25 mM $MgCl_2$ solution, 0.5 μL of 10 mM PCR nucleotide mix, 0.15 μL of tag DNA polymerase (5 U/μL), 0.5 μL of a primer (*invA*), 2.0 μL of template DNA (extracted microbial DNA) and 14.35 μL distilled water (sterile). If necessary, add approximately 50−70 μL of sterile mineral oil. Mineral oil is not added for thermal cycles associated with heat. If volumes of PCR reaction are decreased to 50 μL during heating, the template amount should be decreased to 1.0 μL.

A 1-kb DNA ladder of the *E. coli* strain can be used as a positive control in the PCR test and *E. coli* K12 is used as a negative control.

Temperature cycling. The cycling program is set in the PCR system to denature by heating. PCR conditions in heating for simultaneous amplification of gene fragment, e.g., *invA*, are:

- One cycle heating at 95°C for 5 min;
- Thirty-five cycles heating at 94°C for 30 s (denaturation);
- Heating at 58°C for 1 min (annealing);
- Heating at 72°C for 1 min (extension);
- Final denaturation by heating at 72°C for 5 min; and
- Cooling by holding at 4°C.

After cooling, the reaction mixture is used in electrophoresis for the visualization of PCR products.

PCR products analysis of reaction mixture on agarose gel. Prepare a 1.2%−1.5% agarose gel in 0.5 × TBE (0.9 M Tris-borate, 0.02 M EDTA, pH 8.3) containing 0.5 μg ethidium bromide. Agarose gel is melted using a microwave and allows the gel to solidify. Mix 10 μL of PCR products (reaction mixture) with 2.0 μL 6 × gel dye and load onto gel submerged in 1 × TBE. Apply a constant 100 V/cm voltage for 60 min and allow fragments to migrate in the gel until the separation of the appropriate band. Use a short-wave UV light to visualize bands depending on the molecular weight marker. Determine gene fragment lengths. Use the gel photographs to indicate the results by a comparable gel documentation system.

2. *Genome sequence identification of E. coli*

Microbial whole-genome resequencing is a critical tool to complete the genomes of known microorganisms, as well as to compare multiple genomes or to map genomes of new microorganisms. Sequencing of entire genomes of microorganisms is used to generate accurate reference genomes, to do microbial identification, and in other comparative genome studies. A whole-genome sequencing (WGS) data platform is used in the prediction of isolates. In the WGS analysis, extract genomic deoxyribonucleic acid (DNA) with a full automating purification from 1 mL of bacterial culture. Extract bacterium DNA using a DNA kit protocol. In this extraction, harvest 0.7 mL of overnight bacterium culture in a deep well plate. Lyse bacterial cells in the 220 μL of buffer and 20 μL protease and incubate the mixture while shaking at 56°C for 30 min. Add 4 μL of RNase (100 mg mL^{-1}) to the lysed cells and reincubate at 37°C for 15 min. This step increases the purity of the DNA for further downstream sequencing. DNA extraction is continued by elution in 100 μL of sterile distilled water. After extraction, detect DNA concentration. The pooled amplicon library (PAL) is performed by combining equal volumes of normalized library, quantitated and diluted in hybridization buffer. This results in the formation of a diluted amplicon library (DAL). Heat denatures DAL and load it onto the reagent cartridge for sequencing. Sequence the extracted DNA with a standard base protocol on the appropriate instrument. Perform a sequence reading with bases removed from the trailing end. If the length posttrimming is less than 50 bp, discard the read and its pair. The read pipeline is used to compare isolated sequences against one or more sets of reference genomes (public genome databases) to identify the isolate.

13.6.3.3 Diarrheagenic E. coli *strains genotypic identifications*

Identification of enterotoxigenic E. coli (ETEC). When the number of *E. coli* is higher than 10^4 cells per g or mL of foods, colony hybridization analysis is performed on the isolates using gene probes coding ETEC heat-resistant and heat-labile toxins. PCR and genomic analysis techniques can also be used in the identification of ETEC as indicated in Technique 13.5.2.1 and 13.5.2.2, respectively, with specific isolated gene and primers (Appendix A: Gene Primers in Table A-1).

Identification of enteroinvasive E. coli (EIEC). The invasive characteristic of the isolates can be detected to indicate EIEC. This can be performed by the mouse (or Guinea pig Sereny) or keratoconjunctivitis test. Invasive characteristic of the bacterium can be determined by the HeLa tissue culture cell assay technique or by staining intracellular bacteria in HeLa cells in the in vitro staining technique using acridine orange dye. PCR and genomic analysis techniques can also be used in the identification of EIEC as indicated in Techniques 13.5.2.1 and 13.5.2.2, respectively, with the use of specific isolated genes and primers (Appendix A: Gene Primers in Table A-1).

Identification of enteropathogenic E. coli (EPEC). EPEC isolates can be identified depending on three key traits: attachment and effacing lesion (A/E), localized adherence on cells and the absence of Shiga toxin (Stx). Phenotypically, lesions and localized adherences are tested using HeLa or Hep-2 tissue cells. PCR and genomic analysis techniques can be used in the identification of EPEC as indicated in Technique 13.5.2.1 and 13.5.2.2, respectively, with specific isolated genes and primers (Appendix A: Gene Primers in Table A-1).

Identification of enterohemorrhagic E. coli (EHEC). In direct EHEC isolation from foods, an EHEC enrichment broth can be performed and followed with plating on selective medium TCSMAC agar. All *E. coli* strains grow in this medium but EHEC is identified depending on non-sorbitol fermentation. Both the enrichment and the selective medium contain different antimicrobial agents that inhibit the growth of normal microflora of food except *E. coli.* Add 25 g of solid food into 225 mL of sterile EE broth in the blender (or stomacher plastic bag). Homogenize the sample with blending for 1 min (or stomaching for 2 min). Pour homogenate into a sterile Erlenmeyer flask. In the case of liquid food, add 25 mL of liquid foods into 225 mL of sterile EE broth in the sterile Erlenmeyer flask and homogenize by gently shaking. Incubate the homogenized sample at

35°C with shaking in an incubator for 24 h. After incubation of the flask, one loopful of the enriched sample is streak plated to TCSMAC agar plate. Incubate the inverted Petri plates for 24 h at 35°C. After incubation, analyze the characteristic colonies on the Petri plates. Sorbitol-fermenting bacteria appear as pink to red colonies on TCSMAC agar and EHEC can't ferment sorbitol. EHEC typical colonies on TCSMAC agar are neutral/gray or colorless with a smoky center. Atypical sorbitol nonfermenting EHEC strains along with other sorbitol nonfermenting bacteria (such as *Morganella* and *Hafnia*) may be similar to the EHEC colonies on TCSMAC agar. Therefore, confirmation tests (phenotypic and serological) are performed to differentiate EHEC isolates from other bacteria. PCR and genomic analysis techniques can be used in the identification of EHEC as indicated in Technique 13.5.2.1 and 13.5.2.2, respectively, with specific isolated genes and primers (Appendix A: Gene Primers in Table A-1).

Identification of EAEC. The identification of EAEC depends on the characteristic "stacked-brick" aggregative adherence when culturing in LB broth at 35°C for 3 h in HEp-2 cells. Serotyping of EAEC is a problem due to their aggregative phenotype, many of the strains autoagglutinate as nontypeable or as O-rough. The isolate is identified by the antiaggregate protein transporter gene by PCR. PCR and genomic analysis techniques can be used in the identification of EAEC, as indicated in Technique 13.5.2.1 and 13.5.2.2, respectively, with the use of specific isolated genes and primers (Appendix A: Gene Primers in Table A-1).

PRACTICE

Isolation and counting of *Enterococcus*

14

14.1 Introduction

All streptococci of fecal origin (fecal streptococci) producing Lancefield group D antigens are considered to be *Enterococcus*. They are Gram-positive, nonspore forming, nonmotile, coccus or coccobacillus, facultative anaerobic, and catalase negative. Some species of *Enterococcus* are *Enterococcus faecalis* (its varieties are *liquefaciens*, *faecalis* and *zymogenes*), *Enterococcus faecium*, *Enterococcus avium*, *Enterococcus casseliflavus*, *Enterococcus hirae*, *Enterococcus durans*, *Enterococcus malodoratus*, *Enterococcus gallinarum*, and *Enterococcus mundtii*. They usually grow in 6.5% NaCl, at a temperature range from 10°C to 45°C, and at pH 9.6. *Enterococcus* is associated with cheeses, fermented sausages, processing equipment, environments, water, and mud. *E. faecalis* and *E. faecium* have proteolytic and hemolytic characteristics. They are heat-resistant thermoduric species and may survive at pasteurization temperature. Most *Enterococcus* spp. are relatively resistant to freezing. *Enterococcus* resistance to antibiotics can cause serious illness in humans as opportunistic pathogens. In general, the levels of *Enterococcus* can be used as a good sanitary index bacterium to indicate sanitation conditions in food processing and proper holding conditions of equipment.

KF streptococcal agar is a selective/differential medium that contains sodium azide as a selective agent and triphenyltetrazolium chloride as a differential agent. Azide inhibits *Streptococcus bovis*, and possibly some *Enterococcus* spp. *Enterococcus* can be identified from foods with growth on fluorogenic gentamicin-thallous-carbonate (fGTC) agar as a differential/selective medium. *Enterococcus* counts from foods may be two or more times higher on fGTC agar than on KF streptococcus agar. An excellent confirmatory test is the ability of *Enterococcus* to grow on bile esculin azide agar. All *Enterococcus* species tolerate bile salt and hydrolyze esculin. Phenotypic and foodomics techniques are used in the identification of *Enterococcus* spp. Serological group D reaction can also be used in the identification of *Enterococcus*.

14.2 Isolation and counting techniques

Sampling, sample preparation, and all experiments should be performed under aseptic conditions. Sampling and all analysis should be performed in duplicate, unless otherwise specified. All microbiological media to be used in this application must be sterilized. Again, all solutions and equipment that may cause contamination must be sterile or disinfected.

A selective enrichment isolation and counting flowchart of *Enterococcus* is given in Fig. 14.1.

Microbiological Analysis of Foods and Food Processing Environments. DOI: https://doi.org/10.1016/B978-0-323-91651-6.00025-2

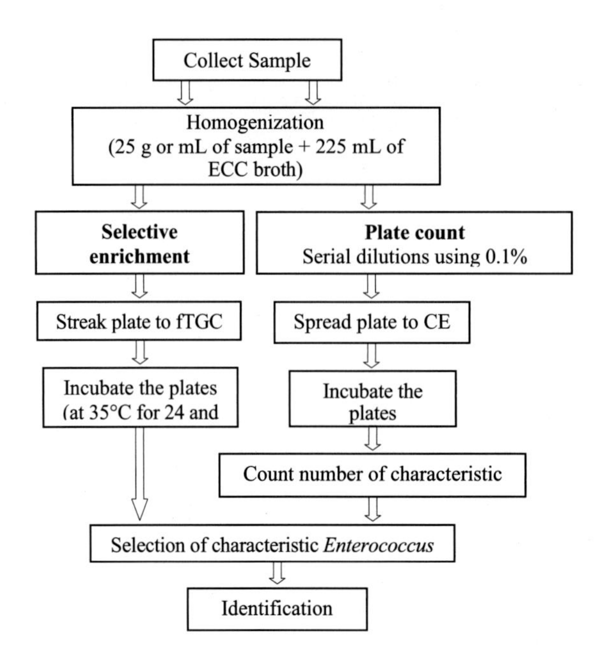

FIGURE 14.1

Enterococcus enrichment isolation and counting flowchart.

14.2.1 Equipment, materials, and media

Equipment. Blender (or Stomacher), cover slip, Erlenmeyer flask, incubator, long-wave (365 nm) ultraviolet lamp, microscope, microscope glass slide, pipette discard equipment (e.g., a graduated cylinder) containing 10% bleach solution and a piece of cloth at bottom, pipettes (1 mL) in the pipette box, and vortex.

Materials. Cinnamaldehyde reagent (0.01%), D-group antiserum, Gram stains, H_2O_2 (3%), immersion oil, 0.1% peptone water, PYR reagents, and triphenyl-tetrazolium chloride (TTC, 1% aq.).

Media. Brain heart infusion (BHI) broth, BHI agar slant, BHI broth containing 6.5% NaCl and 0.0016% bromocresol purple, BHI broth with pH 9.6, Bile esculin (BE) agar, chromocult enterococci (CE) agar, enterococcosel (ECC) broth, fluorogenic gentamicin-thallous-carbonate (fGTC) agar, glucose azide (GA) broth, M-E agar, PYR broth, and trypticase soy (TS) broth.

14.2.2 Selective enrichment isolation technique

Use the enrichment technique in the isolation of *Enterococcus* from a sample that contains a low number of *Enterococcus*. Refer to Technique 2.2.2 for details of sample preparation and dilutions. Add 25 g of sample into 225 mL of sterile ECC broth in the blender (or stomacher plastic bag). The content is mixed by swirling the flask and it is allowed to remain at 4°C for 60 min to extract microbial cells. Then blend for 1 min (or stomaching for 2 min) to homogenize the sample. Pour the homogenate into an Erlenmeyer flask. In the case of liquid sample, add 25 mL sample into

225 mL of sterile ECC broth in the sterile Erlenmeyer flask. The content is mixed by gently shaking the flask and is allowed to remain at 4°C for 60 min. Mix again by shaking. Incubate flasks at 35°C for 24 and 48 h. After incubation, examine the flask for growth (turbidity) and color changes (black color formation). Streak plate one loopful of culture onto fTGC agar and incubate inverted plates at 35°C for 24 and 48 h. After incubation, examine plates for characteristic colony formation on agar medium. *E. faecium* hydrolyzes starch and its colonies surrounded by a zone of bright bluish fluorescence. *E. faecalis* and *E. ovium* cannot hydrolyze starch and cannot produce fluorescence. Confirm the *Enterococcus* isolates as indicated in Technique 14.3.

14.2.3 Counting techniques

14.2.3.1 Plate count technique

Refer to Technique 2.2.2 for details of sample preparation and dilutions. Add 25 g of sample into 225 mL of 0.1% peptone water in the blender (or stomacher plastic bag). The content is blended for 1 min (or stomached for 2 min) to homogenize the sample. Moist and liquid samples (25 mL) are directly added into 225 mL of 0.1% peptone water in the Erlenmeyer flask and homogenized by gently shaking the flask. Add 1 mL of diluted sample into sterile Petri dishes in duplicate. Pour 17–19 mL of melted CE agar (at 47°C) into Petri dishes. Mix agar and 1 mL of sample by thoroughly rotating the plates by drawing a figure of eight while lying on the table tap, and allow the agar to solidify in the plates (within 10 min). Incubate inverted Petri plates at 35°C for 24 h. After incubation, examine the formation of colonies and color changes on the medium. If there is no color change nor visible growth continue the incubation up to 48 h. Count all characteristic colonies (red colonies) from plates containing 25–250 colonies. Confirm the presumptive isolates using identification tests (Technique 14.3). Calculate the number of *Enterococcus* by multiplying the average count by the dilution rate and dividing by the inoculation amount. Record the results as the number of *Enterococcus* colony forming units (cfu) per g or mL of sample.

14.2.3.2 Most probable number count technique

Refer to Technique 14.2.3.1 for sample preparation and dilutions and refer to Technique 4.2 for MPN count. Inoculate 1 mL of diluted sample into each of 3-MPN tubes of GA broth. If very low numbers of *Enterococcus* are expected, add the sample into equal volumes of double strength medium. After inoculation of diluted sample, MPN tubes (or bottles) are incubated at 35°C for 24–72 h. After incubation, examine tubes for growth (turbidity) and color changes (yellow color). The formation of acid with growth indicates a positive tube(s). Record the number of positive tubes for each dilution. Confirm the presumptive *Enterococcus* isolates, as indicated in Technique 14.3. Refer to MPN Table 4.1, calculate the number of MPN and report the results as the *Enterococcus* MPN per g or mL of sample.

Thermoduric *Enterococcus*. Inoculate 1 mL of diluted sample into 3-MPN tubes of GA broth. Incubate tubes at 45°C ± 0.2°C in a thermostatic shaking water bath up to 48 h. After incubation, the tubes are examined, and the formation of the yellow color indicates the presence of thermoduric *Enterococcus* spp. Indicate the number of positive MPN tubes showing positive results at 45°C. Confirm the presumptive *Enterococcus* isolates as indicated in Technique 14.3. Refer to MPN

Table 4.1 and calculate the number of MPN. Report the results as the thermoduric *Enterococcus* MPN per g or mL of sample.

14.2.3.3 Membrane filter count technique

Perform the membrane filter (MF) method as explained in Technique 5.2. M-E agar is a selective medium for *Enterococcus*. BE agar and EL agar are used to identify *Enterococcus* hydrolyzing esculin (to esculitin) and tolerating bile salt. The esculitin reacts with the iron salt (ferric citrate) in the medium to produce a dark brown–black color on the ground colony.

Mostly water sanitary quality is determined by the MF technique. MF permits to recover the *Enterococcus* from a large volume of water and foods. Place a sterile MF (0.22 μm pore size) on the filter base while grid side up and place the funnel on the base; MF is now held between the funnel and the base. A liquid sample is homogenized by gently shaking about 25 times, and 100 mL of sample (diluted or nondiluted) is added into the sterile funnel. The sample is filtered under vacuum and the funnel sides are rinsed at least twice by 20–30 mL of sterile 0.1% peptone water. After filtration, the vacuum is turned off and the funnel is removed from the filtration apparatus. The MF is removed from the filtration apparatus by sterile forceps and rolled to the surface of M-E agar plate while avoiding bubble formation between the filter and the surface of agar. If bubbles occur, filtration should be repeated. The Petri plates are incubated at 42°C for 24–48 h. After incubation, MF from the plate is examined for the formation of characteristic *Enterococcus* colonies. Transfer MF from M-E agar plate and place onto an EL agar plate. The EL agar (or BE agar) plates are retained at room temperature for 20–30 min, and then the plates are incubated at 42°C for 20 min. After incubation, count the number of colonies on MF containing pink to red colonies with black or reddish-brown precipitation on the medium. Calculate the number of *Enterococcus* using the formula:

$$Enterococcus cfu \text{ per } 100 \text{ mL} = (\text{average colony count})(\text{DF})$$

where DF is the dilution rate.

Confirm the *Enterococcus* isolates as indicated in Technique 14.3. Report the results as the number of *Enterococcus* cfu per 100 mL of sample.

14.3 Identification of *Enterococcus*

Obtain a pure culture by the pure culture techniques (Technique 13.4.1) using BHI broth and fTGC agar in the purification. Carry out inoculation into BHI broth, BHI agar slant, and BHI agar plate from pure colonies, and incubate them at 35°C for 18 h. Use cultures in phenotypic and foodomics identification techniques. The *Enterococcus* identification flowchart is given in Fig. 14.2.

14.3.1 Morphological identification

Gram staining. Prepare Gram stain from the BHI agar slant culture, as given in Appendix B under the topic "Gram Staining Reagents." A Gram stain of *Enterococcus* will produce Gram-positive, cocci, forming pairs and occasionally short chains (Fig. 14.3).

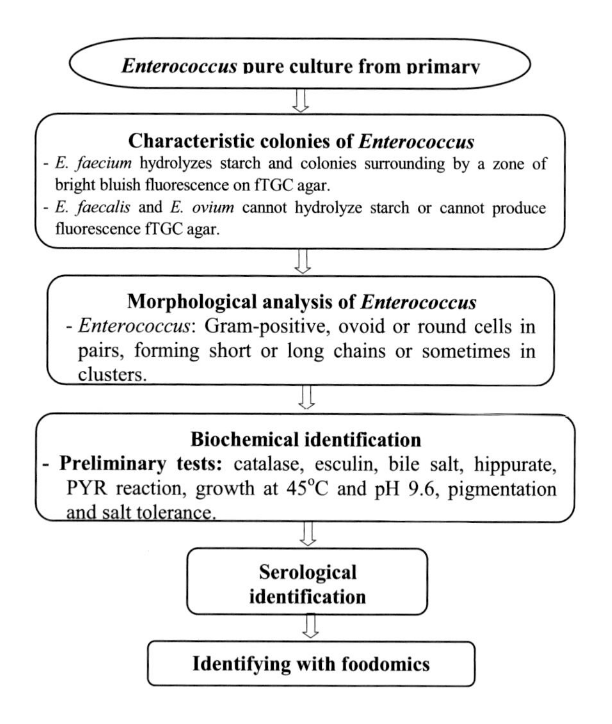

FIGURE 14.2

Enterococcus identification flowchart.

Motility. Refer to Technique 13.4.2 for a motility test with wet mount preparation. *Enterococcus* species are not motile (except for *E. casseliflavus* and *E. gallinarum*).

14.3.2 Biochemical identification

Biochemical characteristics of *Enterococcus* are given in Table 14.1.

14.3.2.1 Bile esculin agar

Inoculate one loopful of BHI broth culture to BE agar slant and incubate slant at 35°C for 24 h. *Enterococcus* hydrolyzes esculin in the BE agar slant with the formation of a dark color (Fig. 14.4). Alternatively, the test can also be performed on a BE agar plate. Streak one loopful of BHI broth culture to the BE agar slant and incubate inverted plates at 35°C for 24 h. *Enterococcus* produces dark brown or black halo colonies on BE agar. Intestinally *Enterococcus* hydrolyzes the glycoside esculin to produce dextrose and esculetin. Esculetin produces an olive green to black color complex with iron(III) ions.

14.3.2.2 Salt tolerance test

Inoculate one loopful of BHI broth culture to TS broth containing 6.5% NCl and bromocresol purple. Incubate the tube at 35°C for 24 h. Examine tubes for growth (turbidity) and color change after

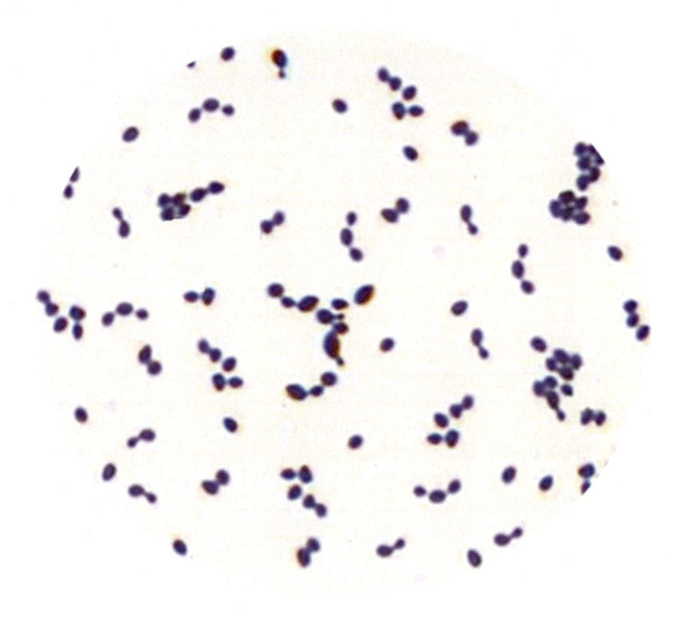

FIGURE 14.3

Staining appearance of *Enterococcus faecalis*.

24 h and if growth cannot appear, incubate tube again for an additional 24 h and again examine tubes. Visible turbidity (growth) in the broth with a color change from purple to yellow indicates tolerance to salt. No turbidity and no color change after 72 h indicates a negative salt-tolerant test. *Enterococcus* tolerates 6.5% NaCl. It is used to differentiate *Enterococcus* from nonsalt-tolerant bacteria.

14.3.2.3 Pyrrolidonyl arylamidase test

Enterococcus species produce the enzyme L-pyrrolidonyl aminopeptidase which hydrolyzes the substrate L-pyroglutamic acid-beta-naohthylamide (PYR) with the formation of a free beta-naphthylamine. The amine then combines with the cinnamaldehyde reagent that is added to form a bright red end product. For this test, emulsify two to three morphologically similar colonies from BHI agar plate in the small amount of PYR broth in the test tube with a loop. Incubate the tube at 35°C for 4 h. After incubation, add one drop of cinnamaldehyde reagent and observe the color change. The reaction should be read 1 min after the addition of the reagent. A bright cherry red color formation within a minute after adding the reagent indicates a positive result. A yellow or orange color indicates a negative result. *Enterococcus* and *Streptococcus pyogenes* are PYR positive; and *S. bovis* and *Streptococcus anginosus* are PYR negative.

14.3.2.4 Other tests

Yellow pigmentation. Inoculate one loopful of BHI broth culture into TS agar and incubate at 35°C for 24 h. *E. casseliflavus* and *E. mundtii* (variable) produce a yellow pigment on this medium.

Table 14.1 Phenotypic characteristics of *Enterococcus* spp.

Characteristic	*Enterococcus avium*	*Enterococcus casseliflavus*	*Enterococcus durans*	*Enterococcus faecalis*	*Enterococcus faecium*	*Enterococcus gallinarum*	*Enterococcus hirae*	*Enterococcus malodoratus*	*Enterococcus mundtii*	*Streptococcus bovis*	*Streptococcus pneumoniae*
PYR/H_2S	+/v	+/ −	+/ −	+/ −	+/ −	+/ −	+/ −	+/+	+/ −	− / −	v/ −
Catalase, oxidase, indole, citrate, urease	−	−	−	−	−	−	−	−	−	−	−
Nitrite reduction	+	+	+	+	+	+	+	+	+	+	+
Mr/VP	− /v	− /v	− /+	− /+	− /+	− /v	− /+	− /v	− /+	− /+	− /+
Growth											
At 4°C	−	−	−	−	+	−	−	−	−	−	−
At 10°C and 45°C	+	+	+	+	+	+	+	+	+	−	−
6.5% NaCl	+	+	+	+	+	+	+	+	+	−	−
pH 9.6	+	+	+	+	+	+	+	+	+	−	−
Hydrolyzing											
Arginine	−	v	+	+	+	+	+	−	+	−	−
Esculin/ hippurate	+/v	+/ −	+/v	(+)/v	+/(−)	+/(+)	+/(−)	+/v	+/ −	+/ −	− / −
Fermentation of											
Lactose/ malonate	+/+	+/+	+/ −	+/+	+/+	+/+	+/+	+/+	+/+	+/+	+/+
Gluconate/ glycerol	+/v	+/v	− /(+)	+/ −	+/ −	+/ −	− / −	+/ −	+/ −	+/ −	+/ −
Arabinose/ sorbitol	+/+	+/v	− / −	− /(+)	+/v	+/ −	− / −	− /+	+/v	v/v	− / −
Melezitose/ sorbose	+/+	− / −	− / −	(+)/ −	− /	(−)/ −	v/ −	− /+	− / −	− / −	− / −
Melibiose/ tagatose	v/+	+/(+)	v/ −	− /+	v/(+)	+/v	(+)/v	+/ −	+/ −	+/ −	
Motility/ Yellow color	− / −	+/+	− / −	− / −	− / −	+/ −	− / −	− / −	− /v	− / −	− / −

+, Positive; − , negative; v, variable; (), weak result.

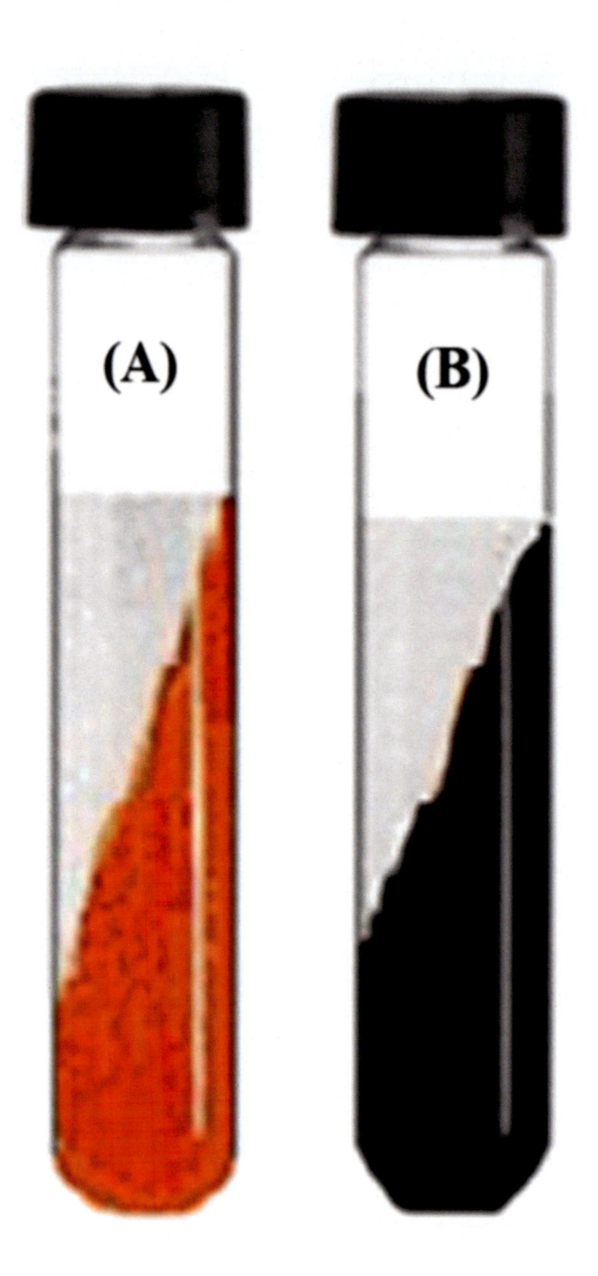

FIGURE 14.4

Growth of *Enterococcus* on bile esculin agar. (A) esculin nonhydrolysis (*red color*) and (B) esculin hydrolysis (*black color*).

Growth at 45°C. Inoculate one loopful of BHI broth culture into BHI broth, and incubate at 45°C for 24 h. After incubation, examine the BHI broth for growth (turbidity). *Enterococcus* grows at that temperature (except *E. malodoratus*).

Growth at pH 9.6. Inoculate one loopful of BHI broth culture into BHI broth (with pH 9.6) and incubate tubes at 35°C for 24 h. *Enterococcus* grows at pH 9.6.

Hippurate hydrolysis. Refer to Technique 17.4.2.8 for the hippurate hydrolysis test. *Enterococcus* hydrolyzes hippurate.

Carbohydrate fermentation. Refer to Technique 15.3.2.2 for carbohydrate fermentation. Carbohydrates are used in the identification of *Enterococcus* species, as indicated in Table 14.1.

Decarboxylase test. Refer to Technique 15.3.2.1 for the decarboxylase test. *Enterococcus* species are variable in the decarboxylation of arginine.

Catalase test. Refer to Technique 13.4.3.10 for the catalase test. Use BHI agar slant culture in the catalase test. *Enterococcus* species are catalase negative bacteria. Caution: Do not test catalase on azide-containing media, For example, KF streptococcal agar.

14.3.3 Serological identification of *Enterococcus*

Refer to Technique 13.5.4 for serological tests with the use of D group antigens. *Enterococcus* species have D group antigen.

14.3.4 Identification of *Enterococcus* species by foodomics techniques

14.3.4.1 MALDI-TOF Ms identification of Enterococcus

Enterococcus can be rapidly identified with the matrix-mediated laser desorption ionization-time-of-flight-mass spectrometry (MALDI-TOF Ms) technique, which is based on the principle of ionizing a specific protein profile of microbial cells. *Enterococcus* species can be easily identified by comparing these profiles with the reference spectrum. MALDI-TOF Ms identification has been explained in Technique 13.6.3.1.

14.3.4.2 Molecular identification of Enterococcus

Polymerase chain reaction (PCR) is a molecular genotypic technique and is used in the identification of *Enterococcus* species. Gene coding specific *Enterococcus* characteristics can be detected by PCR assays. Some *Enterococcus* specific primer sets that can be used in the identification of *Enterococcus* are given in the Appendix A: Gene Primers (Table A.1). Using *Enterococcus*-specific techniques and materials, the bacterium can be identified by PCR analysis, as described in Technique 13.6.3.2.

PRACTICE

Isolation and counting of *Salmonella* 15

15.1 Introduction

Salmonella is a member of Enterobacteriaceae, and is motile with peritrichous flagella, Gram-negative, nonspore forming, rod-shaped, facultative anaerobic, catalase positive, and oxidase negative. *Salmonella* contains two species: *S. enterica* and *S. bongori*. *S. enterica* is further divided into six subspecies and *S. enterica* subsp. enterica includes over 2600 serotypes according to biochemical and genomic characteristics. *Salmonella* nomenclature and some examples of serotypes are given in Fig. 15.1. At present, some of *S. enterica* subsp. *enterica* serotypes are ser. Typhimurium, ser. Typhi and ser. Enteritidis.

Salmonella causes systemic and enteric disease. *S.* typhimurium, *S. enteritidis*, and *S. virchow* predominantly cause gastrointestinal disease (enteric). *S. typhi*, and *S. paratyphi* A, B, and C cause systemic disease. A wide range of animals, e.g., rodents, birds, reptiles, and insects, carry *Salmonella* in their gastrointestinal tract. Meat and milk products are the primary, and poultry and eggs the most frequent vehicles of *Salmonella*. Cross-contamination of undercooked foods can occur by direct contact with kitchen equipment and utensils. *Salmonella* strains are sensitive to gastric acidity in the stomach, therefore the consumption of more viable cells together with foods allow the cells to pass into the small intestine and to cause the foodborne disease (salmonellosis). Ingestion of 10^6 viable *Salmonella* cells per g or mL of foods is necessary for salmonellosis. Following the ingestion, the symptoms occur within 24–36 h and last about 2–3 days. The principal symptoms are nausea, vomiting, diarrhea, mild fever, and abdominal cramps. Salmonellosis can be fatal to infants and elderly people. A small number of *Salmonella* (10–100 viable cells per g or mL) can also cause foodborne disease with high protein and fat foods (e.g., chocolate, hamburgers, ice-cream, potato chips, peanut butter, and salami sticks).

The isolation techniques in the analysis of foods for *Salmonella* include preenrichment, selective enrichment, and selective/differential plating and identification. Relatively small numbers of *Salmonella* are usually present in foods, compared to another competing microflora. Subjecting the foods to a process (e.g., heat, desiccation, preservatives, high osmotic pressure, high carbon dioxide pressure, and changes in pH) can also cause sublethal damage (injury) to *Salmonella*. Preenrichment broth increases the recovery rate of *Salmonella*. The usual ratio of sample size for adding to preenrichment broth would be 1:10. Selective enrichment is followed to increase the number of *Salmonella* cells and to prevent the further unwanted microbial growth in the medium. The selective enrichment broth culture is streak plated on selective/differential agar medium. Unprocessed raw foods or highly contaminated products may contain a high number of *Salmonella*; therefore, the sample is directly analyzed using the selective/differential media. Some of the selective/differential plating media are bismuth sulfite agar, SS agar, deoxycholate citrate agar, lactose sucrose urea agar, XLD agar, and brilliant green agar. The addition of brilliant green into media

Microbiological Analysis of Foods and Food Processing Environments. DOI: https://doi.org/10.1016/B978-0-323-91651-6.00041-0

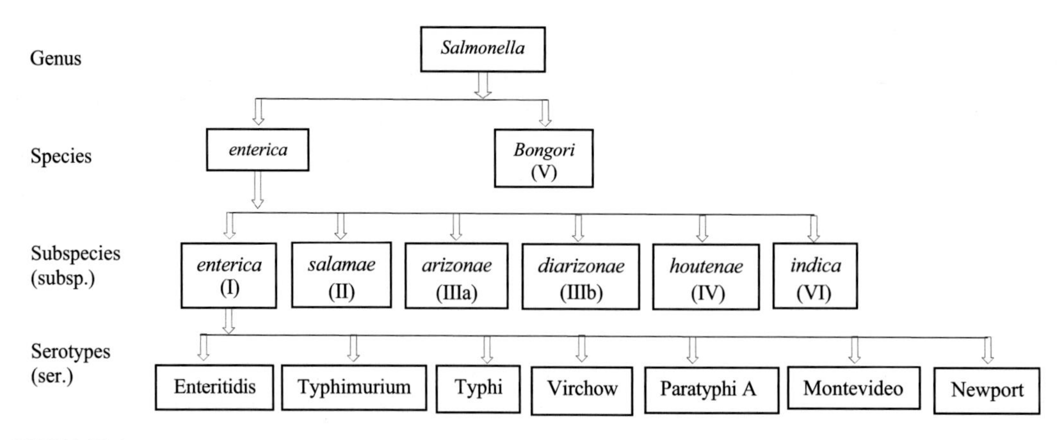

FIGURE 15.1

Current *Salmonella* nomenclature and examples of *S. enterica* subsp. *enterica* serotypes.

inhibits *Pseudomonas* and coliforms, but brilliant green can also inhibit *S.* Typhi. Isolated *Salmonella* are identified by phenotypic, serological, and foodomics techniques.

15.2 Isolation and counting techniques

Sampling, sample preparation, and all experiments should be performed under aseptic conditions. Sampling and all analysis should be performed in duplicate, unless otherwise specified. All microbiological media to be used in this application must be sterilized. Again, all solutions and equipment that may cause contamination must be sterile or disinfected.

15.2.1 Equipment, materials, and media

Equipment. Blender (or Stomacher), bottle (500 mL), cover slip, Erlenmeyer flask (500 mL), incubator, microscope, microscope glass slide, pipette discard equipment (e.g., a graduated cylinder) containing 10% bleach solution and a piece of cloth at bottom, pipettes (1 mL) in the pipette box, vortex, and water bath (at 50°C).

Materials and solutions. Brilliant green (BG) stain (1%, aq.), Gram stains, Liefson's flagella stain, 0.1% peptone water, tergitol anionic-7, Formalized physiological saline (FPS) and triton X-100.

Media: Bismuth sulfite (BS) agar, Brain heart infusion (BHI) broth, BHI agar slant, brilliant green (BG) agar, buffered peptone (BP) broth, carbohydrate fermentation (CF) broth (containing glucose, lactose, sucrose and dulcitol separately), decarboxylase broth (containing lysine, arginine and ornithine separately), Dey-Engley (DE) broth, Hektoen (HE) agar, lactose broth, lysine iron (LI) agar slant, malonate broth, motility test (MT) medium, nutrient broth, Rappaport-Vassiliadis (RV) broth, tetrathionate (TT) broth, triple sugar iron (TSI) agar slant, trypticase (tryptic) soy (TS) agar slant, TS-ferrous sulfate (TS-Fe) broth, trypticase broth, universal preenrichment (UP) broth, and xylose lysine deoxycholate (XLD) agar.

15.2.2 Isolation techniques of *Salmonella*

15.2.2.1 Homogenization of sample and preenrichment

The following homogenization and preenriching techniques can be used in the isolation of *Salmonella*. Frozen samples should not be thawed if they are not analyzed immediately. Growth of unwanted microorganisms during thawing can be minimized by thawing as quickly as possible. To reduce damage to *Salmonella* cells, they should be thawed within 15 min with continuous shaking in a thermostatic shaking water bath below 45°C or thawed at 2–5°C within 18 h.

1. *Dried whole egg, dried egg, egg yolk, skim milk, whole milk, powders (cake, bread, biscuit, cookie and doughnut), and infant foods.* (a) Weigh 25 g or mL of nonpowdered sample into an Erlenmeyer flask containing 225 mL of lactose broth. Mix the sample well in the broth by gently shaking the flask. (b) For powdered sample, weigh 25 g of sample into a sterile Erlenmeyer flask, add 15 mL of lactose broth, and stir using a sterile glass rod or spoon to obtain a smooth suspension. Add 10, 10, and 190 mL portions of lactose broth in that order to provide a total of 225 mL. After each lactose broth portion addition, mix the sample well in the broth by gently shaking the flask.

   After keeping the solution at room temperature for 60 min, mix the homogenate well by gently shaking the flask. Measure the pH using an amount of homogenate sample, if necessary, adjust the pH of the homogenate to 6.8 with a sterile base (1 N NaOH) or acid (1 N HCl). Incubate the flask at 35°C for 24 h. After incubation, proceed as in Technique 15.2.2.2 for the selective enrichment of *Salmonella*.
2. *Eggs*
   a. *Shell eggs.* Wash the eggs' surface by rubbing with a hard brusher. After washing, water should be drained. Eggs should be disinfected by keeping in 200 ppm chlorine solution containing 0.1% sodium dodecyl sulfate for 30 min and the surface left to dry. Each shell eggs sample should contain 20 eggs and therefore remove 50 samples per poultry house. Break the eggs aseptically into a sterile beaker, and the egg yolks and whites are thoroughly mixed with a sterile spoon. Aseptically, weigh 25 g of liquid whole eggs sample into an Erlenmeyer flask containing 225 mL of TS-Fe broth. Mix the sample thoroughly in the broth by gently shaking the flask. After keeping the homogenate at room temperature for 60 min, mix the homogenate well by gently shaking the flask. Measure the pH using an amount of homogenate sample, if necessary, adjust the pH of the homogenate to 6.8. Incubate the flask at 35°C for 24 h. Proceed as in Technique 15.2.2.2 for the selective enrichment of *Salmonella*.
   b. *Liquid whole eggs.* Weigh 25 g of sample into an Erlenmeyer flask containing 225 mL of TS-Fe broth. Mix the sample thoroughly in the broth by gently shaking the flask. After keeping the homogenate at room temperature for 60 min, mix the homogenate well by gently shaking the flask. Measure the pH using an amount of homogenate sample, if necessary, adjust the pH of the homogenate to 6.8. Incubate the flask at 35°C for 24 h. Proceed as in Technique 15.2.2.2 for the selective enrichment of *Salmonella*.
3. *Nonfat dry milk and dry whole milk.* Weigh 25 g of sample on a clean paper funnel and add the dry milk sample slowly to 225 mL of sterile BG stain solution in an Erlenmeyer flask while shaking. After keeping the homogenate at room temperature for 60 min, mix the

homogenate well by gently shaking the flask. Incubate the mixture in the loosely capped flask at 35°C for 24. Proceed as in Technique 15.2.2.2 for the selective enrichment of *Salmonella*.

4. *Casein*
   a. *Lactic casein*. Weigh 25 g of sample on a clean paper funnel and add the sample slowly over the surface of 225 mL of lactose broth in the Erlenmeyer flask. Mix the sample thoroughly in the broth by gently shaking the flask. After keeping the homogenate at room temperature for 60 min, mix the homogenate well by gently shaking the flask. Incubate loosely the capped flask without shaking at 35°C for 24 h. Proceed as in Technique 15.2.2.2 for the selective enrichment of *Salmonella*.
   b. *Rennet casein*. Weigh 25 g of sample on a clean paper funnel and add the sample slowly over the sterile 225 mL of lactose broth in the Erlenmeyer flask while shaking. Mix the sample thoroughly in the broth by gently shaking the flask. After keeping the homogenate at room temperature for 60 min, mix the homogenate well by gently shaking the flask. Incubate the mixture in a loosely capped flask at 35°C for 24 h. Proceed as in Technique 15.2.2.2 for the selective enrichment of *Salmonella*.
   c. *Sodium caseinate*. Aseptically weigh 25 g of the sample into a sterile Erlenmeyer flask and add 225 mL of lactose broth slowly into flask. Mix the sample thoroughly in the broth by gently shaking the flask. After keeping the homogenate at room temperature for 60 min, mix the homogenate well by gently shaking the flask. Measure the pH using an amount of homogenate sample, if necessary, adjust the pH of the homogenate to 6.8. Incubate the flask at 35°C for 24 h. Proceed as in Technique 15.2.2.2 for the selective enrichment of *Salmonella*.
5. *Soy flour*. Weigh 25 g of sample on a clean paper funnel and add the sample slowly over the sterile 225 mL of lactose broth in the Erlenmeyer flask while shaking. After keeping the homogenate at room temperature for 60 min, mix the homogenate well by gently shaking the flask. Incubate the loosely capped flask at 35°C for 24 h. Proceed as in Technique 15.2.2.2 for the selective enrichment of *Salmonella*.
6. *Food products containing egg (egg rolls, spaghetti, macaroni, noodles), salads (from tuna, egg, chicken, ham), dough, cheese, fruits and vegetables (fresh, frozen or dried), fish, crustaceans (crab, shrimp), and nut meats*. Weigh 25 g of the sample into a sterile blender containing 225 mL of lactose broth and blend for 1 min. After keeping the homogenate at room temperature for 60 min, mix the homogenate well by gently shaking the flask. Measure the pH using an amount of homogenate sample, if necessary, adjust the pH of the homogenate to 6.8. Incubate the flask at 35°C for 24 h. Proceed as in Technique 15.2.2.2 for the selective enrichment of *Salmonella*.
7. *Inactive and active dried yeast*. Weigh 25 g of sample into an Erlenmeyer flask containing 225 mL of trypticase broth. Mix the sample thoroughly in the broth by gently shaking the flask. After keeping the homogenate at room temperature for 60 min, mix the homogenate well by gently shaking the flask. Measure the pH using an amount of homogenate sample, if necessary, adjust the pH of the homogenate to 6.8. Incubate the flask at 35°C for 24 h. Proceed as in Technique 15.2.2.2 for the selective enrichment of *Salmonella*.
8. *Topping and frosting mixes*. Weigh 25 g of sample into 225 mL of nutrient broth in the Erlenmeyer flask containing. Mix the sample thoroughly in the broth by gently shaking the flask. After keeping the homogenate at room temperature for 60 min, mix the homogenate well by gently shaking the flask. Measure the pH using an amount of homogenate sample, if

necessary, adjust the pH of the homogenate to 6.8. Incubate the flask at 35°C for 24 h. Proceed as in Technique 15.2.2.2 for the selective enrichment of *Salmonella*.

9. *Spices*
   a. *Pepper (black, red, or white), paprika, celery seed or flakes, sesame seed, cumin, thyme, rosemary, and vegetable flakes*. Weigh 25 g of sample into 225 mL of TS broth in the sterile Erlenmeyer flask. Mix the sample thoroughly in the broth by gently shaking the flask. After keeping the homogenate at room temperature for 60 min, mix the homogenate well by gently shaking the flask. Measure the pH using an amount of homogenate sample, if necessary, adjust the pH of the homogenate to 6.8. Incubate the flask at 35°C for 24 h. Proceed as in Technique 15.2.2.2 for the selective enrichment of *Salmonella*.
   b. *Onion and garlic flakes and onion powder*. Weigh 25 g of sample into 225 mL of TS broth in the Erlenmeyer flask with added K_2SO_3 solution before sterilization (mix 5 g K_2SO_3 per 1000 mL of TS broth before sterilization; resulting in a final broth with 0.5% K_2SO_3). Mix the sample thoroughly in the broth by gently shaking the flask. Measure the pH using an amount of homogenate sample, if necessary, adjust the pH of the homogenate to 6.8. Incubate the flask at 35°C for 24 h. Proceed as in Technique 15.2.2.2 for the selective enrichment of *Salmonella*.
   c. *Cloves, cinnamon, allspice, and oregano*. Weigh 1 g of sample into 100 mL of TS broth in the Erlenmeyer flask. Mix the sample thoroughly in the broth by gently shaking the flask. After keeping the homogenate at room temperature for 60 min, mix the homogenate well by gently shaking the flask. Measure the pH using an amount of homogenate sample, if necessary, adjust the pH of the homogenate to 6.8. Incubate the flask at 35°C for 24 h. Proceed as in Technique 15.2.2.2 for the selective enrichment of *Salmonella*.
10. *Candy and candy products (including chocolate)*. Weigh 25 g of sample into 225 mL of TS broth in the blender and homogenize by blending for 1 min. Aseptically transfer homogenate into a sterile Erlenmeyer flask. After keeping the homogenate at room temperature for 60 min, mix the homogenate well by gently shaking the flask. Measure the pH using an amount of homogenate sample, if necessary, adjust the pH of the homogenate to 6.8. Add 0.45 mL BG solution and mix the sample thoroughly by gently shaking the flask. Incubate the flask at 35°C for 24 h. Proceed as in Technique 15.2.2.2 for the selective enrichment of *Salmonella*.
11. *Coconut*. Weigh 25 g of sample into 225 mL of lactose broth in the Erlenmeyer flask and mix well by gently shaking the flask. Add 2.25 mL of steamed (for 15 min) tergitol-7 (or use 15 min steamed triton X-100) into mixture and mix well again by gently shaking. Incubate flask at 35°C for 24 h. Proceed as in Technique 15.2.2.2 for the selective enrichment of *Salmonella*.
12. *Food dyes and coloring substances*. For dyes with pH 6.0 or above, use technique described for dried whole egg-1 (powdered product) above. For dyes with pH below 6.0, aseptically weigh 25 g of sample into 225 mL of tetrathionate (TT) broth in the Erlenmeyer flask. Mix the sample thoroughly in the broth by gently shaking the flask. After keeping the homogenate at room temperature for 60 min, mix the homogenate well by gently shaking the flask. Measure the pH using an amount of homogenate sample, if necessary, adjust the pH of the homogenate to 6.8. Add 2.25 mL of 0.1% BG stain to the mixture and mix well again by

gently shaking. Incubate the flask at 35°C for 24 h. Proceed as in Technique 15.2.2.2 for the selective enrichment of *Salmonella*.

13. *Gelatin*. Weigh 25 g of sample into 220 mL of lactose broth in the Erlenmeyer flask and 5 mL papain solution (5%). Mix the sample thoroughly in the broth by gently shaking the flask. After keeping the homogenate at room temperature for 60 min, mix the homogenate well by gently shaking the flask. Measure the pH using an amount of homogenate sample, if necessary, adjust the pH of the homogenate to 6.8. Incubate the flask at 35°C for 24 h. Proceed as in Technique 15.2.2.2 for the selective enrichment of *Salmonella*.
14. *Meat and meat products, meals (meat, fish, bone), and animal substances*. Weigh 25 g of sample into a blender containing add 225 mL of lactose broth and blend for 1 min. If sample is powder or ground, blending is omitted. Aseptically transfer homogenate into a sterile Erlenmeyer flask. After keeping the homogenate at room temperature for 60 min, mix the homogenate well by gently shaking the flask. Measure the pH using an amount of homogenate sample, if necessary, adjust the pH of the homogenate to 6.8. Add 2.25 mL of steamed (15 min) tergitol-7 (or 15 min steamed triton X-100) and mix well again by gently shaking. Incubate the flask at 35°C for 24 h. Proceed as in Technique 15.2.2.2 for the selective enrichment of *Salmonella*.
15. *Gum*. Weigh 25 g of sample into 225 mL of lactose broth in the Erlenmeyer flask and add 2.25 mL of sterile cellulase solution (1% aq.). Mix the sample thoroughly in the broth by gently shaking the flask. After keeping the homogenate at room temperature for 60 min, mix the homogenate well by gently shaking the flask. Incubate the homogenate without pH adjustment at 35°C for 24 h. Proceed as in Technique 15.2.2.2 for the selective enrichment of *Salmonella*.
16. *Orange and apple juices, and apple cider*. Add 25 mL of sample into 225 mL of lactose broth in an Erlenmeyer flask. Mix the sample thoroughly in the broth by gently shaking the flask. After keeping the homogenate at room temperature for 60 min, mix the homogenate well by gently shaking the flask. Mix well again by gently shaking and do not adjust pH. Incubate the flask at 35°C for 24 h. Proceed as in Technique 15.2.2.2 for the selective enrichment of *Salmonella*.
17. *Rabbit carcasses*. Place rabbit carcass into sterile plastic bag. Place bag in an appropriate container. Add sterile lactose broth at a 1:9 carcass-to-broth (g per mL) ratio to cover carcass. Mix the sample thoroughly in the broth by gently shaking the flask. After keeping the homogenate at room temperature for 60 min, mix the homogenate well by gently shaking the flask. Measure the pH using an amount of homogenate sample, if necessary, adjust the pH of the homogenate to 6.8. Incubate the flask at 35°C for 24 h. Proceed as in Technique 15.2.2.2 for the selective enrichment of *Salmonella*.
18. (a) *Melon and watermelon*. From cut fruit, weigh 25 g of sample into 225 mL of UP broth in the blender and blend for 2 min. Transfer the homogenized mixture to a sterile Erlenmeyer flask and let it stand undisturbed for 60 min at room temperature and do not adjust pH. Incubate the flask at 35°C for 24 h. Proceed as in Technique 15.2.2.2 for the selective enrichment of *Salmonella*.

    (b) *Whole melon and water melon*. Place the cantaloupe into a sterile plastic bag in a container. Add enough UP broth to cover the fruit. Loosely tighten the open end of the plastic bag (not airtight). Let it stand undisturbed for 60 min at room temperature. Lightly mix by gently shaking the flask and do not adjust the pH. Incubate the flask at 35°C for 24 h. Proceed as in Technique 15.2.2.2 for the selective enrichment of *Salmonella*.

19. (a) *Tomato juices*. Weigh 25 g of sample into 225 mL of UP broth in the blender and blend for 2 min. Aseptically transfer the homogenized mixture to a sterile Erlenmeyer flask. After keeping the homogenate at room temperature for 60 min, mix the homogenate well by gently shaking the flask. Measure the pH using an amount of homogenate sample, if necessary, adjust the pH of the homogenate to 6.8. Mix well again by gently shaking. Incubate the homogenate in the flask at 35°C for 24 h. Proceed as in Technique 15.2.2.2 for the selective enrichment of *Salmonella*.
    (b) *Whole tomatoes*. Place the tomato into a sterile plastic bag in a container. Add enough UP broth into plastic bag to cover the tomato. Loosely tighten the open end of the plastic bag (not airtight). Let it stand undisturbed for 60 min at room temperature. Lightly mix by gently swirling and do not adjust the pH. Incubate the flask at 35°C for 24 h. Proceed as in Technique 15.2.2.2 for the selective enrichment of *Salmonella*.
20. *Environmental sample*. Remove the sample from environmental surfaces with sterile swabs. Place the swab/sponge in a sterile bag or tube containing Dey-Engley (DE) broth to cover the swab. If samples are not analyzed immediately, refrigerate at 4°C. Start the sample analysis within 48 h of collection. Place the swab containing broth into 225 mL of lactose broth in an Erlenmeyer flask. Mix the sample thoroughly in the broth by gently shaking the flask. After keeping the homogenate at room temperature for 60 min, mix the homogenate well by gently shaking the flask. Measure the pH using an amount of homogenate sample, if necessary, adjust the pH of the homogenate to 6.8. Mix well again by gently shaking. Incubate the flask at 35°C for 24 h. Proceed as in Technique 15.2.2.2 for the selective enrichment of *Salmonella*.
21. *Dried beans*. Weigh 25 g of dried beans into 225 mL of lactose broth in an Erlenmeyer flask and mix by gently shaking the flask. After keeping the homogenate at room temperature for 60 min, mix the homogenate well by gently shaking the flask. Measure the pH using an amount of homogenate sample, if necessary, adjust the pH of the homogenate to 6.8. Mix well again by gently shaking. Incubate the flask at 35°C for 24 h. Proceed as in Technique 15.2.2.2 for the selective enrichment of *Salmonella*.
22. *Fresh vegetables, herbs, and sprouts*. Weigh 25 g of sample into 225 mL of UP broth in an Erlenmeyer flask (for cabbage, use modified BP water) and mix the contents by gently shaking the flask. Do not adjust pH. Incubate the flask at 35°C for 24 h. Proceed as in Technique 15.2.2.2 for the selective enrichment of *Salmonella*.
23. *Animal feed*. Weigh 25 g of sample into 225 mL of BP water in an Erlenmeyer flask. Mix the sample thoroughly in the broth by gently shaking the flask. After keeping the homogenate at room temperature for 60 min, mix the homogenate well by gently shaking the flask. Measure the pH using an amount of homogenate sample, if necessary, adjust the pH of the homogenate to 6.8. Mix well again by gently shaking. Incubate the flask at 35°C for 24 h. Proceed as in Technique 15.2.2.2 for the selective enrichment of *Salmonella*.

15.2.2.2 Selective enrichment of Salmonella

The *Salmonella* selective enrichment isolation and counting flowchart is given in Fig. 15.2 Gently shake the incubated preenriched homogenate from Technique 15.2.2.1 and carry out inoculation

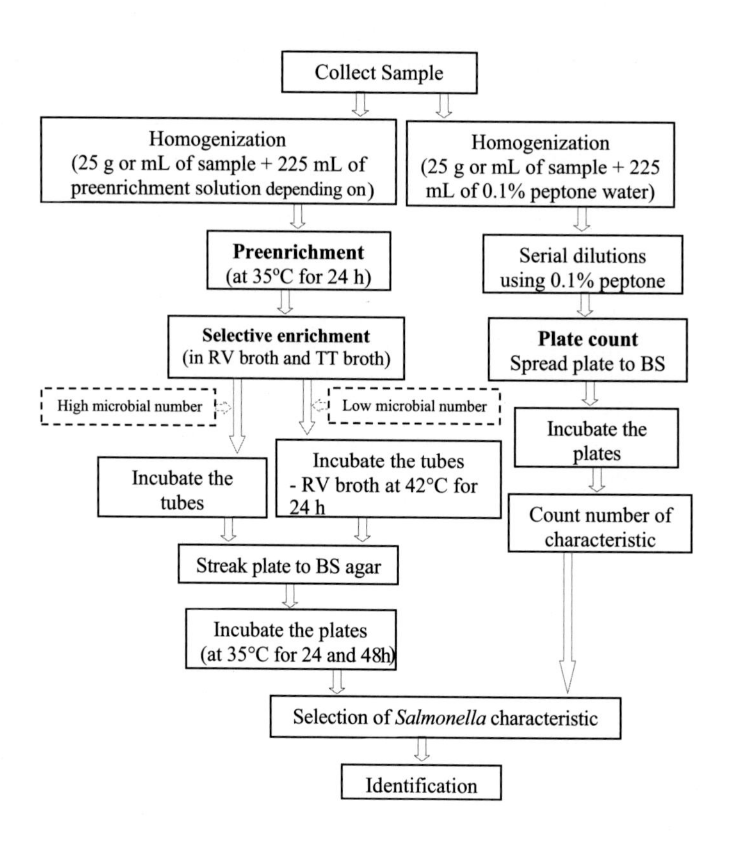

FIGURE 15.2

Salmonella selective enrichment isolation and counting flowchart.

into the selective enrichment broth. For this, inoculate 0.1 mL of preenriched homogenate into 10 mL of RV broth and 1 mL into 10 mL of TT broth. Vortex the tubes to mix. Incubate the selective enrichment broths as follows:

1. *Foods with a high number of microorganisms.* Incubate RV broth and TT broth at 42°C for 24 h in the thermostatic shaking water bath.
2. *Foods with a low number of microorganisms.* Incubate RV broth at 42°C in the thermostatic shaking water bath and TT broth at 35°C for 24 h.

15.2.2.3 Selective isolation

Prepare BS agar plates daily before streaking and store plates in the dark at room temperature until use. After the incubation of a selective enrichment broth (Technique 15.2.2.2), examine the broth for growth (turbidity) and color changes. Vortex the TT broth to mix. Streak plate a loopful of TT broth culture to BS agar (or XLD agar, HE agar, BG agar). Repeat streaking from the RV broth culture to BS agar plates. Incubate the inverted Petri plates at 35°C for 24 h. After incubation of the BS agar plates, examine the plates for the typical *Salmonella* colonies as indicated below. The interpretation of results with typical colonies is as follows:

1. Typical *Salmonella* colony morphology on selective/differential agar media
   Bismuth sulfite (BS) agar. Salmonella produces blue to blue-green colonies with or without a black center on this agar medium. The medium around the colony usually appears brown at first, may turn to black with increasing incubation (Fig. 15.3).
   Hektoen enteric (HE) agar. Salmonella produces blue to blue-green and black-centered large colonies with or without a black center on this agar medium.
   Xylose lysine deoxycholate (XLD) agar. Salmonella produces pink, large colonies with or without black centers and may appear as completely black colonies on this medium.
   Brilliant green (BG) agar. Salmonella produces colorless, pink, and translucent to opaque colonies with pink to red color with surrounding medium. Some *Salmonella* produce transparent green colonies. On the other hand, lactose- or sucrose-fermenting bacteria produce yellow-green colonies.
2. *Atypical Salmonella colony morphology on selective/differential agar media*
   When typical *Salmonella* colonies are not detected on the selective/differential media, atypical *Salmonella* colonies can be indicated as follows:
   BS agar. If typical *Salmonella* colonies are not present on this medium after 24 h, reincubate the plates for an additional 24 h. Atypically *Salmonella* produces green colonies with little or

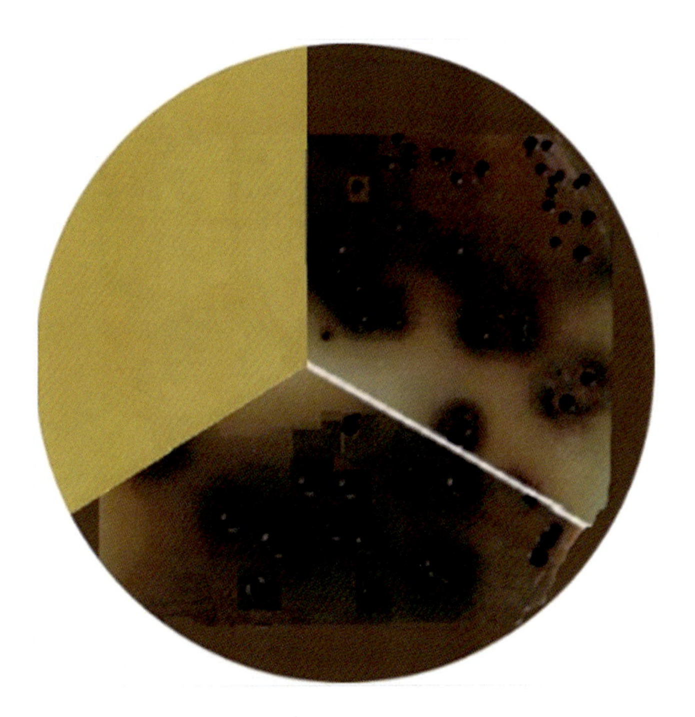

FIGURE 15.3

Salmonella typhimurium colonies (right bottom) and *S. typhi* colonies (right top) on BS agar.

no darkening of the surrounding medium. Select two or more atypical colonies per BS agar plate for interpretation of results.

HE agar and XLD agar. If typical *Salmonella* colonies are not present on the medium after 24 h and reincubate the plates an additional 24 h. Atypically *Salmonella* produces yellow colonies with or without black centers on HE and XLD agar. Select two or more atypical *Salmonella* colonies per plate for the interpretation of results.

3. *Interpretation of results from TSI agar and LI agar slants*

Perform interpretation of results on TSI agar and LI agar slants on the typical and atypical isolates from selective/differential agar plate cultures.

a. With a sterile needle, the bacterial cells are taken by twice touching onto colony center, and the TSI agar slant is inoculated by stabbing into agar and then streaking on the slant surface. Without flaming the needle, the LI agar slant is inoculated by stabbing into agar butt twice and then streaking on the slant surface.

b. Incubate the tubes at 35°C for 18 h. After incubation, record the characteristic changes from each medium (Table 15.1). (i) TSI agar slant: *Salmonella* positive cultures show alkaline (red) slant (not utilize lactose) and yellow butt (acid from utilization of glucose), with H_2S (blacking of the agar surface) or without H_2S, and with and without gas (raising medium) (Fig. 15.4A). *S. arizonae* may produce yellow slant due to late fermentation of lactose. With further incubation, TSI agar may show the appearance of coliform type growth with yellow butts and slants. (ii) LI agar slant: *Salmonella* typically produce an alkaline (purple) butt and slant with H_2S production (Fig. 15.4B). Consider typical results only a distinct yellow butt. Permanently yellow color throughout the agar as a negative test for *Salmonella.* If H_2S is produced, the butt of the LI agar is blackened.

Table 15.1 Cultural characteristics of *Salmonella* on triple sugar iron (TSI) agar/lysine iron (LI) agar slants.

Salmonella strain[a]	Butt	Slant surface	H_2S
Salmonella typhi	A/A	NC/NC	+/+
Salmonella entericia	AG/NC	NC/NC	+/+
Salmonella pullorum	AG/NC	NC/NC	v/v
Salmonella gallinarum	A/NC	NC/NC	v/v
Salmonella paratyphi A	AG/NC	NC/NC	-/-
Salmonella paratyphi B	AG/NC	NC/NC	+/+
Salmonella enteritidis	AG/NC	NC/NC	+/+
Salmonella typhimurium	AG/NC	NC/NC	+/+
Salmonella *choleraesuis*	AG/NC	NC/NC	+/+
Salmonella arizonae	AG/NC	NC/NC	+/+
S. bongori	AG/NC	NC/NC	+/+

A, *acid (yellow);* G, *gas;* NC, *no color change (red);* +, *H2S production;* -, *no H2S production.*
[a]Salmonella *enterica subsp. enterica ser. (e.g.,* S. enterica *subsp. enterica ser. Choleraesuis, S. enterica subsp. arizona).*

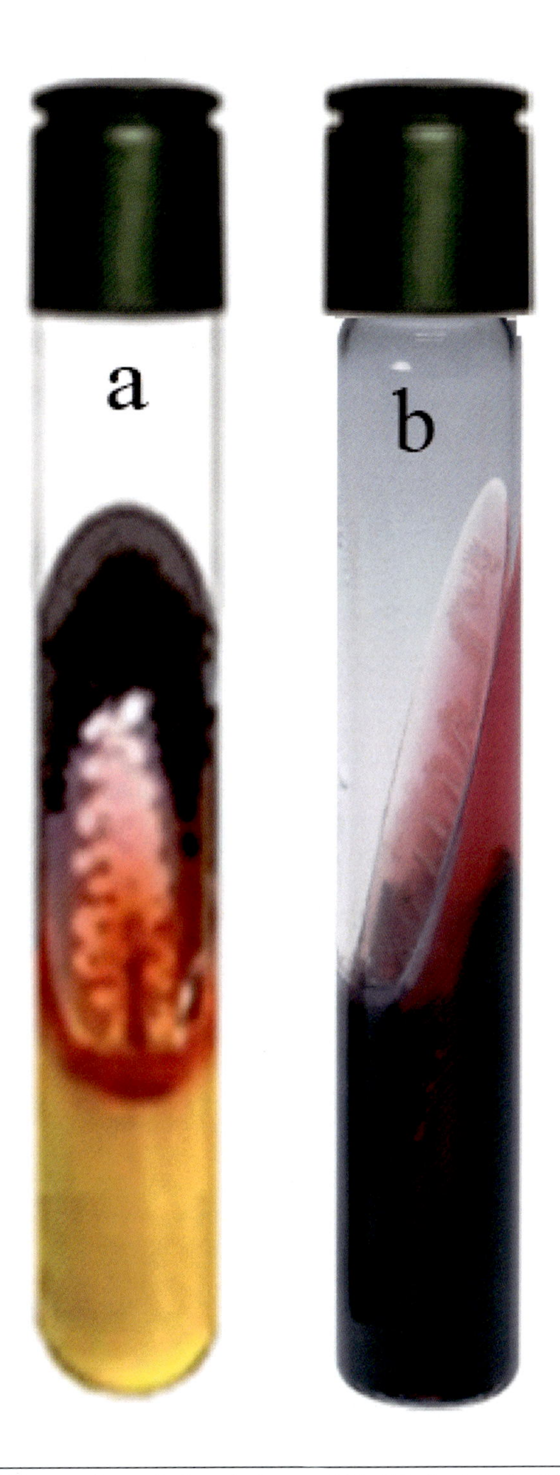

FIGURE 15.4

(A) *Salmonella* growth on TSI agar slant: alkaline slant, an acid butt (yellow) and a blackening of the agar slant (H_2S). (B) *Salmonella* growth on LI agar slant: alkaline (purple) slant and butt, and a blackening of the butt (H_2S).

TSI agar cultures with yellow slants cannot be excluded from further analysis and they should be considered as presumptive positive TSI agar cultures if the corresponding LI agar gives typical *Salmonella* results. LI agar is a useful medium in the detection of lactose or sucrose fermenting *Salmonella* (atypical).

1. (a) The cultures giving alkaline butt in LI agar slant, regardless of TSI agar slant reaction, should be retained as potential *Salmonella* isolates and submitted to biochemical and serological tests. (b) The cultures giving an acid butt in LI agar slant, and an alkaline slant and acid butt in TSI agar slant should also be considered potential *Salmonella* isolates and they should be submitted to biochemical and serological tests. (c) The cultures giving an acid butt in LI agar slant, and an acid slant and butt in TSI agar slant may be discarded as not being *Salmonella*. (d) If the TSI agar slant cultures cannot give typical reactions for *Salmonella* (alkaline slant and acid butt), pick colonies on selective/differential agar plate and repeat test by inoculation to TSI agar and LIA agar slants.
2. On characteristic isolates, apply phenotypical, serological, and foodomics identification techniques.

15.2.3 Counting of *Salmonella*

Refer to Technique 2.2.2 for details of sample preparation and dilutions. Weigh 25 g of the sample into a blender (or stomacher plastic bag) containing 225 mL of 0.1% peptone water. Homogenize the sample by blending for 1 min (or stomaching for 2 min). In the case of a liquid sample, add 25 mL of the sample into a sterile Erlenmeyer flask containing 225 mL of 0.1% peptone water and homogenize the sample by gently shaking the flask. Prepare further dilutions using 0.1% peptone water. Spread plate 1 mL of diluted sample to surface area of predried BS agar in duplicate (Technique 2.2.4). After spreading, allow the inoculum to dry on the surface of the medium and incubate inverted plates at 35°C for 24 h. After incubation, examine the plates for typical *Salmonella* colony formation (Technique 15.2.2.3). Count characteristic colonies from the plates containing 25–250 colonies. Calculate the number of *Salmonella* by multiplying the average count by the dilution rate and dividing by the inoculation amount. Confirm the presumptive isolates (Technique 15.3) and record the results as the number of *Salmonella* colony forming units (cfu) per g or mL of sample.

15.3 Identification of *Salmonella*

Obtain a pure culture as described in Technique 13.4.1, if necessary, using BHI broth and BS agar in the purification of *Salmonella* colonies. Inoculate a pure culture into BHI broth and BHI agar slant, incubate tubes at 35°C for 24 h, and use the cultures in identification tests. The *Salmonella* identification flowchart is given in Fig. 15.5.

15.3.1 Morphological tests

Gram staining. Prepare a Gram stain from 18 h BHI agar slant cultures as explained in Appendix B under the topic "Gram Staining Reagents." A Gram stain of *Salmonella* will produce a Gram-negative, rod-shaped, nonspore-forming, single or paired cell appearance.

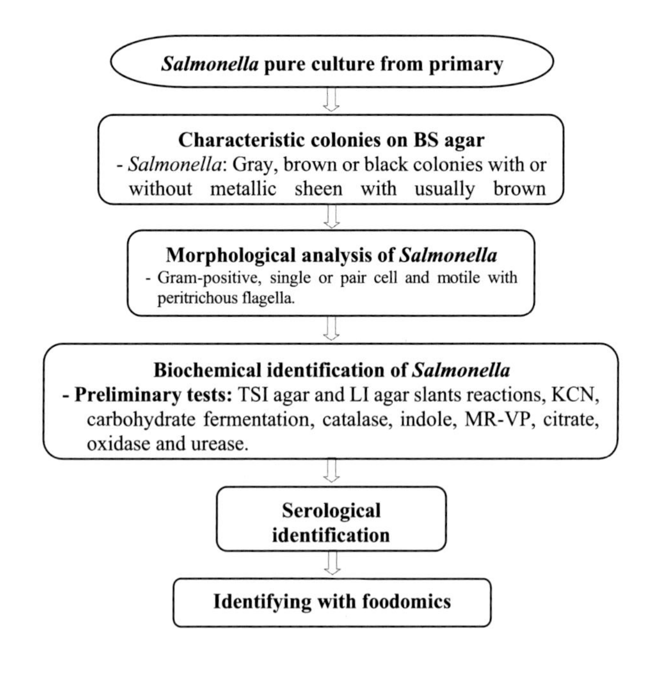

FIGURE 15.5

Salmonella identification flowchart.

Motility test. Refer to Technique 13.4.2 for the MT technique. *Salmonella* species are motile with peritrichous flagella.

Liefson's flagella staining. Prepare a Liefson's flagella stain from 18 h BHI agar slant cultures, as explained in Liefson's flagella stain in Appendix B under the topic "Liefson's Flagella Stain."

The staining appearance of *S*. Typhimurium is given in Fig. 15.6.

15.3.2 Biochemical tests

15.3.2.1 Decarboxylase test

Inoculate one loopful of BHI broth culture into a series of four tubes of Moller's decarboxylase broth. The series should consist of control without amino acid and with lysine (LD broth), ornithine (OD broth), and arginine (AD broth) separately. Add about 1–2 mL of sterile 2% agar carefully over the inoculated media to cover the surface area without disturbing broth. Incubate tubes at 25°C and examine broth daily for up to 7 days. A color change in broth from yellow to purple indicates a positive result. *Salmonella* can grow in all broths and can cause an alkaline reaction with purple (violet) color throughout the broth. A yellow color throughout the medium indicates a negative test for *Salmonella*. If the broth is discolored (neither purple nor yellow), add a few drops of bromocresol purple stain (0.2%) into the broth culture and reread the color change in the tube.

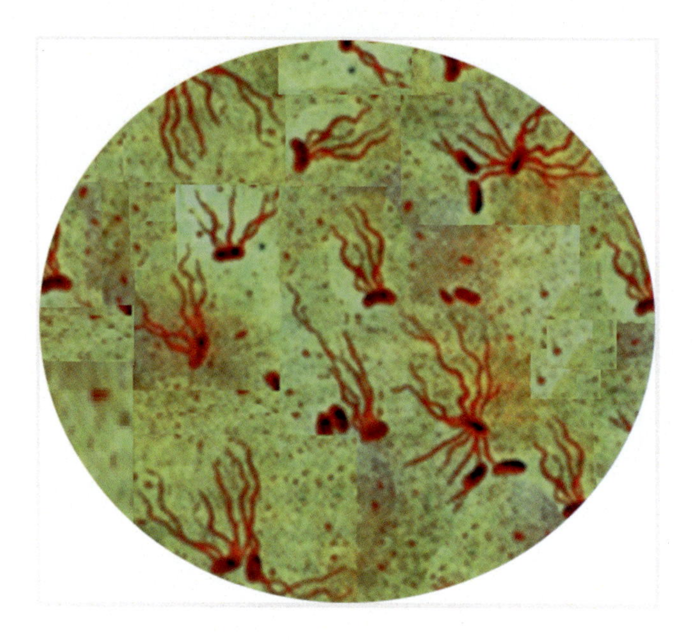

FIGURE 15.6

Staining appearance of *Salmonella typhimurium*.

15.3.2.2 Carbohydrate fermentation

Use carbohydrate fermentation (CF) broth containing sugar separately.

1. *CFL (CF lactose) broth.* Inoculate a loopful of BHI broth culture into CFL broth (containing lactose). Incubate tubes at 35°C up to 48 h. After incubation, examine the broth for growth (turbidity), color changes, and gas production. A positive reaction consists of acid production (yellow color) and gas collection in the inverted Durham tube. A negative reaction consists of a red (with phenol red) or purple (with bromocresol purple) color throughout the medium without gas production. Most *Salmonella* cannot ferment lactose.
2. *CFS (CF glucose) broth.* Follow the procedure described for CFL broth using CFS broth (containing sucrose). Most *Salmonella* cannot ferment sucrose.
3. *CFG broth.* Follow the procedure described for CFL broth using CFG broth (containing glucose). *Salmonella* ferments glucose by gas formation in the inverted Durham tube and yellow color (acidity) in the broth. A negative test is indicated by the absence of gas and yellow color (acidity) of the broth.
4. *CFD (CF dulci) broth.* Follow the procedure described for CFL broth using CFD broth (containing dulcitol). Most *Salmonella* ferment dulcitol by gas formation in the inverted Durham tube and yellow color (acid) in the broth. A negative test is indicated by the absence of gas and yellow color throughout the CFD broth.

15.3.2.3 Malonate utilization test

Inoculate one loopful of tryptone broth culture into malonate broth in duplicate and incubate the tubes at 35°C for 48 h. The uninoculated malonate broth may turn to blue (positive test) on standing, therefore include uninoculated malonate broth as a control during incubation. Most *Salmonella* give a negative test (retain green color). A positive test (alkaline reaction) is shown by a blue color. *S. arizona* is malonate positive.

15.3.2.4 Other tests

Urease test. Refer to Technique 13.4.3.5 for the urease test. *Salmonella* give a negative urease test on urea agar medium (retain purple red color). Include uninoculated urea agar slant as a control.

Simmons citrate test. Refer to Technique 13.4.3.4 for the citrate test. A color change (from green to blue) with growth on the slant indicates a positive citrate test. Most *Salmonella* are citrate-positive. No growth or little growth without a color change (retains green color) indicates a negative citrate test.

KCN test. Refer to Technique 13.4.3.7 for the KCN test. Most *Salmonella* cannot grow (no turbidity) in the KCN broth.

Indole test. Refer to Technique 13.4.3.2 for the indole test. Most *Salmonella* give indole negative result (lack of deep red color at surface of broth). Indicate intermediate orange and pink color as (±) result.

Mr-VP test. Refer to Technique 13.4.3.3 for Mr-VP test. Most *Salmonella* are VP-negative without color change (no deep red color) throughout the test Mr-VP broth culture. Most *Salmonella* give a Mr-positive result by a diffuse red color in the test Mr-VP broth culture. A distinct yellow color indicates a negative test.

If required, perform further biochemical tests, as indicated in Table 15.2.

15.3.3 Serological identification of *Salmonella*

1. *Polyvalent flagellar (H) antigen test*

   Inoculate one loopful BHI broth culture into BHI broth (5 mL) and TS broth (5 mL), separately. Incubate the tubes at 35°C for 4–6 h until the appearance of turbidity. After the appearance of turbidity, add 2.5 mL of materials and solutions (FPS) solution into both broth cultures.

   Place 0.5 mL of appropriately diluted *Salmonella* polyvalent flagellar (H) antiserum into a 13 × 100 mm tube. Add 0.5 mL culture (antigen) into the test tube and mix. Prepare the saline control by mixing 0.5 mL FPS solution with a 0.5 mL culture. Incubate mixtures at 49°C in the water bath. Observe at each 15 min interval and read the final results after 1 h. Report the results as follows:

   Positive. Agglutination in test mixture and no agglutination in control.
   Negative. No agglutination in both test mixture and control.
   Nonspecific. Agglutination in both test mixture and control.
   Salmonella is positive for flagellar antiserum.
2. *Polyvalent somatic (O) antigen test*

Using a waterproof pencil, label two parts about 1 × 2 cm each under the microscopic slides. Place one drop of tryptose broth culture on the upper portion of each labeled part. Add one drop of PS solution to the lower part of the first part. Alternatively add 1 mL of PS solution onto 24–48 h TS agar slant colonies on the slide and emulsify colonies on the slide using a flamed loop. Add one

Table 15.2 Phenotypic characteristics of *Salmonella*.

Characteristic	*S. enterica*	*S. salamae*	*S. houtenae*	*S. indica*	*S. arizonae*	*S. diarizonae*	*S. bongori*	*S. typhimurium*	*S. paratyphi*	*S. typhi*
Motility	+	+	+	+	+	+	+	+	+	+
H_2S/ Catalase	+/+	+/+	+/+	+/+	+/+	+/+	+/+	+/+	-/+	+/+
Oxidase/Urease	-/-	-/-	-/-	-/-	-/-	-/-	-/-	-/-	-/-	-/-
Mr/VP	+/-	+/-	+/-	+/-	+/-	+/-	+/-	+/-	+/-	+/-
Indole/Citrate	-/v	-/+	-/+	-/+	-/+	-/+	-/+	-/+	-/+	-/+
Galacturonate	-	+	+	+	-	+	+	-	-	-
MUG/ONPG	v/-	v/-	-/-	v/v	+/+	v/+	-/-	v/-	v/-	v/-
KCN broth	-	-	+	-	-	-	+	-	-	-
Nitrate reduction	+	+	+	+	+	+	+	+	-	+
Decarboxylation of										
ornithine/ arginine	v/v				+/+	+/+				
lysine	+	+	+	+	+	+				
phenylalanine	-	-	-	-	-	-				
Fermentation of										
glucose/lactose	+/-	+/-	+/-	+/-	+/v	+/v	+/-	+/-	+/-	+/-
dulcitol/salicin	v/-	+/-	-/v	v/-	-/-	-/-	+/-	v/-	+/-	-/-
malonate/sorbitol	-/+	+/+	-/+	-/-	+/+	+/+	-/-	-/+	-/+	-/+
sucrose/mannitol	-/+	-/+	-/+	-/+	-/+	-/+	-/+	-/+	-/+	-/+

+, positive; -, negative; v, variable.

drop of *Salmonella* polyvalent somatic (O) antiserum to the lower part of second Technique of slide of second slide. With flamed loop, homogenize the culture in PS solution on one part with flamed loop, repeat the homogenization for the other part of the slide containing the antiserum and culture. Tilt the mixtures in a back-and-forth motion while holding the slide in one hand for 1 min and observe against a dark background in good illumination. Any degree of agglutination on the slide is considered as a positive test. Record the polyvalent somatic (O) test results as follows:

Positive. Agglutination in test mixture and no agglutination in saline control.

Negative. No agglutination in both test mixture and control.

Nonspecific. Agglutination in test and in control mixtures.

15.3.4 Identification of *Salmonella* species by foodomics techniques

15.3.4.1 Matrix-mediated laser desorption ionization-time-of-flight-mass spectrometry identification of Salmonella

Salmonella can be rapidly identified with the matrix-mediated laser desorption ionization-time-of-flight-mass spectrometry (MALDI-TOF Ms) technique, which is based on the principle of ionizing a specific protein profile of microbial cells. *Salmonella* can be easily identified by comparing these profiles with the reference spectrum. MALDI-TOF Ms identification has been explained in Technique 13.6.3.1.

15.3.4.2 Molecular identification of Salmonella

Polymerase chain reaction (PCR) is a molecular genotypic technique and is used in the identification of *Salmonella* species. Gene-coding of specific *Salmonella* characteristics can be detected by PCR assay. Some *Salmonella* specific primer sets that can be used in the identification of *Salmonella* are given in Appendix A: Gene Primers; Table A-1. Using *Salmonella*-specific techniques and materials, the bacteria can be identified by PCR analysis, as described in Technique 13.6.3.2.

15.4 Interpretation of results

Discard *Salmonella* cultures giving positive KCN and VP tests, and negative Mr test. Isolates giving negative results from phenotypic tests are discarded. In recent years, 90% of foodborne infections have been caused by *Salmonella*. The spread of *Salmonella* is encouraged by the inadequate cooling of foods. The infection dose of pathogens can be quickly reached through rapid multiplication in food. If only a few pathogens are ingested, the immune system is normally able to destroy the germs without manifesting any symptoms of disease. The type of animal husbandry, slaughter, and marketing also influences the incidence of *Salmonella*. The major food vehicles of transmission are animal-derived foods. Plant foods also may act as vehicles, following environmental contamination.

PRACTICE

Isolation and counting of *Listeria monocytogenes*

16

16.1 Introduction

Listeria is a Gram-positive, facultative anaerobic, catalase positive, oxidase negative, nonspore-forming, and rod-shaped bacterium. *Listeria* multiplies at a temperature range from 3°C to 45°C with an optimum from 30°C to 37°C but can also able to grow from 7°C to 10°C. *Listeria monocytogenes* is considered to be a psychrotrophic foodborne pathogen. It produces a blue-green sheen colony on tryptose agar. The genus *Listeria* includes the following species: *L. monocytogenes*, *L. innocua*, *L. ivanovii*, *L. seeligeri*, *L. grayi*, and *L. welshimeri*. Each of *L. ivanovii* and *L. grayi* contain two sub-species. Only *L. monocytogenes* commonly is associated with human listeriosis. Listeriosis from *L. ivanovii* and *L. seeligeris* is rare in humans. Pathogenic and nonpathogenic *Listeria* spp. can be differentiated from each other depending on hemolysin production and β-hemolysis formation on blood agar. *L. monocytogenes* produces hemolysin and listeriosin O as virulence factors. *L. monocytogenes* invades different body tissues, grows in the cells, and releases toxin. The toxins can cause death on the host cells. The infective dose is about 100−1000 viable cells per g or mL of food. In foodborne disease, the mortality rate is approximately 30%. Symptoms of listeriosis occur within 1−7 days after the consumption of food together with living *L. monocytogenes*. The symptoms start with enteric disorders: abdominal cramps, nausea, vomiting, fever, headache, and diarrhea. Then the pathogen invades different organ tissues (including the central nervous system) and blood to cause systemic disease. In pregnant women, *L. monocytogenes* may invade fetus tissues and organs. Systemic symptoms are bacteremia, meningitis, encephalitis, endocarditis, etc. The fatality rate is very high among infected newborn infants, fetuses, and immunocompromised individuals. The foods commonly associated with *L. monocytogenes* are leafy vegetables, raw and cold meats, sausage, chicken, eggs, seafoods, potatoes, ready-to-eat meats, raw and pasteurized milk, soft cheeses, and vegetables. *L. monocytogenes* may have environmental sources, e.g., silage, soil, decaying, and feces. Isolation of *L. monocytogenes* involves a two-step enrichment technique: preenrichment and selective enrichment. Culture from enriched broth is then subcultured on selective/differential esculin-based agar and chromogenic selective agar. *L. monocytogenes* can be counted using L.mono chromogen agar. *L. monocytogenes* can be identified using phenotypic and foodomics techniques.

16.2 Selective enrichment isolation and counting techniques of *L. monocytogenes*

Sampling, sample preparation, and all experiments should be performed under aseptic conditions. Sampling and all analysis should be performed in duplicate, unless otherwise specified. All

Microbiological Analysis of Foods and Food Processing Environments. DOI: https://doi.org/10.1016/B978-0-323-91651-6.00048-3

microbiological media to be used in this application must be sterilized. Again, all solutions and equipment that may cause contamination must be sterile or disinfected.

The *L. monocytogenes* selective enrichment isolation and counting flowchart is given in Fig. 16.1.

16.2.1 Equipment, reagents and media

Equipment. Blender (or Stomacher), cover slip, Erlenmeyer flask, incubator, loop, microscope, microscope glass slide, pipette discard equipment (e.g., a graduated cylinder) containing 10% bleach solution and a piece of cloth at bottom, pipettes (1 mL) in the pipette box, and vortex.

Reagents. Gram stains, Liefson's flagella stains, *Listeria* polyvalent O and H antiserum, and 0.1% peptone water.

Media. Buffered listeria enrichment (BLE) broth, Colombia blood (CB) agar, CHROMagar, HL agar, L.mono chromogen (LMC) agar, lithium chloride phenylethanol moxalactam (LPM) agar, motility test (MT) medium, nutrient agar, polymyxin-acriflavidine-lithium chloride-ceftazidime-

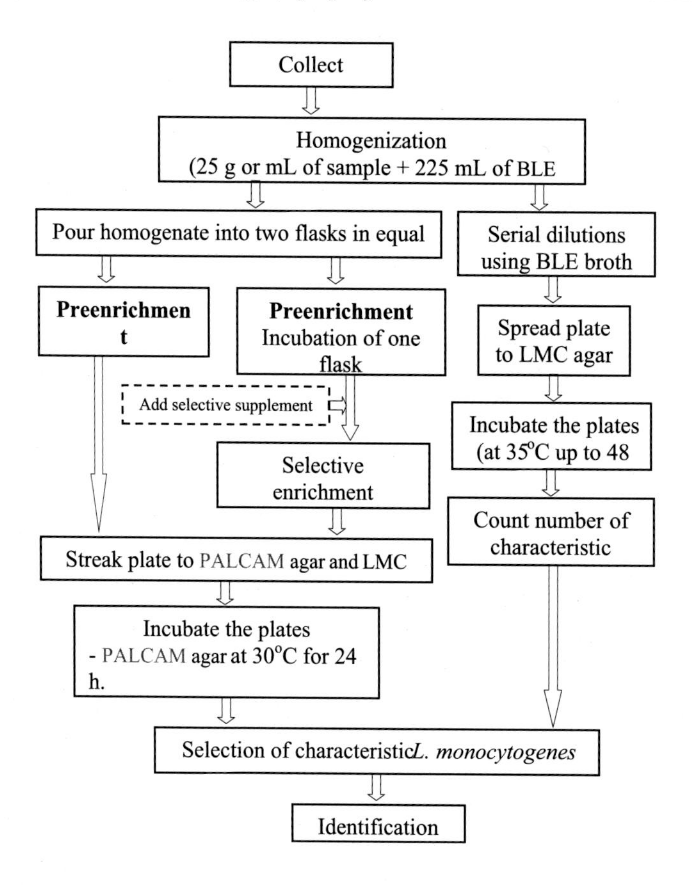

FIGURE 16.1

Listeria monocytogenes isolation and counting flowchart.

aesculin-mannitol (PALCAM) agar, triple sugar iron (TSI) agar slant, trypticase soy yeast extract (TSYE) agar, and TSYE broth.

16.2.2 Sample preparation

Refer to Technique 2.2.2 for sample preparation and dilutions. Analysis of the sample should be performed as soon as possible after receiving the sample. If the analysis of sample is performed later, frozen samples are stored (below −20°C). Nonperishable, canned, or low-moisture foods can be stored at room temperature. Unfrozen perishable foods are stored at 4°C. Weigh 25 g of sample into a blender (or stomacher plastic bag) containing 225 mL of BLE broth (without selective supplement), homogenize the sample by blending for 1 min (or stomaching for 2 min) and pour the homogenate into a sterile Erlenmeyer flask. In the case of a liquid sample, add 25 mL of the sample into an Erlenmeyer flask containing 225 mL of BLE broth (without selective supplement) and homogenize the sample by gently shaking the flask.

16.2.3 Preenrichment and selective enrichment isolation

Use the preenrichment technique to isolate low numbers of *L. monocytogenes* from foods. Preenrichment can be performed at two temperatures. Divide the homogenized sample in equal amount into two sterile Erlenmeyer flasks. Incubate one flask at 30°C for 24 and 48 h and the other at 4°C for 4 h. After incubation at 4°C, add selective antibiotic supplements (present in BLE broth) into flask. Mix the homogenate by gently shaking and continue to incubate at 30°C for up to 48 h for selective enrichment. After incubation, examine the flasks for growth (turbidity) and color changes (acid production) (Fig. 16.2). Streak plate one loopful of the preenriched BLE broth culture and selectively enrich both cultures separately onto esculin base PALCAM agar in duplicate and one chromogenic base LMC agar in duplicate (Technique 2.2.4). Incubate PALCAM agar plates at 30°C for 24 h and LMC agar plates at 37°C for 24 h. After incubation, examine the plates for *Listeria*'s characteristic colony formation.

After 24-h incubation *Listeria* produces gray to black color colonies with 1 mm diameter surrounded by a black halo (zone) with a background color on the bile esculin containing PALCAM agar (Fig. 16.3). *L. monocytogenes* hydrolyzes esculin to produce colonies with black zones due to the formation of black iron phenolic compounds on PALCAM agar. Gram-negative and most Gram-positive bacteria are completely inhibited or suppressed on this medium, but some *Enterococcus* species may grow and show a weak esculin reaction. After 24 h incubation of PALCAM agar, *Listeria* produces a blue-green colored colony with hydrolyzing chromogenic substrate with an enzyme specific to *Listeria* species on the chromogenic-based LMC agar. *L. monocytogenes* and *L. ivanovii* produce enzyme and the enzyme hydrolyzes chromogenic phospholipid substrate in LMC agar to produce 1−3 mm diameter blue-green colonies surrounded by an opaque, white zone (Fig. 16.3). Additionally, *L. ivanovii* colonies can be surrounded with a yellow halo region and heavy growth. *L. ivanovii* can turn the entire colony to a yellow color with further incubation. This color change causes *L. monocytogenes* to appearing like *L. ivanovii*. Other *Listeria* species can also grow and produce smooth, convex white colonies with or without a yellow zone on the LMC agar.

Select least five characteristic *L. monocytogenes* colonies per PALGAM agar and LMC agar. Perform identification tests with the isolates for the differentiation of *Listeria* species (Technique 16.3).

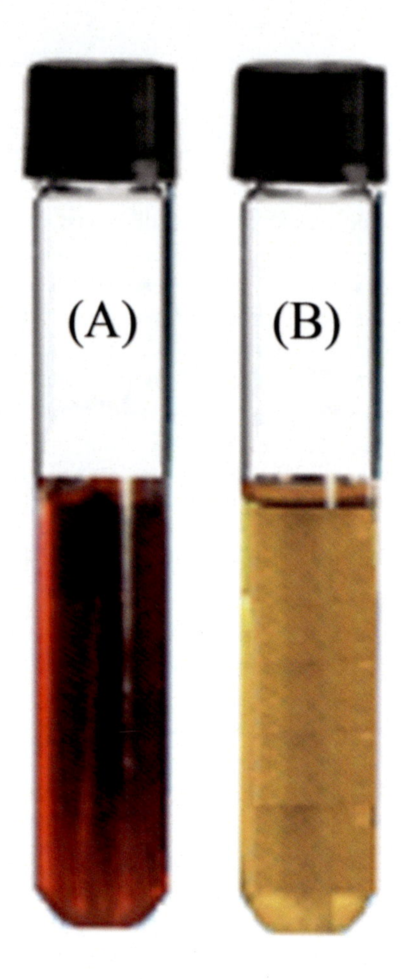

FIGURE 16.2

Growth characteristics of *Listeria* in BLE broth. (A) uninoculated broth (control) and (B) growth of *Listeria* in BLE broth with yellow color.

16.2.4 Counting of *L. monocytogenes*

16.2.4.1 Plate counting technique

Refer to Technique 16.2.2 for details of the sample preparation. Prepare serial dilutions using BLE broth (without selective supplement) from homogenate. Spread plate 1 mL of diluted sample over the surface of the predried LMC agar in duplicate (Technique 2.2.4). Incubate the inverted Petri plates at 35°C for 24–48 h. After incubation, examine the characteristic colony formation on the LMC agar. Count *L. monocytogenes* characteristic colonies on the plates containing 25–250 colonies. Calculate the number of *L. monocytogenes* by multiplying the average count by the dilution rate and dividing by the inoculation amount. Select at least five characteristic colonies from each plate and use in the identification tests. Perform further identification tests as indicated in

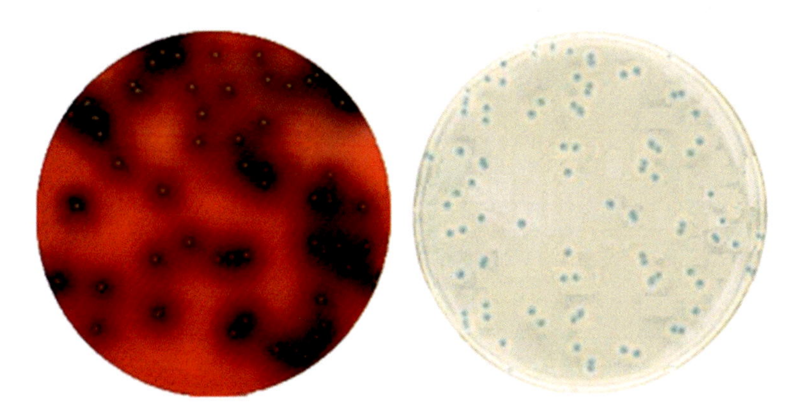

FIGURE 16.3

Typical *Listeria monocytogenes* colony appearance on PALCAM agar (left) and LMC agar (right).

Technique 16.3. Record the results as the number of *L. monocytogenes* colony forming units (cfu) per g or mL of sample.

16.2.4.2 MPN counting technique

Refer to Technique 16.2.2 for details of sample preparation. Prepare serial dilutions using BLE broth from the homogenate. Inoculate 1 mL of each diluted sample into 3-MPN tubes of BLE broth. Incubate tubes at 30°C for up to 48 h. After incubation, examine the tubes for growth (turbidity) and color change. Indicate the number of positive tubes showing growth for each dilution. Confirm all presumptive positives tubes by the identification tests (Technique 16.3). Refer to MPN Table 4.1 and calculate MPN numbers and record the result as the number of *L. monocytogenes* MPN per g or mL of sample.

Streak plate one loopful of culture from each presumptive positive MPN tube on LMC agar and incubate at 35°C for 24 h. Select at least five characteristic colonies per plate and use in identification tests, as indicated in Technique 16.3.

16.3 Identification of *L. monocytogenes*

Obtain a pure culture, as explained in Technique 13.4.1 using TSYE broth and LMC agar in the purification, and finally inoculate pure culture to TSYE agar slant and TSYE broth. The *L. monocytogenes* identification flowchart is given in Fig. 16.4.

16.3.1 Morphological identification

Gram staining. Prepare a Gram stain from 18 h TSYE agar slant culture, as indicated in Appendix B under the topic "Gram Staining Reagents." *Listeria* will appear as Gram-positive, short rods with rounded ends, singly or in pairs, and may resemble "coryneform" or diplococci (Fig. 16.5).

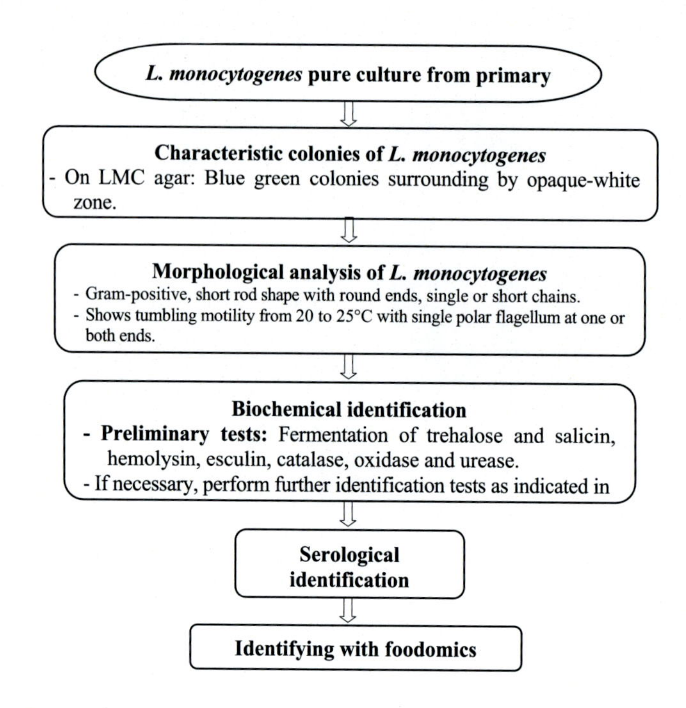

FIGURE 16.4

Listeria monocytogenes identification flowchart.

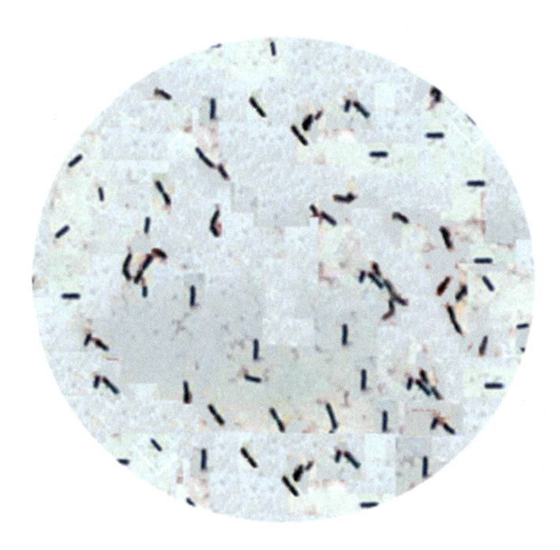

FIGURE 16.5

Staining appearance of *Listeria monocytogenes*.

Motility test. Refer to Technique 13.4.2 for complete details of the motility test. Inoculate MT medium with BHI broth culture, and incubate at 25°C for up to 7 days and observe growth daily. *Listeria* is motile and forms a typical umbrella-shaped growth pattern in MT medium (Fig. 13.6). This pattern of motility is best seen when the culture is incubated at room temperature.

Motility can also be confirmed by a wet mount or hanging drops from overnight *L. monocytogenes* TSYE broth culture (Technique 13.4.2). Observe characteristic darting, zigzag, tumbling motility at 20°C–25°C but not above 30°C. Examine a wet mount or hanging drop immediately, since motility decreases with time.

Flagella staining. Motility can also be performed by flagella staining. Flagella can be demonstrated by staining with flagella staining as given in Appendix B under the topic "Liefson's Flagella Stain." *L. monocytogenes* has five to six peritrichous flagella.

16.3.2 Biochemical identification

16.3.2.1 Forty five degrees transillumination

Streak plate three loopfuls of TSYE broth culture onto CHROMagar and LPM agar plates and incubate the inverted plates at 35°C for 24 h. After incubation, examine colonies from plates for 45 degrees transillumination. A more stable 45 degrees transillumination can be achieved by using the low-power objective of a microscope. *L monocytogenes* gives a blue-green sheen on these media.

16.3.2.2 β-Hemolysis

Streak plate a loopful of TSYE broth cultures onto HL agar plates. Incubate the inverted Petri plates at 35°C for 24 h in 5%–10% CO_2 incubator. *L. monocytogenes* produces smooth, translucent colonies with a characteristic ground glass appearance with hazy beta-hemolysis. Some strains of *L. monocytogenes*, *L. seeligeri*, and *L. ivanovii* produce hemolytic colonies (Fig. 16.6A), and the other *Listeria* species are nonhemolytic (Fig. 16.6B). The characteristic blue color *Listeria* colonies can be observed by exposing the HL plate at an angle to the light sources.

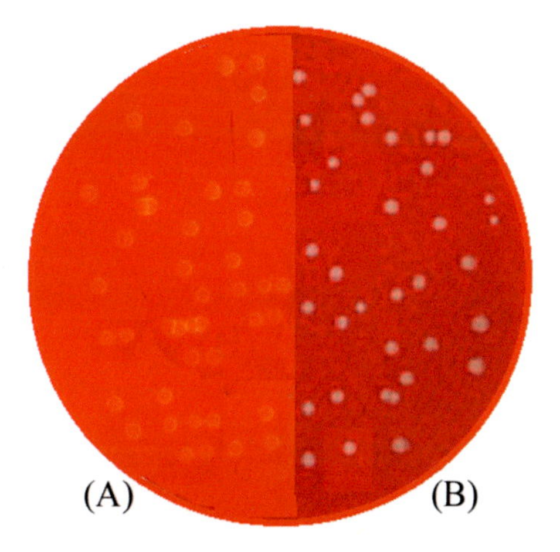

FIGURE 16.6

Listeria (A) hemolytic colonies and (B) nonhemolytic colonies on HL agar.

16.3.2.3 CAMP test

Streak plate one loopful of TSYE broth cultures onto CB agar from one edge of the plate to another as a line. Carry out another streaking by *Staphylococcus aureus* (or *Rhodococcus equi*) as a line that is perpendicular to the test culture line. Incubate the inverted plates at 35°C for 24 h. After incubation, examine the plates for CAMP reaction. CAMP reaction or a zone of enhanced beta-hemolysin interaction between bacteria indicates a positive CAMP test (Fig. 16.7). *L. monocytogenes* is positive by *S. aureus* and negative by *R. equi*. *L. innocua* is negative by *S. aureus* and positive by *R. equi*. Other *Listeria* spp. are negative for CAMP test.

16.3.2.4 Nonpigmented colony formation

Streak plate one loopful of TSYE broth culture onto nutrient agar and incubate the inverted plates at 25°C for 24 h. *L. monocytogenes* produces nonpigmented 1–2 mm diameter colonies with a caramel/sour butter smell and a ground glass appearance.

16.3.2.5 Other tests

Use TSYE broth culture in the following tests.

Catalase test. Refer to Technique 13.4.3.10 for the catalase test. Most *Listeria* species are catalase positive but some species may be weak or negative.

Oxidase test. Refer to Technique 21.3.2.7 for the oxidase test. *Listeria* species are oxidase negative.

Urease test. Refer to Technique 13.4.3.5 for the urease test. *Listeria* species are urease negative.

TSI agar slant reactions. Refer to Technique 13.4.3.1 for the TSI agar slant reactions. *Listeria* species produce yellow slant and butt without gas or H_2S in the TSI agar slant.

Bile esculin test. Refer to Technique 14.3.2.1 for the bile esculin test. *Listeria* species grow on BE agar and blackening the BE agar.

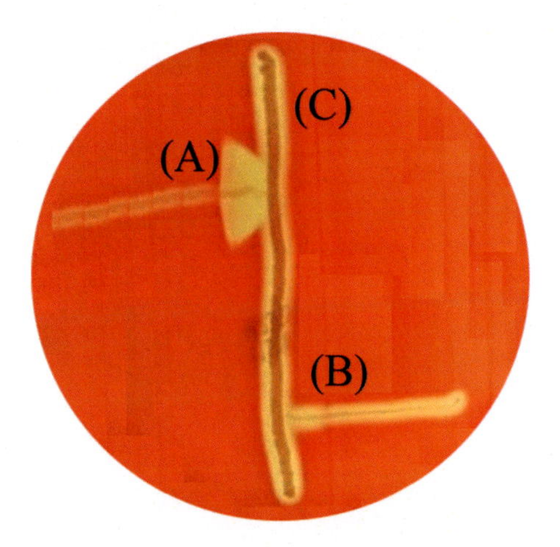

FIGURE 16.7

CAMP test result. (A) positive test result for *Listeria monocytogenes*, (B) negative test result for *Listeria innocua* and (C) *Staphylococcus aureus* growth.

Mr-VP test. Refer to Technique 13.4.3.3 for the Mr-VP tests. *Listeria* species are positive for both of Mr and VP tests.

Nitrate test. Refer to Technique 19.3.2.11 for the nitrate test. *Listeria* species are nitrate negative (except *L. murrayi* and *L. dinitrificans*).

Carbohydrate fermentation. Refer to Technique 15.3.2.2 for the carbohydrate fermentation tests. *Listeria* species ferment glucose with acid production without gas and variable for other carbohydrates fermentation.

Indole test. Refer to Technique 13.4.3.2 for the indole test. *Listeria* species are indole negative.

Refer to Table 16.1 for other useful identification phenotypic characteristics of *Listeria* species.

16.3.3 Serological identification of *L. monocytogenes*

Refer to Technique 13.5.4 for the complete details of serological identification tests. *L. monocytogenes* can be serotyped using polyvalent *Listeria* O and H antiserum.

16.3.4 Identification of *Listeria* species by foodomics techniques

16.3.4.1 MALDI-TOF Ms identification of Listeria

L. monocytogenes can be rapidly identified with the matrix-mediated laser desorption ionization-time-of-flight-mass spectrometry (MALDI-TOF Ms) technique, which is based on the principle of ionizing a specific protein profile of microbial cells. *L. monocytogenes* can be easily identified by comparing these profiles with the reference spectra. MALDI-TOF Ms identification has been explained in Technique 13.6.3.1.

16.3.4.2 Molecular identification of Listeria

Polymerase chain reaction (PCR) is a molecular genotypic technique and is used in the identification of *L. monocytogenes* species. Gene coding of specific *L. monocytogenes* characteristics can be detected by PCR assay. Some *L. monocytogenes*-specific primer sets that can be used in the identification of *L. monocytogenes* are given in the Appendix A: Gene Primers (Table A.1). Using *L. monocytogenes*-specific techniques and materials, the bacterium can be identified by PCR analysis, as described in Technique 13.6.3.2.

16.4 Interpretation of results

L. monocytogenes, as a foodborne pathogenic bacterium, is a major causative agent for the serious disease listeriosis. Milk and dairy products arethe main sources of *Listeria* spp. Accurate detection of *L. monocytogenes* in milk and dairy products, vegetables, meat, poultry, and seafood products is needed to prevent its passage through the food chain. When contaminating foods with this pathogen, antibiotic resistance can occur after exposure to preservatives, antibiotics, and stress conditions. This is a major problem for public health, which indicates the need for special attention to its control along the food chain and disease management in the patients. Two basic phenotypic and foodomics techniques are used in the identification of *Listeria* following enrichment and primary

Table 16.1 Phenotypic characteristics of *Listeria* species.

Characteristic	*Listeria monocytogenes*	*Listeria ivanovii*	*Listeria innocua*	*Listeria welshimeri*	*Listeria seeligeri*	*Listeria murrayi*	*Listeria grayi*	*Listeria fleshmanii*	*Listeria floridensis*	*Listeria aquatica*	*Listeria newyorkensis*	*Listeria cornellensis*	*Listeria rocourtiae*	*Listeria weihenst.*	*Listeria grandensis*	*Listeria riparia*	*Listeria booriae*	*Listeria marthii*
Motility	+	+	+	+	+	+	+	−	−	−								+
TSI agar (slant/butt)	A/A	A/A	A/A	A/A	A/A	A/A	A/A	A/A	A/A	−/A	A/A	(A)/A	A/A	v/a	−/A	A/A	A/A	A/A
Acid from fermentation of																		
Glucose	+	+	+	+	+	+	+	+	+	+	+	+	+	+	+	+	+	+
Mannitol/ xylose	−/−	−/+	−/−	−/−	−/+	+/−	−/−	v/+	−/+	−/+	+/+	−/+	+/+	+/+	−/+	v/+		
Rhamnose	+	−	v	v	−	v	−	+	+	+	v	−	+	+	−	+		
Melibiose	v	v	v	−	−	−	v	−	−		−	−	+	−	−	v		
Beta-hemolysis	+	+	−	−	(+)	−	−	−	−	−	−	−	−	−	−	−	−	−
Catalase/ Indole	+/−	+/−	+/−	+/−	+/−	+/−	+/−	+/−	+/−	+/−	+/−	+/−	+/−	+/−	+/−	+/−	+/−	+/−
Growth at 4°C	+	+	+	+	+	+	+	−	−	−	+	+	+	+	+	+	+	+
Esculin hidrolyze	+	+	+	+	+	+	+	+	+	+								+
Nitrate reduction	−	−	−	−	−	+	v	+	−	+	+	+	+	+	+	+	+	−
Mr/VP	+/+	+/+	+/+	+/+	+/+	+/+	+/+	+/+	+/+	+/+	+/−	v/−	v/−	v/−	+/−	v/−	+/−	+/+

+, Positive; −, negative; A/A, acid/acid; v, variable; (), weak reaction.

isolation from food and environmental samples. *L. monocytogenes* remains as the major challenge in food industries. It can survive under an adverse environmental condition, overcome various types of stress like heat treatment and application of antimicrobials, and can persist for a long time in the food industry by attaching to food-contact surface. The application of good manufacturing practice in food industries, such as improvement in food products, methods of storage, shipping, and handling, along with the application of food safety training programs, especially for food industry employees and staff in restaurant or distribution centers, should be taken into consideration.

PRACTICE

Isolation and counting of *Campylobacter jejuni*

17

17.1 Introduction

Campylobacter species are nonperforming, catalase and oxidase positive, Gram-negative, very small, curved, thin, spiral shaped rods (1.5–5 μm diameter), and corkscrew motile-bacterium. They often appear as zigzag shapes. They cannot ferment or oxidize carbohydrates and are microaerophilic (growing best in an atmosphere containing 5% O_2, 8% CO_2, and 87% N_2), and grow easily at temperature ranging from 30°C to 47°C. They are sensitive to drying, acids, and heat treatment (e.g., pasteurization). *Campylobacter jejuni* can survive in low oxygen and low moist environment for 2–4 weeks at 4°C and grow within a few days at room temperature. Therefore, this is sufficient to reduce the shelf life of foods. Environmental stresses (e.g., oxygen, heating, drying, low pH, freezing, and long-term storage) prevent growth of *Campylobacter*. Old and stressed cells gradually become cocoidal and difficult to culture. Enteric *Campylobacter* species are *C. jejuni*, *Campylobacter coli*, *Campylobacter laridis*, *Campylobacter fetus* subsp. *fetus*, *Campylobacter fenneliae,* *Campylobacter cinaedi*, *Campylobacter intestinalis*, *Campylobacter pylori*, and *Campylobacter upsaliensis*. *C. pylori* associates with the gastric and peptic ulcer diseases in humans. *C. jejuni* produces a thermolabile enterotoxin that is responsible for the enteric symptoms in food poisoning. Toxin production by *Campylobacter* spp. is plasmid linked. *Campylobacter* produces an invasive factor that enables the cells to invade epithelial cells of the small and large intestines. *C. jejuni* causes acute gastroenteritis and is highly infective in humans. The infective dose ranges from 500 to 10,000 viable cells per g or mL of sample. The symptoms of the campylobacteriosis appear within 2–5 days after ingestion of food containing viable *Campylobacter* cells and generally last 2–3 days. *Campylobacter* can cause severe abdominal pain, watery mild to severe diarrhea, often followed by bloody diarrhea, malaise, fever, headache, and chills. Vomiting is less common. Mesophilic (optimum 35°C) *C. jejuni* is occasionally invasive and thermoduric (optimum 42°C). *Campylobacter* commonly associates with raw meat, water, poultry, eggs, raw milk, milk products, shellfish, mushrooms, fruits, and vegetables. *C. jejuni* stays viable best on foods with a long cooling temperature and is highly susceptible to freezing. *Campylobacter* species are carried in the intestines of many domestic and wild animals, especially birds. Consumption of food and drinking water that are contaminated with untreated animal or human wastes cause about 70% of *Campylobacter* diseases each year. The isolation of *Campylobacter* involves the following techniques: two-step enrichment (preenrichment and selective enrichment) and selective/differential isolation of bacterium. *C. jejuni* can also be counted using selective/differential medium. Pure culture of *Campylobacter* isolates can be identified by phenotypic and foodomics techniques.

Microbiological Analysis of Foods and Food Processing Environments. DOI: https://doi.org/10.1016/B978-0-323-91651-6.00035-5

17.2 Isolation and counting techniques

Sampling, sample preparation, and all experiments should be performed under aseptic conditions. Sampling and all analysis should be performed in duplicate, unless otherwise specified. All microbiological media to be used in this application must be sterilized. Again, all solutions and equipment that may cause contamination must be sterile or disinfected.

17.2.1 Equipment, reagents, and media

Equipment. Blender (or stomacher), centrifuge, centrifuge tube, cheesecloth, Erlenmeyer flask, filtration apparatus, filter paper, lead acetate paper, microaerophilic incubator (5% O_2, 8% CO_2, 87% N_2), pipette discard equipment (e.g., a graduated cylinder) containing 10% bleach solution and a piece of cloth at bottom, pipettes (1 mL) in the pipette box, plastic bag, platinum loop, screw cap tube (13 × 100 mm), swab, and vortex.

Reagents. Cephalothin (30 μg), Gram stains, H_2O_2 (3%), lead acetate paper strip, nalidixic acid (30 μg), ninhydrin solution (3.5%), saline solution (0.85% NaCl), sodium hippurate (1%) solution, 0.1% sterile peptone water, and tetramethyl-p-*phenylenediamine* dihydrochloride (1% aq.) solution.

Media. Abey-Hunt-Bark (AHB) agar, Bolton selective enrichment (BSE) broth, Brucella broth containing 1% glycerol, Brucella broth, Brucella-FBP agar, Campy-BAP agar, Campylobacter charcoal differential (CCD) agar, CCD-blood-free selective (CCDA-Preston) agar, Campylobacter modified (CM) agar, CM broth, CM broth-glycine, CM broth-NaCl (3.5%), CM broth-nitrate, carbohydrate fermentation broth containing 0.5% glucose, Cary-Blair transport (CBT) medium, cystine agar slant, motility test (MT) medium, Preston enrichment (PE) broth, trimethylamine-*N*-oxide (TMAO) broth, and triple sugar iron (TSI) agar slant.

17.2.2 Sampling, enrichment, and isolation

Appropriate sample handling and transporting must be emphasized when attempting to isolate *C. jejuni* from foods and water. *C. jejuni* is sensitive to oxygen and storage at room temperature. Meat sample storage at 4°C with an equal amount of CBT medium results in little change in *Campylobacter* viability. Viability during 14 days of storage in this medium is retained and it may be used in the transport and storage of sample. CCDA-Preston and Campy-BAP agars are commonly used selective/differential media for *C. jejuni* isolation from foods. After the media are poured into sterile Petri dishes, they should not be excessively exposed to light, and moisture should be dried to limit moisture on the agar before streak plating. Dry plates can be obtained by holding plates overnight at room temperature in the absence of light. If Petri plates are used on the same day, plates are placed in a 42°C incubator for several hours. Do not dry plates with lids open. Even very brief drying of the agar surface will inhibit *Campylobacter* growth.

The following techniques are used in the enrichment isolation of *Campylobacter* from samples.

1. *Most common samples.* Weigh 25 g or mL of sample (50 g of vegetables) into an Erlenmeyer flask containing 100 mL BSE broth in duplicate and mix well by gently shaking for 5 min. Incubate flasks at 37°C for 4 h or at 30°C for 3 h and then 37°C for 2 h. The preenriched mixture is enriched at 42°C for 20 and 44 h microaerobically (5% O_2, 8% CO_2, and 87% N_2).

After incubation, examine the mixture for growth (turbidity) and color change. Streak plate one loopful of enriched culture to selective/differential CCDA-Preston agar (Technique 17.3).

2. *Water*. Filter 2–4 L of water through 0.25 μm filter in duplicate. After filtration, place filter paper into a sterile beaker. Add 100 mL BSE broth to cover the filter in the beaker. Preenrich at 30°C for 3 h and then selectively enrich at 42°C for 21 and 45 h microaerobically. After incubation, examine the broth for growth (turbidity) and color change. Streak plate enriched culture to selective/differential CCDA-Preston agar (Technique 17.3).
3. *Shellfish*. Add 25 g of sample into an Erlenmeyer flask and add 225 mL of BSE broth in duplicate. Mix well by gently shaking the flask. Preenrich BSE broth at 35°C for 4 h. Then enrich at 42°C for 20 and 44 h microaerobically. After incubation, examine mixture for growth (turbidity) and color change. Streak plate selectively enriched culture to selective/differential agar CCDA-Preston agar (Technique 17.3).
4. *Liquid milk products*. Adjust sample pH to 7.6. Centrifuge 50 mL of sample (12,000 × g for 40 min) and discard supernatant fluid. Mix pellet with 10 mL of BSE broth and then add 90 mL of BSE broth and mix well by gently shaking. Incubate mixture at 37°C for 4 h and then continue to selectively enrich at 42°C for 20 and 44 h microaerobically. After incubation, examine mixture for growth (turbidity) and color change. Streak plate enriched culture to selective/differential CCDA-Preston agar (Technique 17.3).
5. *Cheese products*. Weigh 50 g of sample into 50 mL of 0.1% peptone water in an Erlenmeyer flask and completely homogenize by gently shaking. Centrifuge the homogenized sample (12,000xg for 40 min) and discard supernatant fluid. Mix the pellet with 10 mL of BSE broth in an Erlenmeyer flask and then add 90 mL BSE broth and mix by gently shaking the flask. Incubate the flask at 35°C for 4 h and then continue to selectively enrich at 42°C for 20 and 44 h microaerobically. After incubation, examine the mixture for growth (turbidity) and color change. Streak plate from selectively enriched culture to selective/differential CCDA-Preston agar (Technique 17.3).
6. *Surface rinse sample*. This procedure is useful for sampling from the poultry carcasses. Place the large sample (e.g., poultry with 1–2 kg) in a sterile plastic bag containing 225 mL of selective PE broth and rinse the surface by shaking and massaging. Filter the rinse suspension through two layers of disinfected cheesecloth. Centrifuge the filtrate at 16,000 × g for 10 min at 4°C. Discard the supernatant fluid and resuspend the pellet in 3 mL of PE broth. Add the suspended pellet equally into three 10 mL of PE broth. Incubate PE broth at each of 25°C, 35°C, and 42°C for 24–48 h microaerobically. Streak plate each selectively enriched culture to selective/differential CCDA-Preston agar (Technique 17.3).
7. *Swab sample*. This procedure is useful for sampling the surface area of very large animal carcasses and equipment. Dip a sterile swab into PE broth and remove excess moisture. Take a representative sample by wiping the surface of the sample with the moistened swab (Technique 1.3.7). Return the swab into the PE broth or directly inoculate on modified CCDA-Preston agar plates by rotating the swab over a 3–5 cm^2 area of plate and streak plate with a flamed loop to isolate bacterium. Inoculate to three PE broths and three dried modified CCDA-Preston agar plates. Incubate one broth and plate at each of 25°C, 35°C, and 42°C microaerobically at 24 and 48 h. After incubation, examine tubes for growth and formation of individual colonies on plates. Streak plate each selectively enriched culture to one selective/differential CCDA-Preston agar (Technique 17.3). Incubate CCDA-Preston agar plates at 35°C for 24 and 48 h. After

incubation of selective/differential CCDA-Preston agar plate, examine the Petri plates for characteristic colony formations. *C. jejuni* will produce confluent growth on CCDA-Preston agar, making it difficult to obtain single colonies. Individual colonies can be obtained by cultivating the culture on dry plates. Protect plates from light. Confirm presumptive isolates as indicated in Technique 17.4.

8. When the sample contains high number of *C. jejuni*, the direct isolation of *C. jejuni* without enrichment is possible. Streak plate two or three loopful homogenized sample to the surface of selective/differential three sets of CCDA-Preston agar (or AHB agar) plates in duplicate. Incubate one set of plates at each of 25°C, 35°C, and 42°C microaerobically for 24 and 48 h. After incubation, examine tubes for growth and formation of individual colonies on plates. Streak plate each selectively enriched culture to selective/differential CCDA-Preston agar (Technique 17.3).

The *C. jejuni* selective enrichment isolation and counting flowchart is given in Fig. 17.1.

17.2.3 Counting of *Campylobacter jejuni*

Refer to Technique 17.2.2 for sample preparation and homogenization techniques without preenriching. Prepare further dilutions from an homogenized sample using BSE broth. Spread plate (Technique 2.2.4) to two sets of predried selective/differential Campy-BAP agar (or CCD-Preston agar) plates using 1 mL of diluted sample in duplicate. Incubate one set of inverted plates at 35°C and the other at 42°C microaerobically (5% O_2, 10% CO_2, and 85% N_2) up to 48 h. After

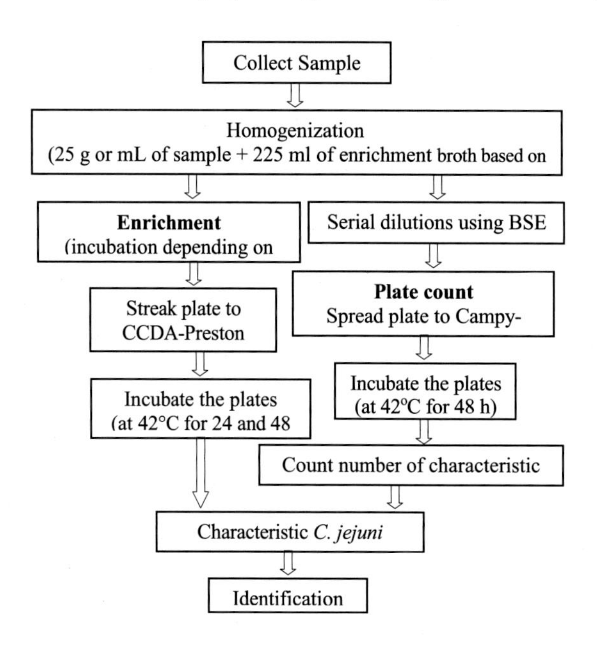

FIGURE 17.1

C. jejuni selective enrichment isolation and counting flowchart.

incubation, examine the plates for characteristic colony formations. Count smooth, convex, and glistening with flat, shiny, translucent, spreading, and watery colorless to grayish or light cream colonies from Campy-BAP agar plates containing 25–250 colonies. Growth may be confluent without distinct colonies. Individual colonies can be obtained by cultivating the culture on dry plates. Calculate the number of *C. jejuni* by multiplying the average count by the dilution rate and dividing by the inoculation amount. Confirm the presumptive isolates (Technique 17.3) and record the results as the number of *C. jejuni* colony forming units (cfu) per g or mL of sample.

17.3 Preliminary identification of *Campylobacter jejuni*

Pure culture. Select least three typical *C. jejuni* colonies from each CCDA-Preston agar plate. With a loop, the bacterial cells are taken by twice touching the center of the colony, and streak plate two sets of modified CCDA-Preston agar (surface dried). Protect plates from light. Incubate one set of inverted plates at 35°C (*C. fetus*) and the other at 42°C microaerobically (5% O_2, 10% CO_2, and 85% N_2) for up to 48 h. After incubation, examine the plates for characteristic colony formation. *Campylobacter* produces round to irregular colonies with smooth edges showing thick translucent white growth to spreading, film-like transparent growth. If bacterium colonies are not pure, restreak to modified CCDA agar without antibiotics to obtain a pure culture. Incubate plates at 37°C for up to 48 h. Use pure culture in identification tests.

Wet mount. Pick one typical colony and emulsify the colony in 0.1 mL of stain (50–50 mix of Gram's crystal violet). After 3–5 min (not drying), cover preparation with a cover slip and magnify under first 20 × low-power objective and then by 40 × high-power objective of light microscope. *Campylobacter* cells are curved, usually in chains similar to zigzag shapes with any length and "wiggly" motility from agar medium and show rapidly swimming corkscrew motion. Stressed *Campylobacter* cells show a coccus shape.

Biochemical tests. Perform catalase (Technique 13.4.3.10) and oxidase (Technique 17.4.2.10) tests on colonies from modified CCDA agar. *Campylobacter* spp. are catalase and oxidase-positive.

Inoculate a pure colony into two sets of CM broth (without supplements). Incubate tubes at 37°C for up to 48 h. After incubation, examine the broth for growth (turbidity). Streak plate one loopful of CM broth culture to two sets of CM agar and two sets of CM broth. Incubate one set of inverted plates and broth at 35°C, and the other set at 42°C for up to 48 h microaerobically. Use cultures in identification tests (Technique 17.4).

The *Campylobacter* identification flowchart is given in Fig. 17.2.

17.4 Identification of *Campylobacter jejuni*

17.4.1 Morphological identification

Gram staining. Prepare Gram stain from the CM broth (or CM agar plate) culture, as indicated in Appendix B under the topic "Hucker Modified Gram Staining." *C. jejuni* is Gram-negative, showing curved, s-shaped, gull-winged, or spiral rods (Fig. 17.3).

Motility test. Prepare a wet-mount as indicated in Technique 13.4.2. Confirm the MT with flagella staining.

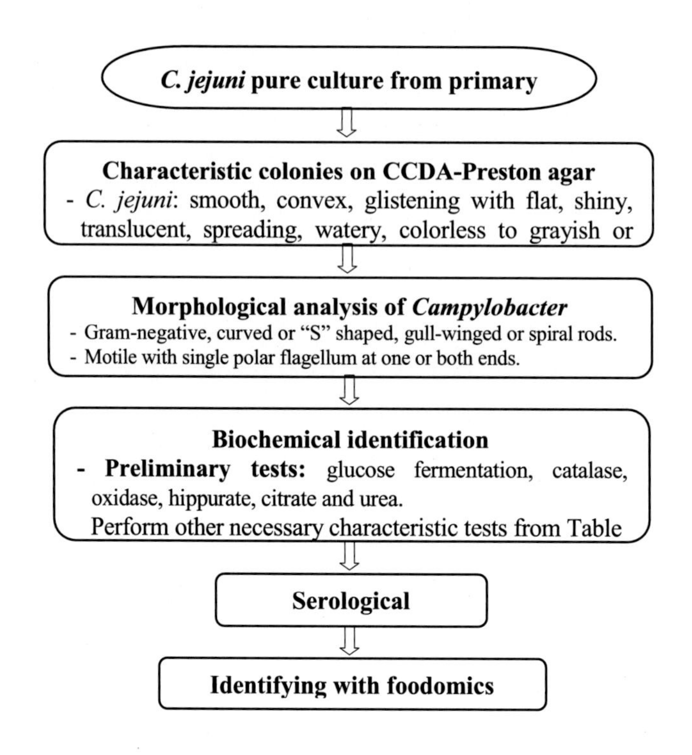

FIGURE 17.2

C. jejuni identification flowchart.

Flagella staining. Prepare a flagella stain from CM broth culture, as indicated in Appendix B under the topic "Liefson's Flagella Stain." *C. jejuni* cells are motile with a single polar flagellum at one or both ends.

17.4.2 Biochemical identification

17.4.2.1 Nitrate reduction and glycine utilization tests

Inoculate one loopful of CM broth culture into each of CM broth-nitrate (2 mL) broth and CM broth-glycine broth. Inoculate culture into a control medium without nitrate or glycine. Incubate the tubes in a microaerophilic atmosphere at 35°C for up to 3 days in a loose-capped tube. After incubation, examine tubes for growth (turbidity). *Campylobacter* species will grow in a medium containing nitrate and reduce nitrate to nitrite without the reduction of nitrite to gases (e.g., NH_3 and N_2). *Campylobacter* species will grow in a medium containing glycine.

17.4.2.2 Carbohydrate fermentation

Inoculate carbohydrate fermentation broth containing glucose with one loopful of CM broth culture. Incubate the tubes in a microaerophilic atmosphere at 35°C (or 42°C) for 3 days with the caps loose. After incubation, examine tubes for growth (turbidity) and color change. Growth with red

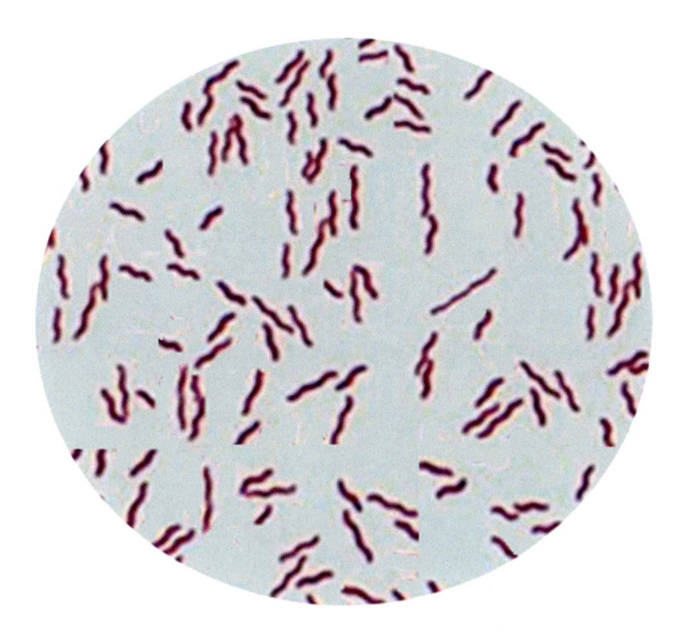

FIGURE 17.3

Staining appearance of *C. jejuni*.

color (no acid production) indicates oxidation in the broth with phenol red indicator. Growth with yellow color (acid production) indicates fermentation in the broth with phenol red indicator. When tubes are removed from the microaerobic atmosphere, carbohydrate fermentation appears as a yellow color due to the absorption of CO_2. The oxidation fermentation reactions are read after the control tubes become neutral or alkaline in color after standing at room temperature (usually within 1 to 2 hours). *Campylobacter* species do not ferment nor oxidize glucose or other sugars and cannot show a change in the tube.

17.4.2.3 Triple sugar iron agar slant reactions

Inoculate a loopful of CM broth culture to TSI agar slant by stabbing the butt and streaking the slant once. Incubate the tubes in a microaerophilic atmosphere at 35°C for up to 3 days. After incubation, examine tubes for growth (colony formation) on the slant and butt. A red color on slants and in butt indicates the culture is negative for lactose and glucose utilization. No black color on the slant indicates the test to be negative for H_2S production. Most (80%) *C. coli* and some *C. lari* produce H_2S on this medium but *C. jejuni* cannot produce H_2S. All *Campylobacter* species produce alkaline reactions on the slant and butt of medium. Use slants no more than 7 days after preparation.

17.4.2.4 Lead acetate cystine agar slant

Inoculate one loopful of CM broth culture to cystine agar slant by stabbing the butt and streaking the slant once. Place a lead acetate paper strip over the slant to detect H_2S production. Incubate the tubes in a microaerophilic atmosphere at 35°C for up to 3 days. After incubation, examine the slant for growth. Brownish-black coloring of the strip indicates H_2S production. Record the degree of blackening from trace to 4 + . Lead acetate strips are more sensitive than sulfide indicators in the medium. Some strains of *Campylobacter* can grow on this medium producing a brownish-black color with H_2S formation as a positive for lead acetate.

17.4.2.5 NaCl tolerance

Inoculate one loopful of CM broth culture into CM broth-3%NaCl in duplicate. Incubate the tubes in a microaerophilic atmosphere at 35°C for up to 3 days with loose-capped tube. After incubation, examine the tubes for growth (turbidity). Any growth (turbidity), usually in the top 10 mm of the medium, is considered to be NaCl tolerance. *C. jejuni*, *C. coli*, *C. laridis*, and *C. fetus* subsp. *fetus* will grow in the presence of 3.5% NaCl and others will not grow.

17.4.2.6 Temperature tolerance

Streak plate three sets of CM broth in duplicate from CM broth culture and incubate each set of plates at 25°C, 35°C, and 42°C in a microaerophilic atmosphere for up to 3 days. *C. jejuni*, *C. coli*, and *C. laridis* will grow at both 35°C and 42°C but not at 25°C. *C. fetus* subsp. *fetus* and *C. fetus* subsp. *venerealis* grow at both 25°C and 35°C but not at 42°C.

17.4.2.7 Antimicrobial test

Inoculate the entire surface of a BHI agar-RB plate with a swab soaked with 4 h CM broth culture (at appearance of turbidity) and allow to dry the surface area of plate. Place the disks of nalidixic acid (30 μg) and cephalothin (30 μg) on the plate and incubate the plates in a microaerophilic atmosphere at 35°C for up to 24 h. After incubation, examine plates for the formation of the zone around antibiotic disks. Any zone of inhibition around disks indicates the bacterium to be sensitive to the antimicrobial agent. Measure the inhibition zone and record the diameter of the antibiotic.

17.4.2.8 Trimethylamine-N-oxide test

Inoculate one loopful of the CM broth culture into TMAO broth. Incubate the tubes in a microaerophilic atmosphere at 35°C for up to 7 days with the cap loose. After incubation, dispersed growth in the broth is considered positive. Compare with the controls. *C. jejuni* cannot grow in this broth.

17.4.2.9 Hippurate hydrolysis

Prepare sodium hippurate solution (1%) in sterile distilled water, add 0.4 mL of solution into 13 × 100 mm screw cap tubes in duplicate, and freeze at −20°C until use. Emulsify a loopful culture from CM agar plate in the thawed sodium hippurate solution (0.4 mL). Incubate the suspension at 35°C in a water bath for 2 h. After incubation, overlay 0.2 mL of ninhydrin solution (1:1 mixture of acetone and butanol; and store solution in the dark at room temperature) and reincubate the tube in the water bath at 35°C for 10 min. Violet color development in the mixture indicates a positive result. A colorless test is considered to be negative. *C. jejuni* is positive for hippurate hydrolysis but hippurate negative strains may be present, and the other *Campylobacter* species are negative.

17.4.2.10 Oxidase test

Add oxidase reagent (1% tetramethyl-p-paraphenylenediamine dihydrochloride) onto a filter paper until moist. Remove colonies from the CM agar plates and heavily smear the surface area of the filter paper using a sterile toothpick or platinum or plastic loop (do not use nonplatinum loops). Observe this for a color change within 30 s. Examine colonies for color formation. The formation of a dark purple color indicates a positive oxidase test and no color change indicates a negative oxidase test. *Campylobacter* species are oxidase positive.

17.4.2.11 Catalase tests

After performing all other tests from CM agar plates, add two to three drops of H_2O_2 (3%) to the CM agar plate surface. Examine the colonies within 1–10 min for bubble formation due to the generation of O_2 in the presence of the enzyme catalase. *C. jejuni* and *C. coli* are catalase positive.

Other phenotypic characteristics of *Campylobacter* species are given in Table 17.1.

17.4.3 Serological identification of *Campylobacter* species

C. jejuni, *C. laridis*, and *C. coli* strains can be serotyped. Perform the soluble heat-labile antigen by slide or tube agglutination test (Technique 13.5.4). Also perform the serological tests for O and H antigens.

17.4.4 Identification of *Campylobacter* species by foodomics techniques

17.4.4.1 MALDI-TOF Ms identification of Campylobacter

Campylobacter can be rapidly identified with the matrix-mediated laser desorption ionization-time-of-flight-mass spectrometry (MALDI-TOF Ms) technique, which is based on the principle of ionizing a specific protein profile of microbial cells. *Campylobacter* can be easily identified by comparing these profiles with the reference spectrum. MALDI-TOF Ms identification has been explained in Technique 13.6.3.1.

17.4.4.2 Molecular identification of Campylobacter

Polymerase chain reaction (PCR) is a molecular genotypic technique and is used in the identification of *Campylobacter* species. Gene coding of specific *Campylobacter* characteristics can be detected by PCR assay. Some *Campylobacter* specific primer sets that can be used in the identification of *Campylobacter* are given in Appendix A: Gene Primers; Table A.1. Using *Campylobacter*-specific techniques and materials, the bacterium can be identified by PCR analysis, as described in Technique 13.6.3.2.

17.5 Stock culture maintenance

Inoculate a heavy loopful of the CM agar slant culture into a semisolid Brucella broth in a screw cap test tube to prepare a stock culture. Loosen the cap to allow exchange of the atmosphere. Incubate the culture for approximately 24 h at the respective temperature in microaerophilic

Table 17.1 Phenotypic characteristics of *Campylobacter* species.

Campylobacter species	Oxidase	Catalase	Motility	H_2S on lead acetate	H_2S on TSI	Nalidixic acid	Cephalothin	Nitrate reduction	1% Glycine	3.5% NaCl	1% Bile salt	Growth at 25°C	Growth at 35°C	Growth at 42°C	Hippurate hydrolysis	Glucose fermentation	Indole	Urease	TMAO
C. jejuni	+	+	+	+	−	S	R	+	+	−	+	−	+	+	+	−	−	−	−
C. coli	+	+	+	+	v	S	R	+	+	−	+	−	+	+	−	−	−	−	−
C. laridis	+	+	+	+	−	R	R	+	+	−	+	−	+	+	−	−	−	−	+
C. fetus subsp. *fetus*	+	+	+	+	−	R	S	+	+	−	+	+	+	(−)	−	−	−	−	V
C. fetus subsp. *veneralis*	+	+	+	+	−	R	S	+	−	−	+	+	+	−	−	−	−	−	−
C. pylori	+	+	+	+	−	R	S	v	v	−	v	−	+	+	−	−	−	+	−
C. sputorum var. *sputorum*	+	−	+	+	+	(S)	S	+	+	−	+	−	+	+	−	−	−	−	−
C. sputorum var. *bubulus*	+	−	+	+	+	R	S	+	+	+	+	−	+	+	−	−	−	−	−
C. sputorum var. *faecalis*	+	+	+	+	+	R	S	+	−	−	+	−	+	+	−	−	−	−	−
C. concisus	+	−	+	+	+	R	R	+	−	−	+	−	+	−	−	−	−	−	−
C. intestinalis	+	+	+	+	+	R	S	+			+	(+)	+	+	−	−	−	−	−
C. upsaliensis	+	(−)	+	+	−	S	S	+			+	−	+	+	−	−	−	−	−
C. cinaedi	+	+	+	+	−	S	V	+			+	−	+	−	−	−	−	−	−
C. fennelliae	+	+	+	+	−	S	S	−			+	−	+	−	−	−	−	−	−
C. cryaerophila	+	+	+	+	−	v	R	+			+	+	+	−	−	−	−	−	−
C. nitrofigilis	+	+	+	+		S	S	+			+	+	+	−	−	−	−	+	−
C. mucosslis	+	−	+	+	+	R	S	+			+	−	+	+	−	−	−	−	−

+, Positive; −, negative; v, variable; (), most positive or negative; S, susceptible; R, resistant.

atmosphere. Store cultures for up to 1 month at 4°C in a microaerophilic atmosphere or in a vacuum. Semisolid Brucella medium with 10% sheep blood and 0.15% agar can be used for storing the bacterium at 4°C for 3 months. For long-term storage, *C. jejuni* grows on Brucella-FBP broth in a microaerophilic atmosphere at 42°C for 24 h, concentrate by centrifugation and resuspend the pellet in Brucella broth with 10% glycerol. Small quantities (1–2 mL) of bacterium suspension can be stored for several years at −70°C. Cultures can also be lyophilized in skim milk and stored indefinitely at −20°C.

17.6 Interpretation of results

Campylobacteriosis is an important foodborne infection. Foods of animal origin, mostly poultry and meat, play an important role in the infection. Barbecues appear to present special hazards for infection, because they permit easy transfer of *Campylobacter* from raw meats to hands and other foods. Milk is can also contaminated with *Campylobacter* and the consumption of raw milk can cause campylobacteriosis. Inadequate cooking and the consumption of poorly chlorinated drinking water or unpasteurized milk are other infection sources of *Campylobacter*. *Campylobacter* can survive in fresh cheese for only a short period of time. *Campylobacter* can contaminate shellfish. *Campylobacter* is sensitive to high temperatures, dry environments, and oxygen. A number of preventive measures are needed to reduce the incidence of campylobacteriosis.

PRACTICE

Isolation and counting of *Yersinia enterocolitica* 18

18.1 Introduction

Yersinia enterocolitica is Gram-negative, nonsporeforming, appears as short rods (young culture may be coccoid), is motile at 25°C with peritrichous flagella and not motile at 35°C, is lactose negative, produces acid from glucose without gas, and is urea positive. It is a psychrotrophic bacterium and will grow between 0°C and 45°C with the optimum being at 25°C. *Y. enterocolitica* strains can be identified serologically with their heat-stable somatic antigens. All strains of *Y. enterocolitica* cause yersiniosis. The serogroups that predominantly cause foodborne diseases are O:3, O:8, O:9, and O:5,25.

Y. enterocolitica often presents in the environment and the intestinal tract of different animals. *Y. enterocolitica* is associated with milk, spring and stream water, beef, poultry, crabs, oysters, and shrimp. Refrigerated foods are potential sources of *Y. enterocolitica* because contamination can occur with foods at the production site or at home before storage. The most common transmission route of pathogenic *Y. enterocolitica* is thought to be fecal–oral via contaminated food. This bacterium can survive and multiply in foods stored at cold temperature. Both pathogenic and nonpathogenic strains produce heat-resistant toxins. The human illness is associated with the consumption of food and water contaminated with *Y. enterocolitica*. The consumption of a high number of cells allows pathogen to survive stomach acidity and pass to the intestines. The infective dose is about 10^7 viable cells per g or mL of food. *Y. enterocolitica* produces an invasive factor that allows pathogenic strains to colonize the intestinal epithelial cells and lymph nodes. Colonized *Y. enterocolitica* starts to grow and produces heat-stable enterotoxin. Enterotoxin causes yersiniosis. The symptoms appear within 24–30 h after consumption of the bacterium together with food or water and last about 2–3 days. The typical symptoms of yersiniosis are abdominal pain in the abdomen lower quadrant, chills, headache, fever, nausea, vomiting, diarrhea often mixed with blood, mucous, and pus. The disease can be fatal in rare cases.

Virulence tests can be used to differentiate pathogenic *Y. enterocolitica* biotypes from nonpathogens. *Y. enterocolitica* strains produce heat-stable toxin below 30°C. While many *Y. enterocolitica* strains produce this toxin, some virulent *Y. enterocolitica* strains cannot produce it.

Cold enrichment can be used in the isolation of *Y. enterocolitica* from a sample with phosphate-buffer (PBS) solution at cold temperature. The treatment of enrichment broth with alkali after incubation increases the recovery of *Y. enterocolitica*. The selective enrichment media are irgasan-sorbitol-bile salt chlorate (ITPC) broth and peptone-sorbitol-bile-salt (PSB) broth. Cold enrichment in a selective enrichment media prevents the growth of unwanted bacteria, reduces the competition of microflora, and facilitates the growth of *Y. enterocolitica*. A selective/differential agar medium (CIN agar) is used in the isolation of *Y. enterocolitica* from enrichment culture. *Y. enterocolitica*

Microbiological Analysis of Foods and Food Processing Environments. DOI: https://doi.org/10.1016/B978-0-323-91651-6.00031-8

can be counted using selective/differential medium CIN agar. Isolates can be by identified phenotypical, serological, and foodomics techniques.

18.2 Isolation and counting techniques

Sampling, sample preparation, and all experiments should be performed under aseptic conditions. Sampling and all analysis should be performed in duplicate, unless otherwise specified. All microbiological media to be used in this application must be sterilized. Again, all solutions and equipment that may cause contamination must be sterile or disinfected.

18.2.1 Equipment, materials, and media

Equipment. Incubator, microscope, microscope glass slide, Petri dishes (15 × 100 mm), pipette discard equipment (e.g., a graduated cylinder) containing 10% bleach solution and a piece of cloth at the bottom, pipettes (1 mL) in the pipette box, thermostatic water bath, vial (5 mL), and Whatman 541 filter paper.

Reagents. Crystal violet (CV) solution (0.085 μg per mL, aq.), ferrous ammonium sulfate (1%), ferrous chloride solution (10%), Gram stains, 0.5% KOH (prepare with 0.5% saline solution), NaH_2PO_4 (0.666 M), phosphate buffer saline (PBS) solution, and physiological saline (PS) solution.

Media. Anaerobic egg yolk (AEY) agar, arginine decarboxylase (AD) broth, Brain heart infusion (BHI) agar slant, BHI broth, bile esculin (BE) agar, cefsulodin-irgasan-novabiocin (CIN) agar, Congo red-brain heard infusion agar, irgasan-ticarcillin-chlorate (ITC) broth, MacConkey agar, motility test (MT) medium containing 2,3,5-triphenyl tetrazolium (1%), Mr-VP broth, peptone-sorbitol-bile-salt (PSBS) broth, phenylalanine deaminase (PAD) agar slant, preenrichment medium, pyrazine-amidase (PA) agar slant, triple sugar iron (TSI) agar slant, tryptose-sulfite-cycloserine (TSC) agar, and urea agar slant.

18.2.2 Enrichment isolation of *Yersinia enterocolitica*

The *Yersinia* selective enrichment isolation and counting flowchart is given in Fig. 18.1. The following simplified procedure is recommended for *Yersinia* isolation from food, water, and environmental samples (e.g., soil).

1. Samples are promptly analyzed after being received or refrigerated at 4°C until analysis. Freezing of sample before analysis is not recommended. Aseptically, weigh 25 g of the sample into each of two sterile blenders (or stomacher plastic bag) containing 225 mL of PSB broth and blend for 1 min (stomach for 2 min) to homogenize the sample. For a liquid sample, add 25 mL of the sample into each of two sterile Erlenmeyer flasks containing 225 mL of PSB broth and mix well by gently shaking. Similarly add 25 g or mL of the sample into each of two 225 mL of ITC broth and follow as per PSB broth. Incubate one set of broth at 10°C for up to 10 days and the other set at 25°C for up to 3 days.
2. a. After 3 days of incubation of preenrichment broth at 25°C, remove broth from incubator and mix well by gently shaking. Streak plate one loopful of the broth culture on to a CIN agar plate

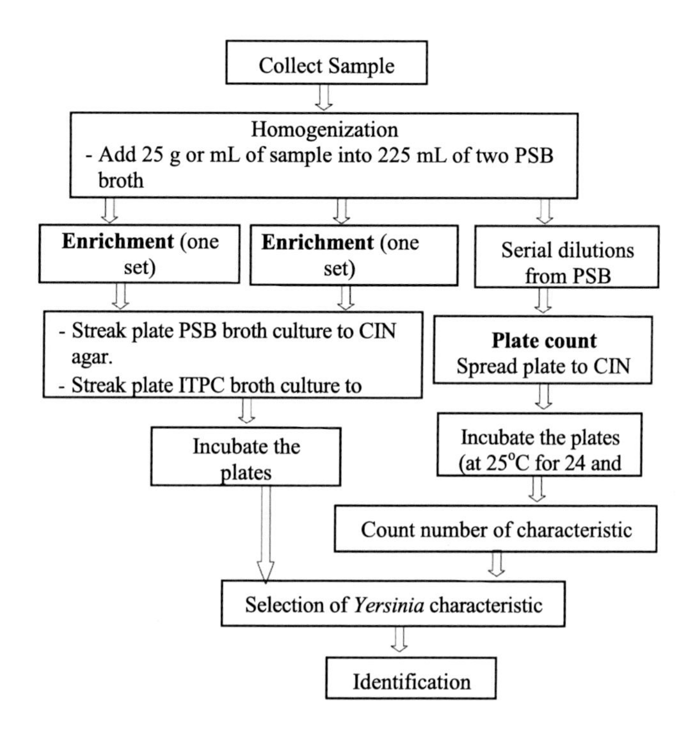

FIGURE 18.1

Yersinia selective enrichment isolation and counting flowchart.

and the ITC broth culture on to a MacConkey agar plate. Incubate the inverted plates at 25°C for up to 3 days.

b. After 10 days of incubation of preenrichment broth at 10°C, remove broth from incubator and mix well by gently shaking. Streak plate the ITPC broth culture to MacConkey agar and the PSB culture to CIN agar. Transfer 0.1 mL of the TSB broth culture to 1 mL of KOH (0.5%) and mix by gently shaking for 10 s. Streak plate one loopful mixture to each of the MacConkey agar and CIN agar plates. Transfer 0.1 mL of TSB broth culture into 1 mL of saline solution (0.5%) and mix well by gently shaking for 10 s and streak plate one loopful of mixture to each of the MacConkey agar and CIN agar plates as above. Incubate inverted plates at 25°C for 24 h.

After incubation, examine the characteristic colony formation on MacConkey agar and CIN agar plates. *Y. enterocolitica* produces small (1–2 mm in diameter), flat, colorless, or pale pink colonies on MacConkey agar. Other bacteria cannot produce red or mucoid colonies on MacConkey agar. *Y. enterocolitica* produces small colonies (1–2 mm diameter) with a smooth surface with a deep red center "bullseye" and translucent edge colonies surrounded by a clear, colorless precipitate bile zone (Fig. 18.2A) on CIN agar. Other bacteria produce large colonies with diffuse pinkish centers and opaque outer zones on CIN agar. *Citrobacter freundii*, *Serratia liquefaciens*, and *Enterobacter agglomerans* may produce similar colonies to *Y. enterocolitica*. *Y. enterocolitica* can be differentiated from these bacteria by phenotypic characteristics tests (Technique 18.3).

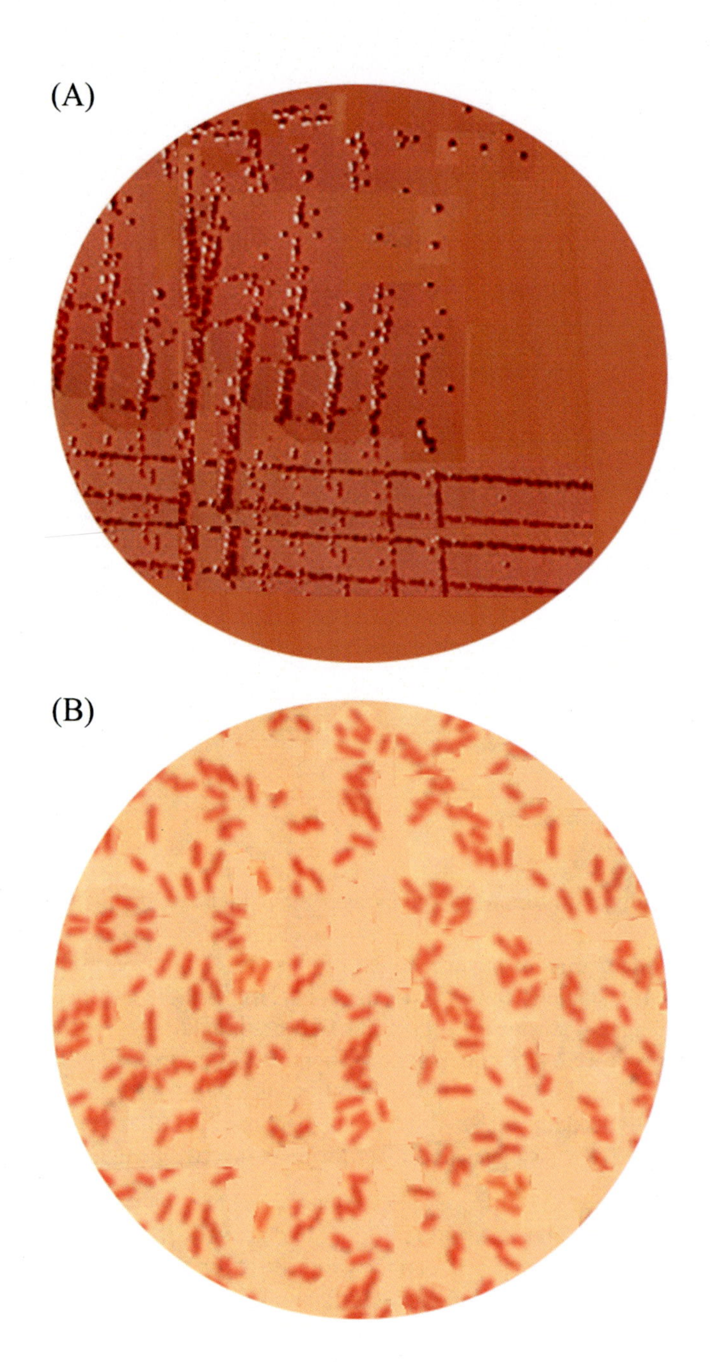

FIGURE 18.2

Yersinia enterocolitica: (A) characteristic colonies on CIN agar and (B) staining appearance.

18.2.3 Counting of *Yersinia enterocolitica*

Refer to Technique 2.2.2 for details of sample preparation and dilutions and gently shake the flask to homogenate the sample. Add 25 g or mL of the sample into 225 mL of PBS solution in the sterile Erlenmeyer flask. Prepare further dilutions in PBS solution. If high numbers of *Yersinia* are expected from the food, spread plate 1 mL of serially diluted sample to each of the MacConkey agar and CIN agar in duplicate. Transfer 1 mL of homogenate to 9 mL of KOH (0.5%; prepared by 0.5% saline), mix the mixture for several seconds, and spread plate 1 mL of the sample mixture to each of the MacConkey agar and CIN agar plates (Technique 2.2.4). Incubate the inverted Petri plates at 25°C for up to 3 days. After incubation, examine the plates for characteristic colony formation. Count the number of characteristic colonies from the plates containing 25–250 colonies. Calculate the number of *Y. enterocolitica* by multiplying the average count by dilution rate and dividing by inoculation amount. Record the results as the number of *Y. enterocolitica* colony forming units (cfu) per g or mL of sample. Confirm the presumptive isolates as described in Technique 18.3.

18.3 Identification of *Yersinia enterocolitica*

Obtain a pure culture, as described in Technique 13.4.1, using CIN agar and BHI broth and incubate at 25°C. Inoculate the pure culture to BHI agar slant and BHI broth, and incubate at 25°C for 24 h.

The *Y. enterocolitica* identification flowchart is given in Fig. 18.3.

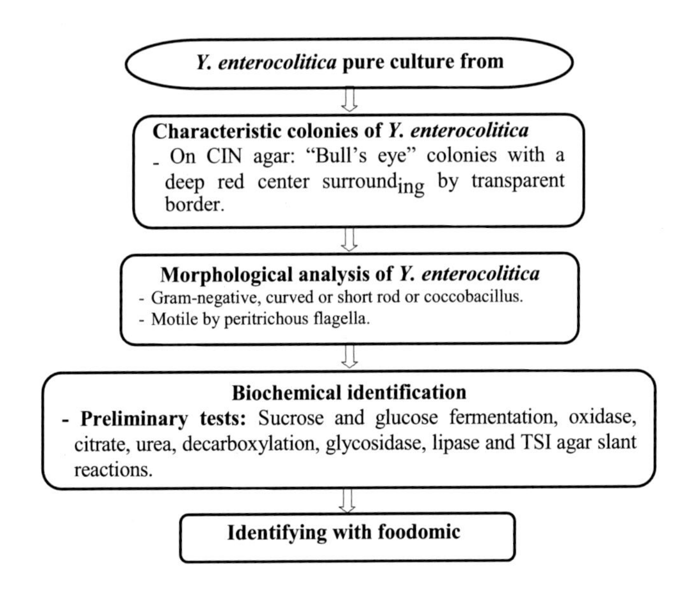

FIGURE 18.3

Yersinia enterocolitica identification flowchart.

18.3.1 Morphological identification

Gram staining. Prepare a Gram stain from fresh BHI broth culture as given in Appendix B under the topic "Gram Staining Reagents." A Gram stain of *Y. enterocolitica* will produce Gram-negative, curved or short rods or coccobacillus shapes (Fig. 18.2B).

Flagella staining. Perform flagella staining from fresh BHI broth culture, as given in Appendix B under the topic "Liefson's Flagella Stain." *Y. enterocolitica* contains peritrichous flagella and is a motile bacterium.

Motility test. Inoculate two tubes MT medium from BHI broth culture. Incubate one MT medium at 25°C and the other at 35°C for 24 h. Growth only along the inoculation line indicates that the bacterium is nonmotile. A circular growth through the medium from the line of the stab inoculation indicates that the bacterium is motile (Technique 13.4.2). *Y. enterocolitica* is motile at 25°C and not at 35°C.

18.3.2 Biochemical identification

18.3.2.1 Phenylalanine deamination

With a sterile needle, the bacterial cells are taken by twice touching onto the colony center, and inoculate to PAD agar slant by stubbing the butt and streaking slant once. Incubate at 25°C for up to 3 days. After incubation, pour two to three drops of 10% freshly prepared ferric chloride solution onto colonies on the PDA agar slant. Green color formation indicates positive phenylalanine deaminase test. *Y. enterocolitica* cannot produce phenylalanine deaminase and the test is negative.

18.3.2.2 Pyrazinamidase test

Inoculate one loopful of BHI broth culture to PA agar slant by stubbing the butt and single streaking the slant. Incubate tubes at 25°C for up to 3 days. After incubation, pour 1 mL of 1% fresh ferrous ammonium sulfate solution to the colonies on the slant. The color of the colonies turns to pink within 15 min to indicate a positive test and a positive test indicates the presence of pyrazinoic acid formation by the bacterium with the production of pyrazinamidase enzyme. Two biotypes of *Y. enterocolitica* produce pyrazinamidase enzyme and give positive results for this test.

18.3.3 3-β-D-glucosidase test

Weigh 0.1 g of 4-nitrophenyl-beta-D-glucopyranoside into 100 mL of 0.666 M NaH_2PO_4 (pH 6.0). Dissolve the mixture and filter-sterilize. Emulsify BHI agar slant culture in a PSB solution in the sterile tube (13 × 100 mm) up to McFarland Turbidity Standard No. 3. Add 0.75 mL of emulsified culture into sterile tube (13 × 100 mm) containing 0.25 mL of test solution and mix by gently shaking the tube. Incubate tube at 30°C for 18 h. Formation of a distinct yellow color indicates a positive test. *Y. enterocolitica* biotype 1A is positive for this test.

18.3.3.1 Lipase test

Streak plate one loopful BHI broth culture to AEY agar (Technique 13.4.1) and incubate inverted plates at 25°C for up to 3 days. After incubation, a positive reaction for lipase is indicated by oily, iridescent, pearl-like colonies surrounded by a precipitation ring and the ring's outer side appears as a clear zone. *Y. enterocolitica* biotypes 1A and 1B are positive in the lipase test.

18.3.3.2 Other tests

Use BHI broth culture in the following tests.

TSI agar slant reaction. Refer to Technique 13.4.3.1 for details of the TSI agar slant reactions and incubate inoculated TSI agar slant at 25°C for up to 3 days. *Y. enterocolitica* will give the following typical reactions: a red (alkaline) slant and yellow (acid) butt, without black color on slant (no H_2S) and little or no gas.

Citrate utilization. Refer to Technique 13.4.3.4 for details of the citrate utilization test. Incubate inoculated citrate agar slant at 25°C for up to 3 days. After incubation, examine the citrate agar slant for growth and color changes. Positive test: formation of purple slant/purple butt (alkaline) by black color due to H_2S production (left); and negative test: purple slant/yellow butt (right) without H_2S production. *Y. enterocolitica* is a citrate negative bacterium.

Decarboxylase test. Inoculate a loopful of BHI broth culture to AD broth (Technique 15.3.2.1) and incubate at 35°C for up to 3 days. A color change from yellow to purple indicates the decarboxylation of arginine and the test is positive. *Y. enterocolitica* decarboxylates arginine.

Esculin hydrolysis. Refer to Technique 14.3.2.1 for details of esculin hydrolyzes. Streak plate one loopful of BHI broth culture to an area of BE agar plate. Incubate at 25°C for up to 3 days. Virulent *Y. enterocolitica* biotype 1A hydrolyzes esculin with black coloring on the surface of the medium (Fig. 18.4A) and avirulent strains of *Y. enterocolitica* do not.

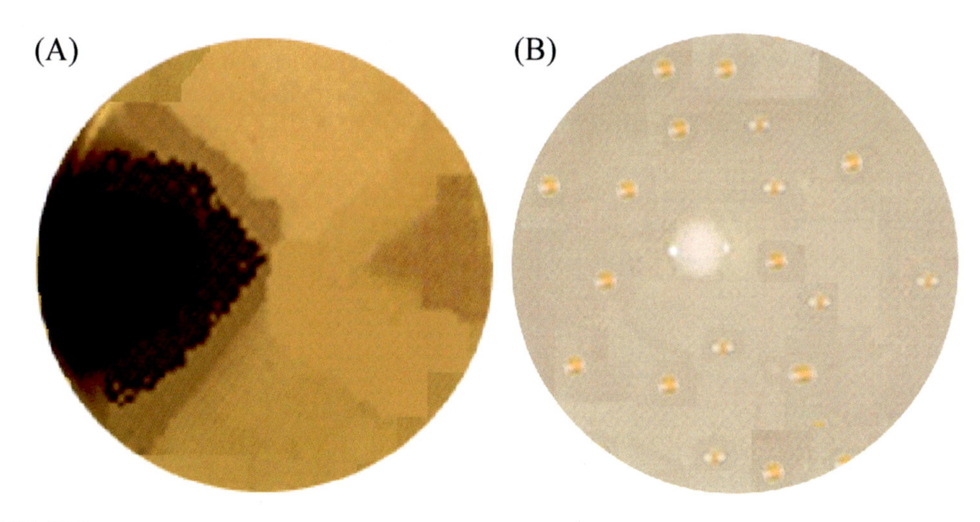

FIGURE 18.4

(A) Virulent *Yersinia enterocolitica* strain hydrolyzes esculin with black coloring on the surface of the medium (*left*), and avirulent strain of *Y. enterocolitica* does not hydrolyze esculin without coloring (*right*). (B) Plasmid-bearing colony is pinpoint convex, red, opaque; and plasmidless colony is large, irregular flat, and translucent (at center).

Urease test. Refer to Technique 13.4.3.5 for details of the urease test. Incubate the inoculated urea agar slant at 25°C for up to 3 days. Most *Y. enterocolitica* biotypes are urease positive. Urease negative nonpathogenic strains of *Y. enterocolitica* are also present.

Catalase test. Refer to Technique 13.4.3.10 for details of the catalase test. *Y. enterocolitica* is catalase positive.

Oxidase test. Refer to Technique 17.4.2.10 for details of the oxidase test. *Y. enterocolitica* is an oxidase negative bacterium.

Indole test. Refer to Technique 13.4.3.2 for details of the indole test. *Y. enterocolitica* is indole variable.

Mr-VP tests. Refer to Technique 13.4.3.3 for details of the Mr-VP tests. *Y. enterocolitica* is positive for Mr and variable positive for VP.

Carbohydrate fermentation test. Refer to Technique 15.3.2.2 for details of the carbohydrate fermentation test. *Y. enterocolitica* shows variable carbohydrate utilization results (Table 18.2).

For further phenotypic identification refer to Tables 18.1 and 18.2.

18.3.4 Pathogenicity testing

Autoagglutination test. Inoculate one loopful BHI broth culture into the Mr-VP broth containing *Y. enterocolitica* agglutinin and incubate at room temperature for 24 h. Growth of *Y. enterocolitica* should show turbidity in the broth. *Y. enterocolitica* gives a positive result for the virulence plasmid with agglutination (clumping) of bacterium along the walls and/or bottom of the Mr-VP broth tube with a clear supernatant fluid.

Virulence plasmid test. Inoculate one loopful of BHI broth culture into the BHI broth and incubate at 25°C for 24 h. After incubation, dilute *Y. enterocolitica* culture with PS solution to obtain 10 cells per mL. Spread plate 1 mL of the diluted culture onto two Congo Red agar plates. Incubate inverted one of the plates at 35°C and the other plate at 25°C up to 48 h. After incubation,

Table 18.1 Phenotypic identification of *Yersinia enterocolitica* biotypes.

Biochemical test	Biotype reaction						
	1A	1B	2	3	4	5	6
Lipase/indole	+/+	+/+	−/(+)	−/−	−/−	−/−	−/−
Nitrate reduction	+	+	+	+	+	−	
Ornithine decarboxylase	+	+	+	+	+	+	+
Sorbose/xylose (acid production)	+/+	+/v	+/+	+/+	+/−	−/v	−/+
Inositol (acid production)	+	+	+	+	+	+	+
Trehalose	+	+	+	+	+	−	+
Esculin/VP	+/+	−/+	−/+	−/+	−/+	−/+	−/−
Pyrazine amidase	+	−	−	−	−	−	+
β-D-Glucosidase	+	−	−	−	−	−	−

v, Variable; +, positive; −, negative; (), delayed reaction.

Table 18.2 Phenotypic characteristics of *Yersinia* species.

Test (at 25°C)	*Y. enterocolitica*	*Y. pseudotuberculosis*	*Y. pestis*	*Y. kristensenii*	*Y. frederiksenii*	*Y. intermedia*	*Y. aldovae*	*Y. rohdei*	*Y. mollaretii*	*Y. bercovieri*	*Y. ruckeri*
Fermentation of (acid from)											
Glucose/sucrose	−/+	v/−	v/−	v/−	v/+	+/+	−/−	v/+	−/+	−/+	−/−
Mannitol/rhamnose	+/−	+/+	−/−	+/+	+/+	+/+	+/+	+/−	+/−	+/−	+/−
Melibiose/raffinose	+/+	+/+	+/+	+/+	−	+	−	−			
Maltose/cellobiose	+/+	+/−	+/−	+/+	+/+	+/+	+/−	+/+	+/+	+/+	+/−
Sorbose/sorbitol	−/+	−/−	−/v	−/+	−/+	−/+	−/+	−/+	−/+	−/+	−/−
Fructose	+		−	v							
Decarboxylation of											
Lysine/arginine	−/−	−/−	−/−	−/−	−/−	−/−	−/−	−/−	−/−	−/−	−/−
Ornithine	+	−	−	+	+	+	+	+	+	+	+
Nitrate reduction	+	+	+	+	+	+	+	+	+	+	+
Lipase/pyrazineamidase	v/v	−/−	−/−	v/+	v/+	v/+	v/+	−/+	−/+	−/+	−/+
Oxidase/catalase	−/+	−/+	−/+	−/+	−/+	−/+	−/+	−/+	−/+	−/+	−/+
Esculin/inositol	v/v	+/−	+/−	−/v	+/v	+/v	+/v	−/−	(+)/−	v/−	+/−
H_2S/indole	−/−	−/−	+/+	−/−	−/−	−/−	−/−	−/−	−/−	−/−	−/−
Mr/VP	+/v	−/−	−/−	v/v	v/+	v/+	v/+	v/v	−/v	−/v	−/v
Urease/citrate	+/−	+/−	−/−	+/−	+/v	+/v	+/−	+/+	+/−	+/−	−/+

+, *Positive;* −, *negative;* v, *variable.*

examine the plates for the characteristic colony formation. When *Y. enterocolitica* cells carry the characteristic virulent plasmid, they will appear as dot, round, convex, and opaque colonies (Fig. 18.4B). Plasmid-free *Y. enterocolitica* cells will appear as large, irregular flat, and translucent centered colonies (Fig. 18.4B at center).

Crystal violet binding test. This rapid screening test can be used to distinguish virulent *Y. enterocolitica* cells. Tubes containing BHI broth are incubated at 25°C for 18 h with shaking in a thermostatic shaking water bath. After incubation, the culture is diluted to 1000 cells per mL in PS solution. Spread plate 1 mL of the diluted culture on two BHI agar plates (Technique 2.2.4). One of the plates is incubated at 25°C and the other plate at 35°C for 30 h. After incubation, 8 mL of CV solution is added onto the colonies in each plate and allowed to remain for 2 min. Then the CV solution is slowly drained from plate. The plates are analyzed to detect colonies that bind to CV. The cells in the colonies containing plasmid at 35°C will bind to CV but not the colonies formed at 25°C. Plasmid-free cells in the colonies growing at either temperature should not bind to CV.

18.3.5 Serological identification of *Campylobacter* species

Fifty-seven somatic heat-stable O factors are used for serotyping (biotyping) *Y. enterocolitica* and related species. Perform the serological tests as described in Technique 13.5.4.

18.3.6 Identification of *Campylobacter* species by foodomics techniques

18.3.6.1 MALDI-TOF Ms identification of Campylobacter

Campylobacter can be rapidly identified with the matrix-mediated laser desorption ionization-time-of-flight-mass spectrometry (MALDI-TOF Ms) technique, which is based on the principle of ionizing a specific protein profile of microbial cells. *Campylobacter* can be easily identified by comparing these profiles with the reference spectra. MALDI-TOF Ms identification has been explained in Technique 13.6.3.1.

18.3.6.2 Molecular identification of Campylobacter

Polymerase chain reaction (PCR) is a molecular genotypic technique and is used in the identification of *Campylobacter* species. Gene coding of virulence factors of *Campylobacter* can be detected by PCR assay. Some *Campylobacter*-specific primer sets that can be used in the identification of *Campylobacter* are given in "Appendix A: Gene Primers (Table A.1)." Using *Campylobacter*-specific techniques and materials, the bacterium can be identified by PCR analysis, as described in Technique 13.6.3.2.

18.4 Interpretation of results

Y. enterocolitica produces acid from sucrose and glucose, is oxidase and citrate negative, and is urea positive. *Yersinia* biotypes (except 1A) do not hydrolyze esculin or ferment salicin. *Y. enterocolitica* biotype 6 ferments sucrose but *Y. kristensenii* cannot. *Y. enterocolitica* causes gastroenteritis with a variety of clinical symptoms in humans. Consequently, the presence of this bacterium in food and water is potentially hazardous to human health. Any food (not to be subsequently cooked) and water containing *Y. enterocolitica* should be considered unfit for human consumption. Cross-contamination between raw and ready-to-eat foods should be avoided. Foods, particularly meats, should be thoroughly cooked to destroy the bacteria. *Y. enterocolitica* in raw meat at 7°C may grow to more than 10^9 cells per g within 10 days. It is sensitive to pasteurization and cooking temperature.

PRACTICE

Isolation and counting of *Bacillus cereus*

19

19.1 Introduction

Bacillus cereus is Gram-positive, motile, facultative anaerobic, hemolytic, spore-forming and rod-shaped, arranged in pairs or chains with round or square ends. It cannot ferment xylose, mannitol, and arabinose; produces lecithinase and acetylmethylcarbinol; does not produce acid from glucose; reduces nitrate to nitrite; and decomposes L-tyrosine. *Bacillus* usually has a single endospore and spores do not swell the vegetative cell. The endospores are central and generally oval, sometimes round or cylindrical. The endospores are very resistant to adverse conditions. It grows optimally from 28°C to 35°C. It is not sensitive to cold or frozen storage. It most commonly is associated with foods, for example, spices, pudding, dry milk, cream, dry potatoes, sauces, spaghetti, and rice. Food product contamination mostly occurs before cooking. If food is maintained from 30°C to 50°C during processing and serving, endospores can germinate and *B. cereus* starts to grow.

As a sporeformer, *B. cereus* is distributed widely in the environment and is present commonly in soil and water. The emetic (vomit) and diarrheal food poisoning are caused by different distinct enterotoxins of *B. cereus*. Different types of foods, including puddings, sauce, vegetables, salads, inadequately pasteurized milk and cream, and meat-based foods, can associate with *B. cereus* diarrheal-type illnesses. The emetic type illnesses can associate with starchy foods (e.g., rice, pasta dishes, macaroni, etc.) and small meat dishes. In the diarrheal form, the infective dose is about $10^6 - 10^7$ viable cells per g or mL of food. The symptoms appear within 6–12 h after the consumption of food containing viable cells. The disease lasts 24 h. The principal symptoms are abdominal pain, and watery diarrhea without vomiting or fever. In the emetic type of foodborne disease, the symptoms appear within 1–5 h after the consumption of food containing preformed enterotoxin by *B. cereus* and last 24 h. The principal symptoms are abdominal pain, nausea, vomiting, and diarrhea.

Bacillus species can be divided into three groups depending on spore and sporangium morphology. They are:

Group I: Gram-positive, produce central or terminal, ellipsoidal or cylindrical spores that cannot cause swelling of vegetative cells. It includes two subgroups:

- Large cell: *B. cereus, Bacillus anthracis, Bacillus mycoides*, *Bacillus megaterium* and *Bacillus thuringiensis*.
- Small cell: *Bacillus pumilus, Bacillus subtilis*, and *Bacillus licheniformis*.

Group II: Gram variable, produces central or ellipsoidal spores and cells swell with endospore: *Bacillus coagulans* and *Bacillus circulans. Bacillus alvei, Bacillus macerans*, and *Bacillus brevis*.

Group III: Gram variable, cells swell with endospore, produces terminal or subterminal spores: *Bacillus sphaericus*.

Microbiological Analysis of Foods and Food Processing Environments. DOI: https://doi.org/10.1016/B978-0-323-91651-6.00030-6

The selective enrichment technique can be used in the isolation of *B. cereus* from foods containing low number. The selective/differential media used in the isolation and counting of *B. cereus* are mannitol egg-yolk polymyxin (MYP) agar, Kim-Gofer agar containing polymyxin, *B. cereus* agar, and polymyxin-pyruvate-egg yolk-mannitol-bromothymol blue (PEMB) agar. Low levels of peptone in PEMB agar promote sporulation and pyruvate reduces the colony size. MYP agar can be used to recover thermally injured *B. cereus*. *B. cereus* isolates can be identified by phenotypic and foodomics techniques.

19.2 Isolation and counting techniques

Sampling, sample preparation, and all experiments should be performed under aseptic conditions. Sampling and all analysis should be performed in duplicate, unless otherwise specified. All microbiological media to be used in this application must be sterilized. Again, all solutions and equipment that may cause contamination of microorganisms must be sterilized or disinfected.

19.2.1 Equipment, reagents, and media

Equipment. Blender (or homogenizer), cover slip, incubator. microscope, microscope glass slide, Petri dishes (15 × 100 mm), pipette discard equipment (e.g., a graduated cylinder) containing 10% bleach solution and a piece of cloth at bottom, pipettes (1 mL) in the pipette box, and test tube.

Reagents. Basic fuchsine (0.5% aq.), ethanol (70% aq.), Gram stains, KOH (40% aq.), lipid globule stain (Burdon), Lugol's iodine solution, malachite green stain (5% aq.), nitrate reagents, phosphate buffer saline (PBS) solution, safranin stain (0.5% aq.), spore stains (Ashby), Sudan black (0.5% in 70% ethanol), xylene, and sterile distilled water,

Media. *B. cereus* (BC) agar, BC broth, Brain heart infusion (BHI) agar slant, BHI broth, carbohydrate fermentation (CF) broth containing glucose, gelatin agar, Kim-Geopfert (KG) agar containing polymyxin, mannitol egg yolk polymyxin (MYP) agar, modified Mr-VP broth, motility test (MT) medium, nitrate broth, nutrient agar slant, nutrient agar, nutrient broth (containing 0.001% lysozyme), polymyxin pyruvate egg yolk mannitol bromothymol blue (PEMB) agar, starch agar, trypticase soy polymyxin (TSP) broth, trypticase soy-sheep blood (TSSB) agar, and tyrosine agar slant.

19.2.2 Selective enrichment isolation of *Bacillus cereus*

The *B. cereus* selective enrichment isolation and counting flowchart is given in Fig. 19.1. Aseptically, weigh 25 g of sample into a blender (or stomacher plastic bag) containing 225 mL of sterile TSP broth. Homogenize by blending for 1 min (or stomaching for 2 min) and pour the homogenate into an Erlenmeyer flask. In the case of a liquid sample, add 25 mL of the sample into a sterile Erlenmeyer flask containing 225 mL of TSP broth and mix by gently shaking the flask. Incubate the flask at 30°C for up to 48 h. After incubation, examine growth (turbidity) and color changes. Streak plate one loopful of enriched culture to PEMB agar and TSSB agar. Incubate inverted plates at 35°C for 18 h. After incubation, examine plates for *B. cereus* characteristic colony formations. Identify colonies by colonial morphology and the presence or absence of β-hemolysis. Characteristic *Bacillus* colonies on these agar media are as follows:

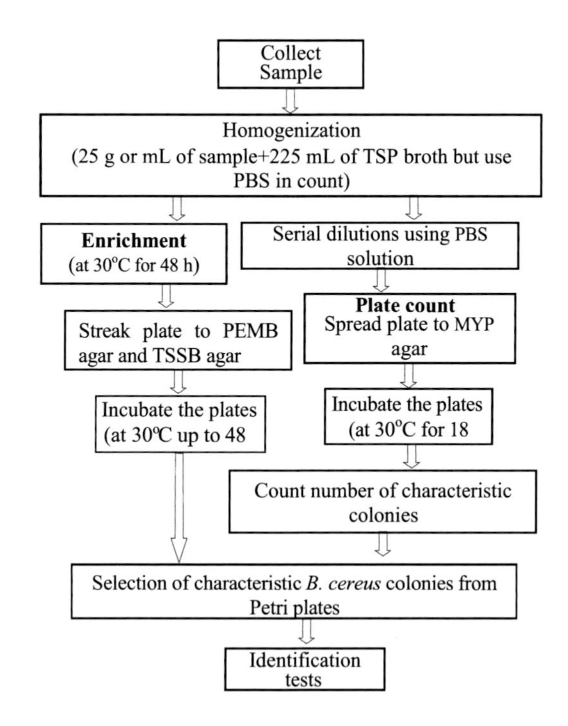

FIGURE 19.1

Bacillus cereus selective enrichment isolation and counting flowchart.

***Bacillus* spp.**	**Hemolysis**	**Characteristics of colonies**
Bacillus anthracis	Nonhemolytic (may be weakly hemolytic)	*TSSB agar*: *B. anthracis* colonies are flat and irregular, gray/white color with a ground glass appearance. Colonies will be pulled up and stay upright on teasing with a loop.
Bacillus cereus group (*B. cereus*, *B. thuringiensis*, *B. weihenstephanensis*, *B. mycoides*, *B. pseudomycoides*)	β-Hemolytic	*TSSB agar*: *B. cereus* colonies are cream to white or gray and have a slight green tinge. *B. mycoides* colonies are rhizoid or hairy. Colonies of others are similar to *B. anthracis*. *PEMB agar*: Colonies of *B. anthracis* and *B. cereus* are crenated, turquoise to peacock blue with a zone of egg yolk precipitation.
Bacillus subtilis	β-Hemolytic	*TSSB agar*: Colonies of *B. subtilis* are large with a frosted-glass appearance and opaque. Some strains may produce mucoid or smooth or raised wrinkly colonies. *PEMB agar*: Colonies of *B. subtilis* are cream to yellow color without egg yolk precipitation zone.

Perform identification tests as indicated in Technique 19.3.

19.2.3 Counting of *Bacillus cereus*

19.2.3.1 Plate counting technique

Refer to Technique 19.2.2 for details of sample preparation without enrichment but use PBS solution in the homogenization of sample. Prepare further dilutions using PBS solution. Spread plate 1 mL of diluted sample to MYP agar (or KG agar) plates in duplicate (Technique 2.2.4). Allow the plates to absorb the liquid by medium at room temperature (within 10 min). Incubate inverted plates at 30°C for 18 h. After incubation, examine the formation of characteristic colonies on plates. Count the number of typical colonies from the Petri plates containing 25–250 colonies. Calculate the number of *B. cereus* by multiplying the average count by the dilution rate and dividing by the inoculation amount. Confirm (Technique 19.3) the presumptive isolates and record the results as the number of *B. cereus* colony forming units (cfu) per g or mL of sample.

Growth on MYP agar. MYP agar differentiates *B. cereus* from other *Bacillus* species by its nonfermenting mannitol and production of lecithinase. Acid produced by *Bacillus* spp. diffuses throughout the MYP agar, the color of medium turns to yellow, thus distinguishing mannitol-fermenting species from nonfermenting *B. cereus*. *B. cereus* sporulates poorly on MYP agar. *B. cereus* produces pink colored colonies surrounded by a precipitation zone on MYP agar. Precipitation indicates lecithinase production. Extended incubation increases the intensity of pink color colonies. If the pink color on the medium is not clear, incubate the Petri plates for an additional 24 h and count the colonies again after incubation.

Growth on KG agar. This medium is equally sensitive and selective but is used much less frequently than MYP agar. *B. cereus* produces free spores within the 24 h incubation period on KG agar. *B. cereus* produces round, flat, and dry colonies with a ground glass, translucent or creamy white appearance.

19.2.3.2 Most probable number counting technique

Most probable number (MPN) technique is suitable for counting *B. cereus* from foods that are expected to contain less than 10 *B. cereus* per g or mL of sample. Perform the MPN technique as described in Technique 4.2. Refer to Technique 19.2.2 for sample preparation and dilutions but use PBS solution in the homogenization of the sample. Prepare further dilutions using PBS solution. Inoculate 1 mL of sample into each 3-MPN tubes of TSP broth and incubate the tubes at 30°C for up to 48 h. After incubation, examine the tubes for dense growth (turbidity) and color change. Indicate the number positive tubes for each dilution. Confirm each positive tube on selective/differential agar medium and phenotypic tests (Technique 19.3). Streak plate presumptive positive MPN tubes on MYP agar for confirmation. Incubate Petri plates at 35°C for 24 h. After incubation, examine Petri plates for characteristic colony formation. Refer to MPN Table 4.1 to find the number of *B. cereus* MPN per g or mL of sample.

19.3 Identification of *Bacillus cereus*

Obtain a pure culture as described in Technique 13.4.2, if necessary, using MYP agar BC broth. Inoculate pure cultures to KG agar plates (for spore staining), BHI agar slant, and BHI broth. Incubate plates and tubes at 35°C for 24 h. Use pure cultures in phenotypic and molecular identification tests. The *B. cereus* identification flowchart is given in Fig. 19.2.

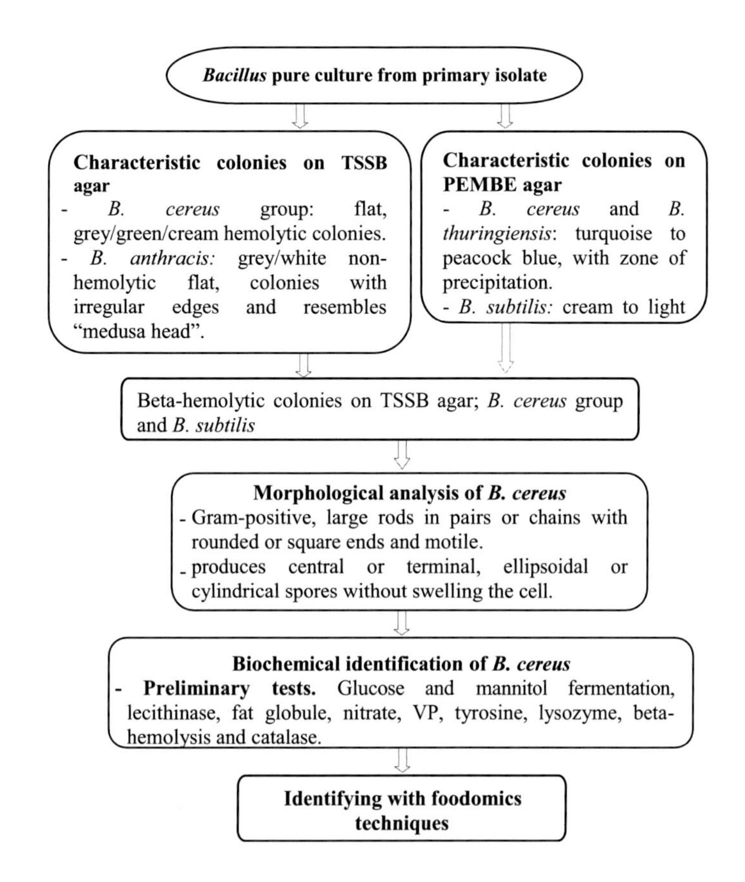

FIGURE 19.2

Bacillus cereus identification flowchart.

19.3.1 Morphological identification

Gram staining. Prepare a Gram stain from BHI agar slant (or BHI broth) culture, as explained in Appendix B under the topic "Gram Staining Reagents." A Gram stain of *B. cereus* will produce Gram-positive rods, appearing singly or in pairs (Fig. 19.3).

Spore staining. Remove a small amount of growth from the center of a colony on KG agar culture and place near a small drop of sterile distilled water on the microscope glass slide. Emulsify in water with a loop and spread on the small area of the slide. Allow the slide to air dry, fix, and perform spore staining, as explained in Appendix B under the topic "Spore Stain." Examine the stained smear under a microscope for the formation of spores within cells. *B. cereus* produces spores within the vegetative cell without swelling.

Motility test. Inoculate the MT medium in tube by stabbing down the medium center with a 3 mm loopful of overnight BHI broth culture. The tubes are incubated at 30°C for 24 h. After incubation, analyze growth through the medium. A motile *B. cereus* bacterium produces diffuse growth throughout the medium. Nonmotile *Bacillus* spp. produce growth only along the stab line.

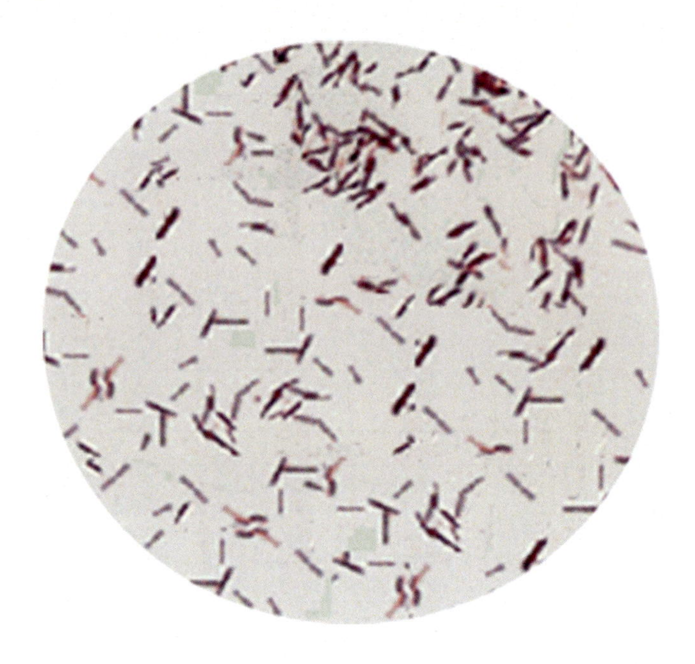

FIGURE 19.3

Staining appearance of *Bacillus cereus.*

Confirmation can be performed by the direct microscopic technique: add 0.2 mL of 0.1% peptone water to the BHI agar slant surface, shake the liquid on the slant and inoculate one loopful of this washed liquid to the BHI agar slant. Incubate the slant at 30°C for 6–8 h. Remove a loopful of BHI agar slant culture from liquid base of slant and place on a microscope glass slide. Place the cover slip onto the culture, replace the slide to a microscope (wet-mount preparation) and examine immediately for motile and nonmotile cells. *B. cereus* and *B. thuringiensis* cells are motile with the help of peritrichous flagella. *B. anthracis* and all other *Bacillus* spp. except a few *B. mycoides* strains are nonmotile.

19.3.2 Biochemical identification

19.3.2.1 Hemolytic test

A TSSB agar plate is marked into 6–8 equal parts (each 4 cm^2 area) at the bottom with a waterproof pen. Remove three or four loopfuls of culture from BHI broth culture and streak culture on a premarked area of TSSB agar plate by gently touching the loop to the medium's surface. Incubate inverted Petri plates at 35°C for 24 h in an incubator containing 5% CO_2 (or in candle jar). After incubation, examine the plates for hemolysis around colonies. *B. cereus* is a strongly hemolytic bacterium and produces a 2–4 mm zone of complete beta-hemolysis around the colony. Most *B. thuringiensis* strains and *B. mycoides* strains are weakly beta-hemolyztic. *B. anthracis* strains are usually nonhemolytic after 24 h incubation.

19.3.2.2 Anaerobic glucose fermentation

Inoculate one loopful of BHI broth culture into CF broth containing glucose. Incubate tubes at 35°C for 24 h in an anaerobic incubator or anaerobic jar. After incubation, examine the CF broth by shaking the tube vigorously, and examine for growth (turbidity) and color change. A change from red to yellow indicates acid production anaerobically from glucose. A pH reduction in the control tube with medium exposure to CO_2 in the GasPak anaerobic jar may be possible. Therefore, use appropriate positive and negative controls for the distinct between "positive" and "false-positive" reactions. *B. cereus* utilizes glucose and produces acid anaerobically.

19.3.2.3 Gelatin hydrolysis

Inoculate gelatin agar deep by using a needle to stab the gelatin medium with a BHI broth culture in duplicate. Incubate the tubes at 32°C for 24 and 48 h. After incubation, place the tubes of gelatin agar into the refrigerator and allow them to remain for 15 min. If gelatin agar is not liquid after 24 h incubation, incubate for an additional 24 h and analyze again by placing into a refrigerator for 15 min Gelatinase activity is indicated when the medium remains liquid after refrigeration due to the gelatinase production of bacterium to hydrolyze gelatin. *B. cereus* hydrolyzes gelatin. When the medium is not liquefied and retains a solid form, the test is negative.

19.3.2.4 Rhizoid growth

Allow nutrient agar plates without inverting to dry at room temperature for 24 h. Streak plate one loopful of the BHI broth culture onto dried nutrient agar by gently touching onto the center of plate and streak to the small area. Incubate inverted plates at 30°C for 48–72 h. After incubation, examine plates for rhizoid growth by extending long hair or root-like colonies that extend several centimeters from the inoculation point of culture. *B. cereus* strains cannot show rhizoid growth and produce rough galaxy-shaped colonies and *B. mycoides* produces typical rhizoid growth.

19.3.2.5 Tyrosine decomposition

Inoculate a tyrosine agar slant surface with one loopful of BHI broth culture and incubate tubes at 35°C for up to 3 days. Medium clearing near growth indicates the decomposition of tyrosine. This clearing progresses to 3 or 4 mm depth after 3 days of incubation. *Bacillus* spp. (except *B. anthracis*) readily decompose tyrosine. Incubate a tyrosine negative slant for a total of 7 days before considering it as negative.

19.3.2.6 Lysozyme resistance

Inoculate nutrient broth containing 0.001% lysozyme with a loopful of BHI broth culture in duplicate and incubate tubes at 35°C for 48 h. Growth in the broth indicates the resistance of bacterium to lysozyme enzyme. *B. cereus* and the other members of *Bacillus* are resistant to lysozyme enzyme.

19.3.2.7 Toxin crystals detection

Inoculate a nutrient agar slant using a sterile needle from BHI broth culture. Incubate slants at 30°C for 24 h, and then hold at room temperature for 2 or 3 days to permit sporangiolysis. Prepare a smear on a microscope glass slide. Place one drop of water onto a slide and place one loopful of

BHI agar slant culture near the water. Homogenize the colonies in water. Air dry the slide, and lightly heat-fix the smear by passing it one or two times through a Bunsen burner flame. Further fix the smear by flooding the slide with methanol for 30 s, pour off the methanol and air dry the slide. Stain the smear by flooding the slide with basic fuchsine (0.5%) and gently heat the slide until steam appears. After 1 or 2 min, heat the slide again until steam appears again, hold the stain for 30 s. Pour additional stain to prevent drying and cool the slide, rinse thoroughly in tap water, and dry without blotting. Examine the stained smear using oil-immersion objective of a microscope for the free spore presence and darkly stained tetragonal toxin crystals. Free toxin crystals usually appear within 3 days of incubation on BHI agar but they are detected after staining with the lysis of cells. *B. thuringiensis* produces endotoxin crystals. *B. cereus* cannot produce crystals.

19.3.2.8 VP test

Inoculate one loopful of BHI broth culture to modified Mr-VP broth (5 mL) and incubate the tubes at 35°C for 48 h. Test for the production of acetyl-methyl-carbinol by pipetting 1 mL of Mr-VP culture into a sterile test tube and add 0.6 mL of alpha-naphthol solution and 0.2 mL KOH (40%). Shake the tube and add a few creatine crystals. Observe results after holding for 1 h at room temperature. Pink or violet color formation indicates the test is positive. *B. cereus* is positive for VP test.

19.3.2.9 Test for psychrotolerant Bacillus *spp*

Streak one loopful of BHI broth culture to two TS agar slants in duplicate. Incubate one of the slants at 6°C for 28 days and the other slant at 43°C for 4 days. Psychrotrophic *B. weihenstephanensis* grows at 6°C but cannot grow at 43°C.

19.3.2.10 Lipid globule formation

Colonies from KG agar are used directly for this test. Prepare the smear from the center of 24 h colonies or from the edge of 2-day-old colonies from an MYP agar plate. Place one drop of sterile distilled water onto a microscope glass slide and place one loopful of culture near to the water drop. Homogenize the culture on the slide. Air dry the smear and heat fix with minimal flaming by passing one or three times through the Bunsen burner flame. Allow the slide to cool for 30 s. Place the slide on a U glass rod and place the glass rod over a beaker of boiling water and immediately flood with malachite green for 2 min. Drying of the slide should be prevented with the addition of dye during staining. After 2 min, cool and wash the slide with distilled water and blot dry. Add Sudan black B (0.3% in 70% ethanol) and allow it to stand for 20 min. Pour off the stain from the slide, wash with distilled water, and wash with xylene for 5–10 s. Rinse the slide with distilled water immediately. Pour on counter stain safranin (0.5% aq.) for 20 s, wash slide with distilled water, and blot dry the slide. Examine the stained slide using a microscope for the presence of lipid globules that appear black within the cytoplasm. Stained spores are usually pale to mid-green, central or paracentral in position, and don't swell the cell. The presence of intracellular lipid globules and typical spores for *B. cereus* is easily indicated from day old culture on KG agar.

19.3.2.11 Nitrate reduction

Inoculate one loopful of BHI broth culture to nitrate broth in tubes in duplicate and incubate tubes at 35°C for 24 h. After incubation, add 0.25 mL of nitrate reagent-A (sulfanilic acid solution) and

0.25 mL of nitrate reagent-B (alpha-naphthol solution) into nitrate broth culture. There are three possible results in this test:

1. The red (violet) color formation indicates presence of nitrites with reduction of nitrate to nitrite by bacterium (Fig. 19.4A). This indicates that the bacterium is nitrate reduction positive. This test is based on the detection of nitrite in the medium after incubation with a bacterium.
2. If there is no red (violet) color formation in medium after adding reagents, this only indicates the absence of nitrite in the medium. There may be two explanations for this observation.
   a. The nitrate may not be reduced; the bacterium is nitrate reduction negative; or
   b. The nitrate may be reduced to nitrite which has then been completely reduced to nitric oxide, nitrous oxide, or nitrogen; the bacterium is nitrate reduction positive.

If the medium does not change color after the addition of reagents (Fig. 19.4B), a small amount ("knife point") of zinc dust is added to the medium. The zinc dust will catalyze the reduction of nitrate to nitrite chemically. Thus, if the nitrate is not reduced by the bacterium (nitrate-negative), nitrate will be reduced to nitrite by the zinc dust and a red color will develop in the incubated medium within 15 min. If no color develops in the incubated medium after the addition of reagents and zinc dust, the bacterium has not only reduced nitrate to nitrite, but has reduced nitrite to nitrogenous gases; the bacterium is also nitrate reduction positive. Nitrogen gas is observed as bubbles in the Durham tube (Fig. 19.4C).

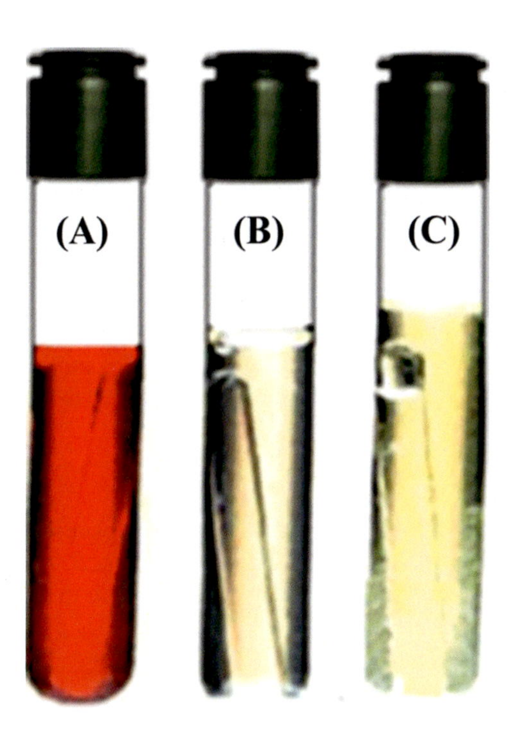

FIGURE 19.4

Nitrate reduction: (A) nitrate reduction (red color) to nitrite; (B) no nitrate reduction (no color change); (C) nitrate reduction to nitrite (no color change) and nitrite to nitrogen gas (space in Durham tube).

19.3.2.12 Sensitivities to gamma-bacteriophage

Inoculate BHI broth with BHI broth culture and incubate tubes at 35°C for about 4 h (up to the appearance of turbidity). Inoculate the BHI agar plate by swabbing BHI broth culture and allow the plate to stand to absorb liquid from swabbing. Label the plates for each sample and draw a circle 0.5 cm in diameter. Spot inoculate a loopful (3 mm platinum loop) of diluted gamma-bacteriophage culture (10^{-6} phage; plaque) onto the circle in the BHI agar (refer to Technique 29.3.4 for the preparation of bacteriophage culture). Allow the plate to stand for a time period for the medium to absorb liquid from the phage culture. Incubate the inverted plates at 35°C for 16 h. Examine plates during the incubation period for the formation of plaques (clear zone). *B. cereus* is resistant to gamma-bacteriophages and plaque cannot form around bacteriophage inoculum on the agar medium.

19.3.2.13 Sensitivities to penicillin

Inoculate BHI broth with BHI broth culture and incubate tubes at 35°C for about 4 h (up to appearance of turbidity). Inoculate the BHI agar plate by swabbing the BHI broth culture and allow the plate to stand to absorb liquid from the swabbing. Label the plates for each sample. Place the penicillin antibiotic disk on the labeled area. Incubate the inverted plates at 35°C for 16 h. Examine plates for the formation of a clear zone around disks. The formation of a zone around a disk indicates the sensitivity of the bacterium to antibiotic and nonzone formation around the disk indicates the resistance of the bacterium to antibiotic. *B. cereus* is resistant to penicillin.

19.3.2.14 Starch hydrolysis

Bacteria can hydrolyze starch (amylose and amylopectin) to simple sugars with the production of extracellular α-amylase and oligo-1,6-glucosidase enzymes. Streak plate one loopful of BHI broth culture on starch agar and incubate inverted plates at 35°C for 24 h. After incubation, examine plates for colony formation. Pour Lugol's iodine solution onto colonies. Iodine changes the color of a colony from yellow-brown to blue-black in the presence of starch within 10 min. The appearance of a blue-black color around colonies indicates starch is not hydrolyzed and starch is present around the colonies. The appearance of a clear zone around the colonies indicates the hydrolysis of starch and the test is positive. *B. cereus* produces amylase enzymes to hydrolyze the starch in the medium.

19.3.2.15 Egg-yolk reaction

Streak plate PEMB agar from a BHI broth culture and incubate plates at 35°C for about 24 h. After incubation, examine the medium for hydrolysis around colonies. *B. cereus* grows on PEMB agar plates and produces an opaque zone of egg yolk precipitation due to lecithin hydrolysis. The zone around colonies indicates the egg-yolk test is positive. *B. cereus* is positive for the egg-yolk reaction test.

19.3.2.16 Other tests

Catalase test. Refer to Technique 13.4.3.10 for details of the catalase test. *B. cereus* is positive for catalase.

For further phenotypic identification refer to Table 19.1.

Table 19.1 Phenotypic differentiation of *Bacillus* species.

Characteristic	*B. cereus*	*B. mycoides*	*B. thuringiensis*	*B. anthracis*	*B. megaterium*	*B. weihenstephan*	*B. sphaericus*
Fermentation (with acid production) of:							
Glucose/mannitol	+/−	+/−	+/−	+/−	+/−	+/−	+/−
Starch	+	+	+	+	+	+	+
Catalase/VP	+/+	+/+	+/+	+/+	+/−	+/+	+/+
Egg yolk reaction	v	v	v	v		+	
Motility/Capsule	+/−	v/−	+/−	−/+	v/−	+/−	+/−
Rhizoid growth	−	+	−	−	−	−	−
Crystal formation	−	−	+	−	−	−	−
Nitrate reduction	v	+	+	+	+	+	v
Tyrosine	+	(+)	+	(+)	v	+	
β-Hemolysis	+	−	+	−	−		−
Lysozyme/Lecithinase	+/+	+/+	+/+	+/+	−/−	+/−	+/−
Penicillin (10 IU)	R	R	R	v	v		R
Pathogenicity	Ent.	RG	EC	AH			

+, *Positive;* −, *negative; (+), weakly positive; Ent., enterotoxin; RG, rhizoid growth; EC, endotoxin crystal; AH, pathogenic; R, resistant; v, variable.*

19.3.3 Identification of *Bacillus cereus* species by foodomics techniques

19.3.3.1 MALDI-TOF Ms identification of Bacillus cereus

B. cereus can be rapidly identified with the matrix-mediated laser desorption ionization-time-of-flight-mass spectrometry (MALDI-TOF Ms) technique, which is based on the principle of ionizing a specific protein profile of microbial cells. *B. cereus* can be easily identified by comparing these profiles with the reference spectra. MALDI-TOF Ms identification has been explained in Technique 13.6.3.1.

19.3.3.2 Molecular identification of Bacillus cereus

Polymerase chain reaction (PCR) is a molecular genotypic technique and is used in the identification of *B. cereus* species. Gene coding of specific *B. cereus* characteristic can be detected by PCR assay. Some *B. cereus*-specific primer sets that can be used in the identification of *B. cereus* are given in "Appendix A: Gene Primers (Table A.1)." Using *B. cereus*-specific techniques and materials, the bacterium can be identified by PCR analysis, as described in Technique 13.6.3.2.

19.4 Interpreting results

Isolated *B. cereus* can be identified by the following results: (1) the appearance of Gram-positive rods and spore within cell without swelling of the cell; (2) the production of lecithinase enzyme; (3) without mannitol fermentation on MYP agar; (4) acid production anaerobically from glucose; (5) positive VP test; (6) decomposition of L-tyrosine; and (7) growth in the presence of 0.001% lysozyme. Some of the characteristics are shared with other *Bacillus* members including the rhizoid growth by *B. cereus* var. *mycoides*, the crystal formation by insect pathogen *B. thuringiensis* and the mammalian pathogen *B. anthracis*. The *B. cereus* group is actively motile and strongly hemolytic, it does not produce rhizoid colonies or protein toxin crystals. The presence of nonmotile and weak hemolytic *B. cereus* strains are rare, and these nonpathogenic *B. cereus* strains can be differentiated from *B. anthracis* by their resistance to gamma bacteriophages and penicillin. Nonmotile and nonhemolytic isolates can be expected to be *B. anthracis*.

PRACTICE

20 Isolation and counting of *Clostridium perfringens*

20.1 Introduction

Clostridium perfringens is Gram-positive, produces heat-resistant endospores, has an anaerobic and fermentative metabolism, and cannot reduce sulfate to sulfide. Its endospore has two characteristics: spores causing a distinct swelling on the cell; and not swelling the cell with small spore formation. *C. perfringens* causes a toxicoinfection type of food poisoning. There are five strains of *C. perfringens*: A, B, C, D, and E. *C. perfringens* produces four types of exotoxins: alpha (α), beta (β), epsilon (ε), and iota (ι). Strain C produces alpha and beta toxins, strain D alpha and epsilon toxin, and strain E alpha and iota toxins. Strain A can cause food poisoning and gas gangrene, produces only α toxin and has lecithinase activity. Strain C causes gastroenteritis; beta-toxin damages the intestinal mucosal membrane and causes necrosis. Strain A produces a chromosomal enterotoxin, while *C. perfringens*' nonfood-borne gastrointestinal (GI) diseases are caused by the toxin of a plasmid-borne gene. *C. perfringens* endospores are extremely resistant to environmental stresses, such as heat, radiation, and toxic chemicals. Endospores are significant agents of food spoilage and food-borne GI diseases. The infective dose is about 10^6-10^8 cells per g or mL of food. The symptoms are caused by heat label protein enterotoxin formation in the gut during sporulation and lysis of a large number of ingested cells. The symptoms appear within 8–24 h following the consumption of food containing the viable cells. The disease lasts 1–2 days. The principal symptoms are diarrhea, intense abdominal pain, gas, less commonly nausea, vomiting and fever. Fatality is rare, although fatality can occur among the very young, elderly and sick people. *C. perfringens* can rapidly grow to high numbers in food products during heating, cooling, and/or rewarming. *C. perfringens* is widespread in the environment (e.g., soil). The vehicles of *C. perfringens* are cooked large meat dishes, cured meats, poultry products, dried foods, herbs, spices, and vegetables. *C. perfringens* foodborne disease may occur when foods are inadequately cooked or inadequately refrigerated before serving. Some *C. perfringens*' strains endospores are resistant to heat treatment up to 100°C for more than 1 h. *C. perfringens* cells may lose viability when foods are frozen or held at prolonged refrigeration temperature. When the food samples are not examined immediately, they can be treated with buffered glycerin-salt solution and stored at low temperature.

A low number of *C. perfringens* can be isolated from sample by an enrichment technique. *C. perfringens* can be counted from samples using plate count and most probable number (MPN) techniques. *C. perfringens* spores can be counted from a heated sample and then pour plating with a suitable agar medium. Selective/differential media commonly used in the isolation and counting of *C. perfringens* are tryptose sulfide iron citrate cycloserine agar and perfringens agar. *C. perfringens* can be identified by phenotypic and foodomics techniques.

Microbiological Analysis of Foods and Food Processing Environments. DOI: https://doi.org/10.1016/B978-0-323-91651-6.00022-7

20.2 Isolation and counting techniques

Sampling, sample preparation, and all experimental techniques should be performed under aseptic conditions. Sampling and all analysis should be performed in duplicate, unless otherwise specified. All microbiological media to be used in this application must be sterilized. Again, all solutions and equipment that may cause contamination of microorganisms must be sterilized or disinfected.

20.2.1 Equipment, reagents and media

Equipment. Anaerobic incubator (or anaerobic jar), blender (or stomacher), container, cover glass, Erlenmeyer flask, microscope, microscope glass slide, Petri dishes (100 × 15 mm), pipette discard equipment (e.g., a graduated cylinder) containing 10% bleach solution and a piece of cloth at bottom, pipettes (1 mL) in the pipette box, and water bath (at 46°C).

Reagents. Buffered glycerin salt (BGS) solution, Gram stains, nitrate reagents, 0.1% peptone water, polyvalent-gas gangrene antitoxin, and spore stains.

Media. Buffered motility nitrate (BMN) broth, cooked meat (CM) medium, fluid thioglycolate (FT) medium (containing 0.4 mg D-cycloserine), lactose egg-yolk milk (LEYM) agar, lactose gelatin (LG) medium, liver broth, modified AE sporulation (MAE) broth, modified Duncan-Strong sporulation broth, modified iron milk (MIM) broth, neomycin blood agar, reinforced clostridial (RC) broth, Spray's fermentation (SF) broth containing carbohydrate, SF broth containing inositol, trypticase peptone glucose yeast extract (TPGY) broth, tyrosine sulfite cycloserine (TSC) agar, and TSC agar without egg-yolk emulsion (EY-free TSC) agar.

20.2.2 Sampling and sample transport

Representative 25 g samples from different food parts are removed. Samples must be transported and analyzed promptly without freezing; if possible, they should be stored at about 10°C until analysis. If analysis cannot be performed within 8 h or if the sample needs shipment to the laboratory, it is treated with sterile BGS solution and stored immediately at −57°C to −68°C. If possible, use a treated sample as soon as available without freezing. If the sample is transported to the laboratory in a pack containing CO_2 gas, the contact of the gas with foods should be prevented, because absorption of CO_2 by the sample can decrease the pH of sample, which this inhibits the viability of *C. perfringens.*

Treatment of sample in BGS solution. Aseptically, weigh 25 g or mL of sample (sliced beef, turkey, etc.) to Erlenmeyer flask (or stomacher) and add 25 mL of BGS solution. Mix the sample well by gently shaking. Store homogenized samples at −57°C to −68°C in freezer. When it is used in analysis, thaw the sample at room temperature and transfer it into a sterile blender.

C. perfringens selective enrichment isolation and counting flowchart is given in Fig. 20.1.

20.2.3 Selective enrichment isolation of *Clostridium perfringens*

Weigh 25 g of sample into a blender (or stomacher plastic bag) containing 225 mL of 0.1% peptone water. Homogenize by blending (at 8000 to 10,000 rpm) for 1 min. In the case of a liquid

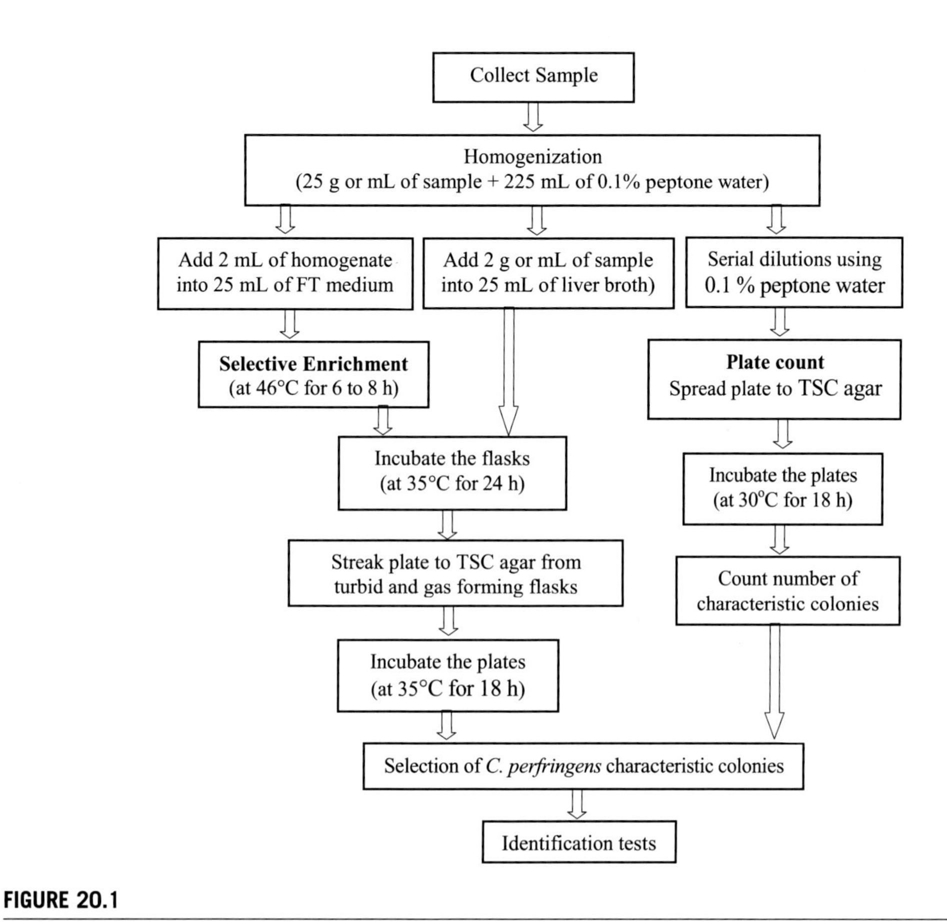

FIGURE 20.1

Clostridium perfringens selective enrichment isolation and counting flowchart.

sample, add 25 mL of the sample to 225 mL of FT medium in the Erlenmeyer flask and slowly mix to homogenize the sample. Add 2 mL of homogenized sample into 25 mL of FT medium (or CM broth) in screw-capped tubes. Incubate tubes in a water bath at 46°C for 6–8 h. (In another enrichment technique, homogenize 2.0 g or mL of sample in 20 mL of liver broth.) Incubate FT medium (or liver broth) at 35°C for 24 h. After incubation, indicate growth (turbidity) and gas production. Record positive tubes. Streak plate one loopful of enriched culture to TSC agar and incubate plates without inverting at 35°C for 20 h in an anaerobic incubator. After incubation, examine plates for characteristic colony formation by placing the plates over a white tissue paper. *C. perfringens* colonies in egg-yolk medium are a black color (due to the reduction of sulfide precipitation from iron sulfide) with a 2–4 mm opaque white zone surrounding the colony as a result of lecithinase activity. Identify presumptive isolates using phenotypic and genotypic techniques (Technique 20.3).

20.2.4 Counting of *Clostridium perfringens*

20.2.4.1 Plate counting

Pour 5–6 mL TSC agar without egg yolk into each sterile Petri dish and spread evenly on the bottom by rapidly rotating the dish and allowing the medium to solidify. Place 25 g of the sample into the blender (or stomacher plastic bag) containing 225 mL of 0.1% peptone water and homogenize by blending for 1 min at low speed. In the case of a liquid sample, place 25 mL of sample into an Erlenmeyer flask containing 225 mL of 0.1% peptone water and mix well by gently shaking the flask. For fecal specimens, homogenize 1 g (or 1 mL liquefied stool) of the sample in 9 mL of 0.1% peptone water and vortex to mix well. Obtain a uniform homogenate with as little aeration as possible. Prepare further serial dilutions using 0.1% peptone water from homogenate. Aseptically, transfer 1 mL of diluted sample onto the center of a sterile Petri plate (containing 5–6 mL of TSC agar) in duplicate, pour additional 12–13 mL TSC agar without egg yolk, and mix sample evenly with an agar medium by rapidly rotating the plate (drawing a figure of eight) on a table top. After the agar has solidified, place the plates without inverting in an anaerobic incubator (or anaerobic jar). Establish anaerobic conditions and incubate the inverted plates at 35°C for 20 h. After incubation, examine the plates for characteristic colony formation. Count characteristic *C. perfringens* colonies from plates containing 25–250 black colonies by placing plates over a white tissue paper. *C. perfringens* colonies on TSC agar containing egg-yolk are a black color with a 2–4 mm opaque white zone as a result of lecithinase activity. Record results as the number of *C. perfringens* colony forming units (cfu) per g or mL of food. Identify presumptive isolates using phenotypic and foodomics techniques (Technique 20.3).

20.2.4.2 Plate counting of Clostridium perfringens *spores*

Pour 5–6 mL of TSC agar without egg yolk into each of sterile Petri dishes and spread evenly on bottom by rapidly rotating dish and allow to solidify. Heat homogenate in a thermostatic shaking water bath at 75°C for 20 min to inactivate vegetative cells and activate spores. After heating, prepare serial decimal dilutions using 0.1% peptone water. Add 1 mL of heated diluted sample into the Petri plate (containing 5–6 mL of TSC agar) in duplicate and pour 12–13 mL of melted TSC agar with egg-yolk (at 47°C). Mix inoculum with medium by gently rotating the plate (drawing a figure of eight) on the table top. After solidifying the agar, incubate the plates without inverting in an anaerobic incubator at 35°C for 18 h.

If neomycin blood agar plates are used (12–13 mL), spread plate the sample on predried agar medium and overlay the plates with a further 5–6 mL of melted neomycin blood agar (at 47°C). After solidifying the agar, incubate the plates without inverting in an anaerobic incubator at 35°C for 18 h.

After incubation, examine the plates for characteristic colony formation on agar medium. On neomycin blood agar, *C. perfringens* produces a narrow zone of complete hemolysis (due to the beta toxin) around colonies or surrounding narrow zone of incomplete hemolysis (due to the alfa-toxin). *C. perfringens* colonies on TSC agar containing egg-yolk are black color with a 2–4 mm opaque white zone as a result of lecithinase activity. Count plates containing 25–250 black colonies. Calculate and record the number of *C. perfringens* cfu per g or mL of sample. Identify presumptive isolates using phenotypic techniques (Technique 20.3).

20.2.4.3 Most probable number counting

MPN technique is suitable for samples containing a low number of *C. perfringens*. Perform the MPN technique as described in Technique 4.2, but use RC broth and homogenize the sample in RC broth. Homogenize the sample in RC broth as indicated in Technique 20.2.4.1 and heat homogenate at 75°C for 20 min. This will destroy vegetative cells but will activate and encourage the germination of spores. After heating the sample, prepare further dilutions from heated homogenate using 0.1% peptone water. Inoculate each heated diluted (or nondiluted) homogenate into 3-MPN tubes of RC broth and incubate the tubes in an anaerobic incubator at 35°C for up to 7 days. Tubes showing blackening should be confirmed for *C. perfringens*, as described in Technique 20.3. Indicate the number of positive tubes and refer to MPN Table 4.1. Calculate the number of *C. perfringens* and record the results as the number of *C. perfringens* MPN per g or mL of sample. For further identifications (Technique 20.3), streak plate one loopful of the culture from each presumptive positive tube to TSC agar and incubate in an anaerobic incubator at 35°C for 24 h.

20.3 Identification of *Clostridium perfringens*

Select at least five representative black colonies from isolates where the sample is analyzed and the presumptive result is obtained. Inoculate a portion of each selected black colony into a tube of TPGY broth. Incubate in a water bath at 46°C for 24 h or at 35°C for 22 h anaerobically. After incubation, analyze the tubes for growth (turbidity) and color changes. Endospores usually are not produced in this medium. Streak plate one loopful of TPGY broth culture onto a TSC agar plate and incubate homogenates anaerobically at 35°C for 24 h. If colonies are not pure, perform another inoculation from single colony into TPGY broth and then streak plate to TSC agar until obtaining pure culture. Then, carry out inoculation from pure single colony (pure culture) into TPGY broth, FT medium, and CM medium. Incubate in an anaerobic incubator at 35°C for 24 h and use in identification tests. The *C. perfringens* identification flowchart is given in Fig. 20.2.

20.3.1 Morphological identification techniques

Inoculate MAE sporulating broth with 1 mL of FT culture and incubate tube at 35°C for 24 h. After incubation, prepare Gram stain and spore stain from MAE sporulating broth culture.

Gram staining. Prepare Gram stain from MAE sporulating broth culture as explained in Appendix B under the topic "Gram Staining Reagents." A Gram stain of *C. perfringens* will produce Gram-positive rods, occurring singly or in pairs (Fig. 20.3).

Spore staining. Perform spore stain from MAE sporulating broth culture as explained in Appendix B under the topic "Spore Stain." *C. perfringens* should show spores within the swelling vegetative cell. This test may be performed directly from colonies on TSC agar culture.

Motility nitrate reduction test. Inoculate one loopful of TPGY broth culture into BMN broth by inserting the loop into a medium. After inoculation, rinse the loop in warm water in the beaker before flaming to avoid splattering. Incubate the tubes in an anaerobic incubator at 35°C for 24 h. Examine BMN medium for growth in the medium. Nonmotile bacterium can show growth along

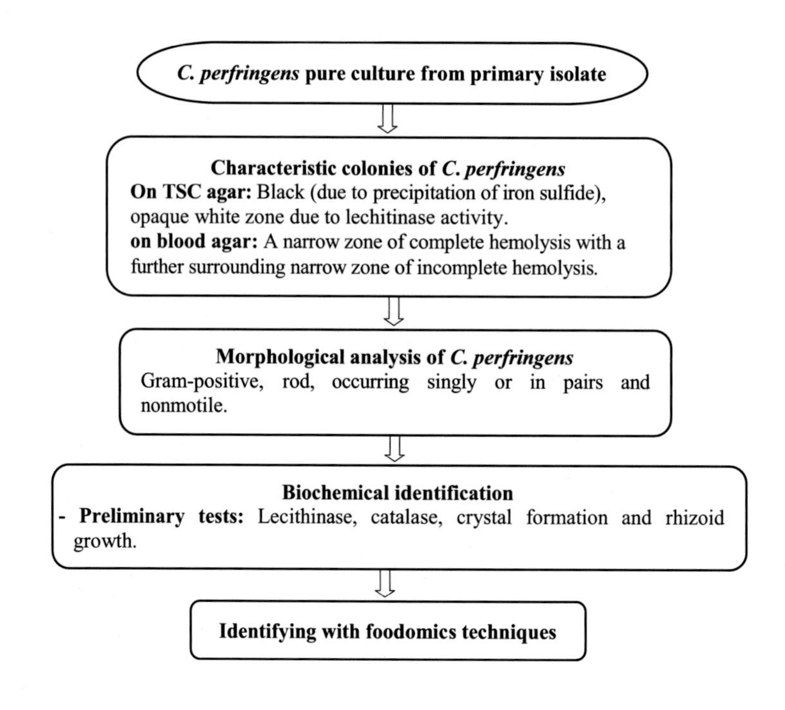

FIGURE 20.2

Clostridium perfringens identification flowchart.

inoculation line. Motile bacteria can show diffuse growth throughout the medium, away from the stab inoculation line. *C. perfringens* is nonmotile.

In the detection of nitrate reduction, add 0.5 mL of nitrate reagent A and 0.2 mL of nitrate reagent B to BMN broth culture. The development of a violet color within 5 min indicates the presence of nitrites and the reduction of nitrate to nitrite. If no color develops, add a few grains of powdered zinc metal and let it stand for a few minutes. A negative test (no violet color) after zinc dust is added indicates that nitrates are completely reduced. A positive test (violet color) after the addition of zinc dust indicates that the bacterium is incapable of reducing nitrates. *C. perfringens* reduces nitrates to nitrites.

20.3.2 Biochemical identification

20.3.2.1 Gelatin hydrolysis

Inoculate a loopful FT culture into an LG medium and incubate anaerobically at 35°C for 24—44 h. After the incubation of tubes for 1 h at 5°C and examine tubes for liquefaction of gelatin. If the medium is not liquefied after chilling and retains gel formation, incubate the medium for an additional 24 h at 35°C and analyze again for gelatin liquefaction. Lactose fermentation is indicated by gas bubbles through medium and acid production by color change from red to yellow. *C. perfringens* liquefies gelatin and ferments lactose within 48 h.

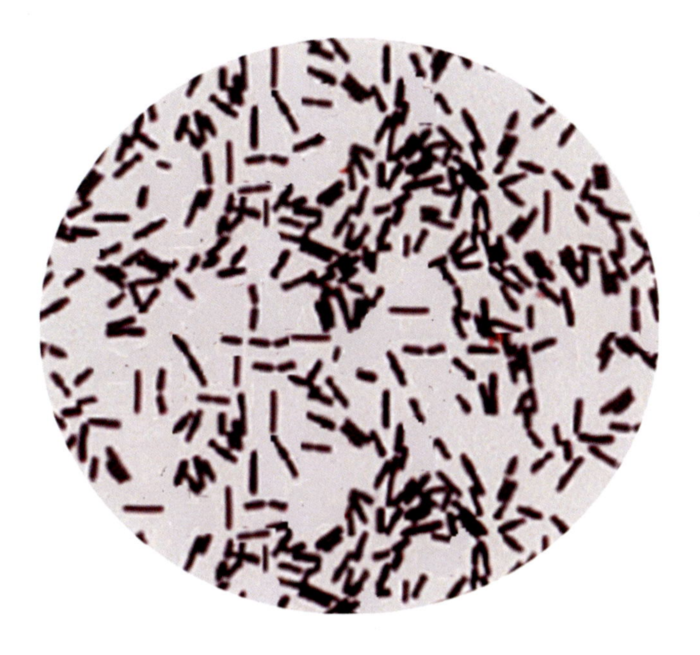

FIGURE 20.3

Staining appearance of *Clostridium perfringens*.

20.3.2.2 Carbohydrate fermentation

Inoculate one loopful of FT culture into a tube of SFS broth containing salicin (1%) and SFR broth containing raffinosae (1%). Incubate the tubes anaerobically at 35°C for 24–72 h. A yellow color indicates acid production and space in the Durham tube indicates gas production. *C. perfringens* ferments salicin and raffinose within 3 days of incubation with acid and gas production.

20.3.2.3 Nagler reaction

Dry the LEYM agar surface by holding the plates at 30°C for 15–30 min with slightly opening the lid near Bunsen burner flame. Divide the plates into two equal parts at the bottom. Spread plate 3–5 drops of *C. perfringens* antitoxin (polyvalent gas-gangrene antitoxin) over one part of the plate and allow the antitoxin to dry. Streak plate one loopful of FT culture over each part from one side to the other separately. Incubate plates at 35°C under anaerobic conditions. Analyze plates for growth from 1 to 3 days. *C. perfringens* produces cream white colonies and a cloudy zone with or without pearl brilliance on the nonantitoxin part of plate; this cannot appear (no cream white colonies) on the antitoxin containing side (Fig. 20.4). Opaque colonies indicate a positive Nagler reaction. A white color shows the production of antitoxin (lechitinase C). The alpha-toxin is lecithinase which hydrolyzes the phospholipid lecithin (a component of cell membrane). *C. perfringens* produces lecithinase, and it will produce the opacity on the egg yolk medium. When the plates are exposed to air by opening the lid, the cream white color turn to pink-red. This color change occurs due to the fermentation of lactose. Therefore, this change differentiates lactose-fermenting Nagler reaction positive *Clostridium* spp. from lactose-nonfermenting Nagler reaction negative species.

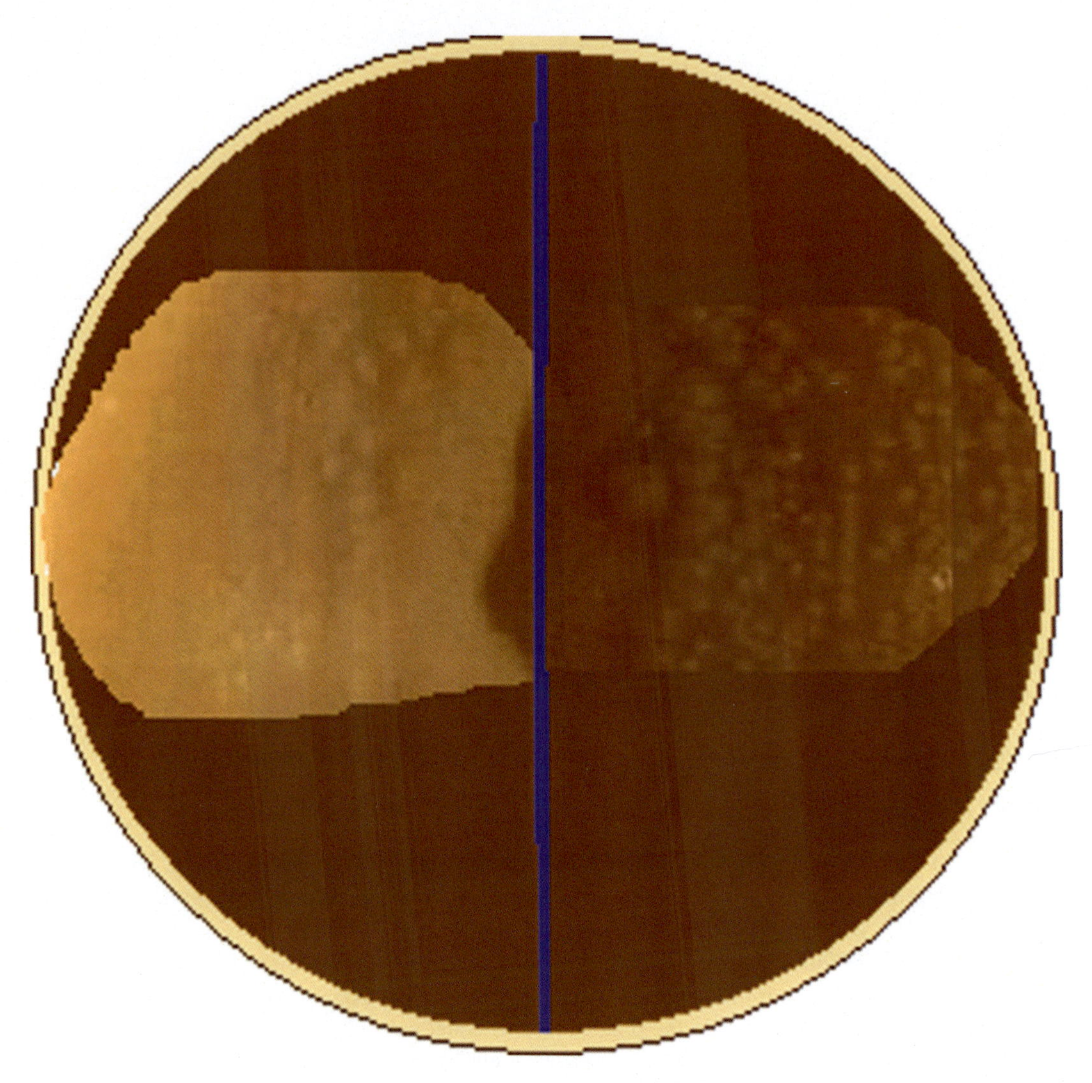

FIGURE 20.4

Nagler's reaction: cream white (opaque) colony on nonantitoxin part (left side) indicates lechitinase (alpha toxin) production and no opaque colony (right side) indicates lechitinase (alpha toxin) inhibition by antitoxin.

20.3.2.4 Stormy fermentation

Use actively growing bacterial culture in this test. Inoculate 1 mL of 24 h FT culture into MIM broth and incubate the medium in a water bath at 46°C. After 2 h incubation, "stormy fermentation" is checked hourly. This fermentation is characterized by rapid milk coagulation with separation of curd into spongy mass and usually raising to surface of medium. Tubes showing stormy fermentation are removed to prevent spilling into the water bath. Short tubes should not be used in the test to prevent the contents spelling out of the tube. *C. perfringens* shows "stormy fermentation" within 6 h. Some strains of *C. baratii* can also give positive result, but *C. perfringens* can be differentiated by its ability to liquefy gelatin.

20.3.2.5 Sporulation

Inoculate one loopful FT culture into freshly reaerated and cooled CM medium. Incubate tubes at 35°C for 24 h and follow an additional incubation at 4 h in a refrigerator. After incubation, mix the CM medium culture by vortexing and inoculate 0.5 mL of this culture into two freshly steamed FT medium in tubes. First heat the tube in a water bath at 75°C for 10 min and after heating, incubate both tubes at 35°C for 18 h. After incubation, the growth in the heated tube indicates spore formation. Incubate this tube at 35°C for 4 h. After incubation add 0.75 mL of FT culture into 15 mL of MAE sporulation broth. Incubate broth anaerobically at 35°C for 24 h. After incubation, prepare spore stain from MAE culture and examine the formation of spores by using a microscope. If fewer than five spores are observed, the process is not considered good sporulation.

20.3.2.6 Anaerobic test

Streak plate one loopful of CM medium culture to each of two PYE broths. Incubate one of the PYE broths aerobically and the other anaerobically at 35°C for 24 h. *C. perfringens* cannot grow on aerobically incubated PYE broth but can grow on anaerobically incubated PYE broth. This test can also be performed using liver broth.

20.3.2.7 Other tests

Urease test. Refer to Technique 13.4.3.5 for details of the urease test. *C. perfringens* is urease negative.

Lipase test. Refer to Technique 18.3.2.4 for details of the lipase test. *C. perfringens* produces lipase enzyme.

Indole test. Refer to Technique 13.4.3.2 for details of the indole test. *C. perfringens* is indole negative.

Esculin hydrolysis. Refer to Technique 14.3.2.1 for details of the bile esculin hydrolyze test. *C. perfringens* is variable in bile esculin hydrolyzes.

For further phenotypic identification refer to Table 20.1.

20.3.3 Identification of *Clostridium perfringens* species by foodomics techniques

20.3.3.1 MALDI-TOF Ms identification of Clostridium perfringens

C. perfringens can be rapidly identified with the matrix-mediated laser desorption ionization-time-of-flight-mass spectrometry (MALDI-TOF Ms) technique, which is based on the principle of ionizing specific protein profile of microbial cells *C. perfringens* can be easily identified by comparing these profiles with the reference spectrum. MALDI-TOF Ms identification has been explained in Technique 13.6.3.1.

Table 20.1 Phenotypic characteristics of *Clostridium* species.

Characteristics	*C. bifermentans*	*C. botulinum ABF*[P]	*C. botulinum BEF*	*C. botulinum CD*	*C. botulinum G*	*C. difficile*	*C. histolyticum*	*C. novyi*	*C. perfringens*	*C. ramosum*	*C. septicum*	*C. sordellii*	*C. sporogenes*	*C. tetani*	*C. absonum*	*C. baritii*	*C. celatum*	*C. paraperfringens*
Fermentation of:																		
Glucose/lactose	+/−	+/−	+/−	+/−	−/−	+/−	−/−	+/−	+	+/+	+/+	+/−	+/−	−/−				
Sucrose/ mannose	−/−	−/−	+/+	−/v	−/−	−/v	−/−	−/−	+	+/+	−/−	−/−	−/−	−/−				
Salicin/ raffinose									AG/A						AG/ −	AG/ −	A/ −	AG/ −
Motility									−						+	−	−	−
Gelatin/esculin	+/v	+/+	+/−	+/−	+/−	+/+	+/−	+/−	+/v	−/+	+/+	+/v	+/	+/−		−	−	−
Milk digestion	+	+	−	v	+	−	+	−	+	−	v	+	+	v				
Lecithinase/ indole	+/+	−/−	−/−	v/v	−/−	−/−	−/−	+/−	+/−	−/−	−/−	+/+	−/−	−/v				
Lipase/urease	−/−	+/−	−/−	−/−	−/−	−/−	−/−	+/−	−/−	−/−	−/−	+/+	+/v	−				
H_2S/nitrate	+/+	+/+	+/+	+/+	+/+	+/+	v/ +	+/+	+/+	v/ +	+/+	+/+	+/+	+/+	v/ +	+/+	+/+	+/+
Nagler's reaction																		
Opaque zone	+	+	+	+	+			+	+			+						
Pearl bright	−	+	+	+	−			+	−			−						
Antitoxin	+	−	−	−	−			−	+			+						

+, Positive; −, negative; v, variable; P, proteolytic; AG, acid and gas.

20.3.3.2 Molecular identification of Clostridium perfringens

Polymerase chain reaction (PCR) is a molecular genotypic technique and is used in the identification of *C. perfringens* species. Gene coding of specific *C. perfringens* virulence factor or the other characteristic can be detected by PCR assay. Some *C. perfringens*-specific primer sets that can be used in the identification of *C. perfringens* are given in "Appendix A: Gene Primers (Table A.1)." Using *C. perfringens*-specific techniques and materials, the bacterium can be identified by PCR analysis, as described in Technique 13.6.3.2.

20.4 Interpretation of results

Culture obtained from presumptive *C. perfringens* with black colonies on selective differential agar are confirmed as *C. perfringens* if it is nonmotile, producing subterminal spores, reduces nitrate, Nagler's reaction positive, ferments lactose, liquefies gelatin within 44 h and produces acid from salicin and raffinose.

C. perfringens can cause a significant public health hazard to consumers of foods that are produced under improper processing conditions or improperly handling at any point before consumption. Factors involved with *C. perfringens* foodborne illness are the contamination of food with either spores or vegetative cells of enterotoxigenic strains of *C. perfringens*; suitable growth temperature, pH, media, oxidation reduction potential; and adequate storage time. With proper processing and handling of foods, the risk of *C. perfringens* foodborne illness outbreaks can be prevented. Both *C. perfringens* cells and spores are widely distributed in the environment such as soil, dust, and vegetation. Soil, as a large habitat, is a direct source of food contamination to foods and processing facilities. Bacterial spores can be detected particularly in foods such as spices, milk powders, and flours (from 10^3-10^4 spores per g). Food processing facilities (equipment, machines, and wash water) can also be the habitat of vegetative cells and spores of the bacterium. This can cause many problems in the food industries. Therefore, it is very important to reduce or inactivate *C. perfringens* cells and spores from contaminated food or to prevent contamination. Oxidizing agents as disinfectants in the washing step can control *C. perfringens* contamination and foodborne illness. Food safety strategies (oral, GMP, and HACCP) should be implemented as a food safety management system to prevent the contamination and disease of *C. perfringens* in foods.

PRACTICE

Isolation and counting of *Staphylococcus aureus*

21

21.1 Introduction

Staphylococcus aureus is nonmotile, Gram-positive, noncapsulated, coccus (spherical to ovoid), a member of Micrococcaceae, catalase positive, oxidase negative, facultative anaerobic, and ferments glucose and mannitol anaerobically and mesophilically. The natural habitats of *S. aureus* are the nose, throat, hair, skin, and the mucous membranes of healthy people. *S. aureus* associates with the nasal tract on 20%–50% of healthy people. The bacterium is also present in air, dust, sewage, and on the surface of food processing equipment. Food contamination generally occurs from these sources, especially from food handlers. *S. aureus* produces enterotoxins if the numbers are higher than 10^4 per g of food. Staphylococcal food poisoning results from the enterotoxin ingestion together with food that is produced by *S. aureus* toxic strains. *S. aureus* has the ability to grow in food at low available water (up to 0.86 water activity). Toxigenic *S. aureus* produces staphylococcal enterotoxins (SET) that are heat stable, i.e., not destroyed by cooking or the canning process. SET also resists dehydration and proteolytic enzymes. Most *S. aureus* strains produce enterotoxin. All *S. aureus* strains in phage group III and some types in phage group II are associated with food poisoning. *S. aureus* of phage group II produces exfoliative toxin and can also cause skin syndrome. *S. aureus* also produces toxic shock syndrome toxin-1. The ingestion of 100 ng of enterotoxins together with foods is sufficient to cause intoxication to consumers. The symptoms appear within 2–8 h after consumption of enterotoxin together with food. The disease lasts 2 days and is rarely fatal. The principal symptoms are salivation, nausea, vomiting, abdominal cramps, diarrhea, sweating, chills, headache, and dehydration. Principal food sources for *S. aureus* are much handled foods such as salads, dairy products especially if prepared from raw milk, custards, cold sweets, and cream-filled bakery products.

Enterotoxigenic *S. aureus* strains produce A, B, C_1, C_2, D, and E serologically distinct enterotoxins. Enterotoxins are heat stable and can withstand heating at 100°C for 30 min, and therefore viable bacterium absence in the food is not proof of safety. *S. aureus* is able to grow in pickled and cured foods but toxin production is inhibited by salt. In the correlation of enterotoxicity (food poisoning), the strains should be examined for colonial morphology on Baird-Parker (BP) agar, coagulase activity, thermostable deoxyribonuclease (TDNase) production, sensitivity to lysostaphin, and the anaerobic utilization of glucose and mannitol. Other properties of *S. aureus* (e.g., pigmentation, hemolysin formation, protein A production, and gelatin hydrolysis) can also be used in combination with the correlation with enterotoxicity testing. *S. aureus* can be lysed by bacteriophages and this permits the classification of *S. aureus* into several groups. The phage typing is a useful technique in epidemiological research.

The basic characteristics of *S. aureus* used in their isolation are (1) the ability to grow in the presence of the toxic chemicals (e.g., media containing 7.5% NaCl and potassium tellurite); (2) its

Microbiological Analysis of Foods and Food Processing Environments. DOI: https://doi.org/10.1016/B978-0-323-91651-6.00044-6

characteristic morphological appearance (grape like clusters); (3) the ability to ferment certain carbohydrates (e.g., mannitol); (4) the ability to produce certain enzymes (e.g., catalase, coagulase and lysostaphin); (5) resistance patterns to certain antibiotics (e.g., penicillin); (6) DNA composition; (7) the presence of certain components in the cell wall; (8) patterns of sensitivity to specific bacteriophages; and (9) production of β-hemolysin.

Two techniques most appropriate for the examination of foods in the isolation and counting of *S. aureus* are direct plating on BP agar and enriching in tellurite mannitol glycine (TMG) broth. Enriched culture is streak plated to milk salt (Ms) agar. Counts of low numbers of *S. aureus* are possible by the most probable number (MPN) technique. Isolated *S. aureus* can be identified by phenotypic and foodomics techniques.

BP agar has the following advantages in the isolation and enumeration of *S. aureus* from foods: (1) selectivity and differentiation; (2) no inhibition of injured *S. aureus*; and (3) easy of recognition of *S. aureus* colonies. *S. aureus* colonies on BP agar are 1.5 mm in size, black to gray colored, circular, smooth, convex with entire margins, moist, and surrounded by an opaque zone and/or clear halo (fibrin halo; coagulase positive) beyond the opaque zone (off-white) (Fig. 21.1).

Coagulase positive and mannitol-fermenting *S. aureus* can be directly isolated from the sample on Vogel-Johnson (VJ) agar. In this medium, *S. aureus* reduces tellurite to metallic tellurium to produce black colonies. *S. aureus* produces black colonies surrounded by a yellow zone due to mannitol fermentation. All *S. aureus* strains with these characteristics are coagulase positive.

21.2 Isolation and counting techniques

Sampling, sample preparation, and all experiments should be performed under aseptic conditions. Sampling and all analysis should be performed in duplicate, unless otherwise specified. All

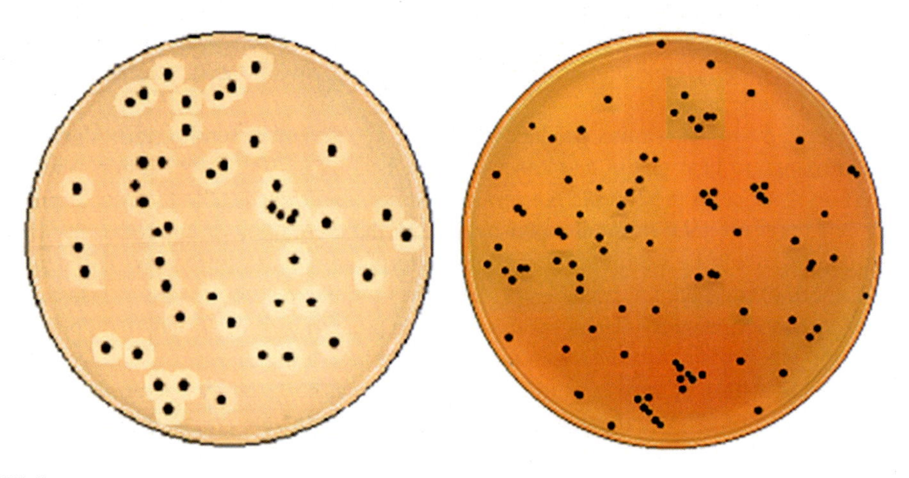

FIGURE 21.1

Staphylococcus aureus colonies on BP agar (*left*) and VJ agar (*right*).

microbiological media to be used in this application must be sterilized. Again, all solutions and equipment that may cause contamination of microorganisms must be sterilized or disinfected.

21.2.1 Equipment, reagents, and media

Equipment. Blender (or Stomacher), centrifuge, Erlenmeyer flask, incubator, microscope, microscope glass slide, pipette discard equipment (e.g., a graduated cylinder) containing 10% bleach solution and a piece of cloth at bottom, pipettes (1 mL) in the pipette box, serological pipette, test tube, vortex, and water bath.

Reagents. Gram stains, HCl (1.0 N), H_2O_2 (10%), human or rabbit plasma (containing 0.1% EDTA), lysostaphin (0.1%), NaOH (0.2 and 1.0 N), nuclease free skim milk, 0.1% peptone water, phosphate buffer saline (PBS) solution (0.1 and 1% NaCl), skim milk (15%), tetramethyl-p-*phenylenediamine* dihydrochloride (1%), and trichloroacetic acid (3.0 M).

Media. 2% Agar, BP agar, Brain heard infusion (BHI) agar slant, BHI blood agar, BHI broth, Ms agar, oxidation-fermentation (OF) broth containing carbohydrate, Staphylococci medium No. 110 (SM), TMG broth, toluidine blue DNA (TB-DNA) agar, trypticase soy (TS) broth, TS broth containing 7% NaCl and 1% sodium pyruvate, and VJ agar.

21.2.2 Preenrichment and selective enrichment isolation of *Staphylococcus aureus*

This technique is recommended for isolating a low number of *S. aureus* from foods and ingredients; and foods expected to contain a high number of competing microorganisms and injured cells. Add 50 g of the sample into a sterile blender (or stomacher plastic bag) containing 50 mL of TS broth (if homogenization of the solid food sample requires more liquid, add a second 50 mL of TS broth into the blender). Homogenize the sample by blending for 1 min (or stomaching for 2 min), transfer the homogenate into an Erlenmeyer flask. In the case of a liquid sample, add 50 mL of the sample into 50 mL of TS broth in the Erlenmeyer flask and mix well by gently shaking the flask. Incubate the flasks at 35°C for 2 h for preenrichment. After incubation, add 100 mL of TMG broth containing 10% NaCl into incubated TS broth and incubate again at 35°C for 18 h for selective enrichment. Examine this TMG broth for growth (turbidity) and color change (black color).

Streak plate one loopful of enriched culture onto BP agar (or VJ agar). Incubate inverted plates at 35°C for up to 48 h. After incubation, examine the plate for characteristic colony formation. Select least three typical *S. aureus* colonies per plate and subject to confirmation tests (Technique 21.3) and report results as the presence or absence of *S. aureus* in the sample.

S. aureus enrichment isolation and counting flowchart is given in Fig. 21.2.

21.2.3 Counting of *Staphylococcus aureus*

21.2.3.1 Plate counting technique

Refer to Technique 2.2.2 for details of sample preparation and dilutions. Weigh 25 g of the sample into a blender (or stomacher plastic bag) containing 225 mL of 0.1% peptone water and blend for

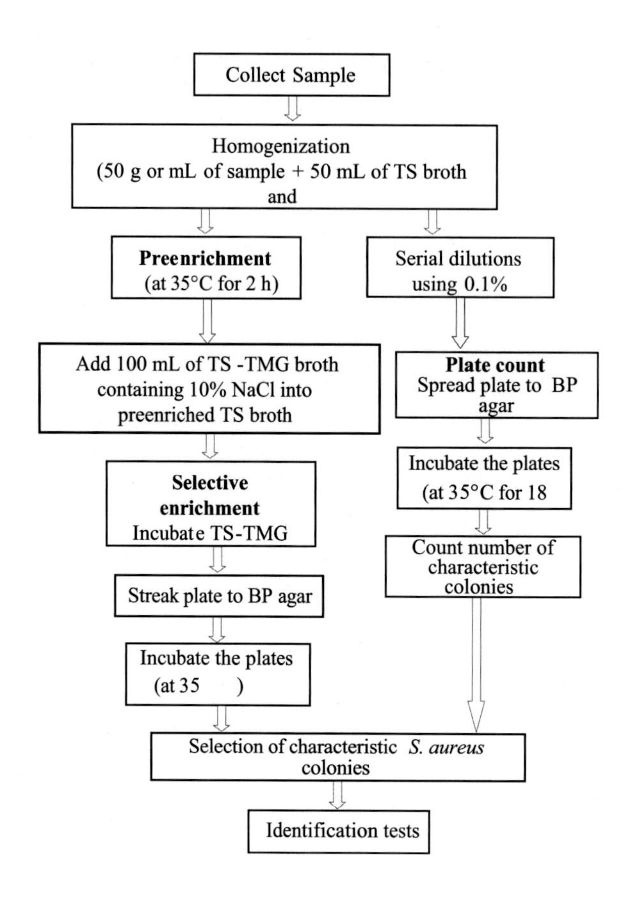

FIGURE 21.2

Staphylococcus aureus enrichment isolation and counting flowchart.

1 min (or stomach for 2 min). For a liquid sample, add 25 mL of the sample into an Erlenmeyer flask containing 225 mL of 0.1% peptone water and mix by gently shaking the flask. Prepare further decimal dilutions using 0.1% peptone water. Spread plate 1 mL of sample on predried BP agar in duplicate (Technique 2.2.4). Retain the plates in an upright position until the inoculum is absorbed by the medium: Place the plates in an incubator in an upright position at 35°C for about 1 h (or slightly opening the lid of the plate near a Bunsen burner flame can also be used). Then incubate inverted plates at 35°C for 24 and 48 h.

After incubation, select plates containing 25–250 colonies. Count characteristic *S. aureus* colonies from BP agar plates as indicated in the introduction of this practice. Select at least three colonies per plate and test for coagulase production. After the coagulase test, add the number of colonies giving a positive coagulase test to result and multiply the count by the dilution rate and dividing by inoculation amount. Report results as the number of *S. aureus* colony forming units (cfu) per g or mL of the sample. Confirm the presumptive isolates by phenotypic techniques (Technique 21.3).

21.2.3.2 Direct counting of coagulase and thermonuclease positive Staphylococcus aureus

Refer to Technique 21.2.3.1 for details of sample preparation and dilutions. Spread plate 1 mL of sample from each dilution over BP agar without egg yolk and then dry the inoculum by placing a plate near a Bunsen burner flame by slightly opening the lid of the plate. After drying the surface area, incubate the inverted plates at 35°C for 24 and 48 h. After incubation, examine the plates for characteristic *S. aureus* colony formation. Select Petri plates containing 25–250 colonies and count the number of characteristic colonies. Reincubate counted plates without inverting at 65°C for 24 h, then cool the plate and overlay each plate with 10 mL of melted (at 45°C) sterile TB-DNA agar, and allow the agar to solidify. After solidification, incubate inverted plates at 35°C for 4 h. Count all colonies showing pink halos as thermonuclease-positive while plates are placed against a blue background and compare with the previous count. Add all colonies that showed both fibrin halos (coagulase-positive) and pink halos (thermonuclease-positive). Calculate the number of *S. aureus* by multiplying by the dilution rate and dividing by the inoculation amount. Report the results as the number of *S. aureus* cfu per g or mL of sample.

21.2.3.3 Counting of Staphylococcus aureus *by most probable number technique*

Refer to Technique 21.2.3.1 for details of sample preparation and dilutions. Inoculate 1 mL of each diluted sample into each 3-MPN tubes of TMG broth (or TS broth-7% NaCl). Incubate tubes at 35°C for 48 h. After incubation, examine TMG broth for growth (turbidity) with the formation of a black color. Indicate the number of positive tubes for each dilution. Refer to MPN Table 4.1 and calculate the number of *S. aureus* MPN per g or mL of sample. Streak plate a loopful of TMG broth culture (from each positive tube) to dried BP agar. Vortex-mix tubes before streaking. Incubate the inverted plates at 35°C for 48 h.

21.3 Identification of *Staphylococcus aureus*

Select at least three representative colonies per Petri plates (BP agar or others) from the isolation and counting technique. Obtain a pure culture as described in Technique 13.4.1 using BHI broth and VJ agar. After obtaining the pure culture, carry out inoculation from pure colonies to TS agar, BHI agar slants, and BHI broth.

The *S. aureus* identification flowchart is given in Fig. 21.3.

21.3.1 Morphological identification

Gram staining. Prepare Gram stain from TS agar slants as explained in Appendix B under the topic "Gram Staining Reagents." A Gram-stained *S. aureus* will appear as Gram-positive cocci, occurring singly or in pairs, most typically in irregular clusters resembling grapes (Fig. 21.4).

Motility test. Perform a motility test with a wet mount preparation as described in Technique 13.4.2. *S. aureus* is a nonmotile bacterium.

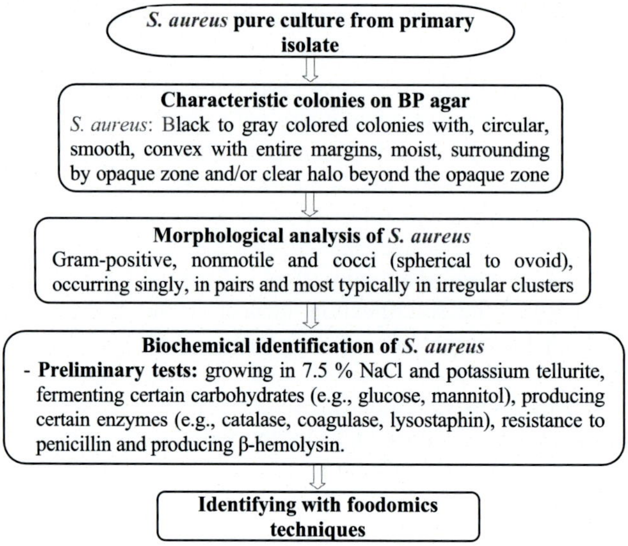

S. aureus **pure culture from primary isolate**

⇩

Characteristic colonies on BP agar
S. aureus: Black to gray colored colonies with, circular, smooth, convex with entire margins, moist, surrounding by opaque zone and/or clear halo beyond the opaque zone

⇩

Morphological analysis of *S. aureus*
Gram-positive, nonmotile and cocci (spherical to ovoid), occurring singly, in pairs and most typically in irregular clusters

⇩

Biochemical identification of *S. aureus*
- **Preliminary tests:** growing in 7.5 % NaCl and potassium tellurite, fermenting certain carbohydrates (e.g., glucose, mannitol), producing certain enzymes (e.g., catalase, coagulase, lysostaphin), resistance to penicillin and producing β-hemolysin.

⇩

Identifying with foodomics techniques

FIGURE 21.3

Staphylococcus aureus identification flowchart.

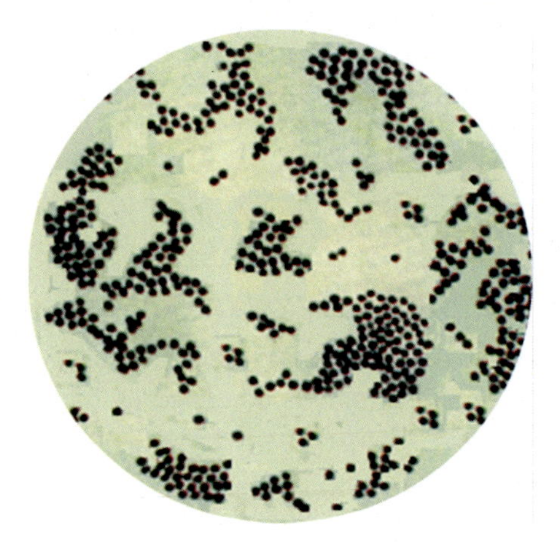

FIGURE 21.4

Staining appearance of *Staphylococcus aureus.*

21.3.2 Biochemical identification

21.3.2.1 Coagulase test

Slide technique. Divide a microscope slide into two parts. Place a loopful of saline solution (0.85% NaCl) on to each part of the slide and place a small amount of 24 h TS agar slant culture near to the saline solution drop and emulsify the culture to obtain a homogenous suspension. Add one drop of human or rabbit plasma (with EDTA) onto one of the suspensions and stir with a flamed and cooled loop for 5 s. Clumps will not reemulsify with gentle shaking of the slide and this indicates positive coagulase test. A second suspension can be served as a control. Do not store dehydrated plasma for longer than 5 days.

Tube technique. Add 0.5 mL of 18 h TS broth cultures into a sterile tube (10 × 75 mm). Add 0.5 mL of plasma onto the culture and mix well thoroughly. Prepare a control tube without plasma. Incubate the tubes at 35°C and periodically examine for clot formation over 6 h intervals. The clothing should be retained after tilting of tube. The clot formation is recorded as positive for *S. aureus* (Fig. 21.5). It is a 3 + or 4 + clot formation. If a clot is not formed after 6 h, the tubes should be incubated at room temperature overnight. A bacterium failing to clot the plasma within 24 h is considered to be coagulase negative and must be confirmed by the slide coagulation

FIGURE 21.5

Tube coagulase test (positive on left and negative on right).

technique if this technique is directly used. Partial, small, or poorly organized clots (2 + and 1 + clot) must be confirmed by the slide technique.

21.3.2.2 Catalase test

Most *S. aureus* strains produce the enzyme catalase aerobically. The catalase test can be performed from solid or liquid culture. (1) From agar culture. One milliliter of H_2O_2 solution is poured over the surface of the colony on the agar plate. Alternatively, place a loopful of the colony from plate or slants onto a microscope glass slide and add one drop of H_2O_2 solution onto colony. (2) From broth culture. Transfer 1 mL of culture into a small clean test tube and add 1 mL of H_2O_2 solution. H_2O_2 (10%) should be freshly prepared each day, and stored in the refrigerator between tests.

After adding H_2O_2 solution onto culture, free oxygen released as gas bubbles indicates the presence of catalase enzyme in the culture and the test is positive for catalase. *S. aureus* produces catalase enzyme and is a catalase positive bacterium.

21.3.2.3 Anaerobic utilization of glucose and mannitol

Inoculate the test microorganism by stabbing a straight wire into two O-F broth tubes containing 1% carbohydrate. Cover the surface of broth in one tube with 20% melted agar (at 45°C) and use the second tube as a control. Incubate O-F broth at 35°C for up to 5 days aerobically (examine periodically). After incubation, examine the tubes for growth (turbidity) and color changes. The result from broths will be one of the following: (1) oxidation of carbohydrates, resulting in acid production in the uncovered tube at the surface of the broth but not in the covered tube, although, on prolonged incubation, the acid reaction may spread through the covered tube (therefore periodic examination is very important). (2) Fermentation of the carbohydrate, resulting in acid production in both covered and uncovered tubes throughout the broth with or without gas production (space in the Durham tube). The reaction is first seen in a part of the broth immediately and spreads rapidly throughout the broth. On prolonged incubation, pH reversal may occur. *S. aureus* will ferment glucose and usually mannitol anaerobically. (3) Inability to utilize carbohydrate; no change in the color of the covered and uncovered broth (alkaline at the surface or throughout the broth). *S. aureus* is a carbohydrate oxidizing and fermenting bacterium. In both cases, acid production takes place throughout the broth.

21.3.2.4 Thermonuclease enzyme production

1. *From culture*. Heat TS broth *S. aureus* culture for 15 min in boiling water bath. Cut 2 mm wells on the TB-DNA agar plates and add 0.01 mL of heated TS broth culture into well using a sterile pipette in duplicate. Incubate the plates without inverting at 35°C for 4 h or at 50°C for 2 h. Formation of a bright pink halo on the agar medium extending at least 1 mm from the periphery of the well indicates the production of thermonuclease enzyme and a positive reaction. *S. aureus* produces thermonuclease enzyme. Include positive and negative controls using *S. aureus* (ATCC 12600) and *S. epidermidis* (ATCC 14990), respectively.
2. *From food*. Sensitivity of this test may be increased by using a higher amount of sample. Add 20 g or mL of food, 5 g of nuclease free skim milk, and 40 mL of sterile distilled water into a sterile blender and blend for 3 min at high speed. (Omit the addition of skim milk when testing milk product, e.g., dry milk, cheese, fruit-filled dairy product, and skim milk). Adjust the pH of the homogenate to 3.8 with HCl. Centrifuge homogenate under refrigeration for 15 min at

7000–10,000 × g (or prechill the suspension to 5°C and centrifuge again at room temperature using a tabletop centrifuge at 7000 × g). After centrifugation, decant the supernatant to discard, and add 0.05 mL of cold 3 M trichloroacetic acid into each mL of original culture, mix and centrifuge as above. Decant and discard the supernatant. Resolubilize the precipitate by adjusting the pH to 8.5 with NaOH (0.2 N). The final volume of this solution should not exceed 2 mL. Boil the solution for 15 min or longer (up to 90 min) if a greater specificity is needed (but prevent loss of volume during heating). Place boiled sample by filling a 2 mm well in TB-DNA agar plate in duplicate. Incubate Petri plates without inverting at 35°C for 4 h. A pink halo extending 1 mm beyond the well is considered as positive for thermonuclease regardless of the actual size of the zone.

21.3.2.5 Hemolysis test

Mark BHI blood agar plate into six or eight equal segments at the bottom of plate and label with the sample name. Inoculate each segment near its center by gently touching the agar surface with a loopful of TS broth culture. After drying the inoculum, incubate inverted Petri plates at 35°C for 24–48 h. During incubation, check for hemolytic activity of bacterium as indicated by a zone of complete hemolysis surrounding the colony. Alpha-hemolysis is the greenish zones with hazy outlines around the colonies (Fig. 21.6A). Beta-hemolysis is the term given to complete clearing around the colonies (Fig. 21.6B). *S. aureus* produces beta-hemolysin on the blood agar. Pathogenic *S. aureus* produces beta-hemolysin zone around colonies.

21.3.2.6 Susceptibility to lysostaphin

1. *From broth culture*. Mix 0.1 mL of TS broth culture with 0.1 mL of 0.1% lysostaphin (prepared using 0.02 M PB-2% NaCl) to produce a final concentration of 25 μg lysostaphin per mL in a narrow small tube. To another portion of 0.1 mL of culture in a narrow small tube, add 0.1 mL

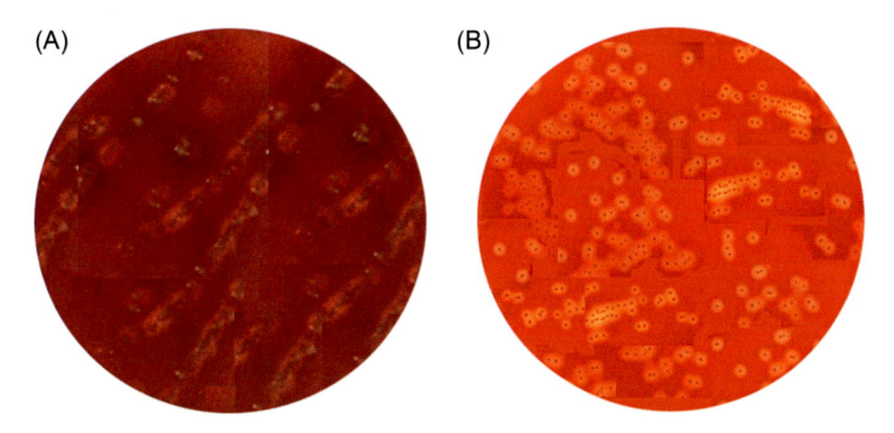

FIGURE 21.6

Hemolytic reactions on BHIB agar. (A) Alfa-hemolysins partially hemolyze red blood cells with the greenish discoloration surrounding the colonies. (B) Beta-hemolysins completely hemolyze red blood cells and clear zones surround the individual colonies.

of PB-NaCl solution (negative control). Use positive control (*S. aureus* ATCC 12600) and negative control (*Micrococcus varians* ATCC 15306) in the assay. Mix the contents of all tubes by gently shaking. Incubate tubes at 35°C up to 2 h. If the turbidity of mixture in the tube clears, this indicates the sensitivity of bacterium to lysostaphin and the test is considered positive. If clearing cannot appear within 2 h, the test is negative. *S. aureus* is susceptible to lysostaphin and the test is positive for *S. aureus*.

2. *From agar culture*. Transfer one half of the selected colony from TS agar plate with loop into 0.2 mL of PBS solution in a narrow small tube and emulsify. The other half of the colony is transferred to another tube and is mixed with sterile PSB solution as a control. Add 0.1 mL of lysostaphin to obtain a concentration of 25 μg lysostaphin per mL. Incubate the tubes at 35°C up to 2 h. If turbidity in the tube clears, the test is considered positive. If turbidity clearing in the tube does not occur within 2 h, the test is considered to be negative.

21.3.2.7 Oxidase test

Technique 1. Pour the reagent (tetramethyl-p-phenylenediamine hydrochloride) over the bacterial colony in the TS agar plate. An oxidase positive colony turns dark–black within 10–30 min.

Technique 2. Add one drop of reagent onto a filter paper in a sterile Petri dish. Prepare a smear with a loopful colony from TS agar culture on filter paper. Use a platinum loop or glass rod (other materials may give false positive results in the preparation of smear). Oxidase positive bacterium turns dark blue within 2 min. Species of staphylococci are oxidase negative (remain colorless). Colonies from other media or colonies older than 24 h will give false results.

21.3.2.8 Bacteriophage sensitivity

Inoculate BHI broth with TS broth culture and incubate tubes at 35°C for about 4 h (up to appearance of turbidity). Inoculate the BHI agar surface in the plate by swabbing BHI broth culture and allow the plate to stand to absorb liquid from swabbing. Label the plates for each sample and draw a circle 0.5 cm in diameter at the bottom of the plate. Spot inoculate a loopful (with 3 mm platinum loop) of diluted bacteriophage culture (10^{-6} phage; plaque) onto the circle on the BHI agar. Prepare the bacteriophage culture as indicated in Technique 29.3.4. Allow the plate to stand for a time period to allow the medium to absorb liquid from the phage culture. Incubate inverted Petri plates at 35°C for 16 h. Examine plates during an incubation period for the formation of plaques (clear zone) on the spot inoculated area. *S. aureus* is sensitive for its bacteriophages and plaque can form on the Petri plate with specific bacteriophages (Fig. 21.7).

Note: $CaCl_2$ (400 μg per mL) must be added into all media to be used for bacteriophage culturing before sterilization.

For further phenotypic identification of isolates refer to Table 21.1.

21.3.3 Identification of *Staphylococcus aureus* species by foodomics techniques

21.3.3.1 *Matrix-mediated laser desorption ionization-time-of-flight-mass spectrometry identification of* Staphylococcus aureus

S. aureus can be rapidly identified with the matrix-mediated laser desorption ionization-time-of-flight-mass spectrometry (MALDI-TOF Ms) technique, which is based on the principle of ionizing

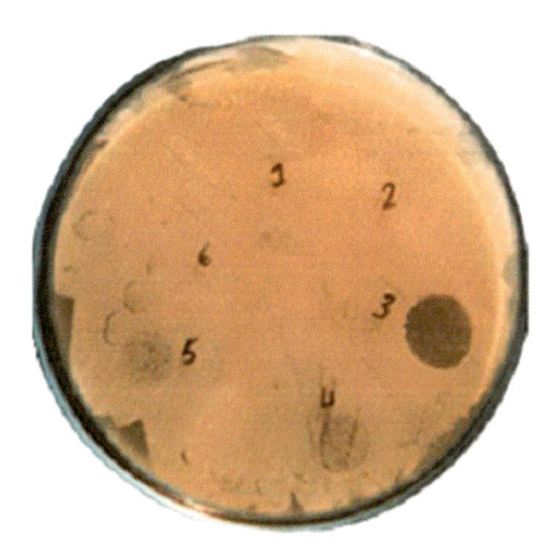

FIGURE 21.7

Bacteriophage typing of *Staphylococcus aureus*: Number 3 indicates the specificity of the bacteriophage with the tested bacterium by formation of plaque.

a specific protein profile of microbial cells. *S. aureus* can be easily identified by comparing these profiles with the reference spectra. MALDI-TOF Ms identification has been explained in Technique 13.6.3.1.

21.3.3.2 Molecular identification of Staphylococcus aureus

Polymerase chain reaction (PCR) is a molecular genotypic technique and is used in the identification of *S. aureus* species. Gene coding of a specific *S. aureus* virulence factor or the other characteristics can be detected by PCR assay. Some *S. aureus*-specific primer sets that can be used in the identification of *S. aureus* are given in Appendix A: Gene Primers; Table A.1. Using *S. aureus*-specific techniques and materials, the bacterium can be identified by PCR analysis, as described in Technique 13.6.3.2.

21.4 Interpretation of results

Healthy people (about 30%–50%) are common carriers of enterotoxigenic *S. aureus* in their nose and throat and on their skin. Thus food handlers can be an important source of food contamination. SET are produced as a by-product during the growth of toxigenic *S. aureus* strains. The toxin production requires the growth of *S. aureus* until the population reaches 10^5 per gram of food. *S. aureus* enterotoxins are highly stable, heat resistant, resistant to proteolytic enzymes, and resistant to environmental conditions, for example, freezing, drying, and low pH. They cannot be destroyed by cooking. An estimated 0.1 μg of *Staphylococcus* enterotoxins can cause food poisoning in human. The microorganism is readily destroyed by heat, but the toxin is heat stable, and still capable of causing staphylococcal foodborne disease outbreaks. Sick employees and poor hygienic

Table 21.1 Phenotypic characteristics of *Staphylococcus* species.

Characteristics	*S. aureus*	*S. epidermidis*	*S. capitis*	*S. warneri*	*S. haemolyticus*	*S. hominis*	*S. auricularis*	*S. saprophyticus*	*S. cohnii*	*S. simulans*	*S. intermedius*
Fermentation of											
Glucose/sucrose	+/+	+/+	+/+	+/+	+/+	+/+	+/v	+/+	+/−	+/+	+/+
Maltose/ mannitol	+/+	+/+	+/+	+/+	+/+	+/+	+/+	+/+	v/+	w/+	w/v
Mannose/ trehalose	+/+	w/−	+/−	−/+	−/+	−/v	−/w	−/+	v/+	v/v	+/+
Lactose	+	v	−	v	v	v	−	v	−	+	+
Growth in											
10% NaCl	+	w	+	+	+	w	+	+	+	+	+
15% NaCl	w	−	−w	w	v	−	w	v	v	w	v
Aerobic/ anaerobic	+/+	+/+	+/w	+/+	+/w	+/w	+/w	+/w	+/v	+/+	+/w
Pigment/ acetone	w/+	−/+	−/v	v/+	v/v	v/v	−/v	v/+	−/v	−/w	−/−
Hyaluronidase	+	v									
Urease/ hemolysis	w/+	+/w	−/w	+/v	−/v	+/w	−/−	+/−	−/w	+/w	+/+
Coagulase/ TDNase	+/+	−/w	−/w	−/v	−/v	−/w	−/w	−/−	−/w	−/w	+/+
Novobiocin	+	−w	−	−	−	−	−	−	−	−w	+

+, *Positive;* −, *negative; v, variable; w, weak.*

practices are the major causes of *Staphylococcus* foodborne diseases. Contaminated equipment and environmental surfaces also can lead to *S. aureus* foodborne diseases. In general, food products that have not been kept hot enough ($>60°C$), cold enough ($<4.5°C$), and/or at room temperatures after preparation are vulnerable to staphylococcus food poisoning. Staphylococcal foodborne diseases can be prevented by implementing an effective HACCP plan, GMPs, and GHPs. Employees should be trained to wash hands with soap and water before entering a food processing facility and handling and preparing food products. People who are infected with *S. aureus* should seek medical attention and stay out of food handling areas. Employees with wounds or skin infection should be strictly prohibited from food-handling areas.

PRACTICE

Isolation of *Clostridium botulinum* 22

22.1 Introduction

Clostridium botulinum is a central or subterminal oval spore former, straight or slightly curved rod with single cells or small chains, and it is an obligate anaerobic. *C. botulinum* is motile by peritrichous flagella and cannot produce capsules. It is distributed widely in soils, and ocean and lake sediments. It is divided into four phenotypic groups: I, II, III, and IV. It multiplies in food and produces neurotoxin, such as exotoxin. Group I is proteolytic and produces A, B, or F neurotoxins; Group II is nonproteolytic and produces B, E, or F neurotoxins; Group III is nonproteolytic and produces C and D neurotoxins; and Group II is weak or nonproteolytic and produces G neurotoxins. *C. botulinum* produces seven serologically distinct types of neurotoxins causing the intoxication known as botulism: A, B, C, D, E, F, and G. Type E botulism is associated with fish or other seafood. Types A and B commonly are associated with soils. Types C and D are not nonproteolytic and cannot digest egg white or meat, and usually are associated with illness in animals and birds. Human botulism commonly is associated with types A, B, and E.

If the pH of the food is above 4.5, heat processing should be designed to destroy spores of *C. botulinum*. The foods commonly associated with botulism are low- and medium-acid canned foods, especially fruits, vegetables, meats, smoked and pickled foods, meat products (usually sausage), vacuum-packed foods, and fish and fish products. The botulism results from the consumption of neurotoxin (a protein) produced in a food. The lethal dose of neurotoxins for adult human is about 1 ng per body weight. The symptoms appear within 12–48 h following consumption of food containing neurotoxins. The disease lasts 1–7 days. The principal symptoms of botulism are vomiting, double vision, urine retention, constipation, difficulty in swallowing and speaking, dry mouth, and respiratory and heart failure. Mortality rate is usually high (20%–50%). *C. botulinum* ingested with food invades the alimentary tract, but is unable to multiply and produce toxin in the intestines. The presence of *C. botulinum* in processed foods indicates underprocessing and postprocess contamination. The inadequate processes occur more frequently in home-canned food production. Swollen containers may contain neurotoxin since *C. botulinum* produces gases during growth. Botulinum toxin associated with canned foods is usually type A and proteolytic type B, since spores of these group are more heat-resistant. The isolation of *C. botulinum* from foods is possible with an enrichment technique. But the isolation of *C. botulinum* is difficult and dangerous, and not to be undertaken in normal quality assurance laboratories. Work must be undertaken only by adequately trained staff (in whom high antitoxin titers are maintained) in specially equipped laboratories. *C. botulinum* can be identified by phenotypic and foodomics techniques. The food can be examined for toxin identification.

Microbiological Analysis of Foods and Food Processing Environments. DOI: https://doi.org/10.1016/B978-0-323-91651-6.00054-9

22.2 Isolation of *Clostridium botulinum*

Sampling, sample preparation, and all experiments should be performed under aseptic conditions. Sampling and all analysis should be performed in duplicate, unless otherwise specified. All microbiological media to be used in this application must be sterilized. Again, all solutions and equipment that may cause contamination must be sterile or disinfected.

22.2.1 Equipment, reagents and media

Equipment. Anaerobic (AB) jars, can opener, centrifuge, clean dry towels, forceps, incubator, microscope, microscope glass slides, Petri dishes, pipette discard equipment (e.g., a graduated cylinder) containing 10% bleach solution and a piece of cloth at bottom, pipettes (1 mL) in the pipette box, sample jars, and tubes.

Reagents. Alcohol, Gram stains, HCl (1 N), monovalent antitoxins, NaOH (1 N), physiological saline (PS) solution, sterile distilled water, and trypsin solution.

Media. AB agar supplemented with 6% horse blood, anaerobic egg yolk (aey) agar, cooked meat (CM) medium, gelatin phosphate buffer (GPB) broth, liver egg yolk (LEY) agar, motility test (MT) medium, trypticase peptone glucose yeast extract (TPGYE) broth, and TPGYE broth containing trypsin.

22.2.2 Sampling

Preliminary examination of can. Refrigerate cans until performing test, except unopened canned food is not refrigerated unless badly swollen. Before opening the can, record the product designation, manufacturer's name, source of sample, container type and size, labeling, production code, and container condition. Examine can for appearance (normal or distortions) at both ends and sides. Record results for the analyzed can.

Opening of canned foods. Sanitize the end of the can with an effective disinfectant. Allow a contact time of a few minutes, then remove the disinfectant and wipe the sanitized area with a sterile dry towel. If the can is swollen, keep the can so that the side seam is away from the analyst. Flame disinfect with a Bunsen burner by directing the flame down onto the can until the visible moisture film evaporates. A container with buckled ends should be flamed with extreme caution to avoid bursting the can. Avoid excessive flaming, indicated by scorching and blackening of the inside enamel coating. Remove a disk of metal from the flamed center area with a flame-disinfected disk cutter.

After opening the can, examine the appearance and odor of the product and liquid. Indicate any evidence of changes. The product is not tested by eating. Record the results for the product and liquid.

Solid foods with little or no free liquid in the can are transferred aseptically into a sterile mortar. An equal amount of GP broth is added and ground with a sterile pestle. Alternatively, small pieces of the product may be added directly into GP broth with sterile forceps. Liquid foods are directly added into GP broth using sterile pipettes. Before the use of GP broth, dissolved oxygen is removed by heating in boiling water for 10–15 min. After heating, cool the broth quickly in cold water to room temperature without agitation before use.

22.2.3 Enrichment isolation of *Clostridium botulinum*

C. botulinum enrichment isolation flowchart is given in Fig. 22.1.

Add 1–2 g or mL of sample from prepared sample into each of two tubes of CM broth and two tubes of TPGYE broth. Introduce the inoculum slowly below the enrichment broth.

Incubate the inoculated one of CM broth and TPGYE broth at 35°C, and the others at 26°C anaerobically for up to 5 days. At the end of incubation, examine each culture for turbidity, production of gas, and digestion of the meat particles (in CM broth). Note the odor in broths. Examine the culture microscopically by a wet mount preparation using a high-power objective and observe the motility of bacteria. Prepare the Gram stain and spore stain. Observe Gram-positive rod-shaped cells and spore location within the cell. Perform further identification tests as indicated in Technique 22.3.

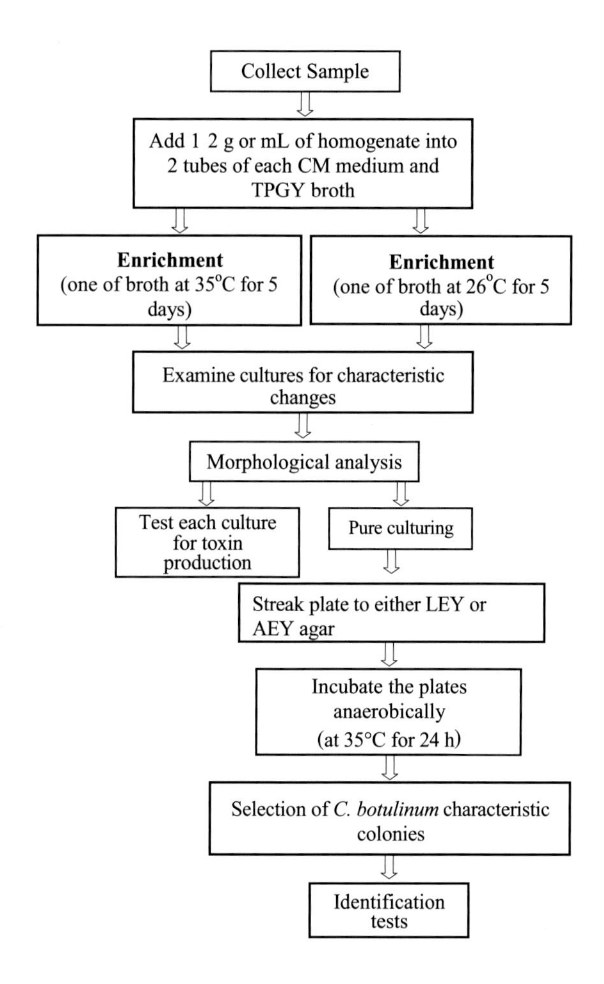

FIGURE 22.1

Clostridium botulinum enrichment isolation flowchart.

22.2.4 Pure culturing

For obtaining a pure culture, gently mix enriched culture, transfer 1–2 mL of the culture into a sterile screw-cap tube and store in a refrigerator until use.

Alcohol treatment. To 1–2 mL of a culture in a sterile screw-cap tube showing some sporulated cells, add an equal volume of filter-sterilized ethanol. Mix the alcohol with the culture and allow the mixture to stand at room temperature for 1 h.

Heat treatment. An alternative method to alcohol treatment is to heat 1–2 mL of the enriched culture to destroy the vegetative cells but not the spores of *C. botulinum*. For a nonproteolytic type (not for proteolytic type), heat at 80°C for 10–15 min.

Plating. To prevent spreading of the colonies, the plates must be well dried. Dry the plates at 35°C for 24 h before use. Streak plate one loopful of the treated culture on either LEY agar or AEY agar (or both) to obtain well separated colonies. Incubate the inverted plates anaerobically at 35°C for 48 h. *C. botulinum* produces flat or raised, rough or smooth colonies, and commonly spreads with an irregular edge. *C. botulinum* types C, D, and E are surrounded by a zone (2–4 mm) of a yellow color due to lecithinase activity. Colonies of types A and B strains generally show a smaller yellow zone.

22.3 Identification of *Clostridium botulinum*

If colonies are not pure on the plates, remove growth from the colony by touching the loop onto the center of the colony and inoculate into TPGYE broth (or AB broth) before incubating in an AB incubator at 35°C for 24 h. After incubation, streak plate onto LEY agar and incubats the inverted plates anaerobically at 35°C for 24 h. After incubation, examine the plates for the formation of pure culture with single colonies (check by Gram staining). Select at least three well-separated typical colonies per plate and inoculate each colony into a tube of TPGYE broth by removing growth from the colony by touching the loop onto the center of the colony. After incubation in an AB incubator at 35°C for 24 h, examine the tubes for growth (turbidity). Streak plate one loopful of TPGYE broth pure culture to LEY agar and into TPGYE broth. Incubate broth and inverted plates anaerobically for 1–7 days at 35°C. After incubation, examine growth in broth and plates. Test the cultures for toxin production and determine the toxin type (Technique 22.3.3). Store the pure culture in sporulated state either under refrigeration, on glass beads or lyophilized.

The *C. botulinum* identification flowchart is given in Fig. 22.2.

22.3.1 Morphological analyses

Gram staining. Prepare a Gram stain from LEY agar culture (or TPGY broth culture), as explained in Appendix B under the topic "Gram Staining Reagents." A Gram stain of *C. botulinum* will produce Gram-positive, rods, occurring singly or in pairs (Fig. 22.3).

Motility test. Stab inoculates each broth culture into MT medium and incubate the medium anaerobically at 35°C for 24 h. *C. botulinum* is a motile bacterium, and produces diffuse growth in semisolid medium (Technique 13.4.2).

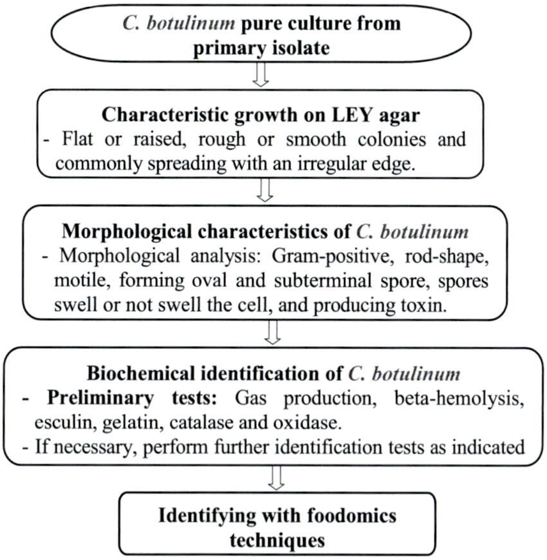

C. botulinum **pure culture from primary isolate**

⇩

Characteristic growth on LEY agar
- Flat or raised, rough or smooth colonies and commonly spreading with an irregular edge.

⇩

Morphological characteristics of *C. botulinum*
- Morphological analysis: Gram-positive, rod-shape, motile, forming oval and subterminal spore, spores swell or not swell the cell, and producing toxin.

⇩

Biochemical identification of *C. botulinum*
- **Preliminary tests:** Gas production, beta-hemolysis, esculin, gelatin, catalase and oxidase.
- If necessary, perform further identification tests as indicated

⇩

Identifying with foodomics techniques

FIGURE 22.2

Clostridium botulinum identification flowchart.

FIGURE 22.3

Staining appearance of *Clostridium botulinum*.

Spore stain. Prepare a spore stain from LEY agar culture, as explained in Appendix B under the topic "Spore Stain." *C. botulinum* produces endospores terminally (Fig. 22.3).

22.3.2 Biochemical tests

Table 22.1 summarizes the necessary phenotypic characteristics useful in the identification of *Clostridium* spp.

Hemolytic test. Streak plate one loopful of culture from LEY agar culture on AB agar. Incubate anaerobically in incubator. *C. botulinum* produces large, irregularly circular, smooth, greyish, translucent colonies with hemolytic zone and colonies may spread. Most strains of *C. botulinum* are β-hemolytic.

Esculin hydrolysis. Refer to Technique 14.3.2.1 for details of esculin hydrolyzes. *C. botulinum* is esculin positive.

Table 22.1 *Clostridium botulinum* groups I and II phenotypic characteristics.

Characteristics	Group I	Group II
Source	Soil	Aquatic environments
Indole/H_2O_2	−/+	−/+
Utilization of proteins	+	−
Lysine, ornithine, VP test, Mr, esculin, gelatin	+	+
ONPG, urease, phenylalanine, nitrate, citrate, malonate utilizations, indole, starch	−	−
Carbohydrate fermentation (positive)	Glucose, maltose, fructose, sorbitol, trehalose salicin	Dextrin, glucose, fructose, maltose, galactose, mannose, sucrose, starch, glycogen, ribose, sorbitol, trehalose
Carbohydrate fermentation (negative)	Mannose, arabinose, xylose	Mannitol
Bacteriocin production	Boticin	Not
Colony morphology on blood agar at 35°C for 48 h	Large, irregularly circular; smooth, opaque, raised center; narrow hemolysis	Irregular with lobate margins (in diameter); opaque to translucent with matt surface; mosaic structure; narrow hemolysis
Min./Opt. growth temp. (°C)	10/35	3/30
Growth limiting min. pH	4.3	5.0
Growth-limiting NaCl (%)	10	5
Spores heat resistance	High resistant ($D_{121°C-110°C}$, 0.15−1.8 min)	Moderate resistant ($D_{85°C}$, 1−98 min)
Source of foods for botulism	*Food-borne botulism*: Meat, vegetables, canned foods. *Infant botulism*: Honey.	*Food-borne botulism*: Meat, fish, minimally processed packaged foods. *Infant botulism*: Honey.

+, Positive; −, negative; D_T, decimal reduction time at temperature T.

Gelatin hydrolysis. Refer to Technique 19.3.2.3 for details of gelatin hydrolyses. *C. botulinum* hydrolyzes gelatin.

Catalase test. Refer to Technique 13.4.3.10 for details of the catalase test. *C. botulinum* is catalase negative.

Oxidase test. Refer to Technique 21.3.2.7 for the oxidase test. *C. botulinum* is oxidase negative.

Carbohydrate fermentation. Refer to Technique 15.3.2.2 for carbohydrate fermentation. Table 22.1 indicates carbohydrate fermentation results for *C. botulinum* groups.

Lipase test. Refer to Technique 18.3.2.4 for details of thelipase test. Lipase positive bacterium produces this zone on egg yolk agar with an iridescent layer around colonies. *C. botulinum* produces lipase.

Lecithinase reaction (Nagler's reaction). Refer to Technique 20.3.2.3 for details of lechitinase activity. *C. botulinum* gives variable results for lecithinase reaction.

22.3.3 Identifying botulinal toxin and typing

22.3.3.1 Detection of toxicity

Preparation of sample. A small food portion is used in toxin identification. The remaining sample is refrigerated. Extract solid food with an equal volume of GPB, homogenize, centrifuge to remove the solids, and use the supernatant fluid for the toxin assay. Wash solid portions of the sample with a few mL of GPB, repeat centrifugation and add supernatant into the previous one.

Trypsin treatment. A portion of the supernatant fluid is treated with trypsin to extract toxins (nonproteolytic toxins require this treatment). At the same time, test another portion of the sample without trypsin treatment, since the fully active toxin of a proteolytic strain, if present, may be degraded by trypsin. TPGY broth culture is not used since this broth contains trypsin.

To trypsinize, adjust an amount of the supernatant to pH 6.2 with NaOH (0.1 N) and HCl (0.1 N). Add 0.2 mL of an aqueous trypsin solution into 8 mL of each supernatant in the sterile small tube and incubate at 35°C for 1 h with gentle agitation.

Preparation of trypsin solution. Place 1 g of 1:250 trypsin into a clean small tube and add 10 mL of sterile distilled water into this tube. Agitate from time to time and keep at room temperature until as much of the trypsin as possible has been dissolved. Check the pH of the trypsin solution and adjust to pH 6.0, if necessary.

Toxicity testing. An untreated supernatant is diluted to 1:2, 1:10, and 1:100 in GP broth. Prepare the same dilutions from trypsinized supernatant. Inject separate pairs of mice intraperitoneally with 0.5 mL of each undiluted and diluted supernatant using a 1 or 2.5 mL syringe. Repeat this for trypsinized supernatant. About 1.5 mL of the untreated supernatant is heated at 100°C for 10 min. Cool heated supernatant and inject 0.5 mL without dilution into each of two mice.

Observe all of the mice during 48 h for botulism symptoms and recover the deaths. Typical botulism symptoms are weakness of the limbs, gasping for breath, and death due to respiratory failure. If all of the mice die with the dilutions used, repeat, using higher dilutions to determine the end point, or the minimum lethal dose (MLD) in the estimation of the toxin amount in sample. The MLD is the highest dilution of sample killing mice. Calculate the amount of MLD per mL.

22.3.3.2 Toxin typing

Use either the untreated or supernatant treated fluid in determining the type of toxin which it is lethal to mice (from Technique 22.2.3.1). Use the supernatant at MLD. Rehydrate the lyophilized antitoxin and dilute the monovalent antitoxins A, B, E, and F in PS to contain one international unit per 0.5 mL. Prepare enough diluted supernatant to inject each of two mice with 0.5 mL of the antitoxin for each dilution. Protect mice for 30–60 min before injecting the toxic preparation. Inject mice with a sufficient dilution to cover a range of at least 10, 100, and 1000 MLD below the previously determined endpoint (MLD) of toxicity. Observe the mice for 48 h for botulism symptoms and record the deaths. If botulinal toxin is present in sufficient quantity, mice receiving the nonneutralized toxin will die, and mice receiving the toxin neutralized by the specific antitoxin will survive. If the toxin is not neutralized, repeat the test using monovalent antitoxins.

22.3.4 Identification of *Clostridium botulinum* species by foodomics techniques

22.3.4.1 Matrix-mediated laser desorption ionization-time-of-flight-mass spectrometry identification of Clostridium botulinum

C. botulinum can be rapidly identified with the matrix-mediated laser desorption ionization-time-of-flight-mass spectrometry (MALDI-TOF Ms) technique, which is based on the principle of ionizing a specific protein profile of microbial cells. *C. botulinum* can be easily identified by comparing these profiles with the reference spectrum. MALDI-TOF Ms identification has been explained in Technique 13.6.3.1.

22.3.4.2 Molecular identification of Clostridium botulinum

Polymerase chain reaction (PCR) is a molecular genotypic technique and is used in the identification of *C. botulinum* species. Gene coding of the specific *C. botulinum* virulence factor or the other characteristics can be detected by PCR assay. Some *C. botulinum*-specific primer sets that can be used in the identification of *C. botulinum* are given in Appendix A: Gene Primers (Table A.1). Using *C. botulinum*-specific techniques and materials, the bacterium can be identified by PCR analysis as described in Technique 13.6.3.2.

22.4 Interpretation of results

Botulism is caused by the formation of toxins as a result of conditions that allow *C. botulinum* to become infected, multiply, and toxin to build up. Spores can tolerate many conditions, such as heat, and can remain dormant for extended periods of time before germinating when conditions become more favorable. Proteolytic strains of *C. botulinum* do not grow below 10°C. Nonproteolytic strains can grow at cooling temperatures. Control of *C. botulinum* in foods requires the destruction of spores by processing (e.g., high-temperature canned production) or prevention of growth by product formulation (e.g., keeping pH below 4.6, reducing the amount of water available). Failure of one or more measures of food production (e.g., heat or acidification) and storage (such as temperature change) can cause bacteria to multiply in food and produce toxins. Foods that are not acidic (pH above 4.6), contain sufficient water, and contain little or no air are susceptible to

C. botulinum growth. Outbreaks are commonly seen on canned food, oil-jarred foods, some fish products, and meat products. Sometimes it can be responsible for botulism in contamination after heat treatment. Normal complete cooking (pasteurization: 70°C for 2 min or equivalent) kills *C. botulinum* but not its spores. A sterilization process equivalent to 121°C for 3 min is required to kill spores of *C. botulinum*. Botulinum toxin itself is rapidly inactivated (denatured) at temperatures above 80°C.

PRACTICE

Isolation and counting of *Vibrio* 23

23.1 Introduction

Vibrio species are Gram-negative, rod- or curve-shaped, facultative anaerobic motiles with a single polar flagellum, and they are oxidase (except two species) and catalase positive. They ferment glucose without gas production and do not ferment lactose. They can grow at a temperature ranging from 5°C to 42°C with an optimum at 35°C. They are extremely sensitive to drying, freezing, and heating (pasteurization). They commonly are present in marine environments. Most *Vibrio* spp. are human pathogens and cause foodborne disease. *Vibrio* spp. (except *Vibrio cholerae* and *Vibrio mimicus*) cannot grow in media without sodium chloride, and are called halophilic. Four species of *Vibrio* are associated with foodborne illness: *V. cholerae* (O1 and non O1 serogroups), *Vibrio parahaemolyticus*, *Vibrio vulnificus*, and *V. mimicus*. Not all strains of *V. parahaemolyticus* are pathogenic. The pathogenic *Vibrio* spp. can cause hemolysis with the production of a heat-stable hemolysin and are known as Kanagawa positive strain. *V. cholerae* produces heat-sensitive enterotoxin that can cause characteristic cholera symptoms, for example, "rice" water. It is transmitted by the fecal oral route and indirectly by contaminated water. Direct person to person spread is not common. It frequently associates by fish and selfish from coastal water. It is a halotolerant bacterium (grown in the presence or absence of NaCl) and widely distributed in an aquatic environment. Cholera is a main waterborne toxicoinfection. The optimum NaCl for growth is 1%−3%. Serotype O1 and nonserotype O1 (O139, Ogawa, Inaba, and biotype El Tor) can cause cholera.

V. parahaemolyticus is an enteroinvasive pathogen and adheres to human intestinal cells. The infective dose is about 10^3-10^4 viable cells per g or mL. The symptoms appear within 9−25 h after the consumption of contaminated products with *V. parahaemolyticus* and last up to 8 days. The principal symptoms are abdominal pain, profuse watery diarrhea without blood or mucus, vomiting, and fever. *V. parahaemolyticus* foodborne disease is mostly associated with crab, lobster, shrimp, insufficient cooked foods, refrigerated seafoods, and temperature abused food after cooking. It commonly is associated naturally in estuarine waters and coastal environment, and seafoods

V. vulnificus can cause foodborne disease and wound infection with fatality to individuals having liver illness (cirrhosis) or other underlying illnesses for example, diabetes. *V. vulnificus* infection and fatality rate are very high (40%−60%) among people with liver and gastric diseases. It mainly causes gastroenteritis following the consumption of raw (e.g., oyster) or undercooked seafood. *V. mimicus* foodborne diseases are associated with diarrhea. Some strains produce cholera-like toxin. It commonly is associated with estuarine and coastal water.

V. mimicus is a nonsucrose fermenting bacterium. It produces cholera-like toxin to cause gastroenteritis which is characterized by watery diarrhea, nausea, vomiting, abdominal pain, and fever due to contaminated raw fish consumption and other seafoods (e.g., raw oysters, fish, turtle eggs, prawns, squid, and crayfish). It produces several virulence factors, including adhesins, hemolysins,

Microbiological Analysis of Foods and Food Processing Environments. DOI: https://doi.org/10.1016/B978-0-323-91651-6.00012-4

proteases, cytolysins, lipases, and DNAses. It produces a heat-labile cytolytic/hemolytic toxin known as hemolysin.

Vibrio fluvialis, *Vibrio alginolyticus*, *Vibrio hollisae*, *Vibrio metschnikovii*, and *Vibrio furnissii* can cause gastroenteritis and are associated with an estuarine environment together with nonpathogenic *Vibrio* species. *Vibrio cincinnatiensis* and *Vibrio damsela* cannot cause gastroenteritis. *V. damsela*, *Vibrio anguillarum*, and *Vibrio carchariae* are not pathogenic to humans but are pathogenic to fish.

Vibrio species grow in relatively high amounts of bile salts. They grow best in an alkaline medium. Thiosulfate citrate bile salt sucrose (TCBS) agar is an excellent medium for the selective/differential isolation and enumeration of *V. parahaemolyticus*, *V. cholerae*, and *V. vulnificus*. Most nonvibrios bacteria are inhibited on this medium. The sample should be held in a refrigerator (4°C) until being analyzed. There are fewer die-offs of vibriosis when refrigerated than when frozen. *V. hollisae* cannot grow easily on TCBS agar. In the isolation and counting of *V. hollisae*, a differential medium (e.g., blood agar) flooded with oxidase reagent or mannitol-maltose agar can be used. MPN can be used in counting *Vibrio* from seafoods. Different media can be used in the enrichment isolation of *Vibrio*. Phenotypic and foodomics techniques can be used in the identification of *Vibrio* species.

Sampling, sample preparation, and all techniques should be performed under aseptic conditions. Sampling and all analysis should be performed in duplicate, unless otherwise specified. All microbiological media to be used in these applications must be sterilized. Again, all solutions and equipment that may cause contamination of microorganisms must be sterile and disinfected.

23.2 Isolation and counting techniques of *Vibrio cholerae*

The *Vibrio* selective enrichment isolation and counting flowchart is given in Fig. 23.1.

23.2.1 Equipment, reagents and media

Equipment. Blender (or Stomacher), Erlenmeyer flask, loop, Petri dishes, pipette discard equipment (e.g., a graduated cylinder) containing 10% bleach solution and a piece of cloth at bottom, pipettes (1 mL) in the pipette box, and thermostatic shaking water bath.

Reagents. Alkaline peptone water (APW), Gram stains, phosphate buffer saline (PBS) solution, polyvalent and monovalent antiserums: 01, Ogowa and Inaba polyvalent somatic (O), and sodium deoxycholate solution (0.5%).

Media. Arginine glucose (AG) agar slant, Brain heart infusion (BHI) agar, gelatin agar, gelatin phosphate salt (GPS) agar, Kligler iron (KI) agar slant, motility test (MT) medium, Mueller-Hinton agar, T_1N_1 agar, T_1N_1 broth, T_1N_2 broth, thiosulfate citrate bile salt sucrose (TCBS) agar, triple sugar iron (TSI) agar slant, tryptone (Trypticase) soy (TS) agar, TS broth supplemented with NaCl, and TS agar-magnesium sulfate-2%NaCl (TSAMS-NaCl) slant.

Note: Halophilic *Vibrio* spp. grow in the medium containing 1%–3% NaCl. All media should be prepared with 3% NaCl unless otherwise specified. But do not add NaCl to gelatin containing agar.

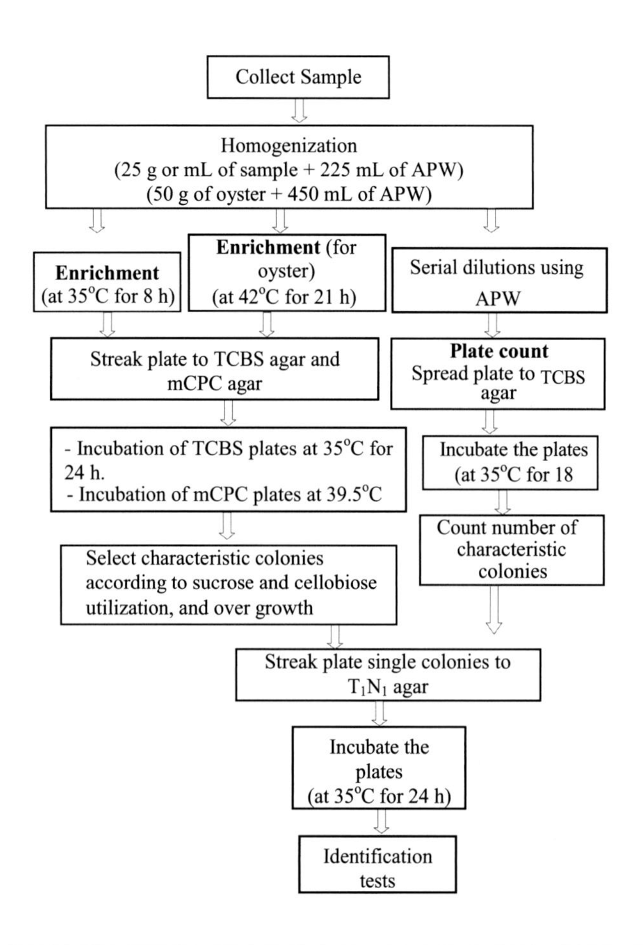

FIGURE 23.1

Vibrio selective enrichment isolation and counting flowchart.

23.2.2 Selective enrichment isolation

Cut large samples (e.g., seafood, vegetables) into smaller pieces with sterile scissor before weighing. Weigh 25 g of sample into a sterile blender (or stomacher plastic bag) containing 225 mL of APW (or weigh 50 g in the case of oyster into 450 mL of APW). Homogenize a sample by blending for 1 min (stomaching for 2 min). Pour the homogenate into a sterile Erlenmeyer flask. In the case of a liquid sample, add 25 mL of the sample into a sterile Erlenmeyer flask containing 225 mL of APW and mix well by gently shaking. Incubate the homogenate at 35°C for 8 h in an incubator and the oyster homogenate at 42°C for 21 h in a thermostatic shaking water bath.

After incubation of the homogenate for 8 and 21 h, streak plate a loopful of enriched homogenate from surface growth (pellicle) separately onto selective/differential agar medium: TCBS agar and mCPC agar (but polymyxin B in mCPC agar may inhibit *V. cholerae*). Incubate inverted TCBS agar plates at 35°C and mCPC agar at 39.5°C for 24 h.

After incubation, examine plates for characteristic colonies as described below. Select three or more selected colonies per plate and inoculate each colony into TS broth containing NaCl. After incubation, streak plate to TCBS agar to obtain a pure culture (Technique 2.2.4). Carry out inoculation to T_1N_1 agar slant, TCBS agar slant, and TS broth. Incubate at 35°C for 24 h and use in identification tests (Technique 23.3).

Thiosulfate-citrate-bile salt-sucrose (TCBS) agar. V. cholerae (classical and El Tor) produces large, yellow (sucrose hydrolyzes), slightly flattened colonies with opaque centers and translucent peripheries (Fig. 23.2). *V. parahaemolyticus* produces round, green or bluish (sucrose-negative) colonies on this medium. *V. vulnificus* produces round, green (sucrose-negative) opaque or translucent colonies. *V. alginolyticus* colonies are large, opaque and yellow. *V. mimicus* produces green (sucrose-negative) colonies. *Vibrio* cannot produce creamy colonies on this medium.

Modified cellobiose-polymyxin B-colistin (mCPC) agar. V. cholerae (classical and El Tor) produces small, smooth, opaque, and green to purple (cellobiose negative) colonies. *V. vulnificus* produces flattened yellow colonies with opaque centers and translucent peripheries. Most other *Vibrio* species cannot grow readily on this medium. But polymyxin cannot be used in the preparation of mCPC in the case of *V. cholerae*, since it inhibits this species. Most *V. parahaemolyticus* strains will not grow on mCPC agar. If growth occurs, colonies will be green-purple in color due to non-cellobiose fermentation.

Gelatin agar or gelatin phosphate salt (GPS) agar. V. cholerae produces transparent colonies usually with a cloudy zone on gelatin agar and the zone becomes more definite after a few minutes of refrigeration. *V. cholerae* on GPS agar produces small and transparent colonies with a cloudy halo. Satellite growth of nongelatinase-producing *Vibrio* colonies may surround *V. cholerae* colonies on these media.

23.2.3 Counting of *Vibrio cholerae*

23.2.3.1 Plate counting technique

Refer to Technique 2.2.2 for details of sample preparation and dilutions. Foods (e.g., seafood or vegetables) are cut into small pieces with sterile scissors. Weigh 25 g of sample (in the case of

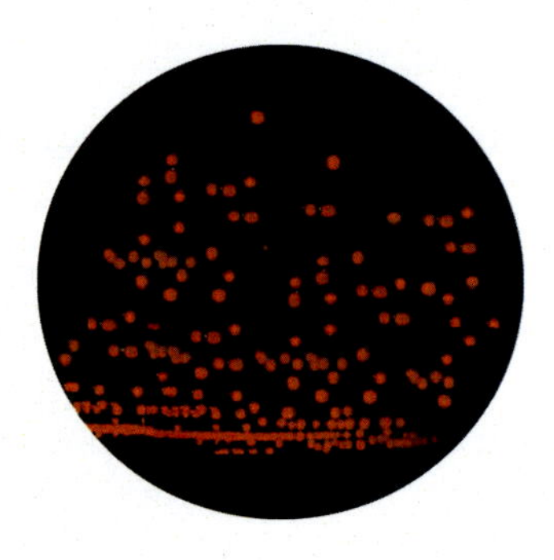

FIGURE 23.2

Vibrio cholerae characteristic colonies on TCBS agar.

oyster, weigh 50 g of sample) into a blender (or stomacher plastic bag) containing 225 mL of APW (for 50 g sample, 450 mL of APW). Homogenize by blending for 1 min (or stomaching for 2 min). In the case of a liquid sample, add 25 mL of the sample into a sterile Erlenmeyer flask containing 225 mL of APW and mix well by gently shaking the flask. Prepare further serial dilutions using APW. Spread plate 1 mL of diluted sample to predried TCBS agar in duplicate (Technique 2.2.4). Retain the plates in upright at room temperature until the inoculum is absorbed by the medium. Incubate inverted plates at 35°C for 48 h. After incubation, examine characteristic colony formation from plates. Count the characteristic *V. cholerae* colonies on plates containing 25–250 colonies. Calculate the number of *V. cholerae* by multiplying the average count by dilution rate and dividing by inoculation amount. Confirm (Technique 23.3) the presumptive isolates and record results as the number of *V. cholerae* colony forming units (cfu) per g or mL of sample.

23.2.3.2 MPN counting technique

Homogenize samples and prepare dilutions as indicated in Technique 23.2.3.1. Inoculate 1 mL of diluted sample into each of 3-MPN tubes of APW. Incubate tubes at 35°C for 24 h. After incubation, examine tubes for growth (turbidity) and color changes. Indicate the numbers of positive tubes for each dilution. From each APW positive tubes, streak plate one loopful of culture to TCBS agar. Incubate inverted plates at 35°C for 24 h. After incubation, indicate the number of positive tubes for each dilution comparing with characteristic colony formation on TCBS agar. Refer to MPN Table 4.1 and calculate the number of *V. cholerae* MPN per g or mL of sample. Perform identification tests to confirm the isolates (Technique 23.3).

23.3 Identification of *Vibrio cholerae*

Obtain pure culture from characteristic colonies as described in Technique 13.4.1 using APW and GPS agar. After the purification of isolates, inoculate pure culture to T_1N_1, agar slant, TS agar slant, and BHI broth, and incubate at 35°C for 24 h.

23.3.1 Morphological identification

Gram staining. Prepare Gram stain from 18 h T_1N_1 agar slant culture as explained in Appendix B under the topic "Gram Staining Reagents." A Gram stain of *V. cholerae* will produce Gram-negative and curved or straight rods (Fig. 23.3).

Motility test. Perform a motility test as described in Technique 13.4.2. Incubate inoculated MT medium at 35°C for 18 h. A spread growth through the medium from the line of inoculation indicates a positive test. *V. cholerae* is a motile bacterium. Confirm the motility of the bacterium with flagella staining.

Flagella staining. Prepare a flagella stain from T_1N_1 agar slant, as explained in Appendix B under the topic "Liefson's Flagella Stain." *V. cholerae* has a single polar flagellum (Fig. 23.3).

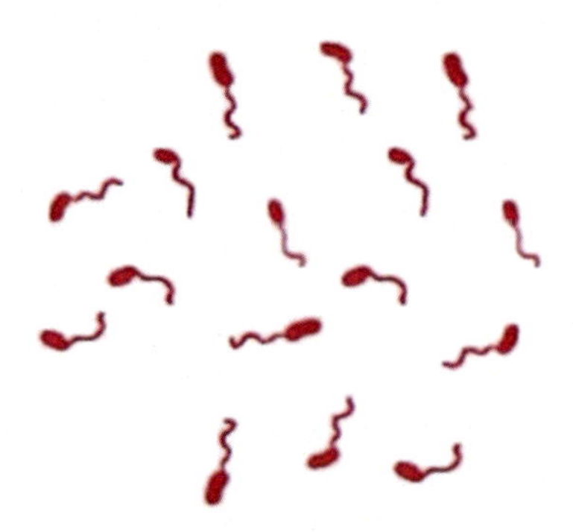

FIGURE 23.3

Staining appearance of *Vibrio cholera*.

Table 23.1 *Vibrio* spp. reactions on/in agar slants.

Vibrio spp.	KI agar		TSI agar		AG agar	
	Slant	Butt	Slant	Butt	Slant	Butt
Vibrio cholera	K	A	A (K)	A	K	A
Vibrio mimicus	K	A	K	A	K	A
Vibrio parahaemolyticus	K	A	K	A	K	A
Vibrio alginolyticus	K	A	A	A	K	A
Vibrio vulnificus	K or A	A	K (A)	A	K	A
Vibrio shigelloides	K or A	A	K or A	A	K	A

A, Acid (yellow); K, basic; (), rare.

23.3.2 Biochemical identification

23.3.2.1 Triple sugar iron agar, Kligler iron agar, and arginine glucose agar slants reactions

Inoculate T_1N_1 agar slant culture with small amount of growth using a sterile needle to each of TSI agar, KI agar, and AG agar slants by stabbing the butt and streaking the slant once. Incubate the tubes at 35°C for 24 h. After incubation, examine slants for characteristic growth (Table 23.1).

On TSI agar slant: *V. cholerae* grows with red slant (not fermenting lactose) and yellow butt (fermenting glucose) without gas and H_2S formation. Use of TSI agar is less desirable since *V. cholerae* ferments sucrose, and results in acid slant and butt, giving false results for lactose and glucose fermentation.

On KI agar slant: *V. cholerae* grows with alkaline (red) slant (not fermenting lactose) and acid (yellow) butt (glucose lactose) without gas and H_2S formation.

On AG agar slant: *Vibrio* species produce alkaline (red) slant and acid (yellow) butt (glucose lactose), as arginine is not hydrolyzed, and cannot produce gas and H_2S.

23.3.2.2 Growth in the presence of salt

Inoculate T_1N_1 agar slant culture by touching a loop onto a colony and into tubes of TS broth containing 0%, 3%, 6%, 8%, and 10% NaCl, before incubating tubes at 35°C for 24 h. *V. cholerae* grows in tryptone broth containing 0% and 3% NaCl, does not grow in 6%, 8%, and 10% NaCl. Reincubate negative tubes for an additional 24 h and record results again.

23.3.2.3 Gelatin hydrolysis

GPS agar plate can be used in the gelatin hydrolyzing test. Divide plates into eight parts. Inoculate a straight line at the center of each part of the agar plate with an isolate. Incubate inverted plates at 35°C for 24 h. After incubation, examine the plates for growth (colony formation). The plate is held above a black surface in the gelatinase reaction analysis. An opaque halo can appear around gelatinase-positive colonies. *Vibrio hallisate* is a gelatinase negative bacterium and other *Vibrio* spp. are positive.

23.3.2.4 Oxidation-fermentation test

Inoculate one loopful of T_1N_1 broth culture into each of two Hugh-Liefson's glucose broth (O-F glucose broth). Add sterile 2% agar (at 45°C) over a depth of 1–2 cm of one of the broths. Incubate broths for up to 2 days or more at 35°C. After incubation, examine broths for growth (turbidity), color change, and gas formation. *Vibrio* spp. cannot utilize carbohydrate oxidatively, cannot produce acid, and cannot cause a yellow color. The purple color of the broth can change to yellow with acid production. *Vibrio* spp. ferment glucose and produce acid. *Pseudomonas* frequently is associated with seafood and utilizes glucose oxidatively.

23.3.2.5 Oxidase test

Place a circle of filter paper into a sterile Petri dish and place a filter paper saturated with oxidase reagent (1% *N,N,N,N′*-tetramethyl-p-phenylenediamine.2HCl) into the Petri dish. Transfer one loopful of T_1N_1 agar slant culture onto the filter paper using a platinum loop (do not use nichrome loop) or a wood stick. Formation of a dark purple color on the colony within 10 s indicates a positive test. *Vibrio* spp. are oxidase positive (except *V. metschnikovii*). Alternatively, a drop of oxidase reagent is added onto the colonies on a T_1N_1 agar slant or T_1N_1 agar plate.

23.3.2.6 Bacteriophage susceptibility

Inoculate one loopful of BHI broth culture to BHI broth and incubate the tube at 35°C for about 4 h (up to appearance of turbidity). Inoculate the surface of the BHI agar plate by swabbing the BHI broth culture and allow the medium in the plate to absorb liquid from swabbing. Label plates (5 min in diameter) at the bottom with name of bacteriophage. Spot inoculate one loopful (3 mm platinum loop) of diluted phage culture (phage group II or IV) onto the labeled area of the BHI agar. Allow the plate to stand for a time period to absorb liquid from phage drop by the medium. Incubate the inverted plates at 35°C overnight and examine the plates during the incubation period for the formation of plaques (Fig. 21.7). Classical *V. cholerae* biotypes are susceptible to these bacteriophages, and the formation of clear plaque (zone) on the agar medium indicates the susceptibility of bacterium to phage. El Tor biotype strains are resistant to these bacteriophages and there will be no clear zone on the agar medium. The same technique can be used to test the sensitivity of El Tor strain to phage III.

23.3.2.7 Polymyxin-B sensitivity

Inoculate one loopful of BHI broth culture to BHI broth and incubate the BHI broth tubes at 35°C for 4 h (up to the appearance of turbidity). Inoculate the surface area of the BHI agar plate by swabbing the BHI broth culture and allow the plate to stand for a time to absorb liquid from swabbing by the medium. Label the plates at the bottom with the name of antibiotics. Place a polymyxin-B antibiotic disk (50 units) on the marked agar surface. Incubate the inverted Petri plates at 35°C for 24 h. After incubation, examine plates for inhibitory zones around antibiotic disks. Classical *V. cholerae* biotype strains produce a complete inhibition zone (10–15 mm in diameter) around the disks on the agar surface or a slight inhibition zone in a 7–8 mm diameter zone.

23.3.2.8 Stretch like thread

Homogenize a colony from a T_1N_1 agar plate culture in a small drop of 0.5% sodium deoxycholate in deionized water in a sterile Petri dish. If the result is positive, the bacterial cells will be lysed by the sodium deoxycholate within 60 s, the suspension will lose turbidity, and DNA will be released from the lysed cells causing the mixture to become viscous. A mucoid "string" is formed when an inoculating loop is drawn slowly away from the suspension (up to 2–3 cm). Most Vibrio species are positive, whereas Aeromonas species are usually negative.

23.3.2.9 Hemolysin test

Add 1 mL (or 0.5 mL) of each BHI broth culture and sheep red blood cells in 5% saline suspension into a sterile tube (13 × 100 mm). Likewise, similar mixtures are prepared in a separate sterile tube with a liquid culture heated at 56°C for 30 min as control. Use hemolytic and nonhemolytic *V. cholerae* strains as controls in the same way. Incubate tubes for 2 h in a water bath at 35°C, then refrigerate the tubes overnight at 4°C. After refrigeration, examine the tubes without shaking for the formation of hemolysis. Most El Tor *V. cholerae* strains hydrolyze and the red color disappears. Hemolysis should not be observed with the heated culture because the hemolysin enzyme is sensitive to heat. Classical *V. cholerae* biotypes and some biotype El Tor strains cannot produce hemolysin enzymes.

The hemolysin test can also be done using an alternative technique. In the Kanagawa phenomenon, spot inoculate one loopful of BHI broth culture on BHIB agar plates (Technique 23.4.4.2). Incubate the inverted plates at 35°C for 24 h. After incubation, analyze plates for hemolysis around the spot inoculum. If hemolysis occurs around the colonies, the test is positive. If a hemolytic zone is not formed around the colonies, the test is negative.

23.3.2.10 Chicken red blood cell agglutination

A thick suspension is prepared with PS solution on a microscope slide from bacterial colonies on TS agar slant incubated for 24 h at 35°C. Add one loopful of washed chicken red blood cells (by 2.5% saline solution) to the slide and mix with the suspended culture. Visible agglutination with red blood cells after mixing indicates a positive test for El-Tor biotype. Use positive and negative controls.

23.3.2.11 Other tests

Perform the following biochemical tests using BHI broth culture.

Mannitol fermentation. Refer to Technique 21.3.2.3 for details of mannitol fermentation. *V. cholerae* ferments mannitol.

Decarboxylase test. Refer to Technique 15.3.2.1 for details of the decarboxylase test. *V. cholerae* decarboxylates lysine and ornithine, but not arginine.

Indole test. Refer to Technique 13.4.3.2 for details of the indole test. *V. cholerae* is indole positive.

Mr-VP tests. Refer to Technique 13.4.3.3 for details of the Mr-VP tests. *V. cholerae* is Mr positive and VP is variable.

Citrate test. Refer to Technique 13.4.3.4 for details of the citrate test. *V. cholerae* ferments citrate.

The *Vibrio* identification flowchart is given in Fig. 23.4.

23.3.3 Serological identification of *Vibrio cholerae*

Refer to Technique 13.5.4 for the complete details of the slide agglutination test. But use T_1N_1 agar slant culture and *V. cholerae* polyvalent somatic (O) antiserum. A positive test is indicated by

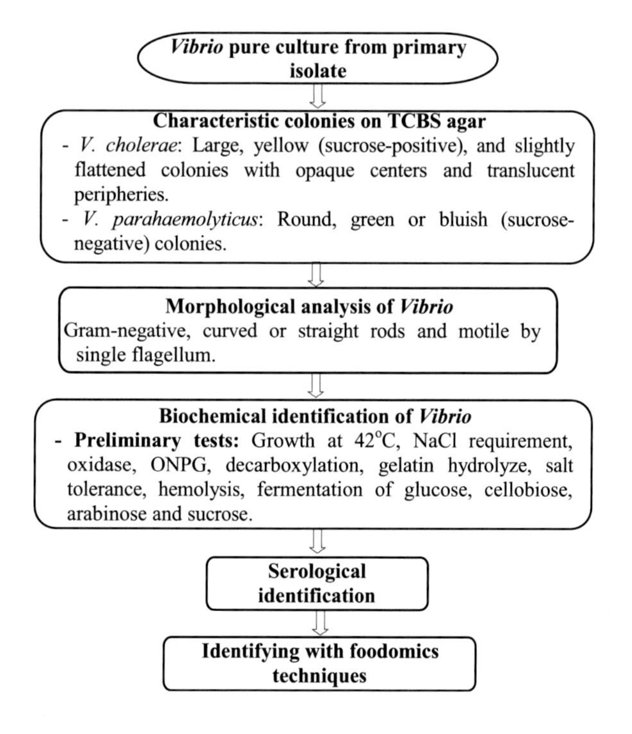

FIGURE 23.4

Vibrio identification flowchart.

a rapid, strong agglutination. If polyvalent O antiserum is positive, perform agglutination tests with monoclonal Inaba, Ogawa, and 01 antiserums.

23.4 Isolation and counting of *Vibrio parahaemolyticus*

All media used in this practice should be prepared with 3% NaCl unless otherwise specified.

23.4.1 Equipment, materials, and media

Equipment. Blender (or stomacher), centrifuge tubes, incubator, microscope glass slide, pipette discard equipment (e.g., a graduated cylinder) containing 10% bleach solution and a piece of cloth at bottom, pipettes (1 mL) in the pipette box, and vortex.

Materials. Alkaline peptone water (APW) containing 3% NaCl, NaCl solution (3%), 3% NaCl-5% glycerol solution, phosphate buffer saline (PBS), and polyvalent and monovalent antiserums (O, K, Inaba, Ogawa, and 01).

Media. Brain heart infusion (BHI) agar slant, BHI broth, cellobiose polymyxin colistin (CPC) agar,

Glucose salt teepol (GST) broth, ONPG peptone water, sodium dodecyl sulfate polymyxin sucrose (SDS) agar, thiosulfate citrate bile salt sucrose (TCBS) agar, triple sugar iron (TSI) agar slant, tryptic soy agar magnesium sulfate-NaCl (TSAMS-NaCl), trypticase soy (TS) agar, TS broth, TS broth containing NaCl (0%, 3%, 6%, 8%, and 10%), and Wagatsuma blood (WB) agar.

23.4.2 Selective enrichment isolation

Refer to Technique 2.2.2 for details of sample preparation and dilutions. Foods (e.g., seafood or vegetables) may be blended or cut into small pieces. Weigh 50 g of a seafood sample into a sterile blender (or stomacher plastic bag) containing 450 mL of APW and blend (or stomach) for 1 min. If the sample is liquid, add 50 mL of the sample into an Erlenmeyer flask containing 450 mL of APW and mix well by gently shaking. Inoculate 10 mL of homogenized sample into each of three 10 mL of double strength GST broth. Inoculate 1 and 0.1 mL of homogenized sample separately into each of three 10 mL of single strength GST broth. Incubate the tubes at 35°C for 18 h. After incubation, examine the tubes for growth (turbidity) and color changes. Streak plate one loopful of GST broth culture onto TCBS agar. Incubate inverted plates at 35°C for 18 h. After incubation, examine the plates for the formation of colonies. *V. parahaemolyticus* appears as round, green, or bluish colonies 2–3 mm in diameter. Confirm the presumptive isolates (Technique 23.4.4).

23.4.3 Counting of *Vibrio parahaemolyticus*

Prepare the sample as indicated in Technique 23.4.2. Spread plate 1 mL of sample on predried TCBS agar in duplicate (Technique 2.2.4). Incubate the inverted plates at 35°C for 48 h. After incubation, examine the plates for characteristic colony formation on the TCBS agar. Count the number of typical colonies on the TCBS agar containing 25–250 colonies. Calculate the number of *V.*

parahaemolyticus by multiplying the average count by the dilution rate and dividing by the inoculation amount. Confirm the presumptive isolates (Technique 23.4.4) and record the results as the number of *V. parahaemolyticus* cfu per g or mL of sample.

23.4.4 Identification of *Vibrio parahaemolyticus*

Obtain pure culture as described in Technique 23.3, if necessary, using GST broth and TCBS agar. Inoculate pure culture into TS agar slant and TS broth.

23.4.4.1 Morphological identification of Vibrio parahaemolyticus

Gram staining. Prepare a Gram stain from 18 h TS agar slant culture as explained in Appendix B under the topic "Gram Staining Reagents." Gram stain of *V. parahaemolyticus* will give Gram-negative, curved or straight rods (Fig. 23.5).

Motility test. Perform the motility test as described in Technique 13.4.2. *V. parahaemolyticus* is motile with polar flagella.

Identification tests are given in Technique 23.3.2 and other characteristics in Tables 23.1 and 23.2.

23.4.4.2 Serological identification

V. parahaemolyticus has three types of antigens: H, O (somatic), and K. All *V. parahaemolyticus* strains contain H antigen. Before O antigen (thermostable) tests, heat bacterial culture for 1 (or 2) h at 100°C to remove the K (capsule) antigen from the bacterial cell, since the capsule masks the O antigen. Serologic tests are not suitable to identify *V. parahaemolyticus* due to cross-reactions with many other marine bacteria. During foodborne outbreak investigations, serologic tests become a valuable epidemiologic tool.

(a) *Determination of O group*. Inoculate two TS agar slants with one loopful of TS broth culture by stubbing butt and streaking slant once. Incubate at 35°C for 24 h. After incubation, wash one slant with 1 mL of a sterile 3% NaCl-5% glycerol solution. Sterilize this cultural suspension at 121°C for 1 h. Aseptically, centrifuge washed and heated culture at 5000 × g and

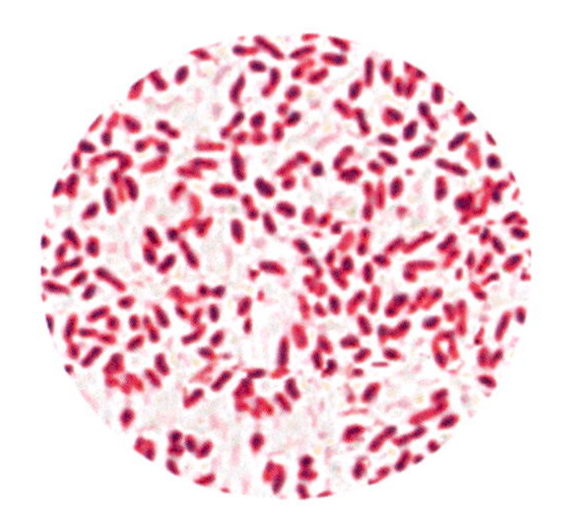

FIGURE 23.5

Staining appearance of *Vibrio parahaemolyticus*.

Table 23.2 Phenotypic characteristics of *Vibrio* species.

Characteristic	*V. cholerae*	*V. cholerae var. eltor*	*V. metschnikovii*	*V. harveyi*	*V. parahaemolyticus*	*V. alginolyticus*	*V. vulnificus*	*V. fluvialis*	*V. furnissii*	*V. alguillarum*	*V. mimicus*	*V. hollisae*
TCBS agar	Y	Y	Y	LG	G	Y	G	Y	Y	NG	G	NG
mCPC agar	P	P	NG		NG	NG	Y	NG	NG	NG	NG	NG
CC agar	P	P	NG		NG	NG	Y	NG	NG	NG	NG	NG
TSI agar (slant/butt)	A/A	A/A			A/A	A/A	A/A		A/A	A/A		
Growth at 42°C	+	+	−	−	−	−	−	−	−	−	−	
Indole/Citrate	+/+	+/+	+/+	+/+	+/+	+/+	v/v	+/+	+/+	+/+		
Mr/VP	+/v	+/v	+/+	+/−	+/−	+	+/−	+/−	+/−	+/−	−/−	−/−
Nitrate reduction	+	+	−	+	+	+	+	+	+	+		
Spread on agar/pili	−/−				+/+	+/+						
Fermentation of												
Arabinose/inositol	+/−	+/−	+/v	v/−	−/−	+/−	−/−	+/−	+/−	v/−	−/−	+/−
Cellobiose/sucrose	−/+	−/+	−/+	v/−	+/−	−/−	v/−	+/+	−/+	−/+	−/−	−/−
Glucose/lactose	−/v	−/−	−/−	v/v	v/−	v/−	−/+	−/−	+/−	−/−		
Mannitol/mannose	+/+	+/+	−/−	v/+	+/+	+/+	v	+/+	−/+	+/+	+/+	−/+
Salicin/starch				−/−		−	v/+					
Growth in:												
0%/3% NaCl	+/+	+/+	v/+	−/+	−/+	−/+	+/+	v/+	−/+	+/+	+/+	−/+
6%/8% NaCl	−/−	−/−	+/v	+/+	+/+	+/+	+/+	+/v	+/+	+/−	−/−	+/−
10% NaCl	−	−	−	−	+	+	−	v	−	−	−	−

Decarboxylation of												
Lysine/ ornithine	+/+	+/+	v/−	+/+	+/+	+/v	+/v	−/−	−/−	−/−	+/+	−/−
Arginine	−	−	+	−	−	−	−	+	+	v	−	−
Enzyme production												
Lipase/ gelatinase	+/+	+/+	+/+	+/+	+/+	+/+	+/+	+/+	+/+	+/+	+/+	+/−
Urease/catalase	+/+	−/+	−/+	+/+	v/+	−/+	−/+	+/+	−/+	−/+	−/+	v/+
β-Galactosidase	+	+	v	−	−	−	+	+	+	+		−
Esculin hydrolysis	−	−	v	−	−	−	−	v	−	−		
Oxidase	+	+	−	+	+	+	+	+	+	+	+	

+, *Positive;* −, *negative; Y, yellow; P, purple; G, green; K, alkali; A, acid; v, variable; NG, no or poor growth.*

resuspend pellets in sterile 3% NaCl solution. Take five microscope glass slides and divide each to three parts with 1 cm^2 at the bottom using a wax pencil. Place a drop of heated culture in each square with a Pasteur pipette. Place a drop of different O group antiserum onto each of two squares representing 12 types of O groups and place a drop of 3% NaCl solution in thirteen squares as a control. Mix all drops with a wire needle, and rock back and forth for at least 2 min. A positive agglutination is produced rapidly. Weak and delayed reactions are considered negative. The control should be negative. Frequently all 12 O groups will be positive or negative. For negative test, the culture must be reported as O group-nontypable.

(b) *Determination of K group*. Wash down the second TS agar slant from previous (a) with 1 mL of 3% NaCl solution to prepare K antigenic suspension. The foregoing O group determination establishes the number and identity of the individual K types that associate with the specific O group. For example, if group O8 is found, one would test against monovalent K20, 21, 22, 39, and 70 antiserums. Thus, five test squares plus one agglutination control square for each O group. A square would be drawn with wax pencil on four microscope glass slides. Place one drop of K antigen onto each of the squares with a Pasteur pipette and place a drop of 3% NaCl solution onto the sample and control square. Add one drop of K antiserum onto each box. Mix all drops with a wire needle, and rock back and forth for at least 2 min. A positive agglutination is produced rapidly and completely. The control should be negative.

23.4.4.3 Determining pathogenicity (Kanagawa test)

A small amount of TS agar slant culture is inoculated into TS broth with a loop and the tubes are incubated at 35°C for 24 h. After incubation, streak plate one loopful of TS broth culture on to well-dried WB agar. Incubate the inverted plates at 35°C for up to 24 h. After incubation, examine the formation of the zone around the colony. A positive test is indicated by beta-hemolysis and transparent clearing of a zone due to the decomposition of red blood cells around the colonies. The special hemolytic response on WB agar is called the Kanagawa phenomenon. *V. parahaemolyticus* is positive for this test and the positive test indicating the bacterium is pathogenic. No observation beyond 24 h is valid.

23.5 Isolation and counting of *Vibrio vulnificus*

All media used in this Technique should be prepared with 3% NaCl, unless otherwise specified.

23.5.1 Equipment, materials and media

Equipment. Blender (or Stomacher), centrifuge tubes, incubator, microscope glass slide, pipette discard equipment (e.g., a graduated cylinder) containing 10% bleach solution and a piece of cloth at bottom, pipettes (1 mL) in the pipette box, and vortex.

Materials. Alkaline peptone water (APW) with 3% NaCl, 3% NaCl-5% glycerol solution, NaCl solution (3%), phosphate buffer saline (PBS) solution, and polyvalent and monovalent antiserums (O, K, Inaba, Ogawa, and 01).

Media. Brain heart infusion (BHI) agar slant, BHI broth, cellobiose polymyxin colistin (CPC) agar, glucose salt teepol (GST) broth, ONPG peptone water, sodium dodecyl sulfate polymyxin sucrose (SDS) agar, thiosulfate citrate bile salt sucrose (TCBS) agar, triple sugar iron (TSI) agar slant, tryptic soy agar magnesium sulfate-NaCl (TSAMS-NaCl), trypticase soy (TS) agar, TS broth, TS broth containing NaCl (0%, 3%, 6%, 8%, and 10%), and Wagatsuma blood (WB) agar.

23.5.2 Selective enrichment isolation

Prepare the sample as indicated in Technique 23.4.2. Incubate the flasks at 35°C for 16 h. After incubation, examine APW for growth (turbidity) and color changes. Streak plate one loopful culture from the 1 cm tap of APW onto each TCBS agar, mCPC agar, and SDS agar plate. Incubate the inverted plates at 35 C for 24 h. After incubation, examine the formation of colonies from plates. *V. vulnificus* produces round, opaque, and green to blue colonies on TCBS agar. On SDS agar, *V. vulnificus* produces round, opaque, blue to brownish colonies with a blue, opaque halo (zone) around each colony. On mCPC agar, *V. vulnificus* produces round, flat, opaque, and yellow colonies. Identification characteristics are given in Technique 23.3 and others in Tables 23.1 and 23.2.

23.5.3 Counting by plate count technique

Refer to Technique 23.4.2 for details of sample preparation and dilutions. Spread plate 1 mL of diluted sample on predried TCBS agar in duplicate (Technique 2.2.4). Incubate inverted plates at 35°C for 48 h. After incubation, examine plates for characteristic colony formation on the TCBS agar. *V. mimicus* produces green (sucrose-negative) colonies. Count the number of typical colonies on the TCBS agar containing 25–250 colonies. Calculate the number of *V. mimicus* multiplying the average count by dilution rate and dividing by inoculation amount. Confirm the presumptive isolates as indicated in Technique 23.3, and Tables 23.1 and 23.2, and record the results as the number of *V. mimicus* cfu per g or mL of sample.

23.5.4 MPN counting technique

Prepare homogenate from sample in APW as indicated in Technique 23.4.2. Inoculate 1 mL of homogenate into each of 3-MPN tubes of GST broth (Technique 4.2). Incubate the tubes at 35°C for 24 h. After incubation, examine APW for growth (turbidity) and color changes. Streak plate one loopful of the culture from positive tubes to each TCBS agar, mCPC agar, and SDS agar plate. Incubate inverted TCBS agar and SDS agar plates at 35°C for 24 h, and mCPC agar at 40°C for 24 h. After incubation, examine characteristic colony formations on the Petri plates and refer to positive tubes according to the presumptive indication on Petri plates. Calculate the number of *V. vulnificus* by referring to MPN Table 4.1. Confirm the presumptive isolates and report the results as the number of *V. vulnificus* MPN per g or mL of sample.

23.6 Identification of *Vibrio* species by foodomics techniques

23.6.1 Matrix-mediated laser desorption ionization-time-of-flight-mass spectrometry identification of *Vibrio* species

Vibrio can be rapidly identified with the matrix-mediated laser desorption ionization-time-of-flight-mass spectrometry (MALDI-TOF Ms) technique, which is based on the principle of ionizing a specific protein profile of microbial cells. *Vibrio* can be easily identified by comparing these profiles with a reference spectrum. MALDI-TOF Ms identification has been explained in Technique 13.6.3.1.

23.6.2 Molecular identification of *Vibrio* species

One of the genotypic identification techniques is polymerase chain reaction (PCR) assays for the identification of bacteria. PCR is a molecular biology technique used in the detection of various bacterial species or strains according to the genes encoding virulence factors or other characteristics of bacterium. Some *Vibrio*-specific primer sets that can be used in the identification of *Vibrio* are given in "Appendix A: Gene Primers (Table A.1)." Using *Vibrio*-specific techniques and materials, the *Vibrio* can be identified by PCR analysis, as described in Technique 13.6.3.2.

23.7 Interpretation of results

Proper cooking and refrigeration of seafood are important factors in the prevention of *Vibrio* gastroenteritis. Consuming only thoroughly cooked shellfish and preventing their recontamination decrease the risk of *Vibrio* foodborne disease. Isolation of *V. cholerae*, *V. parahaemolyticus*, *V. vulnificus*, and *V. mimicus* in foods will be important with respect to public health.

PRACTICE

Isolation and counting of *Shigella dysenteriae*

24

24.1 Introduction

Shigella consists of four species: *Shigella dysenteriae* (subgroup A), *Shigella flexneri* (subgroup B), *Shigella boydii* (subgroup C), and *Shigella sonnei* (subgroup D). *Shigella* species are Gram-negative, facultative anaerobic, catalase positive, oxidase negative, nonspore-forming, nonmotile, and rod-shaped, and within the family Enterobacteriaceae. They cannot decarboxylate lysine, produce acid and cannot produce gas from glucose within 24 h (only one serotype of *S. flexneri* produces gas), and they do not ferment lactose within 48 h. They cannot utilize citrate and malonate. Potassium cyanide (KCN) inhibits some *Shigella* spp. All species of *Shigella* are human pathogens. *S. dysenteriae* can cause severe bacillary dysentery. Only humans are their host. *Shigella* usually spreads via food handlers with poor personal hygiene. The bacteria are either transmitted directly through fecal–oral routes or indirectly through fecal contamination of foods and water. In developing countries, contaminated drinking water is a major cause of shigellosis. Foods are indirectly contaminated by hand contact with fecal material containing *Shigella*. Foods most often associated with *Shigella* are shellfish, raw vegetables, potato, tuna, shrimp, chicken, and many ingredients (e.g., eggs, potatoes, macaroni, and vegetables) used in salads.

S. dysenteriae may spread by food as a result of contamination from food handlers. Shigellosis is a common waterborne disease, but shigellosis can also be a foodborne disease. *Shigella* invades the epithelial mucosa of the small and large intestine to produce exotoxin that has an enterotoxigenic property. The toxin is Shiga toxin. The infective dose is very low, about 10–1000 cells per g or mL of food. Following the ingestion of a food contaminated with *Shigella*, the symptoms occur within 1–3 days and last 5–21 days. The principal symptoms are abdominal pain, fever, chills, headache, and diarrhea often mixed with blood, mucus, and pus.

Hasna's GN broth is used as the enrichment medium in the isolation of *S. dysenteriae*. The isolation of *S. dysenteriae* on solid medium is performed using selective/differential media xylose lysine deoxycholate agar, Hektoen enteric (HE) agar, and Salmonella-Shigella (S-S) agar (a lesser alternative). *Shigella* species are identified by phenotypic and foodomics techniques.

24.2 Isolation and counting techniques

Sampling, sample preparation, and all experimental techniques should be performed under aseptic conditions, and in duplicate, unless otherwise specified. All microbiological media to be used in applications must be sterilized. Again, all solutions and equipment that may cause the contamination of microorganisms must be sterilized or disinfected.

Microbiological Analysis of Foods and Food Processing Environments. DOI: https://doi.org/10.1016/B978-0-323-91651-6.00005-7

Analyze the sample as soon as possible after being obtained. Hold the sample at refrigeration temperature if it needs storage and freeze it if it is to be held longer than 24 h. The *Shigella* selective enrichment isolation and counting flowchart is given in Fig. 24.1.

24.2.1 Equipment, reagents, and media

Equipment. Anaerobic jar with catalyst, blender (or stomacher), incubator, microscope, microscope glass slide, pipette discard equipment (e.g., a graduated cylinder) containing 10% bleach solution and a piece of cloth at bottom, pipettes (1 mL) in the pipette box, and vortex.

Reagents. Alkalescens-Dispar biotypes 1–4, Gram stains, HCl (1 N), Kovacs' reagent, methyl red indicator, NaOH (1 N), novobiocin, physiological saline (PS) solution, polyvalent Shigella antiserums, and Voges-Proskauer reagents.

Media. Acetate agar, acetate differential (AD) agar, carbohydrate fermentation (CF) broth containing carbohydrate (each at a level of 1.0%: adonitol, salicin, rhamnose, glucose, lactose, sucrose, mannitol, raffinose, inositol, xylose, glycerol and dulcitol), Christensen citrate (CC) agar, decarboxylase broth containing lysine or ornithine, Hasna's GN broth, HE agar, malonate broth, motility test

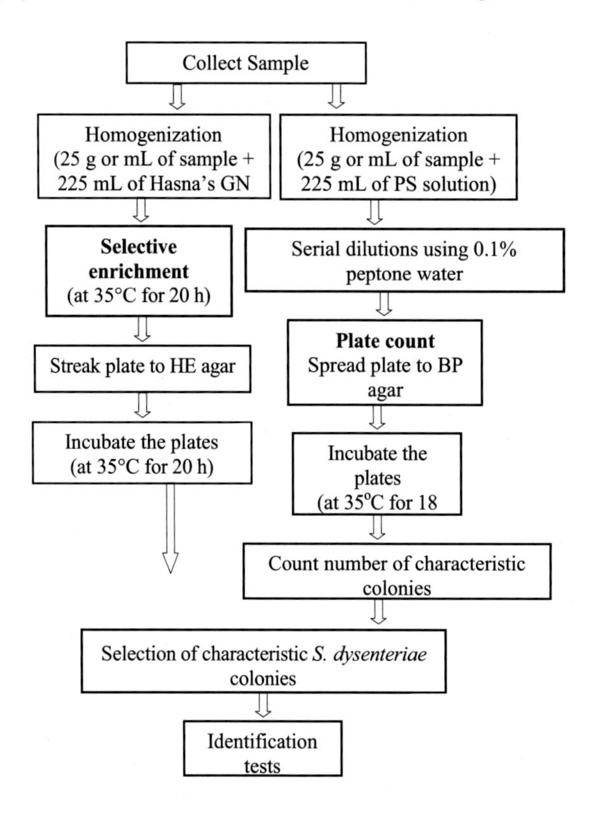

FIGURE 24.1

Shigella dysenteriae selective enrichment isolation and counting flowchart.

(MT) medium, Mr-VP broth, KCN broth, SS agar, Shigella broth containing novobiocin, Shigella broth, triple sugar iron (TSI) agar, trypticase (tryptone) soy (TS) broth, trypticase soy yeast extract (TSYE) broth, urea agar, and veal infusion agar.

24.2.2 Selective enrichment isolation

There are two technique for the isolation of *S. dysenteriae*. The first technique involves the use of a selective enrichment medium, Hasna's GN broth or Shigella broth with novobiocin (0.5 μg mL^{-1}).

The second technique is DNA hybridization. The enzyme DNA gyrase induces a negative reaction. But novobiocin inhibits DNA gyrase. Thus the novobiocin is not added into *Shigella* broth. TSYE can be recommended as enrichment for the DNA hybridization.

Aseptically weigh 25 g or mL of sample into 225 mL of Hasna's GN broth or Shigella broth containing novobiocin (0.5 μg mL^{-1}) in an Erlenmeyer flask. Homogenize the sample by gently shaking the flask. Hold the homogenate for 10 min at room temperature and shake periodically. If necessary, adjust the pH to 7.0 with sterile 1 N NaOH or 1 N HCl. Incubate the mixture at 35°C for 20 h. After incubation, examine the broth for growth (turbidity) and color change. Agitate enrichment culture, streak plate on HE agar (or MacConkey agar) and incubate inverted plates at 35°C for 20 h. After incubation, examine plates for the formation of *Shigella* characteristic colonies. *S. dysenteriae* produces a pink red color with a black center.

24.2.3 Counting of *Shigella dysenteriae*

Weigh 25 g of the sample into a blender (or stomacher plastic bag) containing 225 mL of PS solution, and homogenize it by blending for 1 min. In the case of a liquid sample, aseptically add 25 mL of sample into a sterile Erlenmeyer flask containing 225 mL of PS solution and mix well by gently shaking the flask. Prepare further dilutions using sterile PS solution. Spread plate 1 mL of sample on predried HE agar in duplicate (Technique 2.2.4). Allow plates to absorb liquid from the inoculum. Incubate the inverted plates at 35°C for 24 h. After incubation, examine the plates for the formation of *S. dysenteriae* characteristic colonies. Count typical black center colonies from plates containing 25–250 colonies. Calculate the number of *Shigella* by multiplying the average number by the dilution rate and dividing by the inoculation amount. Verify possible isolates (Technique 24.3) and record the results as the number of *Shigella* colony forming units (cfu) per g or mL sample.

24.3 Identification of *Shigella dysenteriae*

Obtain a pure culture using XLD agar and TS broth as indicated in Technique 13.4.1. Inoculate the pure culture in TS broth and TS agar slant. *S. dysenteriae* can be identified with the morphological and biochemical tests (Table 24.1). The *Shigella* identification flowchart is given in Fig. 24.2.

Table 24.1 Phenotypic differentiations of *Shigella* species.

Characteristic	*Shigella dysenteriae*	*Shigella flexneri*	*Shigella sonnei*	*Shigella boydii*
TSI agar (slant/butt)	K/A	K/A	K/A	K/A
LI agar	−	−	−	−
Glycerol/gelatin hydrolysis	v/ −	v/ −	v/ −	v/ −
KCN/H_2S	−/ −	−/ −	−/ −	−/ −
Mr/VP	+/ −	+/ −	+/ −	+/ −
Nitrate reduction	+	+	+	+
Citrate/indole	−/v	−/v	−/ −	−/v
Phenylalanine deaminase	−	−	−	−
Fermentation of				
Glucose (acid/gas)	+/ −	+/ −	+/ −	+/ −
Lactose/mannitol	v/ −	−/v	+/ −	−/ +
Maltose	−	v	−	−
Mannose/sucrose	+/ −	+/v	+/ +	+/ −
Raffinose/xylose	−/ +	v/ −	v/ −	−/ −
Enzymatic reaction				
Catalase	v	+	+	+
Oxidase/urease	−/ −	−/ −	−/ −	−/ −
β-Galactosidase	v	−	+	−
Decarboxylation of				
Arginine/lysine	−/ −	−/ −	−/ −	−/ −
Ornithine	−	−	+	v

+, *Positive;* −, *negative; v, variable; K, alkaline; A, acid.*

24.3.1 Morphological identification

Gram staining. Prepare a Gram stain from 18 h TS agar slant culture as explained in Appendix B under the topic "Gram Staining Reagents." A Gram stain of *S. dysenteriae* will produce Gram-negative, rod-shaped single cells (Fig. 24.3).

Motility test. Perform the MT as described in Technique 13.4.2 using MT medium. *S. dysenteriae* is nonmotile. Confirm the test using the flagella staining technique.

Flagella staining. Prepare a flagella stain from 18 h TS agar slant culture as indicated in Technique 13.4.2.

24.3.2 Biochemical identification

TSI agar slant reactions. Refer to Technique 13.4.3.1 for details of TSI agar slant reactions. *Shigella* species produce red (alkaline) slant and a yellow (acid) butt, no gas, and no H_2S.

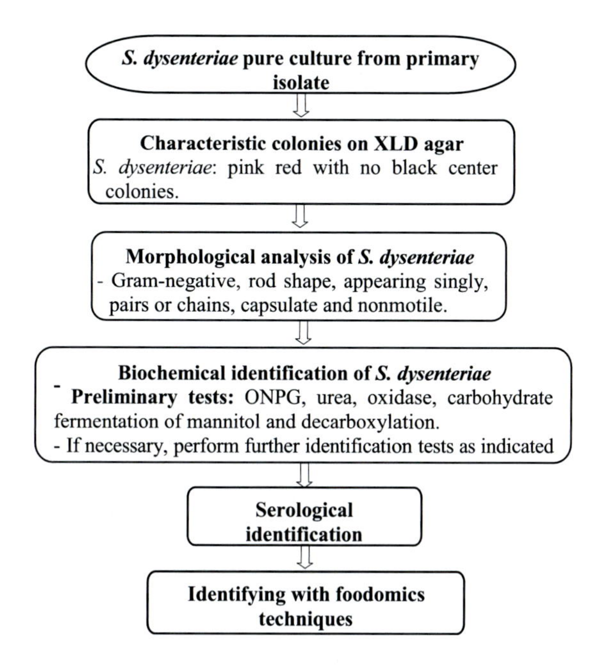

FIGURE 24.2

Shigella dysenteriae identification flowchart.

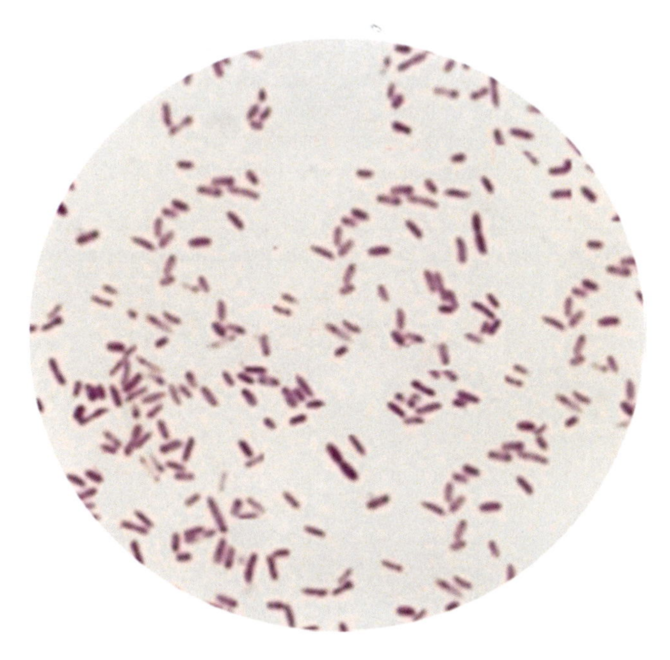

FIGURE 24.3

Staining appearance of *Shigella dysenteriae*.

Urease test. Refer to Technique 13.4.3.5 for details of urease test. *Shigella* species are urease negative.

Citrate test. Refer to Technique 13.4.3.4 for details of citrate test. *Shigella* species are citrate negative

KCN test. Refer to Technique 13.4.3.7 for details of KCN test. *Shigella* gives a negative reaction.

Indole test. Refer to Technique 13.4.3.2 for details of indole test. *Shigella* gives a variable reaction.

Mr-VP tests. Refer to Technique 13.4.3.3 for details of Mr-VP tests. *Shigella* gives a positive result for Mr test and a negative result for VP test.

Acetate differential agar. Streak plate one loopful of TS broth culture to AD agar in duplicate and incubate inverted plates at 35°C for 18 h. *Shigella* cannot grow on this medium.

Carbohydrate fermentation test. Refer to Technique 15.3.2.2 for CF. *Shigella* cannot utilize sucrose (*S. sonnei* utilizes and produces acid after several days) and lactose (some strains of *S. flexneri*, *S. boydii*, and *S. sonnei* produce acid after several days), malonate, adonitol, inositol, and salicin. *Shigella* utilizes mannitol (*S. dysenteriae* and some strains of *S. flexneri* do not utilize it) and glucose (without gas production).

Decarboxylase test. Refer to Technique 15.3.2.1 for the decarboxylase test. *Shigella* does not decarboxylate lysin and arginine but does decarboxylate ornithine.

Perform further phenotypic identification tests as given in Table 24.1, if necessary.

24.3.3 Serological identification of *Shigella dysenteriae*

Transfer 24 h TS agar slant culture into 3 mL of 0.85% saline in sterile tube (13 × 100 mm) and homogenize to prepare McFarland standard no. 5 turbidity. Draw nine 3 × 1 cm circles on three clean microscope glass slides with a waterproof pencil. Add one or two drops of homogenized culture, polyvalent *S. dysenteriae* antiserum, and saline in accordance with the details given in Table 24.2.

The contents are mixed by gently tilting (rock back and forth) the slide through fingers and by continuing to shake the slide for 3–4 min to accelerate agglutination. The extent of agglutination is examined and results are recorded as follows: 0 = no agglutination; 1 + = slight detectable agglutination; 2 + = agglutination with 50% clearing of liquid; 3 + = agglutination with

Table 24.2 Mixing *S. dysenteriae* culture and polyvalent *S. dysenteriae* antiserum and saline on slide.

No	Culture	Polyvalent antiserum								Saline
		A	A_1	B	C	C_1	C_2	D	A-D	
1	+	+								
2	+		+							
3	+			+						
4	+				+					
5	+					+				
6	+						+			
7	+							+		
8	+								+	
9	+									+

75% clearing of liquid; 4 + = visible agglutination with suspending fluid totally cleared. Reexamine the culture with monovalent antiserum for a positive reaction (2 + , 3 + , 4 +). In the case of a negative reaction, heat the culture for 30 min to hydrolyze interfering capsular antigens or other cell components. Reexamine with polyvalent and monovalent antiserums.

24.3.4 Identification of *Shigella dysenteriae* by foodomics techniques

24.3.4.1 Matrix-mediated laser desorption ionization-time-of-flight-mass spectrometryidentification of Shigella dysenteriae

S. dysenteriae can be rapidly identified with the matrix-mediated laser desorption ionization-time-of-flight-mass spectrometry (MALDI-TOF Ms) technique, which is based on the principle of ionizing a specific protein profile of microbial cells. *S. dysenteriae* can be easily identified by comparing these profiles with a reference spectrum. MALDI-TOF Ms identification has been explained in Technique 13.6.3.1.

24.3.4.2 Molecular identification of Shigella dysenteriae

One of the genotypic identification techniques for bacteria is polymerase chain reaction (PCR) assays. PCR is a molecular biology technique used in the detection of various bacterial species or strains according to the genes that encode virulence factors or other characteristics of a bacterium. Some *S. dysenteriae*-specific primer sets that can be used in the identification of *S. dysenteriae* are given in "Appendix A: Gene Primers (Table A.1)." Using *S. dysenteriae*-specific techniques and materials, *S. dysenteriae* can be identified by PCR, as described in Technique 13.6.3.2.

24.4 Interpretation of results

Proper cooking and refrigeration of seafood are important factors in the prevention of *S. dysenteriae* gastroenteritis. Consume only thoroughly cooked shellfish and prevent their recontamination to decrease the risk from bacteria. The isolation of *S. dysenteriae* from foods is important with respect to public health.

PRACTICE

Isolation and counting of *Brucella* 25

25.1 Introduction

Brucella species are Gram-negative, nonmotile, noncapsulated, and short oval rods (coccobacilli) in pairs or single. They are oxidase and catalase positive, grow optimally at 35°C, and live intracellularly or extracellularly in body fluids (facultative parasite). Different species of *Brucella* are animal specific: *Brucella abortus* (cattle), *Brucella melitensis* (sheep and goats), *Brucella ovis* (sheep), *Brucella canis* (dogs), and *Brucella suis* (pigs). Brucellosis (also known as Mediterranean fever, Malta fever) is a zoonosis. Brucellosis is associated with infected animals of farmers, veterinarians, slaughterhouse workers, inadequately heated meat, raw milk, and cream and cheese made from nonpasteurized milk. The incubation period of brucellosis in humans after the consumption of *Brucella* in food is 1–5 weeks. The principal symptoms are abdominal pain, profuse watery diarrhea, rectal tenesmus, and maybe nausea, without fever or vomiting. *B. abortus* requires CO_2 in the first isolation. *B. abortus* and *B. suis* produce H_2S and others do not. *B. canis* and *B. suis* strongly utilize urea (*B. abortus* and *B. melitensis* weakly). *B. abortus* and *B. melitensis* grow in media containing acid fuchsine, others do not. *B. abortus* cannot grow in media containing thionil, but others do grow.

The *Brucella* milk ring test (MRT) can be used to safely identify milk samples that contain antibodies to the bacterium originating from an infected animal. Serological surveys for animal brucellosis are possible by the MRT from pooled bulk milk. This test is very reliable for detecting one infected individual animal in a herd. The isolation of *Brucella* involves selective enrichment isolation with incubation in 10% CO_2 atmosphere for up to 1–2 weeks. A biphasic system with both agar and liquid in the bottle can be used in the selective enrichment isolation of *Brucella* from foods containing low numbers of *Brucella*. Morphological analysis can be performed from enriched broth and *Brucella* species are isolated using selective/differential medium Brucella agar. *Brucella* are identified from isolates using phenotypic and foodomics techniques. *Brucella* species are highly infective bacteria and there is a high risk of infections for laboratory staff working with *Brucella*. The isolation of *Brucella* should be restricted to official specialist public health and clinical laboratories.

25.2 Isolation and counting techniques

Sampling, sample preparation, and all experimental techniques should be performed under aseptic conditions, and in duplicate unless otherwise specified. All microbiological media to be used in applications must be sterilized. Again, all solutions and equipment that may cause the contamination of microorganisms must be sterilized or disinfected.

Microbiological Analysis of Foods and Food Processing Environments. DOI: https://doi.org/10.1016/B978-0-323-91651-6.00018-5

25.2.1 Equipment, reagents, and media

Equipment. Blender (or stomacher), incubator (with 5%–10% CO_2), loop, microscope, microscope glass slide, pipette discard equipment (e.g., a graduated cylinder) containing 10% bleach solution and a piece of cloth at the bottom, pipettes (1 mL) in the pipette box, square bottle (125 mL), test tube (75 × 10 mm), and vortex.

Reagents. Gram stains, monovalent *Brucella* antiserum, phosphate buffer saline (PBS) solution, and stained *Brucella* antigen.

Media. Biphasic media (biphasic Brucella agar and broth), Brucella blood (BB) agar, BB broth, triple sugar iron (TSI) agar slant, Brucella agar containing thionil (1:100,000, 1:50,000, and 1:25,000), Brucella agar containing fuchsine (1:100,000 and 1:50,000), trypticase soy (TS) broth, and sodium citrate Brain heart infusion (SCBHI) broth containing albumin (1%).

25.2.2 Selective enrichment isolation of *Brucella*

A biphasic system with both agar and liquid in the bottle (Fig. 25.1) can be used in the selective enrichment isolation of *Brucella* from foods containing low numbers of *Brucella*. Biphasic selective media (Brucella agar and Brucella blood broth) are used. Refer to Technique 2.2.2 for details of sample preparation and dilutions. If the sample is solid, add 25 g of sample into a blender (or stomacher plastic bag) containing TS broth and homogenize by blending for 1 min (or stomaching for 2 min). Add 5–10 mL of a liquid sample (or homogenized sample) into biphasic media in the bottle. Incubate bottles in the upright position at 35°C in aerobic humidified incubator, but for *B. abortus* and *B. suis* incubate in a humidified microaerophilic incubator (5% CO_2) or in a candle anaerobic jar for 16 days. Humidity may be maintained by placing a pan of water in the bottom of the incubator. During incubation, turbidity formation in broth and tiny colony formation on agar are monitored daily. After 4 days of incubation, look for turbidity every 3-day incubation cycle. The agar should be examined before tilting for visible bacterial colonies. If growth is not observed, the broth is tilted one or two times so that the broth over the agar medium moistens and contaminates invisible growth for colony formation.

Starting from 7 days of incubation, if colonies or turbidity are not observed, subculture by removing 1 mL of broth after each 3-day cycle and spread to selective/differential Brucella agar. Tilt the bottles again and reincubate.

Incubate the inverted plates at 35°C in a humidified incubator or in a humidified microaerophilic incubator (5% CO_2) or in a candle anaerobic jar (for *B. abortus* and *B. suis*) for 3 days. Examine the plates daily for the growth of tiny colonies and evaluate colonies by morphology (wet mount preparation and Gram staining). Fastidious *Brucella* may grow (turbidity) in blood containing broth without visible colonies on agar medium. *Brucella* colonies on Brucella agar appear as raising, convex, white, and pinpoint colonies with a round entire edge, shiny surface, and without a distinct odor (Fig. 25.2). Colonies are 1–2 mm in diameter after 48 h, not mucoid, nonpigmented, and nonhemolytic.

Direct isolation. If the expected number of *Brucella* is high in a sample, perform the direct isolation technique. Add 25 mL of homogenized or liquid sample into Brucella broth in an Erlenmeyer flask and incubate the flask at 35°C for 24 h (microaerobically for *B. abortus* and *B. suis*). After incubation, examine the broth for growth (turbidity) and color changes. Spread plate

FIGURE 25.1

Biphasic media.

FIGURE 25.2

Brucella produces white, nonhemolytic, and pinpoint colonies on Brucella agar.

1 mL of sample (either diluted or nondiluted) to BB agar and incubate inverted plates at 35°C for 3 days. After incubation, examine the plates for *Brucella* characteristic colony formation (Fig. 25.2). Confirm the presumptive isolates by phenotypic techniques (Technique 25.3).

The *Brucella* selective enrichment identification and counting flowchart is given in Fig. 25.3.

25.2.3 Counting of *Brucella*

Add 25 g or mL of the sample to an Erlenmeyer flask (or homogenizer plastic bag) containing an equal volume or 1:10 volume of PBS solution. The sample is homogenized by blending for 1 min (or stomaching for 5 min) and the homogenized sample is diluted using PBS solution. Spread plate one mL of the diluted sample to BB agar in duplicate. The inverted plates are incubated at 35°C (microaerophilic for *B. abortus* and *B. suis*). Examine the plates daily for up to 14 days for *Brucella*-like colonies. *Brucella* appears as raised, convex white colonies (1–2 mm in diameter) with a round entire edge, shiny surface, without a distinct odor, not mucoid, nonpigmented, and nonhemolytic. Count the number of *Brucella* colonies from plates containing 25–30 colonies and record the average count. Calculate the number of *Brucella* by multiplying by dilution rate and dividing by inoculation amount. Confirm the presumptive isolates by phenotypic techniques. Record the results as the number *Brucella* colony forming units (cfu) per g or mL sample.

25.3 Identification of *Brucella*

Obtain a pure culture, as described in Technique 13.4.1, if necessary, using BB agar and BB broth. *Brucella* may not need microaerophilic conditions after the first isolate, therefore microaerophilic

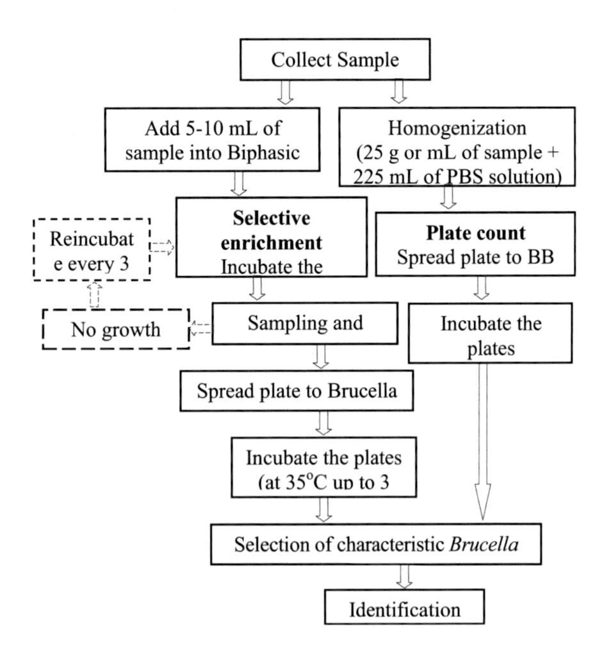

FIGURE 25.3

Brucella selective enrichment isolation and counting flowchart.

conditions cannot be used. Inoculate one loopful of *Brucella* pure culture to BHI broth and BHI agar slant. Incubate the inverted plates and tubes at 35°C for 3 days. Species of *Brucella* can be identified depending on phenotypic and foodomics techniques. The *Brucella* identification flowchart is given in Fig. 25.4.

25.3.1 Morphological identification

Gram staining. Prepare Gram stain from BHI agar slant culture as explained in Appendix B under the topic "Gram Staining Reagents." A Gram stain of *Brucella* species are Gram-negative, short ovoid (small coccobacillus) or rod-shape (Fig. 25.5).

Motility test. Perform motility test with wet mount preparation as described in Technique 13.4.2. *Brucella* species are not motile.

25.3.2 Biochemical identification

25.3.2.1 Dye tolerance

Streak BHI broth culture onto three BB broths containing thionine (1:25,000, 1:50,000 and 1:100,000), and two BB broths containing fuchsine (1:50,000 and 1:100,000). Incubate one set of plates in an air incubator and the other in a 5% CO_2 incubator at 35°C for 3 days. *B. abortus* does not grow in the medium containing thionil, but others do grow. *B. abortus* and *B. melitensis* grow in a medium containing acid fuchsine, but others do not.

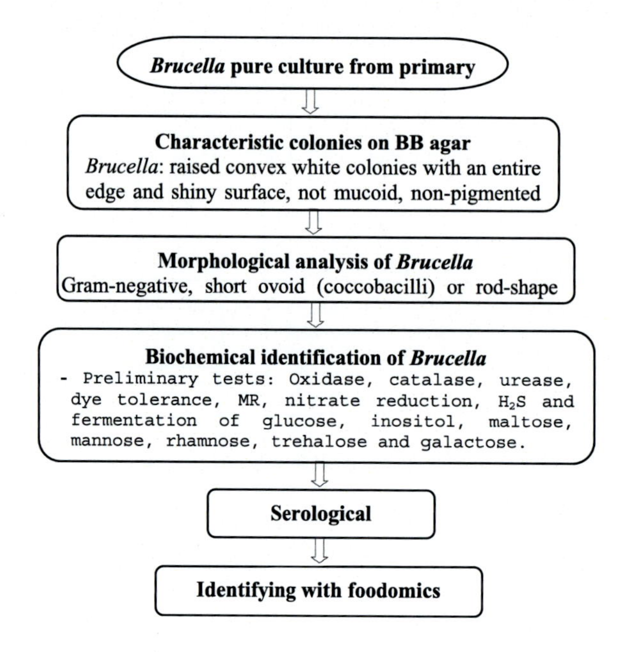

FIGURE 25.4

Brucella identification flowchart.

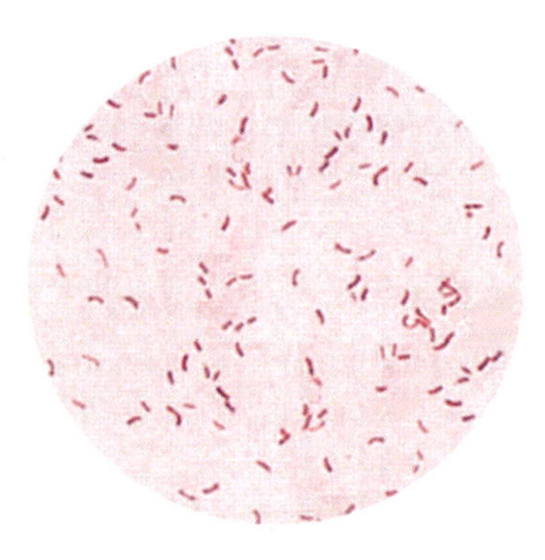

FIGURE 25.5

Staining appearance of *Brucella*.

25.3.2.2 Other tests

TSI agar slant reactions. Refer to Technique 13.4.3.1 for details of TSI agar slant reactions. *B. abortus* and *B. suis* produce H_2S and others do not. *Brucella* ferments glucose (except *B. ovis*) with a yellow color in butt and cannot ferment lactose (no color change) on slant.

Urease test. Refer to Technique 13.4.3.5 for the urease test but use *Brucella* specific urea agar. Inoculate BHI broth culture onto the urea agar slant and incubate tubes at 35°C for up to 3 days. *B. canis* and *B. suis* strongly utilize urea, and *B. abortus* and *B. melitensis* weakly.

Oxidase test. Refer to Technique 21.3.2.7 for oxidase test. *Brucella* species are oxidase positive.

Catalase test. Refer to Technique 21.3.2.2 for the catalase test. *Brucella* species are catalase positive.

Nitrate reduction. Refer to Technique 19.3.2.11 for the nitrate reduction test. *Brucella* species are variable in nitrate reduction.

Mr. Refer to Technique 13.4.3.3 for the Mr test. *Brucella* species are variable in the Mr test.

Perform further phenotypic identification tests, as given in Table 25.1, if necessary.

25.3.3 Serological identification of *Brucella*

Perform the slide agglutination test as described in Technique 13.5.4 using *Brucella* monovalent antiserum.

25.3.3.1 Milk ring test

The MRT is an agglutination reaction used to determine the presence of brucellosis in milk. This test shows the presence of agglutinins (antibodies) in the milk after incubation with the addition of stained inactive *Brucella* into a milk sample. The stained inactive *Brucella* agglutinates with

Table 25.1 Phenotypic characteristics of *Brucella* species.

Characteristic	*Brucella melitensis*	*Brucella abortus*	*Brucella suis*	*Brucella neotoma*	*Brucella ovis*	*Brucella canis*
Thionine (1/50,000)	+	+	+	+	+	+
Basic fuchsin (1/25,000)	+	+	−	−	−	−
Methyl red (1/50,000)	+	+	−	−	+	−
Pyronin (1/100,000)	+	+	−			
Catalase/Oxidase	+/+	+/+	+/+	+/+	+/+	+/+
CO_2 requirement	+	+	−	−	+	−
H_2S/urease	+/v	+/+	+/+	+/+	−/−	−/+
Fermentation of						
Glucose/galactose	+/−	+/+	+/+	+/+	−/−	+/+
Inositol/maltose	−/−	+/−	−/+			
Mannose/rhamnose	−/−	+/+	+/−			
Trehalose	−	−	+			
Decarboxylation of						
Lysine/arginine	−/−	−/−	+/+	+/−	−/−	+/+
Ornithine	−	−	+	−	−	+

+, Positive; −, negative; v, variable.

agglutinins that may be present in cow or sheep milk, the agglutinin-stained antigen complex rises to the surface of the milk with the fat globules. This complex causes deep coloring of the cream layer on the milk's surface. In goat milk, the agglutinin-stained bacterial complex settles in the test tube bottom due to the different creamy properties of the milk.

In performing this test, place a well-mixed milk sample up to 25 mm high in a small, narrow test tube (10 × 75 mm) and a drop of stained Brucella antigen is added into the milk. The tube is mixed by shaking carefully and care should be taken not to foam the milk. The mixture is examined by incubating at 35°C for 60 min. If no result is obtained after that time, it is incubated overnight and examined again. If the cream layer on the surface area is deep colored as a ring and the milk is white or nearly, the test is regarded as positive. This indicates the presence of agglutinins (antibody) in the milk and the milk is obtained from a brucellosis-affected animal. If the cream layer is white and the milk cream layer appears the same color as the milk layer, the test is recorded as negative.

25.3.4 Identification of *Brucella* species by foodomics techniques

25.3.4.1 *Matrix-mediated laser desorption ionization-time-of-flight-mass spectrometry identification of* Brucella

Brucella can be rapidly identified with the matrix-mediated laser desorption ionization-time-of-flight-mass spectrometry (MALDI-TOF Ms) technique, which is based on the principle of ionizing a specific protein profile of microbial cells. *Brucella* can be easily identified by comparing these profiles with a reference spectrum. MALDI-TOF Ms identification has been explained in Technique 13.6.3.1.

25.3.4.2 *Molecular identification of* Brucella

One of the genotypic identification techniques for the identification of bacteria is polymerase chain reaction (PCR) assays. PCR is a molecular technique based on the amplification of specific genomic sequences (such as virulence factors or other characteristics) of the genus, species, or even biotypes of *Brucella*. Some *Brucella*-specific primer sets that can be used in the identification of *Brucella* are given in "Appendix A: Gene Primers (Table A.1)." Using *Brucella*-specific techniques and materials, *Brucella* can be identified by PCR analysis, as described in Technique 13.6.3.2.

25.4 Interpretation of results

For the prevention of brucellosis, undercooked meat and unpasteurized dairy products (e.g., milk, cream, cheese, and ice cream) should not be consumed. Pasteurization of milk destroys *Brucella*. If you are not sure whether the dairy product is pasteurized, do not eat it. People who handle animal tissues (e.g., hunters and animal herdsman) should protect themselves by using goggles, rubber gloves, and gowns or aprons. This prevents the contamination of *Brucella* infection to eyes or inside a cut or abrasion on the skin from infected animals.

PRACTICE

Isolation and counting of *Aeromonas hydrophila* 26

26.1 Introduction

Aeromonas species are Gram-negative, straight rods with round ends (bacilli to coccobacilli shape), motile by a polar flagellum, oxidase and catalase positive, neither salt ($<5\%$) nor acid (minimum pH 6.0) tolerant, do not decarboxylase ornithine, and are noncapsulated. They ferment glucose, lactose, sucrose, and inositol. Psychrotrophic *Aeromonas* spp. grow in foods at chilled temperatures, as low as $-0.1°C$. Their principal reservoir is the aquatic environment, for example, freshwater lakes, streams, and wastewater system; but most commonly are associated with saltwater rather than freshwater. They may present in shellfish and on foods washed with untreated water. They can multiply in a low nutrient level environment such in piped water systems. *Aeromonas* also contaminate red meat and poultry.

Aeromonas hydrophila is mainly associated with a warm climate and fresh or brackish water, and survives in aerobic and anaerobic environments. It is resistant to most common antibiotics and cold temperatures. It is oxidase- and indole-positive. It has a symbiotic relationship with gut flora inside of certain leeches. It is a human pathogen and can be divided into two groups based on the growth temperature. Psychrotrophic *A. hydrophila* strains can grow at an optimum temperature between 15°C and 20°C and can grow as low as 5°C. Mesophilic *A. hydrophila* strains can grow at an optimum temperature of 35°C and can grow at temperatures as high as 45°C.

Aeromonas infections in people are predominantly during warm weather. People are mainly contaminated from water, soil, seafood, or other infected people. Four types of food or waterborne diseases are associated with *Aeromonas* species:

1. wound infection with exposure to water and soil;
2. a general infection with the bacterial spread in the body (septicemia), especially in immunocompromised people and with various other illnesses;
3. gastroenteritis (diarrheal disease); and
4. systemic infection, for example, meningitis, peritonitis, otitis, eye, or urinary tract.

Virulent strains of *A. hydrophila* produce a toxin that can kill tissue cells and lyse blood cells. *A. hydrophila*, *A. sobria*, and *A. caviae* can cause gastroenteritis. *A. hydrophila* gastroenteritis most commonly is associated with children under 5 years old. The infective dose is about 10^5 viable cells per g or mL of food. Two types of gastroenteritis are associated with *A. hydrophila*: a cholera-like illness with a watery (or rice-watery) diarrhea without vomiting; and a dysenteric illness with loose stools containing blood and mucus. Virulent strains of *A. hydrophila* produce heat-labile, β-hemolytic, and cytotoxic enterotoxin. *A. hydrophila* infection is associated with water, poultry, seafoods, raw milk, ice cream, and salad vegetables. It may be an important part of the spoilage flora of chilled meats.

Microbiological Analysis of Foods and Food Processing Environments. DOI: https://doi.org/10.1016/B978-0-323-91651-6.00009-4

A. hydrophila grows on a wide variety of enteric media. Bile salts and ampicillin are selective agents for *the* selective isolation of *A. hydrophila*. Modified XLD agar can be used as a selective/differential medium in the isolation and counting of *Aeromonas*. *A. hydrophila* and *A. sobria* are hemolytic on equine blood agar; producing complete hemolysis and *A. hydrophila* shows a green coloration. *A. caviae* is nonhemolytic but shows a zone of precipitation. Hemolysis may be seen in areas of heavy confluent growth.

26.2 Isolation and counting techniques

Sampling, sample preparation, and all experimental techniques should be performed under aseptic conditions, and in duplicate unless otherwise specified. All microbiological media to be used in applications must be sterilized. Again, all solutions and equipment that may cause contamination of microorganisms must be sterilized or disinfected.

The *A. hydrophila* selective enrichment isolation and counting flowchart is given in Fig. 26.1.

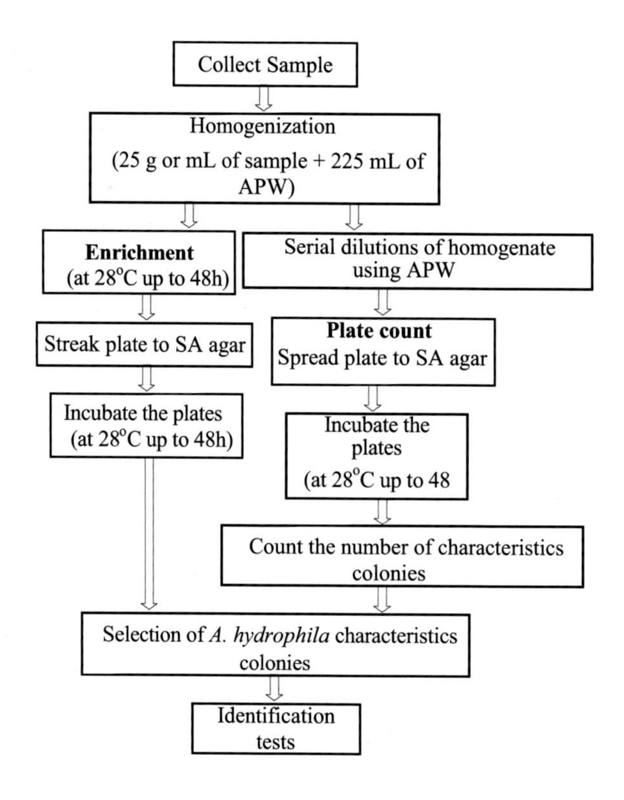

FIGURE 26.1

Aeromonas hydrophila selective enrichment isolation and counting flowchart.

26.2.1 Equipment, reagents, and media

Equipment. Blender (or Stomacher), incubator, loop, pipettes (1 mL) in the pipette box, pipette discard equipment (e.g., a graduated cylinder) containing 10% bleach solution and a piece of cloth at bottom, and vortex.

Reagents. Alkaline peptone water (APW) and Lugol's iodine solution.

Media. Blood ampicillin (BA) agar, Brain heart infusion (BHI) agar slant, BHI broth, modified XLD agar, starch ampicillin (SA) agar, and thiosulfate citrate bile salt sucrose (TCBS) agar.

26.2.2 Selective enrichment and isolation of *Aeromonas hydrophila*

When samples contain a low number of *A. hydrophila*, a selective enrichment isolation technique is used. Refer to Technique 2.2.2 for details of sample preparation and dilutions. Aseptically, weigh 25 g of the sample into a sterile blender (or stomacher plastic bag) containing 225 mL of APW. Homogenize by blending for 1 min and pour the homogenate into a sterile Erlenmeyer flask. In the case of a liquid sample, add 25 mL of the sample into a sterile Erlenmeyer flask containing 225 mL of APW and mix well by gently shaking the flask. Incubate the homogenized sample at 28°C for 18 h. After incubation, examine the homogenate for growth (turbidity) and color change. Streak plate one loopful of enriched broth culture on SA agar (or BA agar, modified XLD agar). Incubate the inverted plates at 28°C for up to 48 h. After incubation, examine the plates for *A. hydrophila* characteristic colony formation. Confirm the primary isolates as indicated in Technique 26.3.

26.2.3 Counting of *Aeromonas hydrophila*

Refer to Technique 26.2.2 for details of sample preparation and dilutions. Prepare necessary dilutions from the homogenate using APW. Spread plate 1 mL of diluted sample onto SA agar (or BA agar) in duplicate (Technique 2.2.4). Retain the Petri plates in an upright position until the absorption of the inoculum by the medium. Incubate the inverted plates at 28°C for up to 48 h. After incubation, examine the plates for characteristic colony formation. Count the number of typical colonies surrounded by a light halo against a blue background from the plates containing 25–250 colonies. Calculate the number of *A. hydrophila* by multiplying the average count by the dilution rate and dividing by the inoculation amount. Confirm the presumptive isolates (Technique 26.3) and record the results as the number of *A. hydrophila* colony forming units (cfu) per g or mL of sample.

26.3 Identification of *Aeromonas hydrophila*

Obtain a pure culture from enriched or counted plates, as indicated in Technique 13.4.1, using modified XLD agar and BHI broth. After obtaining a pure culture, propagate pure cultures in BHI agar slant and BHI broth, and use in phenotypic and foodomics identification techniques. The *A. hydrophila* identification flowchart is given in Fig. 26.2.

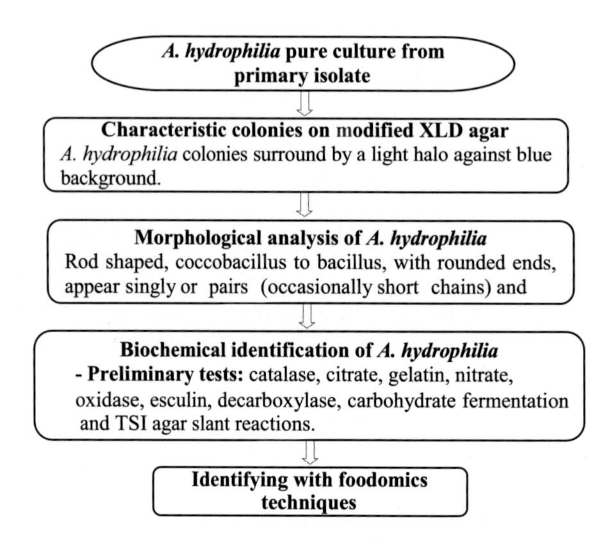

FIGURE 26.2

Aeromonas hydrophila identification flowchart.

26.3.1 Morphological identification

Gram staining. Prepare a Gram stain from BHI broth culture as explained in Appendix B under the topic "Gram Staining Reagents." A Gram stain of *A. hydrophila* appears as Gram-negative, straight rods with round ends (bacilli to coccobacilli shape) (Fig. 26.3).

Motility test. Refer to Technique 13.4.2 for the motility test. *A. hydrophila* is motile by a polar flagellum. Also perform flagella staining as explained in Appendix B under the topic "Liefson's Flagella Stain."

26.3.2 Biochemical identification

TSI agar slants. Refer to Technique 13.4.3.1 for TSI agar slant reactions. *A. hydrophila* produces acid at the butt and slant of TSI agar, and produces gas without H_2S.

Decarboxylation. Refer to Technique 15.3.2.1 for the decarboxylation tests. *A. hydrophila* can decarboxylate arginine and lysine but cannot decarboxylate ornithine.

Nitrate reduction. Refer to Technique 19.3.2.11 for the nitrate reduction test. *A. hydrophila* reduces nitrate.

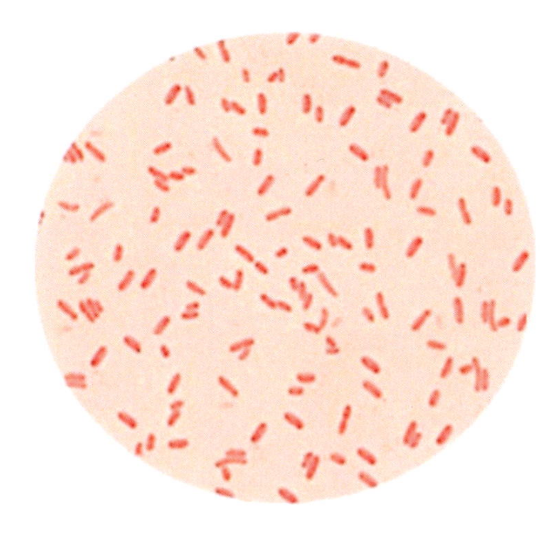

FIGURE 26.3

Staining appearance of *Aeromonas hydrophilia*.

Carbohydrate fermentation. Refer to Technique 15.3.2.2 for carbohydrate fermentation. *A. hydrophila* ferments glucose, sucrose, and inositol, and is variable for lactose.

Esculin hydrolysis. Refer to Technique 14.3.2.1 for the esculin hydrolysis test. *A. hydrophila* hydrolyzes esculin.

Urease test. Refer to Technique 13.4.3.5 for the urease test. *A. hydrophila* cannot hydrolyze urea.

Citrate utilization. Refer to Technique 13.4.3.4 for the citrate utilization test. *A. hydrophila* utilizes citrate.

Perform further phenotypic identification tests, as given in Table 26.1, if necessary.

26.3.3 Identification of *Aeromonas hydrophila* by foodomics techniques

26.3.3.1 Matrix-mediated laser desorption ionization-time-of-flight-mass spectrometry identification of Aeromonas hydrophila

A. hydrophila can be rapidly identified with the matrix-mediated laser desorption ionization-time-of-flight-mass spectrometry (MALDI-TOF Ms) technique, which is based on the principle of ionizing a specific protein profile of microbial cells. *A. hydrophila* can be easily identified by comparing these profiles with a reference spectrum. MALDI-TOF Ms identification has been explained in Technique 13.6.3.1.

26.3.3.2 Molecular identification of Aeromonas hydrophila

One of the molecular identification techniques for the identification of *Aeromonas* species is the polymerase chain reaction (PCR) assay. PCR is a molecular technique based on the amplification of specific genomic sequences (such as virulence factors or other characteristics) of the *Aeromonas* species. Some *A. hydrophila*-specific primer sets that can be used in the identification of

Table 26.1 Phenotypic characteristics of *Aeromonas* species.

Biochemical test	*Aeromonas hydrophila*	*Aeromonas caviae*	*Aeromonas sobria*
Esculin/gelatin hydrolyzes	+/+	+/+	−/+
Nitrate reduction	+	+	+
KCN/pectinase	+/+	+/+	−/+
H_2S/indole	−/+	−/+	−/+
Beta-hemolysis	+	−	+
IMViC	++++	++ − +	+ − v +
Oxidase/catalase	+/+	+/+	+/+
Urease	−	−	−
TSI agar slant	K/A,G	K/A,G	K/A,G
Phenylalanine hydrolyzes	−	−	−
Fermentation of			
Glucose/sucrose	+/+	+/+	+/+
Inositol/arabinose	−/+	−/+	−/−
Mannitol/salicin	+/+	+/+	−/−
Decarboxylation of			
Arginine/lysine	+/+	+/−	−/−
Ornithine	−	−	−

+, Positive; −, negative; v, variable; A, acid; G, gas; K, alkaline.

A. hydrophila are given in "Appendix A: Gene Primers (Table A.1)." Using *A. hydrophila*-specific techniques and materials, *A. hydrophila* can be identified by PCR analysis as described in Technique 13.6.3.2.

26.4 Interpretation

Seafoods have a high possibility to be contaminate with *Aeromonas* spp. due to their growth and survival in water and fish. Fish carry this pathogen in their gut. *Aeromonas* spp. should be inactivated with adequate food processing techniques. *Aeromonas* contamination can be prevented by (1) reducing the potential growth conditions by the storage of food in chilled temperature; (2) maintaining good hygienic practice in the food processing plant; (3) reducing the cross-contamination risk by keeping raw and processed food products separately; and (4) the application of handling and packaging practices that will prevent the possible contamination of processed products.

PRACTICE

Isolation and counting of *Plesiomonas shigelloides* 27

27.1 Introduction

Plesiomonas shigelloides is Gram-negative, motile with 2–7 flagella, nonsporulating, rod-shaped, and a member of the Plesiomonadaceae family. It is facultative anaerobic; catalase and oxidase positive; gelatinase, citrate, and urea negative; produces acid without gas from glucose, mannitol, sucrose, salicin, inositol, and lactose (not from arabinose and xylose); decarboxylates ornithine and lysine; and reduces nitrate to nitrite. It is associated with the intestinal contents of humans and animals. It grows between 8°C and 45°C with an optimum at 25°C–35°C. It is sensitive to pasteurization. It does not grow in foods held in refrigeration. *P. shigelloides* gastroenteritis commonly is associated with seafood (e.g., oysters, fish, shellfish, and crabs), drinking water, and freshwater (streams, rivers, ponds, etc.). The usual foodborne disease is associated with the ingestion of contaminated raw shellfish and water. It produces cholera-like, thermostable, and thermolabile enterotoxins in the intestinal tract. It secretes a β-hemolysin, which may be the major cell-associated virulence factor. It causes gastroenteritis. The incubation period is 48 h after the consumption of contaminated food or water containing *P. shigelloides* and lasts about 1–7 days. The typical symptoms are watery diarrhea without mucoid or blood, nausea, abdominal pain, vomiting, and fever. In severe cases, diarrhea may be greenish-yellow, foamy, and bloody. All people are susceptible to *P. shigelloides* infection. *P. shigelloides* can be isolated from food and water using an enrichment isolation technique. It can be counted using a selective/differential medium and identified by phenotypic and foodomics techniques.

27.2 Isolation and counting techniques

Sampling, sample preparation, and all experimental techniques should be performed under aseptic conditions, and in duplicate, unless otherwise specified. All microbiological media to be used in applications must be sterilized. Again, all solutions and equipment that may cause contamination of microorganisms must be sterilized or disinfected.

The *P. shigelloides* selective enrichment isolation and counting flowchart is given in Fig. 27.1.

27.2.1 Equipment, reagents, and media

Equipment. Blender (or homogenizer), incubator, pipette discard equipment (e.g., a graduated cylinder) containing 10% bleach solution and a piece of cloth at bottom, pipettes (1 mL) in the pipette box, and vortex.

Reagents. Alkaline peptone water (APW) and bile peptone water (BPW).

Microbiological Analysis of Foods and Food Processing Environments. DOI: https://doi.org/10.1016/B978-0-323-91651-6.00028-8

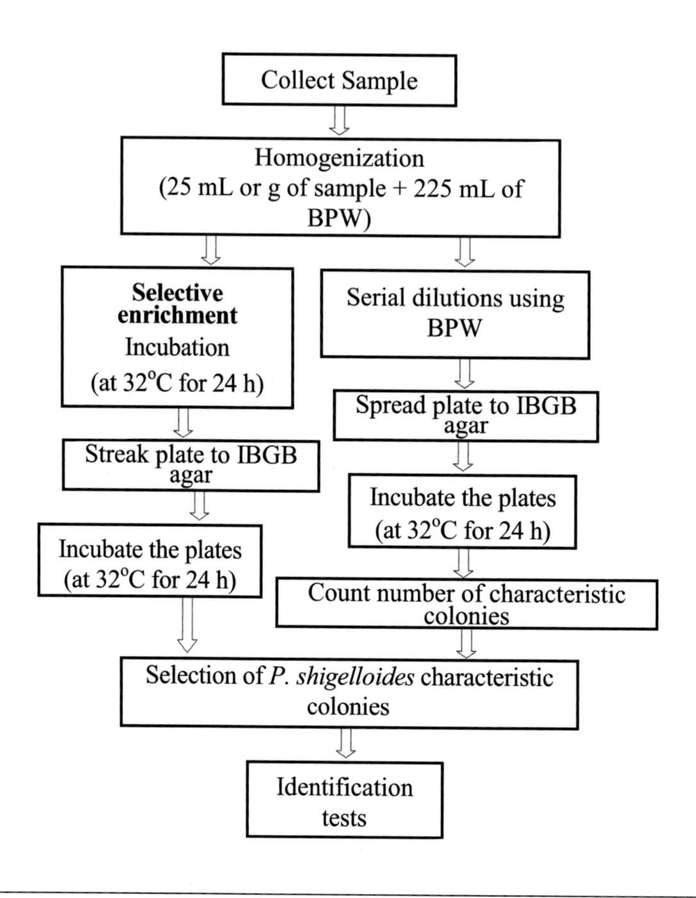

FIGURE 27.1

Plesiomonas shigelloides selective enrichment isolation and counting flowchart.

Media. Inositol brilliant green bile salt (IBGB) agar, IBGB broth, Inositol gelatin (IG) deep, starch ampicillin (SA) agar, tetrathionate broth containing salt (0, 3 and 6%), and triple sugar iron (TSI) agar slant.

27.2.2 Selective enrichment isolation of *Plesiomonas shigelloides*

Refer to Technique 2.2.2 for details of sample preparation and dilutions. Weigh 25 g of the sample into a sterile blender (or stomacher plastic bag) containing 225 mL of BPW. Homogenize by blending for 1 min (or homogenizing for 2 min) and pour the homogenate into an Erlenmeyer flask. In the case of a liquid sample, add 25 mL of sample into a sterile Erlenmeyer flask containing 225 mL of BPW and mix well by gently shaking the flask. Incubate the homogenate at 32°C for 24 h. After incubation, examine the BPW for growth (turbidity) and color changes. Streak plate one loopful of BPW culture onto selective/differential IBGB agar (or SA agar). Incubate the inverted plates at 32°C for 24 h. Examine the plates for characteristic colony formation on agar medium. *P. shigelloides* appears as whitish or slightly pink colonies on IBGB agar (Fig. 27.2), while coliforms are generally

FIGURE 27.2

Plesiomonas shigelloides colonies on IBGB agar.

greenish in color. However, some coliforms can also appear as pink and yellow colonies. It appears as yellow colonies on SA agar. Isolates must be further identified as indicated in Technique 27.3.

27.2.3 Counting of *Plesiomonas shigelloides*

Refer to Technique 27.2.2 for details of sample preparation and dilutions. Prepare necessary dilutions using sterile BPW. Spread plate 1 mL of diluted sample onto predried IBGB agar (or SA agar) (Technique 2.2.4). Retain the plates in an upright position until the inoculum is absorbed by the medium (about 10 min on properly dried plates). Invert the plates and incubate at 32°C for 24 h. After incubation, examine the plates for characteristic colony formation. Count the number of typical whitish or slightly pink colonies on IBGB agar (yellow colonies on SA agar) plates containing 30–300 colonies. Calculate the number of *P. shigelloides* by multiplying the average count by the dilution rate and dividing by the inoculation amount. Confirm the presumptive isolates (Technique 27.3) and record the results as the number of *P. shigelloides* colony forming units (cfu) per g or mL of sample.

27.3 Identification of *Plesiomonas shigelloides*

Obtain a pure culture as described in Technique 13.3.4.1 using IBGB agar and IBGB broth in the purification of *P. shigelloides*. Propagate a pure culture using BHI agar slant and BHI broth and use in identification techniques. The *P. shigelloides* identification flowchart is given in Fig. 27.3.

27.3.1 Morphological identification

Gram staining. Prepare a Gram stain from a BHI broth culture as explained in Technique 13.4.2. A Gram stain of *P. shigelloides* will produce a Gram-negative nonspore-forming rod shape (Fig. 27.4).

Motility test. Perform the motility test as described in Technique 13.4.2. *P. shigelloides* is a motile bacterium with lophotrichous flagella. Confirm the motility of *P. shigelloides* with flagella staining.

Flagella staining. Prepare a flagella stain using the BHI agar slant culture as explained in Technique 13.4.2.

27.3.2 Biochemical identification

Oxidase test. Colonies from enteric agar media cannot be tested directly for oxidase because acid production from carbohydrate fermentation causes a false-negative reaction. Use nonselective nutrient agar medium in the oxidase test. Refer to Technique 21.3.2.7 for oxidase test. *P. shigelloides* is oxidase positive.

Catalase. Refer to Technique 13.4.3.10 for the catalase test. *P. shigelloides* is catalase positive.

Decarboxylase. Refer to Technique 15.3.2.1 for the decarboxylase test. *P. shigelloides* decarboxylases arginine, lysine and ornithine.

Urease test. Refer to Technique 13.4.3.5 for the urease test. *P. shigelloides* is urease negative.

Citrate test. Refer to Technique 13.4.3.4 for the citrate test. *P. shigelloides* is citrate negative.

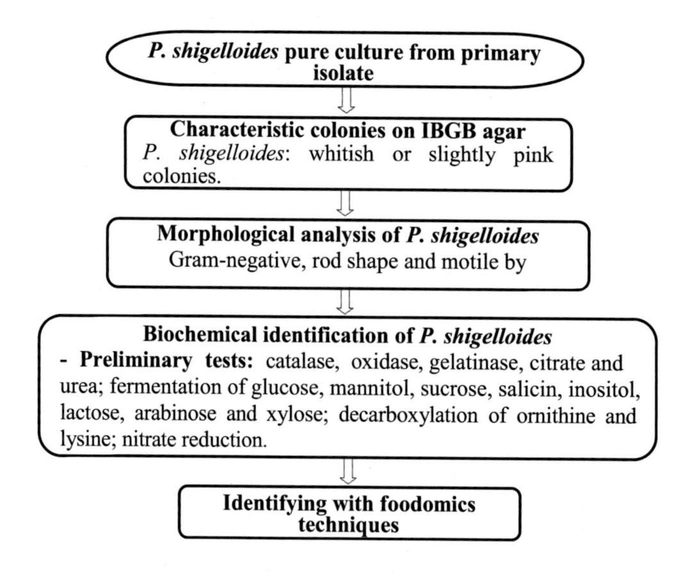

FIGURE 27.3

Plesiomonas shigelloides identification flowchart.

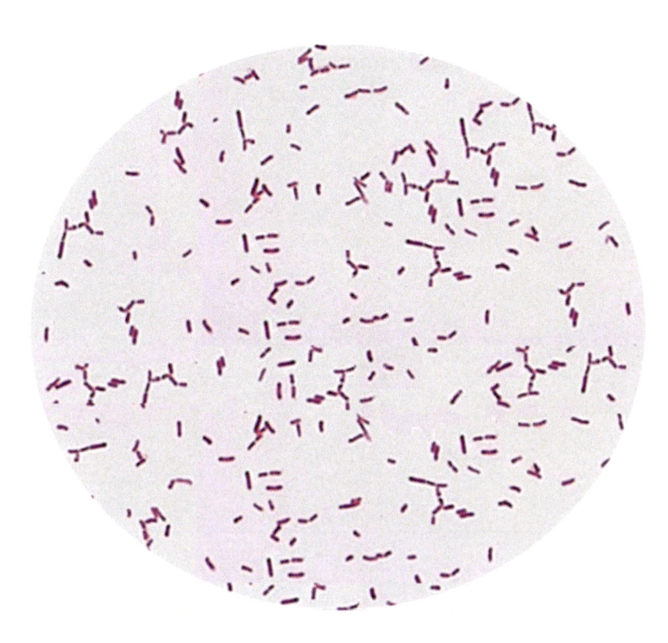

FIGURE 27.4

Staining appearance of *Plesiomonas shigelloides*.

Salt requirement. Inoculate one loopful BHI broth culture into each tetrathionate broth containing 0%, 3%, and 6% NaCl and incubate tubes at 35°C for 24 h. After incubation, examine the tubes for growth (turbidity) in the broth. *P. shigelloides* can grow in tetrathionate broth containing 0% and 3% NaCl, and cannot grow in 6% NaCl.

Growth temperature. Inoculate a loopful culture to tetrathionate broth and incubate tubes at 42°C for 24 h. After incubation, examine the tubes for growth (turbidity) in the broth. *P. shigelloides* can grow at 42°C.

Perform other phenotypic tests for *P. shigelloides* as indicated in Table 27.1.

27.3.3 Serological identification of *Plesiomonas shigelloides*

An antigenic characteristic can be used in the differentiation of *P. shigelloides* strains with the use of somatic (O) and flagellar (H) antigens. Perform serological tests as described in Technique 13.5.4 using *P. shigelloides* BHI broth culture and monovalent antiserum.

27.3.4 Identification of *Plesiomonas shigelloides* by foodomics techniques

27.3.4.1 Matrix-mediated laser desorption ionization-time-of-flight-mass spectrometry identification of Plesiomonas shigelloides

P. shigelloides can be rapidly identified with the matrix-mediated laser desorption ionization-time-of-flight-mass spectrometry (MALDI-TOF Ms) technique, which is based on the principle of

Table 27.1 Phenotypic characteristics of *Plesiomonas shigelloides*.

Characteristic	Reaction
H_2S from sistine	+
Oxidase/catalase	+/+
Nitrate reduction	+
Gelatin hydrolysis	−
Urease/citrate	−/−
Esculin/sodium acetate	−/+
Indole	−
Fermentation of	
Glucose (A/G	+/−
Arabinose/inositol	−/+
Lactose/mannitol	v/−
SUCROSE/xylose	−/−
mannose	+
Decarboxylation of	
Arginine/lysine	+/+
Ornithine	+

+, Positive; −, negative; v, variable; A, acid; G, gas.

ionizing a specific protein profile of microbial cells. *P. shigelloides* can be easily identified by comparing these profiles with a reference spectrum. MALDI-TOF Ms identification has been explained in Technique 13.6.3.1.

27.3.4.2 Molecular identification of Plesiomonas shigelloides

One of the genotypic identification techniques for the identification of *Plesiomonas* species is the polymerase chain reaction (PCR) assay. PCR is a molecular biology technique used in the detection of various *Plesiomonas* species or strains according to the genes that encode virulence factors or other characteristics of *Plesiomonas*. Some bacterium-specific primer sets that can be used in the identification of *P. shigelloides* are given in "Appendix A: Gene Primers Table A.1." Using *P. shigelloides*-specific techniques and materials, *P. shigelloides* can be identified by PCR analysis, as described in Technique 13.6.3.2.

27.4 Interpretation of results

P. shigelloides shows little strain variation in its phenotypic tests. The TSI reaction is coupled with positive oxidase, lysine, ornithine, and arginine decarboxylase, and inositol fermentation can be used to differentiate it from other microorganisms. It is positive for indole and ONPG (*o*-nitrophenyl-β-D-galactopyranoside); ferments glucose, trehalose, and maltose; and reduces nitrate. Citrate, urea, Voges-Proskauer, and potassium cyanide tests, and gas production from any carbohydrate are uniformly negative. Lactose and salicin reaction results are variable. *P. shigelloides* does not require salt for growth but can grow in salt concentrations up to 4%. Hemolysis is usually not produced on plates with 5% sheep blood agar. In controlling the infection caused by the consumption of *P. shigelloide*-contaminated water and raw or insufficiently cooked seafood, especially crustaceans, it is recommended not to consume fishery products raw, to undergo adequate heat treatment, to provide cold storage, to carry out appropriate disinfection of drinking water, and to pay attention to environmental and personnel hygiene.

SECTION

Detection of toxigenic fungi, viruses, and parasites

When molds colonize with grains, there is a high possibility of contamination of grains with mycotoxins. Mycotoxins are toxic secondary metabolic chemical products of molds. Different species of Alternaria, Aspergillus, Fusarium, *and* Penicillium *can produce mycotoxins. Toxigenic molds in crops can be divided into two groups: "field molds," which grow on grains and produce their toxins before the harvesting of grains; and "storage" molds, which contaminate grains after harvest.* Aspergillus flavus *is a field mold associated with infections of grain crops in the field; it can also contaminate stored grains when abiotic factors (e.g., temperature and water activity) are favorable for its growth. Interactions among environmental stress factors (e.g., water activity and temperature) may effect mold growth and mycotoxin production. Good postharvest practices, avoiding high temperatures, and rapid drying of grain can increase the grain quality and reduce the health risk to humans due to mold growth and toxin contamination. Poor postharvest practice can lead to mold contamination, rapid deterioration in food quality, and toxin production. Viruses require a living host cell for replication. Viruses can grow in vivo (within a living cell) or in vitro (in an artificial environment, e.g., a test tube, cell culture flask, or agar plate). Bacteriophages can grow in the presence of a bacterial cells in liquid or solid media. Lytic bacteriophages can appear with a clear zone formation called a plaque (on solid media). A parasite lives on or in a host organism and gets its food from its host. There are three main parasite groups causing disease in humans: protozoa, helminths, and ectoparasites. Parasite diseases cause major human health problems throughout the world. During their life, parasites typically go through several growth stages that involve changes not only in structure but also in biochemical and antigenic composition.*

Practice 28. Isolation and Counting of Toxigenic Fungi

Practice 29. Isolation of Foodborne and Waterborne Viruses, and Typing Techniques

Practice 30. Detection of Foodborne and Waterborne Parasites

PRACTICE

Isolation and counting of toxigenic fungi

28

28.1 Introduction

Many strains of molds produce toxic secondary metabolites (known as mycotoxins). Consumption of foods containing mycotoxin causes mycotoxicosis in humans, animals, and birds. Mycotoxins are not proteins or enteric toxins and are carcinogens to human and animals. Some mycotoxins producing by molds are aflatoxins, patulin, ochratoxins, etc. The most important toxigenic molds genera are *Aspergillus*, *Penicillium*, and *Fusarium*. Other toxigenic molds include *Alternaria*, *Stachybotrys*, *Wallemia*, and *Claviceps purpurea*. Mycotoxicosis can also result from the ingestion of toxic mushrooms.

In general, molds survive and grow best in warm and humid environments. The production of mycotoxins is favored by high humidity and water activity (a_w) at about 25°C–30°C. They are obligate aerobes and grow in the presence of air. They can grow at low temperature (e.g., chilling and refrigerator), very low a_w (0.65), and acidic environment (as low as pH 3.5). These conditions are often used in the preservation of many foods. Vacuum packaging prevents the growth of molds. The mold spores mostly associate with soil, dust, and environment. Many foods are contaminated with mold spores or hyphae, especially before processing. If toxigenic molds grow in foods, they can produce mycotoxins. Feeding moldy products (including moldy silage) to food animals can result in indirect contamination with mycotoxins of animal origin foods (milk, eggs, etc.). Many mycotoxins are resistant to the heat that is used in cooking or processing.

Mycological media used in the isolation and counting of molds differ from bacteriological media. Different media are available for the detection of certain toxigenic molds, for example, AFP medium for *Aspergillus flavus* and *Aspergillus paracyticus*, and PRYES agar for *Penicillium viridicatum*.

28.2 Types of mycotoxins

1. *Aflatoxins*. *A. flavus*, and *A. parasiticus* produce aflatoxins. The most important types of aflatoxins are B1, B2, G1, G2, M1, and M2. Foods commonly associated with aflatoxins are wheat, barley, corn, dry beans, flour, bread, cornmeal, popcorn, corn, milk, peanut, nuts, rice, cheese, sausage, moldy meats, peppers, macaroni, refrigerated and frozen pastries, and foods stored in refrigerators and at room temperature.
2. *Ochratoxins*. *Aspergillus ochraceus*, *Aspergillus sulphurus*, *Aspergillus mellus*, *P. viridicatum*, and *Penicillium cyclopium* produce ochratoxins. Foods commonly associated with ochratoxins are barley, wheat, corn, legumes, peppers, surface injured apples, onions, pears, and salted foods.

Microbiological Analysis of Foods and Food Processing Environments. DOI: https://doi.org/10.1016/B978-0-323-91651-6.00055-0

3. *Sterigmatocystin. Aspergillus versicolor*, *Aspergillus nidulans*, and *Aspergillus rugulosus* produce sterigmatocystin. It causes liver and kidney damage, and hepatocarcinogens. Foods commonly associated with sterigmatocystin are wheat, oats, barley, corn, dried beans, soybeans, citrus fruits, nuts, peanuts, moldy tobacco, coffee beans, and cheese.
4. *Citrinin. Penicillium citrinum*, and *P. viridicatum* produce citrinin. It contaminates wheat, rice, rye, oats, moldy bread, and cured hams. It cannot contaminate cocoa and coffee beans, because caffeine inhibits the growth of *Penicillium* spp.
5. *Patulin. Penicillium claviform, Penicillium expansum, Penicillium roquefortii*, and *Penicillium patulum* are common patulin-producing molds. *Penicillium lapidosum, Aspergillus clavatus, Byssochlamys nivea*, and *Byssochlamys fulva* may also produce patulin. Foods commonly associated with patulin are fruits (apple, bananas peas, pineapple, grapes, and peaches), apple juice, moldy bread, sausage, and rice.
6. *Penicillic acid. P. cyclopium, P. viridicatum*, and *Penicillium puberulum* produce penicillic acid. Molds produce penicillic acid on corn, dried beans, and mold fermented sausages.
7. *Rubratoxin. Penicillium rubrum, P. puberulum, P. cyclopium*, and *A. ochraceus* produce rubratoxin. These species are widespread in nature, and present on poultry feed, corn, beans, cheese, and legumes. Rubratoxin causes liver necrosis.
8. *Rice toxins*. Different types of rice toxins are produced by molds. Citreoviridin and citrinin are produced by *P. citrinum*, and cylochlorotin and islanditoxin by *Penicillium islandicum*. Foods commonly associated with rice toxins are rice and corn.
9. *Zearolenon. Fusarium graminearum, Fusarium roseum*, and *Fusarium tricinctum* produce zearolenon. *Fusarium* spp. commonly associate with grains (corn and other grains) in high moisture conditions. Foods commonly associated with zearalenon are corn (especially during heavy rainfall), wheat, oat, barley, and sesame.
10. *Trichothecenes. Fusarium* spp. producing trichothecenes are *Fusarium nivale* and *Fusarium concolor. Trichothecenes* cause hemorrhages on the lips and in the mouth, throat, and gastrointestinal tract. Foods commonly associated with these toxins are grains and rice.
11. *Ergotism*. This poisoning is caused by the mold *C. purpurea*. It produces its fruit bodies on rye. These bodies are not removed before grain is ground; the ergot will poison the corn flour. Foods associated with ergotism are bread and rye.
12. *Alternaria toxins*. Some species of *Alternaria* produce alternaria toxin in apples, tomatoes, blueberries, and fruits. *Alternaria citri, Alternaria alternata, Alternaria solani*, and *Alternaria tenuissima* are toxic species. The alternarial toxins are alternariol, altertoxin, tenuazonic acid, monomethyl ether, and altenuene.

28.3 Isolation and counting of toxigenic molds

Sampling, sample preparation, and all experimental techniques should be performed under aseptic conditions, and in duplicate unless otherwise specified. All microbiological media to be used in applications must be sterilized. Again, all solutions and equipment that may cause the contamination of microorganisms must be sterilized or disinfected.

28.3.1 Equipment, reagents and media

Equipment. Blender (or Stomacher), Erlenmeyer flask, and incubator.

Materials. 0.1% Peptone water and Melzer's solution.

Media. Aspergillus flavus-parasiticus (AFP) agar, potato dextrose (PD) agar, and pentachloronitrobenzene rose Bengal yeast extract sucrose (PRYES) agar.

28.3.2 Isolation and counting techniques

Sample preparation, dilution, and plating techniques are essentially the same as the plate count technique (Technique 2.2). The direct plating technique can be applied to intact particles and whole foods (e.g., grain kernels, seeds, dried beans, nuts, coffee beans, and cocoa beans) and whole spices. Before direct plating of particles, samples should be held in a freezer (at −20°C) for 72 h to kill insects and their eggs.

For a liquid sample, pipette 25 mL of sample into a sterile Erlenmeyer flask containing 225 mL of 0.1% peptone water. Mix well by gently shaking the flame. For a solid sample, weigh 25 g of sample into a sterile blender (or stomacher bag) containing 225 mL of 0.1% of peptone water. Homogenize by blending for 1 min (stomaching for 2 min). But grain samples should not be blended, they should be homogenized by placing them into an Erlenmeyer flask and gently shaking the flask. Prepare further dilutions using 0.1% peptone water. Spread plate 1 mL of the sample on to AFP agar for *A. flavus* and on to PRYES agar for *P. viridicatum* (Technique 2.2.4) in duplicate. Do not invert the plates for incubation. Incubate Petri plates at 30°C for up to 42 h. *A. flavus* and *A. parasiticus* produce insoluble bright orange/yellow colonies on AFP agar and *P. viridicatum* produces violet brown colonies on PRYES agar. Count the number of characteristic mold colonies from plates. *A. oryzae* and *A. niger* can also produce orange/yellow colonies on AFP agar. After 48 h or longer incubation, *A. niger* colonies retain a pale yellow color but begin to produce black conidial heads that can prevent clear the differentiation of colonies. AFP and PRYES inhibit bacteria and rapidly growing fungi due to the inhibitory substances (dicloran, chloramphenicol, etc.). Calculate the number of respective mold species or total number of toxigenic molds multiplying average count by dilution rate and dividing by inoculation amount. Report results as the number of molds cfu per g or mL of sample.

Carry out total mold counts as indicated in Technique 6.3.2. Identify isolated mold colonies by morphological, cultural, and molecular tests as indicated in Technique 6.4. Confirm toxin production of mold species as indicated in Technique 6.4.3.

28.4 Identification of molds

28.4.1 Matrix-mediated laser desorption ionization-time-of-flight-mass spectrometry identification of molds

Molds can be rapidly identified with the matrix-mediated laser desorption ionization-time-of-flight-mass spectrometry (MALDI-TOF Ms) technique, which is based on the principle of ionizing a specific protein profile of microbial cells. Molds can be easily identified by comparing these profiles with a reference spectrum. MALDI-TOF Ms identification has been explained in Technique 13.6.3.1.

28.4.2 Molecular identification of molds

Polymerase chain reaction (PCR) is a genotypic technique used in the detection of various molds species. Some mold-specific primer sets that can be used in the identification are given in Appendix A: Gene Primers in Table A.1. Using mold-specific techniques and materials, the molds can be identified by PCR analysis as described in Technique 13.6.3.1.

28.5 Interpretation of results

The primary protection of humans from mycotoxicosis is to reduce the contamination of food with toxigenic molds. Appropriate packaging can be used to prevent mycotoxin formation in foods and to protect people from mycotoxicosis. Heat treatments to be applied to foods can reduce molds and their spores. Prevention of mold growth in foods and feeds is an important point in reducing the frequency of mycotoxicosis. Special protection against growth of molds and mycotoxin production can be provided by anaerobic packaging, reducing the a_w to 0.60 as much as possible, and freezing storage. In addition, any food or product where molds grow should not be consumed. Foods containing toxigenic mold species may also contain mycotoxin. Some toxigenic mold species are also potential pathogens to humans with spores that are allergic. Spores of *A. flavus* and *A. parasiticus* may also contain low levels of aflatoxins. Therefore, precautions should be taken to prevent inhalation of airborne spores in aerosol sample dusts.

Hazard Analysis Critical Control Point (HACCP) principles must be used to control both contamination of food by molds and growth of molds in food. The counting and identification of molds should be performed as the part of HACCP and quality criteria. The food can be examined directly for the presence of specific mycotoxins by ELISA, high-pressure liquid chromatography, etc. Allowable levels of aflatoxins in some foods are as follows: 20 ppb for food feeds, nuts, peanut, peanut products, and pistachio nuts; and 0.5 ppb for milk. The established upper limit of aflatoxins for some types of foods is 30 $\mu g\ kg^{-1}$ for human consumption.

28.6 Identification of mushrooms

Mushrooms in the Basidiomycota fungi group have umbrella-shaped sporulating structures. Successful identification of a toxic mushroom depends on an accurate botanical identification. The recovery of mushroom toxins involves extraction and quantitation. Chromatographic techniques can detect mushroom toxins, for example, amanitins, muscimol, orellanine, ibotonic acid, psilocybin, muscarine, and gyromitrins. Commercially available kits can also be used for the detection of some types of mushroom toxins. Some of the most common ways to identify poisonous mushrooms are summarized in the followings.

1. *Mushrooms color images*. The overall dimensions, shapes, and colors of mushrooms can be determined by image analysis. In these analyses, different images should be taken from different parts of the mushrooms.

2. *Use of microscope*: Mushroom structures and spores are examined under the microscope. Microscopic analysis always should be started with the low-power objective magnification (10 ×) and increase the magnification when necessary. A wet-mount can be used to analyze the mushroom spores, staining with appropriate stain can improve magnification. Some mushroom can be identified by the shape, size, and color of their spores.

- Some morphological characteristics of toxic mushroom species are given below.

Amanita bisporigera. It is a white, deadly toxigenic mushroom. It has a milk-white color. Macroscopic images will allow for quick identification. The mushroom has a smooth white cap and has a delicate white skirt-like ring near the top. It bears two spores on the basidia.

Amanita muscaria (fly agaric). It produces muscimol and ibotenic acid toxins. It causes sweating, sickness, salivation, weeping, running eyes, diarrhea, hallucination, and possibly coma. When toxigenic *A. muscarine* is consumed, symptoms usually appear after 15 min. It has a large, yellow or orange cap, white gill, white spotted, and is usually red.

Amanita pantherine. It is a panther cap mushroom and produces ibotenic acid. The cap of *A. pantherine* is deep brown. The gills are free to remote, close to crowded, white becoming grayish slowly. It grows under oaks. Symptoms appear after 120 min of consumption with central nervous system depression, ataxia, waxing, religious hallucinations, and hyperkinetic behavior. The toxin is a neurotoxin causing inebriation and delirium.

Amanita phalloides (death cap). It produces amatoxins and phallotoxins. Amatoxins are not destroyed by cooking or drying. Symptoms include severe stomach pain, sickness, sweating, and diarrhea together with intense thirst. Toxins cause kidney and liver failure with death about a week after ingestion. Treatment reduces the mortality rates to 20%. It has a round and hemispherical pale-yellow or olive-green cap, the cap surface is sticky. The smell is faint initially and honey-sweet later, its spores are white and egg-shaped.

Boletus frostii. It is toxic and produces severe gastrointestinal irritation. It is recognized by its bright red color and pores have a moist fruit body from a half sphere to convex. It has amber-colored drops on the pore surface, the edge of the cap is curved inward, its spores are thick walled, smooth, and spindle shaped.

Cortinarius orellanus. It has a concave cap and the cap flattens with age. Its color ranges from orange to brown. The cap surface turns black with potassium hydroxide. The thick gills are colored, changing to a rust-brown with age as the spores mature. It can cause fatal poisonings with the production of orellanin. Orellanin toxicity has an insidiously long latency period and may take 2 days to 3 weeks to cause symptoms after consumption. The toxin causes kidney failure and death without treatment.

Inocybe geophylla. The cap is white or cream-colored with a silky texture. The cap margins may split with age. The crowded gills are cream early, before darkening to a brownish color with the developing spores. The spores are brown. It produces muscarine toxin. The symptoms are similar to muscarine poisoning, namely, greatly increased salivation, perspiration (sweating), and lacrimation (tear flow) within 15–30 min of ingestion. With large doses, these symptoms may be followed by abdominal pain, severe nausea, diarrhea, blurred vision, and labored breathing. Intoxication generally appears within 2 hours after the consumption of mushroom.

Lactarius indigo. It is easily recognized by exuding blue milk appearance. *Lactarius* exudes milk; yellow-brown milk can cause digestive irritation. Its color ranges from dark blue to pale blue-gray. Its spores appear cream to yellow colored, elliptical to nearly spherical in shape.

Macrocybe titans. It is a large mushroom and grows with the formation of clusters, and each mushroom may reach 15–20 inches in diameter, which enables quick visual identification. It has a dark center, is grayish at the margins, and becomes white with age. It has pale a yellowish to brownish colored cap. It has large gills with a creamy to buff-colored mushroom. Younger mushrooms have a more convex surface of the cap and it flattens with age.

Russula emetica. It is recognized by its white color. It can cause nausea when eaten. *Russula* and *Lactarius* are morphologically similar, but *Russula* does not exude milk. It has a cherry red cap with white stem and gills. Its shape ranges from convex to flat. It produces white to yellowish-white spores. Spores are roughly elliptical to egg-shaped.

PRACTICE

Isolation and typing techniques of foodborne and waterborne viruses 29

29.1 Introduction

A virus is on the borderline between living and nonliving things. It has a biological structure mainly composed of nucleic acid within a protein coat. Its size ranges from 0.01 to 0.2 micrometer; they can be seen only with an electron microscope. Viruses only grow in a living cell, either animal, human, or bacteria. They control the host cell's metabolic system to produce energy, and virus-specific proteins and nucleic acid to replicate themselves. After viral components are assembled into viral particles in the infected cell, virus particles are released after the lysis of the host cell. Some viruses do not kill cells but transform them into a cancerous state; some cause illness or survive as lysogenic. Viruses can cause measles, fever, mumps, influenza, poliomyelitis, the common cold, and others. The most important foodborne viruses are hepatitis E virus, hepatitis A virus (HAV), rotavirus, and Norwalk virus. Viruses cannot replicate in foods but can contaminate foods and are transmitted as small, latent particles. Thus the number of infectious virus particles in contaminated foods normally is low. However, the presence of a few infectious particles in foods may be sufficient to cause disease. Ice cream, milk, pastries, salads, shellfish, and other foods consumed raw are the major food vehicles for virus transmission. Foodborne and waterborne viral infections increase due to changes in food processing and consumption patterns. They lead to the worldwide availability of high-risk food. Virus diseases may occur due to the contamination of food from a single food handler or a single source. Viruses can be fecal–orally transmitted, for example, Norwalk-like caliciviruses (NLV) and HAV. These viruses associate with the greatest risk of foodborne transmission. They are transmitted from person to person and drinking water contamination from feces or vomit.

Viruses are associated with many natural environments. They infect and multiply only in living cells. The basic lytic and lysogenic cycles of bacteriophage (bacterial viruses) are shown in Fig. 29.1. The lytic cycle consists of different stages: (1) attachment (adsorption); a phage attaches to a bacterial cell. (2) Genetic material enters into cell. The phage gene passes through the cell wall and membrane, and enters into the cytoplasm of the bacterial cell. (3) Regulation of cellular DNA. Bacteriophage DNA controls bacterial DNA and the bacterial metabolism works in the direction of the bacteriophage's requirements. Energies are used in the production of bacteriophage-specific components. Adjust cellular activities to phage-producing cell; the bacterial cell starts to synthesize virus components. (4) Synthesis of virus-specific components. Phage-specific nucleic acids, enzymes, and proteins are synthesized in the bacterial cell. (5) Phage particles assembly; phage particles assemble to form a phage in several steps. (6) Production of endolysin; the bacterial cell produces endolysin enzyme and this enzyme lyses the bacterial cell, and the bacterial cell releases the mature bacteriophages. In this practice, the isolation of viruses from foods and environmental

Microbiological Analysis of Foods and Food Processing Environments. DOI: https://doi.org/10.1016/B978-0-323-91651-6.00036-7

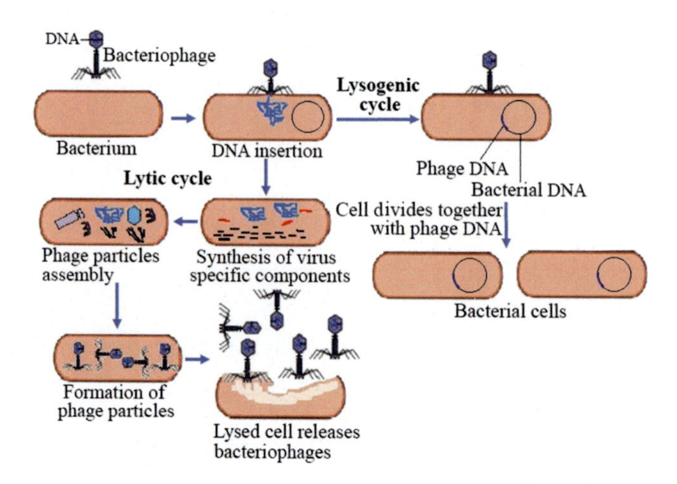

FIGURE 29.1

Lytic and lysogenic cycles of a bacteriophage.

samples are explained, and bacteria and bacteriophage typing techniques have also been included. There is no technique for the direct indication of viruses. The use of bacterial indicators of fecal contamination may indicate the presence of bacteriophages in foods. Infectious virus particles must be separated from food components before identification. Separation is usually accomplished by liquefaction to dissolve food components, and concentration of viral particles from the clarified sample.

Caution. Sampling, sample preparation, and all experimental techniques should be performed under aseptic conditions, and in duplicate, unless otherwise specified. All microbiological media to be used in applications must be sterilized. Again, all solutions and equipment that may cause contamination of microorganisms must be sterilized or disinfected.

29.2 Isolation of foodborne viruses

29.2.1 Equipment, reagents, and media

Equipment. Centrifuge, centrifuge tubes, microscope, incubator, freezer, funnel, pipettes (1 and 10 mL) in pipette box, and pipette discarding equipment.

Reagents. Sterile distilled water, glycine-NaOH buffer (0.09 M; pH 8.8–9.5), glycine-HCl solution (0.05 M, pH 1.5), Na_2HPO_4 (0.15 M, pH 9.0), NaCl solution (0.14 M, pH 7.5 to 9.5), and HCl (1.0 N).

Media. Agar solution (0.75%), neutral red solution (0.0015%), African rhesus monkey kidney cell culture, and viral antiserum.

29.2.2 Sampling and virus extraction technique

Sample sizes may range from 1 to 100 g or mL for the isolation of viruses. Add water, buffers, or media into an Erlenmeyer flask and then homogenize by gently shaking. Figs. 29.2 and 29.3 summarize virus extraction techniques from foods. These include elution–precipitation and adsorption–elution–precipitation techniques. In the adsorption–elution–precipitation technique, the virus is first adsorbed to undissolved food components at acidic pH during liquefaction. Clarification of this sample results in a separation of the virus undissolved food components from the dissolved food components, giving both an initial crude purification and concentration. Viruses are eluted from the undissolved food components by resuspension at alkaline pH. The undissolved food components are removed by clarification, and viruses are concentrated and assayed on the cell culture. In an elution–precipitation technique, viruses are first eluted from food materials by liquefaction at alkaline pH. After removing undissolved food materials by clarification, viruses remaining in suspension are concentrated and assayed on cell culture.

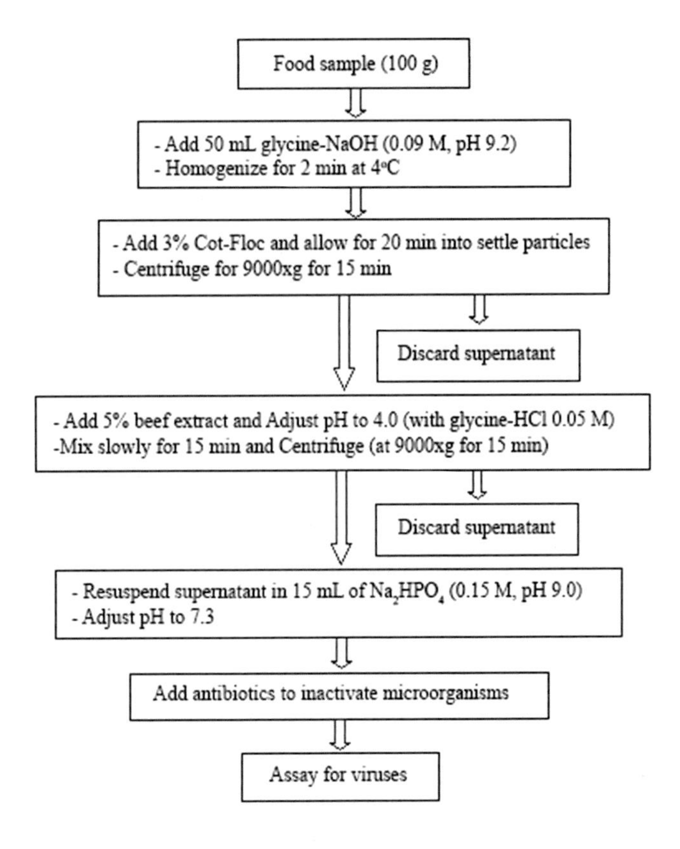

FIGURE 29.2

Extraction of viruses from food sample by elution–precipitation technique.

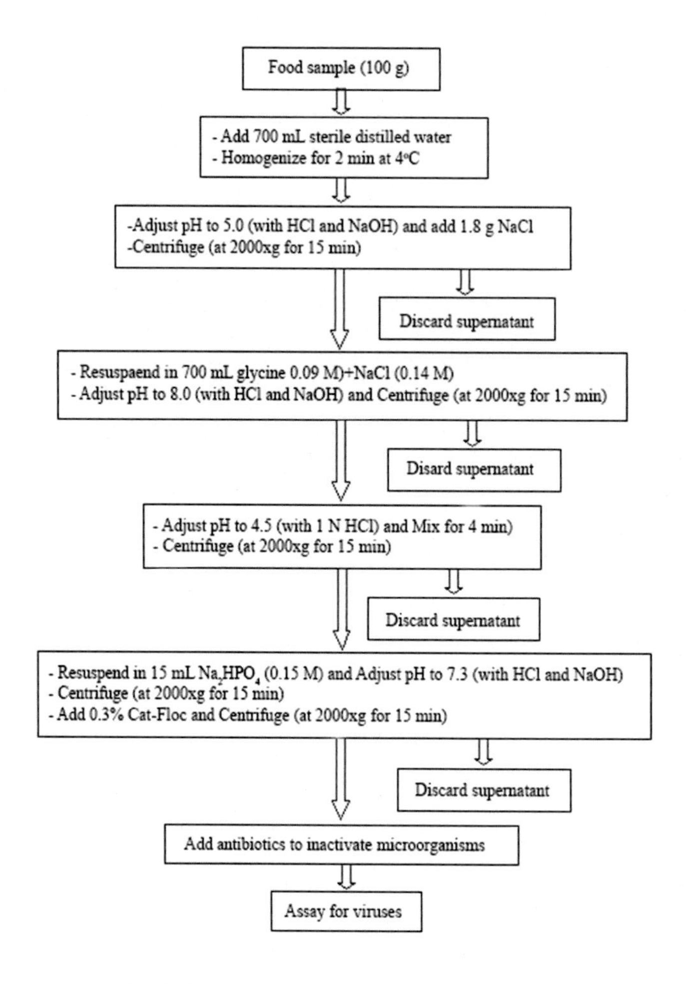

FIGURE 29.3

Extraction of viruses from food sample by adsorption–elution–precipitation technique.

29.2.3 Isolation of enteric viruses

The host cell type is selected based on the type of viruses to be isolated. Cell culture used for food analysis should be susceptible to many types of viruses, not greatly affected by the toxicity of food components, and it should be readily available and easy to maintain in the laboratory. Cultures of African rhesus monkey kidney cells are the most sensitive to enteroviruses.

29.2.3.1 Culturing of viruses

Virus present in food extracts usually can be adsorbed to cell cultures. Viruses should be redistributed over the 90 mL of monolayer cell culture every 15–20 min at 37°C by 0.02 mL per cm^2 during 2 h. At the end of the adsorption period (2 h), the cells are overlaid with maintenance medium containing about 0.75% agar and 0.0015% neutral red for counting. Incubate the inoculated cell culture at 37°C, and look at plaque formation during the 4-week incubation period. Observe periodically every 2 days during the first week of incubation and twice thereafter for up to 4 weeks. The test is considered negative when plaques are not observed after incubation for 4 weeks. Verify both positive and negative results by further passage on tissue culture. The tissue cultures should be observed for a toxic effect on the cells. Cultures developing cytopathic effects are scored as presumptive virus-positive. When 75% of the cell layer is affected by cytopathic effect, the culture is frozen at −70°C. Frozen cultures are thawed and a second passage inoculum is prepared by clarifying the fluid lysate with low-speed centrifugation (2000 × g). The cytopathic effects from the second-round quantal assay are stored at −70°C. Before virus identification, frozen cultures containing viruses are thawed and clarified as above.

29.2.3.2 Virus identification

Identification of virus is done serologically. Using a specific viral antiserum, viruses can be identified depending on the ability of antibodies to block the infectivity of virus particles. Other immunoassays in the identification of viruses are immunoelectron microscopy, immunofluorescence, radioimmunoassay, and enzyme immune-linked assays. In the identification, treating isolated nucleic acid with deoxyribonuclease and ribonuclease, and running the treated material with labeled nucleic acid on agarose gels can determine nucleic acid type. Particle size in identification may be estimated by electron microscopy after filtration through a membrane filter with 100, 50, and 29 nm porosity.

29.3 Bacteriophage isolation and typing

The phages require calcium ions for adsorption and growth on bacteria. Liquid and solid media used for the propagation and typing of bacteriophages should contain 400 μg per mL of $CaCl_2$. But the resulting agar may be slightly cloudy with the addition of calcium. Solid media for propagation and typing should be kept soft and do not dry the plates more than that required to remove surface moisture. For all propagations and typing, use TS agar and broth at pH 7.3 (or use other suitable media depending on microorganism).

29.3.1 Equipment, reagents, and media

Equipment. Centrifuge, centrifuge tubes, incubator, Petri dishes, loop, Erlenmeyer flask, and pipettes (0.1, 1, and 10 mL) in pipette box.

Reagents. Quarter-strength Ringer's solution, agar solution (0.70%), chloroform, and $CaCl_2$ solution.

Media. Tryptone soy (TS) agar, soft TS agar, and TS broth (or other suitable media for microorganisms).

29.3.2 Bacteriophage isolation techniques

29.3.2.1 Preliminary enrichment of bacteriophage

Liquid medium. Add 25 g or mL of food sample into 50 mL of TS broth in an Erlenmeyer flask and homogenize by gently shaking. Use a homogenized or liquid sample in the enrichment of bacteriophage. Add 1 mL of chloroform into 5 mL of the sample, and mix and centrifuge (2000 × g for 15 min). Inoculate 1 mL of 18 h bacterial culture into 5 mL of TS broth and incubate at 35°C for 2–4 h or until the appearance of growth (turbidity) that gives an exponential growth phase. Add 1 mL of the centrifuged supernatant into this TS broth bacterial culture. Incubate TS broth tubes at 35°C for 24 h for enrichment of bacteriophage. After incubation, remove the bacteria by centrifugation at low speed (2000 × g for 15 min) to obtain a clear broth. Inoculate 1 mL of supernatant liquid into 9 mL of bacterial broth culture at exponential phase again and incubate at 35°C for 24 h for the second enrichment. Periodically examine broth for the clearing of turbidity over 24 h. When the inoculated broth is cleared by the supernatant inoculum, this indicates the presence of bacteriophages in the sample and this is used as a bacteriophage culture after centrifugation (2000 × g) to remove remaining bacteria.

Solid medium. Bacteriophages can also be detected on agar medium. Transfer the liquid sample into a sterile centrifuge tube and centrifuge at low speed (2000 × g for 15 min), add 1 mL of chloroform, mix well, and centrifuge again. Insert a swab into an exponential phase bacterial culture and remove the excess fluid by touching the inside of the tube. Spread the swab over the entire surface area of the TS agar plate to obtain a uniform inoculum. As soon as the inoculum on the surface area has been dried, spread 0.5 mL of supernatant bacteriophage culture onto medium. Dry the inoculum again and incubate inverted plates at 35°C for 18 h. After incubation, use a thin, alcohol-flamed rod band at an angle of 90° and funnel for removing agar lumps, and transfer into a sterile Erlenmeyer flask. Add 20 mL of sterile TS broth into the flask. Shake the flask vigorously to obtain a good suspension of the phage. Add a homogenized bacteriophage culture into sterile centrifuge tubes and then centrifuge until the supernatant is clear at low speed (2000 × g for 15 min). The supernatant represents the stock bacteriophage. Alternatively, the whole plate may be frozen at −60°C, thawed at room temperature, the released fluid removed, and centrifuged at low speed. Store the supernatant fluid at 4°C as a bacteriophage stock culture.

29.3.2.2 Isolation of bacteriophage

Spread plate the surface of TS agar with a swab that has been moistened thoroughly in exponential phase bacterial TS broth culture (grown at 35°C). As soon as the inoculum has dried, add 0.1 mL of bacteriophage culture using a sterile pipette onto TS agar and spread bacteriophage stock culture (from 29.3.3.1) over TS agar. Dry the inoculum and incubate at 35°C for 18 h. After incubation, the presence of a phage produces a clear area (plaque) where the bacterial cells have been lysed by the single bacteriophage (Fig. 29.4). Each plaque represents a single bacteriophage. Sometimes, single colonies may appear inside this clear area as phage-resistant mutant colonies. The use of decimal dilutions allows plaques to develop from a single phage particle on the surface area of the agar medium. Use a single plaque without a phage-resistant colony as the pure bacteriophage.

29.3.2.3 Pure bacteriophage culture

Select least five plaques from plates (Technique 29.3.3.2). In the isolation of a pure phage culture, remove the bacteriophage plaque material by spreading the loop over the plaque and transfer each plaque material with a sterile loop to a bacterial TS broth culture in an exponential phase and incubate TS broth

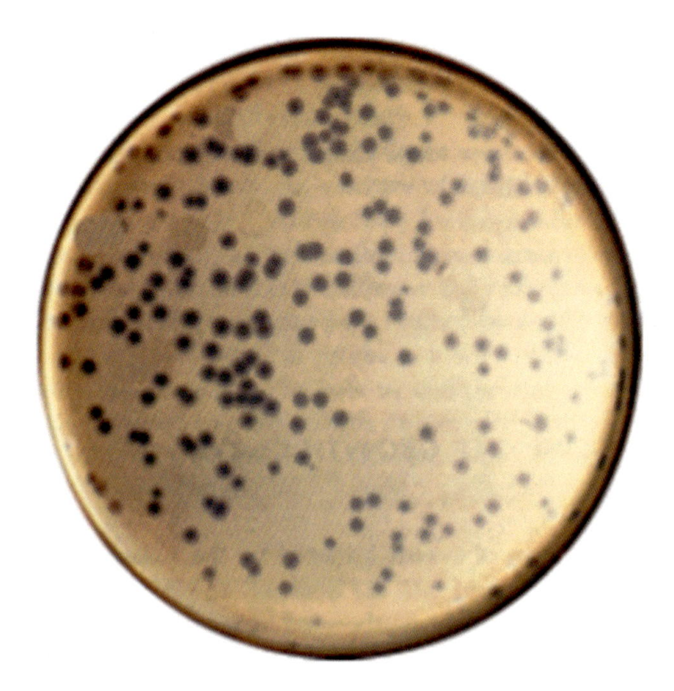

FIGURE 29.4

Bacteriophage culture with plaque formations (confluent lysis).

at bacterial growth temperature until clearing of medium. Include a TS broth inoculated with a bacterium as a control at 35°C. If clearing cannot appear within 24 h it is due to the development of bacteriophage-resistant bacterial mutants. In this case, centrifuge the culture and repeat phage inoculum from supernatant into a fresh bacterial culture at the exponential phase. The pure phage should be free of bacteria and cell debris. If complete clearing occurs in the TS broth culture, this indicates a pure bacteriophage culture. Use this phage culture after centrifugation (2000 × g for 15 min) as a pure bacteriophage culture.

29.3.3 Bacteriophage and bacterium typing

There are two basic bacteriophage typing techniques: (1) liquid medium and (2) solid medium. Propagation (increasing phage particles) in the liquid medium is the simpler technique and yields phage concentrations between 5×10^8 and 1×10^{10} particles per mL. When higher titers (number of bacteriophage) are required, propagation on solid medium should be performed; it can be used as an alternative to the broth propagation technique.

29.3.3.1 Propagation in liquid medium

Transfer 1 mL of the pure bacteriophage culture into 5 mL of the bacterial TS broth culture (in the exponential phase) in duplicate. Incubate tubes at 35°C for 24 h by shaking. After incubation, look for clearing of the bacterial culture. Immediately after clearing, centrifuge (at low speed) culture.

Repeat inoculation from the supernatant fluid into a fresh bacterial TS broth culture by two or three times to increase the number of phage particles in the broth. Finally, centrifuge the incubated culture at low speed, and take the supernatant liquid as bacteriophage culture. Collect the bacteriophage culture in sterile stoppered or screw-capped tubes and store at 4°C.

29.3.3.2 Propagation in soft-agar layer

Add 0.2 mL of bacterial culture (in exponential phase) into a test tube containing 10 mL of soft TS agar (0.7%) and add 0.25 mL of pure bacteriophage culture. Mix the contents of each tube thoroughly and immediately pour into sterile Petri dishes. Incubate plates at 35°C for 18 h. After incubation, use a thin, alcohol-flamed rod band at an angle of 90° and funnel for removing agar lumps, and transfer into sterile centrifuge tubes. Add 20 mL of sterile TS broth into each centrifuge tube. Shake the tubes vigorously to obtain a good suspension and then centrifuge ($2000 \times g$ for 15 min) until the supernatant clears at low speed. Use the supernatant as the stock phage culture. Alternatively, the whole plate may be frozen at −60°C, thawed at room temperature, released fluid removed, and centrifuged at low speed. Collect the bacteriophage culture in sterile stoppered or screw-capped tubes and store at 4°C.

29.3.3.3 Titration of bacteriophage

Prepare serial decimal dilutions from the bacteriophage culture (from Technique 29.3.4.1 or Technique 29.3.4.2) in TS broth using a single 1 mL pipette for each dilution. Spread the swab over the entire surface area of the TS agar plate to obtain a uniform inoculum. Label plates with phage number and draw a circle (1.2 cm in diameter). As soon as the inoculum has been dried, place one loopful (4 mm) or one-drop (0.02 mL) of each diluted bacteriophage onto the labeled area on the TS agar plate. When the drops are dried, incubate the inverted plates at 30°C for 18 h. After incubation, examine the plate for the plaque formation and determine the routine test dilution (RTD). The RTD is the highest dilution that gives confluent (like honeycomb) lysis (RTD includes about 3×10^4 and 2×10^5 particles per mL) on the agar medium, as shown in Fig. 29.5 (from number 1–6) and the RTD value is number 5.

29.3.3.4 Bacteria and bacteriophage typing technique

Only a pure culture of isolated bacterial strain should be submitted to typing. An unknown bacterium is typed with the known bacteriophage at RTD. In the reverse, an unknown bacteriophage can also be identified using kwon bacterial strain in a similar manner. In this technique, only isolated bacteria are typed with a known bacteriophage. A standard 15 cm agar plate can be used for each isolated bacterial culture to be typed. The plates are labeled with an area (1.4 cm in diameter). The agar should be free from surface moisture.

Pour TS agar into sterile Petri dishes and dry at 30°C for 48 h before use. Inoculate 1 mL of TS broth isolated bacterial culture into TS broth and incubate tubes at 30°C for 4–6 h until they show well-defined turbidity (turbidity appears to the naked eye at the exponential phase). Moisten a sterile swab thoroughly in this fresh TS broth culture, remove the excess fluid by touching the inside of the tube, and spread the swab over the entire surface area of the TS agar plate to obtain a uniform inoculum. On a properly prepared plate, the inoculum should be dried within 10 min or less at room temperature. When the inoculum is dried, spot one loopful (4 mm) or a small drop (0.02 mL) of each bacteriophage culture at RTD over the center of a labeled circular area of plate so that each circle corresponds to a particular bacteriophage. If a pipette is used in the transfer of bacteriophage culture, the tip of the pipette should not be touched onto the agar surface. Spread bacteriophage drops over the labeled

FIGURE 29.5

Detection of routine test dilution from 10^{1}–10^{-7} (RTD is number 5).

area with a sterile inoculating loop without disturbing the agar surface. When the drops of the phage have dried within a few min, incubate the inverted plates at 30°C for 18 h. Alternatively, incubate the inverted plates at 35°C for 4–6 h and then hold at room temperature overnight. The plates should not be incubated continuously at 37°C overnight since a heavy growth of the bacteria (or phage-resistant mutant colony) tends to obscure the phage plaques. After incubation, look for lysis (plaque).

When no significant lysis is characterized with the RTD of any phage in the series, repeat the typing of the culture with the same phages in stronger concentrations. The dilution for this purpose should be 1000 times stronger (concentrated) than the RTD or it may be necessary to retype at a 1:10 dilution of bacteriophage.

Reading and reporting the results. Examine the plates with transmitted light against a dark background. When bacteria are typed at RTD, different degrees of lysis from a few discrete lysis to completely confluent lysis indicate the susceptibility of bacteria to the bacteriophage(s) (Fig. 29.6). Secondary phage-resistant growth may occur in the area of confluent lysis. Similarly, reactions ranging from discrete plaques to confluent lysis are also produced when strains are typed at RTD.

The degrees of lysis observed are recorded as follows:

+++ = complete lysis (without secondary growth)
+++ = confluent (honeycomb) lysis (with or without secondary growth)
+ = semiconfluent lysis
− =no lysis

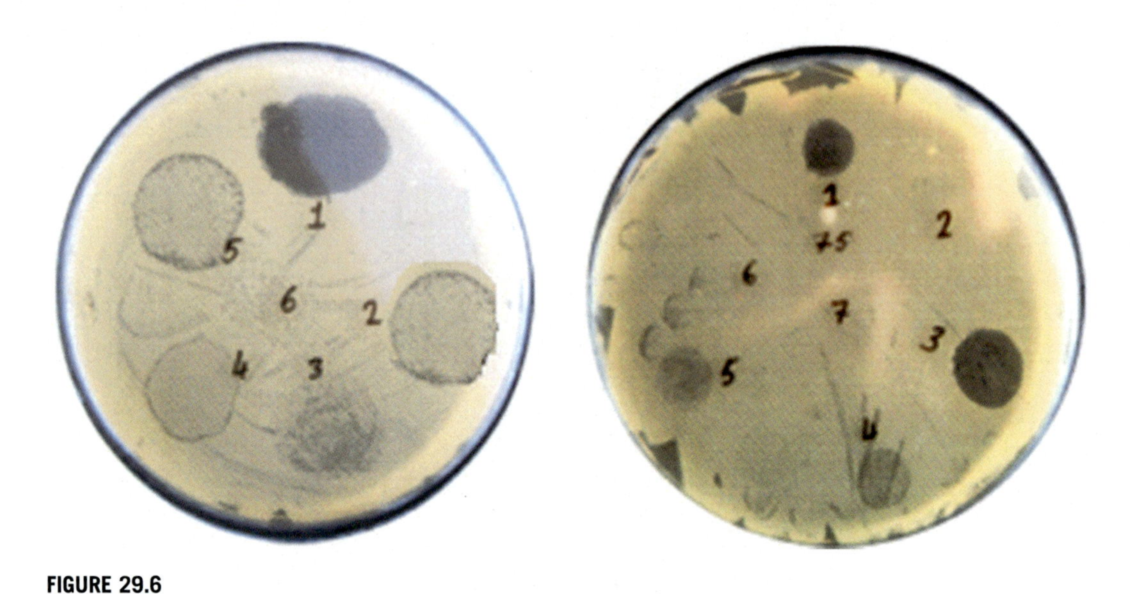

FIGURE 29.6

Typing of bacterium with bacteriophages: typed bacterium with number 1 in (A), and with numbers 1 and 3 in (B).

All lytic reactions with confluent and complete lysis are regarded as "strong" reactions. The results of bacteriophage typing of a bacterial strain are reported in terms of those bacteriophage that produce strong lysis. This is referred to as the *phage type* (Fig. 29.6A) or the *phage pattern* of the bacterial strain (Fig. 29.6B). In this way the bacterial strain is detected with an equivalent phage name. Bacterial strains may be susceptible to more than one bacteriophage (Fig. 29.6B). This is known as a bacteriophage pattern of bacterial strain. The bacteriophage pattern is usually reported as "type 52/52A" or "type 6/47/53/77." Sometimes weak reactions can be reported by placing a + sign after the pattern of significant lysis, for example, "type 6/47/53/77+". The pattern is also reported with RTD with confluent to complete lysis. Lesser degrees of lysis and reactions of inhibition are disregarded.

A similar technique can also be used for the typing of isolated bacteriophages using known bacterial strains. In this case, spread plate the bacteriophage onto an agar medium by swabbing and place the known bacterial strain culture at apparent turbidity onto the labeled area as described for bacterium typing and report.

29.3.4 Storage of bacteriophage cultures

After harvesting and centrifugation at low speed, the phages are rendered free from bacteria by filtration (0.22 μm membrane filter), and are stored in the refrigerator (at 4°C). Filtration involves the possible loss of some bacteriophage by adsorption. Chemical agents can be used instead of filtration, for example, zephran (in low concentration) and thymol. The phage culture can also be stored in glycerol. Lyophilization provides the ideal and probably the most satisfactory preservation technique. Stock cultures of the bacteriophage can also be stored at −10°C.

29.4 Identification of bacteriophage by foodomics

29.4.1 Matrix-mediated laser desorption ionization-time-of-flight-mass spectrometry identification of bacteriophage

Bacteriophage can be rapidly identified with the matrix-mediated laser desorption ionization-time-of-flight-mass spectrometry (MALDI-TOF Ms) technique, which is based on the principle of ionizing a specific protein profile of a bacteriophage. Bacteriophages can be easily identified by comparing these profiles with a reference spectrum. MALDI-TOF Ms identification has been explained in Technique 13.6.3.1.

29.4.2 Molecular identification of bacteriophage

Polymerase chain reaction (PCR) is a genotypic technique used in the detection of various bacteriophages. Some bacteriophage-specific primer sets that can be used in the identification of bacteriophages are given in Appendix A: Gene Primers (Table A-1). Using bacteriophage-specific techniques and materials, the bacteriophage can be identified by PCR analysis, as described in Technique 15.6.3.2.

29.5 Interpretation of results

Viruses are important agents causing foodborne disease, most commonly by minimally processed foods, for example, molluscs, and fresh food produce. Viruses can cause gastroenteritis and common symptoms of viral gastroenteritis include vomiting and diarrhea. Viruses are strict intracellular parasites and cannot replicate in food or water. Therefore viral contamination of food will not increase during processing, transport, and storage, and cannot change the smell and taste of the contaminated foods. Programs for the detection and identification of viruses from foods should be included in the process. Food can be contaminated during different phases of production from: preparation environment, ready-to-eat food by an infected food handler, preparation by a person excreting the virus, drinking or use of waters contaminated with sewage, irrigation water, etc. Many of the foodborne viruses require only a low infectious dose to cause foodborne disease. Viruses may also persist for extended periods (up to 60 days) on several materials. Wastewater can contain viruses around 10^4-10^6 particles per L. Contamination of viruses with persons and surfaces from aerosols can occur during food processing and preparation.

Viruses are recalcitrant to many of the food processing and preservation methods. Standard pasteurization conditions inactivate viruses. Most food or waterborne viruses are resistant to heat, disinfection agents (e.g., chlorine, ozone, high pressure, and UV and ionizing radiation) and pH (as low as 3 to 4). Hand washing reduces viruses but considerable numbers of viruses will remain when hand sanitizers are used instead of proper hand washing. Cleaning and disinfection of surfaces with adequate disinfectant are important in the control of virus transmission during food handling, production, and preparation. However, most surface disinfectants lack efficacy against enteric viruses. Optimization and standardization techniques for detecting foodborne viruses, enhancing laboratory-based surveillance to detect viruses, and developing quality control measures; include the risk of viruses in (Hazard Analysis and Critical Control Points) plans. The most efficient ways to minimize the risks of virus contamination from farm to fork are the application of food safety systems including GAP (Good Agricultural Practices), GHP (Good Hygiene Practices), GMP (Good Manufacturing Practices), and HACCP.

PRACTICE

Detection of foodborne and waterborne parasites

30

30.1 Introduction

The infectious parasites for humans through the ingestion of food and water are protozoa, trematodes (flukes), cestodes (tapeworms), and nematodes (roundworms). All waterborne parasites originate from human or animal fecal contamination of the environment. Vegetables can be contaminated with numerous species of parasites through the use of fertilizers consisting of animal or human feces. Vegetables can also be contaminated during preparation by rinsing with water containing parasites. Shellfish, fish, poultry, beef, goat meat, lamb, pork, and others sometimes contain encysted, encapsulated, or free stages of parasites. These stages are not merely contaminants; they are specific to the developmental life cycle in the animal hosts. All parasites and parasite eggs are temperature sensitive and killed by thorough cooking or freezing. Perhaps the best known of these foodborne parasites are *Trichinella spiralis*, a roundworm, causing trichinosis; and *Toxoplasma gondii*, a protozoa that causes toxoplasmosis in human. A detailed list of the parasites is given in Table 30.1. Visual inspection and microscopic analysis have been indicated for the detection of parasites after recovery techniques.

30.2 Techniques of examination and identification

30.2.1 Equipment and reagents

Equipment. Centrifuge, centrifuge tube, cheesecloth, microscope, microscope glass slide, loop, and cover slip (22 × 40 mm).

Materials. Hematoxylin, iron hematoxylin, physiological saline (0.85% NaCl) solution, NaOH (1.0 N), Lugol's iodine stain, 33% zinc sulfate solution, phosphate buffered formalin (40%) solution, and HCl (1 N).

30.2.2 Sample preparation and analysis

30.2.2.1 Gross analysis

Direct analysis of food by the naked eye or with a hand lens magnification is useful. The large *Taenia* tapeworm larvae (*cysticerci*) found in pork and beef are several millimeters in diameter. The white slender larvae of the broodfish tapeworm *Dipyllobothrium* in fish are visible and several millimeters long.

Microbiological Analysis of Foods and Food Processing Environments. DOI: https://doi.org/10.1016/B978-0-323-91651-6.00019-7

Table 30.1 Pathogenic parasites transmissible through foods.

Parasites	Sources of infection
Protozoa	
Entamoeba histolytica	Food and drink contaminated with cyst
Giardia lamblia	Food and drink contaminated with cyst
Toxoplasma gondii	Meats and milk contaminated with cyst
Balantidium coli	Food and drink contaminated with cysts
Nematoda	
Trichinella spiralis	Meats enclosing larvae
Trichuris trichiura	Food and drink contaminated with eggs
Ascaris lumbricoides	Food and drink contaminated with eggs
Trematoda	
Fasciola hepatica	Vegetables contaminated with metacercariae
Fasciola buski	Vegetables contaminated with metacercariae
Cestoda	
Diphyllobothrium latum	Fish harboring plerocercoids
Taenia solium	Pork containing *cysticerci*
Taenia saginata	Beef containing *cysticerci*
Echinococcus granulosus	Food and drink contaminated with eggs

30.2.2.2 Dissection

The suspected food should be carefully dissected with forceps and dissecting needles, if possible. Small samples should be selected, and animal food material should be tested separately with the aid of a magnifier or microscope. Small agents (e.g., trematode cysts or small nematode larvae) can be viewed by pressing the food between two pieces of glass and holding it against a lamp to permit light to penetrate the material. If no agents are seen on glass analysis or dissection and magnification, the technique described below can be used.

30.2.2.3 Washing procedures

Washing and scrubbing increase the efficiency of the recovery of parasites from foods. Some parasites are stickier than others. An effective technique must be sufficiently forceful to free a significant number of parasites from the food, but not so forceful that it damages the parasites. Scrubbing the surface area recovers *Ascaris lumbricoides* eggs and hookworm larvae from vegetables. This is followed by 10 h of sedimentation with the wash water. The sediment is strained, concentrated further by centrifugation (at least $25000 \times g$ for 20 min), and examined with a microscope. In this technique, fewer parasites are obtained than expected from seeded samples.

30.2.3 Recovery, concentration, and digestion

30.2.3.1 Tissue separation

To obtain parasites that are embedded in foods, it is often necessary to dissect, grind, or homogenize the samples before further processing. Homogenization is appropriate for small pathogens with tough outer membranes, such as bacteria, but damages the larger microorganisms. The larvae of *T. spiralis* in muscle are killed when the meat is homogenized, but not when the meat is ground.

30.2.3.2 Flotation, sedimentation, and centrifugation

1. Use sedimentation or centrifugation techniques for foods, soils, and feces. Dilute a sample with 10 parts of sterile distilled water, filter the suspension through one layer of wet cheesecloth, centrifuge (25,000 rpm for 40 to 60 s), pour off the supernatant, and resuspend the sediment in sterile distilled water. Repeat centrifugation and resuspend three to four times until the supernatant becomes clear. After the last supernatant is discarded, add 3–4 mL of a 33% zinc sulfate solution (specific gravity 1.180) to resuspended packed sediment. After centrifugation, remove the surface film with a loop. Add Lugol's iodine stain and mix to aid in the identification of helminth eggs, larvae, and protozoan cysts. Place this preparation onto a microscope glass slide and examine microscopically for parasites.
2. Use the dilution egg count technique in the identification of parasites. Weigh 4 g of samples (such as feces), and complete to 60 mL with 0.1 N NaOH. Homogenize by gently shaking with glass beads, allow to stand overnight so that the sample disintegrates adequately, and remix. Then add 0.15 mL onto a clean microscope glass slide, cover with a 22×40 mm cover slip, and examine for helminth eggs using microscope. Multiply the total number of helminth eggs in 100 fields to obtain the number per g or mL of sample.

30.2.3.3 Digestion

Add 200 g of the food into liter of sterile distilled water and stir at 37°C for h. This is often used to recover parasites from foods. After dissection, grinding, or homogenization, adjust the pH to 3 with HCl (1 N). This technique serves not only to free parasites from cyst walls and host tissues but also to select the potential parasites of warm-blooded animals. The larvae of *T. spiralis* in muscles classically are recovered from infected meat by such digestion, followed by sedimentation.

Protozoan cysts and helminth eggs are fixed for flotation and sedimentation to concentrate the parasites.

30.2.4 Types of parasites and identification techniques

30.2.4.1 Amoeba

Of the amoeba parasites only *Entamoeba histolytica* is pathogenic to humans. *E. histolytica* is transmitted without a known animal reservoir. The trophozoite stage, a motile form resembling macrophages, presents in the intestine and causes acute colitis, diarrhea, or dysentery. Because the trophozoite is easily destroyed outside the body, transmission usually involves passages of the resistant cyst stage in the stool. This stage is round and surrounded by a tough outer wall. Ingestion of cysts together with contaminated food or water or direct passage from person to person are the usual routes of transmission. Cysts can be transmitted from soiled fingers of infected food handlers who have poor personal hygiene.

Identification of *E. histolytica* from water is based on microscopic examination of sediment obtained either from filters or in the pellet of particulate matter from centrifugation of water. Trophozoites are 20–30 μm, have a thick, clear ectoplasm and granular endoplasm; pseudopods might be visible. The nucleus, unclear in fresh specimens, is distinct when stained with hematoxylin. It has a ring of small peripheral granules and a central dark body (endosome). Cysts are 10–20 μm in diameter, have four nuclei and contain rod-like bodies with rounded ends.

30.2.4.2 Balantidium coli

Balantidium coli is the only species in the genus *Balantidium*. The trophozoite stage in the large intestine can cause ulcerative colitis and diarrhea. The cyst is the infective stage. *B. coli* is a commensal widely distributed in pigs. Water obtained from drainage areas contaminated by human or pig feces is the major source of human infection. The *B. coli* cyst can be identified from water after concentration by filtration or centrifugation. The cyst is large (45–65 μm) and surrounded by a distinct wall and it contains a large nucleus that is often bean-shaped. Both cysts and trophozoites can be identified by microscopic examination of suspect stool specimen mixed with physiological saline. In formalin-fixed samples trophozoites often resemble debris, artifacts, or eggs.

30.2.4.3 Giardia

Giardia spp. are flagellated protozoa of the intestine. The most common species is Giardia lamblia. *Giardia* spp. are responsible for the enteritis and diarrhea. Cysts may be transmitted by the fecal–oral route. Soiled fingers, contaminated drinking water, and fecal contamination of food are the most common routes of human infection. Sand filtration is needed to eliminate cysts from community drinking water because *Giardia* spp. can survive in normal chlorination. *Giardia* cysts are identified by light microscopy in suspect water after concentration by filtration or centrifugation. *Giardia* spp. can be identified as trophozoites or cysts in stool specimens mixed with physiological saline. Trophozoites are motile in samples less than a few hours old. They are 9–21 μm long, 5–15 μm wide, but only about 2 μm thick. When fixed and stained with iron hematoxylin, the trophozoites look like a human face with nuclei for eyes and a median bar for the mouth. The cyst stage contains four nuclei.

30.2.4.4 Toxoplasma

T. gondii is a coccidian protozoon, and has an economic and public health importance. Members of the cat family, and all other vertebrates, including humans, are potential intermediate hosts. Infectious stages of *T. gondii* take place in muscles and various organs, for example, the heart, brain, and liver of chickens, pigs, cattle, goats, and sheep. Toxoplasma infection often is associated with mild influenza-like symptoms; and there is a neonatal form of the infection, especially if the mother becomes infected during the first or second trimester of pregnancy. The presence of oocysts in cat feces or water can be determined by centrifugation. Because oocysts are 10 μm in diameter or slightly greater, the pore size of the filter should be adjusted accordingly. Oocysts can be detected in cat feces by mixing feces in water followed by flotation of oocysts in sucrose solution (or centrifugation) and examination by light microscopy.

30.2.4.5 Trichinella spiralis

Human trichinosis results from eating raw or undercooked meat, most often pork, containing the infective muscle larval stage of *T. spiralis*. Human infection results in many symptoms as the worms develop in the intestine and produce larvae that migrate to and encyst in strait muscles. During the larval migration and encystment, symptoms that appear include fever, myalpia, and periorbital edema. A sample of muscle is obtained from a suspected animal carcass or meat. A carcass sample should be collected from one of the areas where the larvae accumulate in highest density, that is, the tongue, diaphragm, or masseter tissues. The sample is cut into small pieces (thin slices are best) and squeezed between two microscope glass slides until the tissue becomes translucent. The sample is then examined with the aid of a light microscope (at low-power magnification). Muscle larvae in tissue are coiled within a muscle nurse cell or cyst.

30.2.4.6 Taenia solium *and* Taenia saginata

Human taeniasis (tapeworm disease) results from the ingestion of raw or undercooked pork meat containing the infective cysticersus larvae of *Taenia solium*. Humans acquire taenia infection by the ingestion of raw or undercooked pork containing *cysticerci*. Human taeniasis (tapeworm disease) can also result from the ingestion of raw or undercooked beef containing the infective cysticercosis stage of *Taenia saginata*. The life cycle, transmission, and signs of infection are similar to *T. solium*. Human taeniasis can be indicated by the presence of eggs or segments in the stool with microscopic analysis.

30.3 Interpretation of results

Some parasites, for example, *Trichinella*, *Toxoplasma*, and *Giardia*, can be directly or indirectly transmitted between animals and humans through the consumption of contaminated food or drinking water. Parasites can also be transmitted to humans or animals by vectors. Health effects of foodborne parasitic infections vary greatly depending on the type of parasite, ranging from mild discomfort to debilitating illness and possibly death. Contamination of foods by parasites may occur during growing, harvesting, irrigation, spraying water, processing, preparation, and packaging. Different types of parasitic diseases are commonly transmitted to humans from pork, fish, freshwater crustaceans, vegetables, and water. Water serves as an important vehicle among them for the transmission of foodborne parasites.

The keys to providing food safety against the risk of parasite contamination are keeping clean, separating raw and cooked, cooking thoroughly, keeping food at safe temperatures, and using safe water and raw materials in the food production, as well as the safe handling of food and good kitchen hygiene. The risks associated with foods and water can be prevented by the application of good hygiene, farming, and fishing practices, and the promotion of community awareness about the importance of parasites. A number of control practices are used to protect food products from parasites. The traditional practices are the cleaning and cooking of food items prior to consumption. Most parasites are sensitive to heat treatment as low as 56°C–60°C for 1–20 min. Other food processes practices are also effective on parasites, for example, hot smoking, fermentation in brine, drying, cold smoking, and salting. A single chemical treatment of water (such as chlorine) does not inactivate parasites, cysts, and oocysts. Application of cold temperatures (−20°C to −70°C) is effective on parasites. Filtration of water can also improve water quality, although some small-sized parasites may pass through certain types of filters. One micrometer filter is required to exclude some parasites, for example, *Cryptosporidium*. Ozone is a powerful oxidizing agent that can be used as a disinfectant for some parasites depending on temperature, pH, and the amount of organic matter residues on the application area. Irradiation and hydrostatic pressure may be able to decontaminate foods. Surveillance programs should include control measures for foodborne parasites. The standard for water quality is based on a coliform count but this criterion is not a reliable indicator for parasite contamination of water.

SECTION

Identification of foods safety and quality

The microbiological analyses of foods involve the identification and prevention of harmful microbial contamination, which can spoil foods and cause foodborne diseases. The typical reasons for the microbiological analysis of foods are to (1) provide available specifications for raw material, ingredients, and finished product; (2) identify risk factors; (3) verify process and processing conditions; (4) confirm regulatory guidelines; (5) establish prerequisite programs; and (6) produce optimum food quality. Microbiological analysis is one of the components of the food safety system and does not guarantee 100% food safety. Microbiological analysis is a requirement for many industries worldwide where human health is at risk and adversely affected by the presence of biological pathogens and toxins. In some cases, the importance of microbiological testing is a matter of life or death. Microbiological analysis encourages the establishment of programs, for example, Hazard Analysis Critical Control Point (HACCP), Good Manufacturing Practices (GMP), Recall Management, Traceability, and Sanitation Practices. This section describes the most common techniques used to identify microbial contaminants and evaluate microbiological characteristics, microbial significance, and microbial control in different types of foods.

PRACTICE

Analysis of milk and milk products 31

31.1 Introduction

Raw milk from healthy animals normally contains very low numbers of microorganisms. *Micrococcus*, *Staphylococcus*, *Streptococcus*, and *Corynebacterium* are the most common bacteria in milk. Milk may be contaminated with microorganisms from the surface of the animal, the environment, and an unclean milking system. The contaminated raw milk may contain *Bacillus cereus*, *Campylobacter jejuni*, *Escherichia coli*, *Listeria monocytogenes*, *Micrococcus*, *Mycobacterium tuberculosis*, *Salmonella*, *Staphylococcus*, *Streptococcus*, and *Yersinia enterocolitica*. Unclean udders and teats can contribute microorganisms from a variety of sources (e.g., manure, soil, feed, and water). They include lactic acid bacteria (LAB), coliforms, *Bacillus*, *Clostridium*, *Microbacterium*, *Micrococcus*, and *Staphylococcus*. Thermoduric bacteria (e.g., *Bacillus*, *Lactobacillus*, *Micrococcus* and *Streptococcus*) are associated with poorly cleaned milking machines, pipelines, and storage tanks. Pathogenic microorganisms may enter raw milk through infected animals, milking personnel, and the environment. Psychrotrophic microorganisms start to grow in raw milk held at refrigeration temperatures. The most common psychrotrophic bacteria are *Pseudomonas*, *Alcaligenes*, and *Flavobacterium* (e.g., *F. psychrophilum*). The most common acid-forming bacteria in milk are *Lactococcus lactis* and *Lactobacillus acidophilus*. *E. coli* and *Enterobacter aerogenes* are the most common gas-formers in milk. *Bacillus subtilis* and *Enterococcus liquefaciens* are the most common peptonizing bacteria in milk and decompose proteins (principally casein). The most common microorganisms causing ropy milk (a viscous, sticky consistency) are *Alcaligenes viscolactis*, *L. lactis* subsp. *cremoris*, *E. aerogenes*, *Lactobacillus bulgaricus*, and *Micrococcus*. In this practice, different microbiological analysis techniques for milk and dairy products are explained to indicate microbial risks from the desired raw materials and ingredients, identify risk factors, verify the process, and form regulatory guidelines.

31.2 Raw milk analyses

Sampling, sample preparation, and all experimental techniques should be performed under aseptic conditions and in duplicate, unless otherwise specified. All microbiological media to be used in applications must be sterilized. Again, all solutions and equipment that may cause the contamination of microorganisms must be sterilized or disinfected.

31.2.1 Equipment, materials, and media

Equipment. Blender (or stomacher), Erlenmeyer flask, fermentation tube, filter paper, forceps, gasometer, glass rod, incubator, loop, microscope glass slide (in bottle containing alcohol),

Microbiological Analysis of Foods and Food Processing Environments. DOI: https://doi.org/10.1016/B978-0-323-91651-6.00027-6

microscope, penicillin disk, penicillinase, Petri dishes, pipettes (1 and 0.5 mL) in pipette box, pipette discarding pad (containing 10% bleach solution), rubber stoppered test tube, scalpel, spatula (or knife), swab, thermometer, thermostatic shaking water bath, 2,3,5-triphenyltetrazolium chloride (TTC) solution (1%), waterproof pen, and Whatman filter paper (6 mm in diameter).

Materials. Alcohol (95%), alpha-naphthylamine reagent, ammonium sulfate, cold water in a beaker, distilled water, Gram stains, hydrogen peroxide (H_2O_2, 0.2 and 3%), Liefson's methylene blue (LMB) stain, methylene blue (MB) indicator stain, Newman's stain, p-dinitrophenyl phosphate (PP) substrate, 0.1% peptone water, phosphatase reading disk, p-phenyldiamine (2%, p-1,4-benzene diamine), rosalic acid, sodium carbonate (or sodium bicarbonate), sodium hydroxide (NaOH), thallium acetate, tissue paper, xylene, and oil-immersion.

Media. Brain heart infusion (BHI) broth, BHI 5% sheep blood agar, BHI agar, Edward's esculin crystal violet blood (EECVB) agar, neutral red chalk lactose (NRCL) agar, nutrient agar slant, paraffin wax, plate count (PC) agar, skim milk (5%), yeast extract skim milk (YESM) agar, and violet red bile (VRB) agar.

31.2.2 Aerobic plate count

Aerobic plate count (APC) is used more frequently than any other technique in milk. The factors influencing the plate counts include incubation temperature, composition of medium, amount of oxygen, etc. The numbers of colonies appearing on PC agar do not represent all the microorganisms present in the milk. Anaerobic microorganisms do not find favorable conditions in milk for growth. The quality of a milk supply can be determined on the bases of APC.

Label PC agar with dilution rate, sample name, and your initials. Shake the milk sample vigorously at least 25 times to obtain a homogeneous suspension of the microorganisms. Add 1 mL of raw milk into 9 mL of sterile 0.1% peptone water in the Erlenmeyer flask to prepare a 1:10 dilution by a fresh pipette. Transfer 1 mL of 1:10 dilution into a second 9 mL of diluent with the second pipette. Prepare further dilutions in the same manner, if necessary. Transfer 1 mL of diluted sample onto its labeled PC agar and spread plate (Technique 2.2.4). When the agar absorbs water from the sample, incubate the inverted Petri plates at 35°C for up to 48 h. After incubation, count the colonies from plates containing 25–250 colonies. Calculate the number of APC by multiplying the average count by the dilution rate and dividing by the inoculation amount. Report the results as the number of APC colony forming units (cfu) per mL of milk.

Counting other microorganisms. Refer to Technique 13.3.2 for coliforms count and fecal coliforms counts. For counting individual pathogens, for example, *Brucella*, *C. jejuni*, *L. monocytogenes*, and *E. coli*, refer to the corresponding practices of this book.

31.2.3 Direct microscopic count

Refer to Technique 3.2.2 for details of the direct microscopic count (DMC). Take a microscope glass slide and pass over a flame. Draw a 1 cm^2 area under the slide using a waterproof pen. Shake the milk vigorously to obtain a homogeneous suspension. With a sterile 0.5 mL pipette, withdraw a sample of milk slightly above the graduation mark. Wipe the exterior of the pipette tip with clean dry paper tissue. Then absorb the liquid at the tip to give exactly 0.5 mL. Place the tip of pipette at the center of slide and expel the test portion up to 0.4 line from 0.5 line of pipette (0.01 mL). Do not retouch the pipette to the plate. With a clean bent-point needle, spread 0.01 mL of milk uniformly over a 1 cm^2 area. Dry the smear in a dust-free area such as in an incubator. After drying, do not heat fix the slide. Submerge the dried film in a bottle of xylene for at least 1 min. Remove the slide from the xylene and pour LMB stain onto the smear, and allow it to stand for at least 2 min. Drain the stain, wash with water, and air-dry. Examine the slide under oil-immersion objective of the microscope, count at least 25 fields and find the average count per field of microscope. Calculate the number of microorganisms per mL of sample using the microscope factor (MF) of the oil-immersion objective, average number of microorganisms (n) per field, amount of sample (0.01 mL), and the reciprocal of the dilution rate (Dr).

$$\text{DMC per mL of sample} = (\text{MF} \times \text{n} \times \text{DR})/0.01$$

Round off the count to two significant figures and report the result as DMC per mL of sample.

31.2.4 Methylene blue reduction

Add 1 mL of methylene blue indicator stain and 10 mL of milk in sterile rubber-stoppered tubes in duplicate and mix. Incubate tubes at the specified temperature. This technique depends on the ability of microorganisms to change the oxidation–reduction potential of the medium. Bacteria grow in the milk, consume the dissolved oxygen, and reduce the oxidation–reduction potential of the milk. Certain enzymes (e.g., oxidoreductases) produced by bacteria are able to oxidize the substrates by the removal of hydrogen. In this technique, methylene blue acts as a hydrogen acceptor. When microorganisms oxidize the substrate, the oxidized substrates release hydrogen. Methylene blue accepts the hydrogen and becomes reduced; the reduced methylene blue appears colorless (decolorize). The decolorization speed is an indication of the rate of oxidation in the milk. The rate of decolorization (oxidation) depends on the microbial numbers in the milk sample.

There is not always a good agreement between methylene blue reduction test and the plate count. Because (1) some microorganisms fail to grow on plate count agar; (2) a clump of microorganisms records as only one colony, whereas the rate of decolorization is due to the combined effect of each microbial cell in the population; (3) the rate of decolorization of the methylene blue is not the same for all microorganisms; and (4) the test becomes less accurate as the reduction time is increased, freshly drawn milk requires at least 10 h to decolorize methylene blue.

The methylene blue must be prepared weekly and stored in a dark and cold place. Shake the milk sample vigorously to obtain an homogenous suspension. Add 10 mL of milk into a sterile rubber stoppered test tube. Add 1 mL of methylene blue indicator stain solution using a sterile 1 mL pipette into each tube containing milk, and after a lapse of 3 s, blow out the remaining methylene blue indicator stain drops. Do not touch the pipette to the milk. Close the tube with its own rubber stopper and mix well by using a vortex or twice inverting the tube slowly. The same pipette may be used for a series of tubes in the removal of methylene blue indicator stain. Within 5 min, place the tube in a thermostatic shaking water bath or an incubator at 35°C ± 0.5°C and record the time. The water level in the water bath should be 2 cm above the milk in the tubes, and the bath should be closed with a lid to exclude light. Add 10 mL of milk into a sterile stoppered tube, add 1 mL of sterile distilled water, and place in water bath as a control. The control tube will help to determine if decolorization is complete or not. Examine the tubes at frequent intervals (in 15 min) during 5 h of incubation. The milk is regarded as decolorized when the whole column of milk appears completely colorless as compared with the control tube. If tubes are not completely decolorized after each inspection, slowly invert the tubes to mix. Once the cream layer is distributed continue incubation. When the tubes are decolorized, record the decolorization time. Refer to the decolorization (reduction) time table as follows and record the number of bacteria per 1 mL of milk.

Decolorization time limits

1. *Freshly drown milk* requires at least 10 h to decolorize; containing less than 10 microorganisms per mL.
2. *Good milk* cannot decolorize before 5 h; containing less than 1/2 million microorganisms per mL.
3. *Fair milk* decolorizes between 2 and 5 h; containing 4–20 million microorganisms per mL.
4. *Bad milk* decolorizes between 20 min and 2 h; containing 4–20 million microorganisms per mL.
5. *Very bad milk* decolorizes less than 20 min; containing least 20 million microorganisms per mL.

31.2.5 Physical analysis of milk and isolation of physical changes causing microorganisms

Mix the milk thoroughly, determine the pH value, and pour a few mL into a sterile Petri dish to analyze the physical appearance (e.g., precipitation, color change, odor). Place a drop onto a microscope glass slide and analyze microscopically. Microbial faults and taints may occur in milk when particular microorganisms grow. The nature of the fault or taint itself provides information about the type of microorganism. The term "fault" is applied to a change in the physical condition of the milk, whereas the term "taint" indicates normal physical condition but there is a change in flavor or smell.

31.2.5.1 Isolation of fault causing microorganisms

Streak plate one loopful of milk onto the surface of a nonselective medium, for example, YESM agar or any other suitable selective/differential medium. Incubate plates at the milk storage temperature. After incubation, analyze colony formations on plates and check the purity of colonies by restreaking onto YESM agar plate, and incubate at the previously incubated temperature for 24 h. Inoculate each pure culture into pasteurized whole milk and incubate at the previous incubation temperature. When the fault is caused by the pure culture inoculated milk, record the result as the presence of fault-causing microorganisms and indicate the type of fault formation in milk. The following are examples of the types of fault formation in milk.

Sour milk. Milk is coagulated with lactic acid production by lactose-fermenting bacteria (e.g., *Lactococcus* and coliforms). *L. lactis* may induce about 90% of the total flora of freshly soured raw milk. *L. lactis* can be isolated by streaking onto YESM agar (or NRCL agar). Thallium acetate is added at a final concentration of 1:2000 into the medium to inhibit Gram-negative bacteria. Production of acid from lactose results in the formation of yellow colonies. Refer to Technique 33.2 for differential counting and identification of LAB.

Gassiness or frothiness. Gas bubbles produced on the milk surface may be trapped in the cream. The most common microorganisms causing this fault are coliaerogenes bacteria (especially *E. aerogenes* and *E. coli*) and lactose-fermenting yeasts. Refer to the respective practice for the isolation and identification of these microorganisms.

Sweet clotting or sweet curdling. Proteolytic enzymes coagulate milk at neutral pH. The microorganisms responsible for this proteolytic activity are predominantly *Bacillus* spp. The fault frequently develops in heat-treated milk perhaps partly due to the removal of competing bacteria and survival of heat-resistant *Bacillus* spores. Refer to Technique 31.5.2.3 for counting proteolytic microorganisms.

Ropiness or sliminess. Viscous milk can be drawn out into threads with a wire loop. It is caused by the growth of capsule-forming coliaerogenes bacteria (e.g., *Klebsiella pneumoniae* and *A. viscolactis*) and capsule-forming *B. subtilis*, *B. cereus*, and *Micrococcus*. They frequently grow in refrigerated milk. Count ropiness- or sliminess-causing microorganisms using plate count agar as explained in Technique 31.2.5.2. Colonies of rope or slime (due to the production of dextran)-forming bacteria will show a viscous consistency.

Broken cream or bitty cream. This is the breaking of cream into separate particles on the milk's surface, and which do not reemulsify. It appears when milk is poured into hot tea or coffee. The fault is caused by the previous lechitinase production in the milk by *B. cereus* and *B. cereus* var. *mycoides*. Count *B. cereus* from milk as indicated for *Bacillus* (Technique 19.2.3).

31.2.5.2 Isolation of rope- or slime-forming bacteria

Viscous milk positive samples initially are transferred with a 1 mL amount to test tubes containing 10 mL steamed whole milk (at 100°C for 30 min). Incubate the test tubes at 10°C and check daily for rope formation with the appearance of viscous consistency (due to the production of dextran). Streak plate one to three loopfuls of milk from test tubes showing ropiness onto skim YESM agar and incubate the inverted plates at 22°C for up to 3 days. After incubation, streak plate from rope-forming colonies onto YESM agar again to obtain a pure culture and incubate the inverted plates at 22°C for up to 3 days. Rope-forming pure colonies are transferred to YESM agar slants (nutrient agar slant), incubate tubes at 22°C for up to 3 days, then seal the cap end of the screw test tube by immersing into hot paraffin wax to prevent culture dehydration. These tubes are used as stock cultures and stored at 2°C. Use this culture in the identification techniques, if necessary. Refer to practices in this book for the identification of isolated rope-forming microorganisms.

31.2.5.3 Isolation of taint-forming microorganisms

Taint in milk may be easily recognized if the milk is slightly warmed. For the detection of causative microorganisms, transfer about 5 mL of tainted milk to 50 mL of sterile whole milk in a Erlenmeyer flask, and incubate at 20°C or another appropriate temperature and examine twice daily for the development of the taint. The following are examples of the types of taint in milk.

Malty or caramel taint. *L. lactis* var. *maltigenes* can cause malty or caramel taint and produce 3-methylbutanol from the leucine component of casein. *L. lactis* subsp. *lactis* cannot cause these effects.

Carbolic or phenolic taint. Phenol-producing *Bacillus circulans* can cause carbolic or phenolic taint in commercially sterilized bottled milk. The spores of this bacterium may survive the heating process.

31.2.6 Detection of inhibitory substances in milk

31.2.6.1 Detection of antibiotics in milk

Following the treatment of animals with antibiotics for bovine mastitis, the antibiotics can be released into milk. Antibiotics in milk is undesirable for public health due to several reasons: some individuals may show allergic reactions, drug resistance of microorganisms may be encouraged, and residual antibiotics may inhibit lactic acid fermentation in the production of dairy products. The problem arises from milk obtained from animal treatment with antibiotics usually within 72 h, the antibiotic excretion gradually decreases during this time.

1. *Disk assay technique*

   Suspend a sterile swab into an exponential phase BHI broth culture of the test microorganism sensitive to the antimicrobial agent. This culture can be prepared by adding 1 mL of an overnight culture to 4 mL of BHI broth and incubating for 2 h (up to visible turbidity). Insert a sterile swab into the BHI broth culture and press the swab against the tube interior wall with a rotational motion to remove excess culture. Hold the swab handle to make a 50 degrees angle contact with the BHI agar surface. Rub the swab firmly over the surface of the BHI agar. Allow the agar medium surface to dry, and mark the plate at the bottom external surface as required by waterproof pen. Pick up one of the antibiotic assay disks (6 mm diameter, Whatman paper) using clean dry forceps and dip the disk into the well-mixed milk sample. Drain off the excess liquid from the disk by touching to the internal surface of the milk container and then place the disk onto its own labeled plate sector. Rinse, dry, and flame disinfect the forceps before use for another disk. Use a control antibiotic disk (such as penicillin) effective on the tested microorganism and place it onto the labeled plate sector. Incubate the plate at the optimum growth temperature of the test microorganism, for example, at 30°C for *B. subtilis*, at 55°C for *Geobacillus stearothermophilus*, and at 35°C for *Staphylococcus aureus*. After incubation, analyze the plate for the formation of a zone around the disk. The inhibitory zone indicates the presence of antibiotic residues in milk. Compare the inhibition diameter with the control antibiotic disk zone. When the inhibition zone around the disk is present and smaller or greater than the control milk, the milk sample should be retested after heating the milk to destroy any natural antimicrobial agent. Predict whether the heat treatment (e.g., pasteurization) is sufficient to destroy the antimicrobial agent in the milk.

   Prepare a standard curve by placing the known concentrations of antimicrobial disk and their zones onto plates. Fresh milk can be used in the preparation of dilution from 0.001 to 0.1 antibiotic units per mL. Prepare the antimicrobial disk and place onto culture inoculated BHI agar as above. After incubation, measure the zone formation (in mm) for each dilution. Draw a standard curve with a known concentration of the antimicrobial against the zone of inhibition. Indicate the concentration of the antimicrobial agent in milk.

2. *Dye reduction*

   Pipette well-mixed 10 mL of milk sample into each of two sterile labeled test tubes (sample and control) closed with rubber bungs. To the control tube, add 0.2 mL of 1000 IU penicillinase per mL. Add 1 mL of BHI broth *Streptococcus thermophilus* (sensitive to penicillin) culture to the sample and control tubes. Add 1 mL of 1% solution of TTC to each tube. Tightly close with rubber bung and invert to mix. Incubate all tubes in a thermostatic shaking water bath at 44°C for 1.5 h. Invert the tubes to mix (or mix with vortexing), and reincubate at 44°C for a further 1 h. After incubation, analyze the sample for bacterial inhibition by comparison with the control tube. A change in the color in the control tube containing penicillinase and no color change in the sample tube indicates the presence of penicillin in milk. No color change in the sample tube indicates sensitivity of *S. thermophilus* to penicillin.

   All samples should be tested within 36 h of the sampling time.

31.2.6.2 Hydrogen peroxide detection in milk

Hydrogen peroxide (H_2O_2) inactivates LAB when added into milk. So, the milk for fermentation should not contain H_2O_2. Place 10 mL of well-mixed milk into a sterile test tube and add one drop of 2% p-phenyldiamine (p-1,4-benzene diamine). The appearance of a blue color indicates the presence of H_2O_2 in the milk.

31.2.6.3 Carbonate detection in milk

Sodium carbonate or sodium bicarbonate may be added into milk to prevent coagulation during heat treatment. The milk for fermentation should not contain carbonate. Place 5 mL of well-mixed milk into a sterile test tube. The appearance of carbonate should be checked by a control tube: add 5 mL of milk and 0.5 mL of sodium carbonate into the control tube. Add 5 mL of 95% alcohol into both tubes. Mix the contents of the tubes. Add three drops of 1.0% rozalic acid into both tubes and mix. The development of a rose-pink color indicates the presence of carbonate in milk. The control tube should also show rose-pink color formation.

31.2.7 Mastitis test

If lactating cows have mastitis (inflammation of the udder), large numbers of the pathogenic microorganisms may be excreted into the milk and contribute high microbial counts. Many microorganisms cause mastitis; the most importance are *S. aureus*, *E. coli*, *Salmonella* spp., *Streptococcus agalactiae*, *Streptococcus uberus*, *Streptococcus pyogenes*, *Pseudomonas aeruginosa*, and *Micrococcus pyogenes*. The first three of these are also potential human pathogens. A number of other human pathogens also cause mastitis, for example, *L. monocytogenes*, *M. tuberculosis*, and *Mycobacterium bovis*.

Mastitis can cause changes in the characteristics of milk. In milk obtained from an animal with mastitis the following physical and chemical changes can be observed: (1) different milk color, and smell; salty and bad smell; (2) pH of the milk increases from 6.0 to 6.6 over 7.0; (3) fat content decreases by 5%–12% compared to the normal milk; (4) the amount of casein decreases, and lactoalbumin and katoglobulin increase; (5) lactose decreases by 10%–20%; (6) the amount of milk decreases by 20%; (7) calcium, potassium, and phosphor decrease, and sodium and chlorine increase, and can cause a salty taste; (8) B vitamins decrease and E vitamin disappears; (9) catalase, phosphatase, and esterase increase; (10) total microbial count increases; and (11) somatic cells increase.

31.2.7.1 Direct microscopic count of somatic cells

Total somatic cells can be counted from milk using the Breed's smear technique (Section 31.2.3) with trypan blue or a similar stain (e.g., Liefson's methylene blue, Newman's stain). Leucocytes are stained by adding 5–10 drops of trypan blue (0.5% aq.) to an equal volume of milk. Count the leucocytes in 20 successive fields with an oil-immersion objective. Milk obtained from healthy animals contains cells, mainly epithelial cells and small numbers of leukocytes, but the numbers of leukocytes are greatly increased in mastitis milk because of udder inflammation. These cells can be counted by staining techniques. The standards suggesting for somatic cells in milk are as follows:

Individual milk sample	Result
Below 250,000 cells per mL	Negative; from health animal.
250,000–500,000 cells per mL	Suspect; there is a risk of it coming from a diseased animal.
More than 500,000 cells per mL	Positive; from a diseased animal.

31.2.7.2 Whiteside test

This is a simple, rapid test for the detection of mastitis. It indirectly detects the number of somatic cells in milk. When the somatic cells are present in milk, the nucleic acids are released from the cells with the milk titration by an alkaline agent and cause an increase in viscosity.

Mix one part of NaOH (0.1 N) with five parts of milk on a sterile Petri dish and stir for 15 s with a glass rod. Use a control milk from an animal without mastitis. Record the extent of viscosity formation in milk, if any. An increase in viscosity appears with an increase in the number of cells in milk (e.g., leucocyte cells). The absence of a further increase in viscosity indicates a negative test, as well as control milk, and the milk is normal milk obtained from a healthy animal.

31.2.7.3 Isolation and counting of mastitis pathogens

The most satisfactory nonselective medium is a BHI 5% sheep blood agar. The kind of blood used will affect the type of hemolysis production by pathogens. This medium is used to count the total bacterial number in the milk. Prepare dilutions from milk using 0.1% peptone water. Spread plate will not easily differentiate colonies on the surface from a nondiluted milk sample. Therefore, use an appropriate dilution. Pour plate can cause temperature effect on microorganisms. Shake the milk sample vigorously at least 25 times to obtain a homogeneous suspension of the microorganisms. Add 1 mL of raw milk into 9 mL of 0.1% peptone water in the Erlenmeyer flask to prepare 1:10 dilution by a fresh pipette. Transfer 1 mL of 1:10 dilution into a second 9 mL of diluent with the second pipette. Prepare further dilutions in the same manner, if necessary. Add 1 mL of diluted sample onto its labeled BHI 5% sheep blood agar and spread plate (Technique 2.2.4). When the agar has absorbed water from the sample, incubate the inverted Petri plates at 35°C for up to 48 h. After incubation, count the colonies from plates containing 25–250 colonies. Calculate the number of APC by multiplying the average count by the dilution rate and dividing by the inoculation amount. Report the result as the number of APC colony forming units (cfu) per mL of milk.

If required, perform identification tests on pathogens from pure cultures, as indicated in the related practice of this book. Indicate the Gram reaction, cellular morphology and arrangements of the bacteria, catalase test (H_2O_2, 3%) (Technique 31.2.7.4), and coagulase test (Technique 21.3.2.1).

Streptococcus count. EECVB agar is a selective medium for counting or isolation of mastitis-causing *Streptococcus*. Add the selective agent, crystal violet at a concentration (1/500,000) to select *Streptococcus* and to inhibit *Staphylococcus*. Esculin-fermenting bacteria, including *S. uberus*, produce black colonies, while nonesculin-fermenting bacteria (e.g., *S. agalactiae* and *Streptococcus dysgalactiae*) produce colorless colonies. Spread plate 1 mL of sample onto the well-dried EECVB agar in duplicate. Incubate inverted plates at 35°C for up to 2 days. After incubation, count the total number of *Streptococcus* and individual *Streptococcus* spp. colonies. Calculate the number of bacteria or bacterium cfu per mL of sample. The predominant bacteria growing on the EECVB agar plates are of most significance and are presumed to be the mastitis-causative *Streptococcus* spp. Record the colony characteristics (e.g., size, shape, hemolysis, and pigmentation), and prepare a Gram stain to indicate the Gram reaction, and morphology. These characteristics should be sufficient to indicate the probable genus of the bacterium, but further confirmatory tests can be carried out, if required.

For counting other pathogenic microorganisms, for example, *S. aureus*, *Salmonella*, and *E. coli*, refer to the corresponding practices of this book.

31.2.7.4 Catalase test

There is a low level of the enzyme catalase in normal milk. But the level of catalase increases in milk drawn from animals with mastitis. Catalase hydrolyzes H_2O_2 to gas (O_2) and water. The catalase test depends on the detection of the level of gas production after the addition of H_2O_2 into milk. Mix the milk thoroughly and transfer 15 mL of raw milk into a sterile test tube connected to a gasometer (or Smith fermentation tube). Add 5 mL of 1% H_2O_2 and close the tube. Incubate the tube in a water bath at 25°C for 12 h. Read the volume of gas from the space of the fermentation tube (Fig. 31.1). When the level of gas is lower than 1.5 mL, the milk is considered to be normal and the milk has been drawn from a healthy animal. When the level of gas is higher than 1.5 mL, the milk has been drawn from a mastitis animal.

Smith fermentation tube. This is a borosilicate glass tube and traps evolved gas (CO_2) from fermentation. A variety of carbohydrates can be tested for gas production with this tube. A vertical tube is graduated up to 5 mL (space) in 0.1 mL divisions, so that the gas volume can be quantified.

31.2.8 Analysis of pasteurized milk

31.2.8.1 Equipment, materials, and media

Equipment. Erlenmeyer flask, filter paper, incubator, loop, Petri dishes, pipettes (1 mL) in pipette box, pipette discarding pad (containing 10% bleach solution), screw-capped test tube, thermometer, thermostatic shaking water bath, and waterproof pen.

Materials. Ammonium sulfate, Charlette's improved Newman's stain, cold water in a beaker, PP substrate, tissue paper, H_2O_2 (0.2%), 0.1% peptone water, sponge, and p-phenyldiamine (2%, p-1,4-benzene diamine).

Media. Crossley's milk peptone (CMP) broth, Davis's yeast salt (DYS) broth, plate count (PC) agar, tributyrin agar, VRB agar, yeast extract glucose lemco (YEGL) agar, and YESM agar.

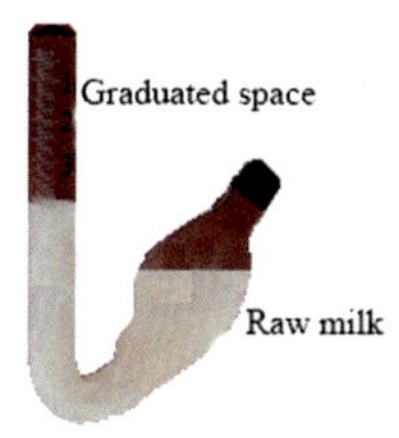

FIGURE 31.1

Fermentation tube.

31.2.8.2 Pasteurization of milk

The destruction of all living things in food is called sterilization. High temperature is required to achieve sterilization. Sterilization of foods may not be possible for two reasons: (1) the cooked flavor is not so pleasant; and (2) heating to such a high temperature might result in a decrease in the vitamin or nutritional value.

As a heat treatment, the destruction of all pathogens and reduction of spoilage microorganisms in milk is called pasteurization. Pasteurization refers to the process of heating milk at 72°C for 15 s.

Commercial sterility of milk is obtained by heating the milk to a high temperature for a short time. This process is known as ultrahigh temperature (UHT) processing and it retains a sufficient quality of the milk. In the UHT process, the milk is heated by injecting steam at high pressure for a rapid temperature increase. In the UHT process, milk is heated to 137°C–140°C for 1–3 s. Following heat treatment and cooling, the milk is packed aseptically. This milk is stored at room temperature, and the UHT milk generally has a 3-month shelf life. However, if psychrotrophic bacterial heat-stable proteinases or lipases are present in the raw milk, UHT cannot inactivate these enzymes. Microbial heat-sensitive toxins will be destroyed, but heat-stable toxins may remain active after commercial sterility.

Adjust the temperature of a thermostatic shaking water bath to 72°C. After the temperature of the water bath reaches 72°C, place the tube containing 10 mL of raw milk sample into the water bath and allow it to remain until the temperature reaches 72°C. Prepare an additional tube containing 10 mL raw milk as a control. Place a thermometer into the control milk to check the pasteurization temperature. After temperature of the milk reaches 72°C, allow the milk to remain for 15 s at 72°C. At the end of pasteurization, remove the tube from the water bath and quickly cool in running cold water. Do not allow the cotton stopper to become wet.

In the following analyses, include pasteurized and UHT milk from the market, laboratory pasteurized milk, and raw milk (as control).

31.2.8.3 Microbial counts from milk

Refer to Technique 31.2.2 for APC by spread plate count to retain heat-injured cells. Refer to Technique 13.3.2 for coliform count by the most probable number (MPN) technique. For counting individual pathogens, for example, *C. jejuni*, *L. monocytogenes*, *Y. enterocolitica*, and *Salmonella*, refer to the corresponding practices of this book.

31.2.8.4 Phosphatase test on heated milk

The aim of the dairy industry is to ensure a supply of milk from healthy animals and then heat-treat the milk in the factory by either pasteurization or UHT heating (commercial sterilization). A number of rapid tests are used to ensure that the milk is sufficiently pasteurized or not. These include the phosphatase test, dye reduction test, turbidity test, and DMC. The phosphatase test depends on the detection of the phosphatase enzyme from the pasteurized milk. Phosphatase is more heat resistant than the most heat-tolerant, vegetative pathogenic bacteria. It is however, destroyed by sufficient pasteurization. Milk containing phosphatase will be detected by hydrolyses of PP substrate releasing p-dinitrophenol. The amount of this product is directly proportional to the amount of phosphatase enzyme in the milk. Under the conditions of the test, the p-dinitrophenol is a yellow-colored compound. The color intensity is proportional to the amount of p-dinitrophenol released in the milk. Milk showing a negative result for phosphatases is considered as properly pasteurized and safe.

Phosphatase test. Pipette 5 mL of PP substrate solution into a sterile test tube. Stopper the tube and place it in a 35°C water bath for 5–10 min to bring the solution temperature to 35°C. Add 1 mL of the milk sample into the tube and replace the rubber bung and mix. Prepare a control tube containing 5 mL of buffer substrate solution and 1 mL of pasteurized milk. Continue to incubate in the water bath for 2 h. After incubation, remove each tube from the water bath and thoroughly mix the contents by inverting the tubes.

Read the intensity of color results by using the appropriate comparator disk. The disk consists of a series of yellow-colored filters each corresponding to a particular concentration of p-dinitrophenol. Place the control tube in the left-hand hole and the test sample in the right-hand hole of the comparator. Match the colors of two tubes by resolving the disk while examining the tubes by daylight or fluorescent light. Note the disk reading when the colors of the two tubes are matched. The reading expresses the μg of p-dinitrophenol per mL of milk. Satisfactorily pasteurized milk contains a concentration of 10 μg or less of p-dinitrophenol per mL of milk.

31.2.8.5 Catalase test on heated milk

This test detects if milk has been heated or not before coming to the factory. Heat treatment from 71°C to 80°C for a second inactivates catalase enzyme, thus heated milk cannot be used in the production of milk products, for example, cheese. Place 10 mL of milk (at 20°C) into a sterile test tube. Add two drops of 0.2% H_2O_2 solution and two drops of 2% p-phenyldiamine (p-1,4-benzene diamine). When a blue color is observed within 30 s, the test is considered positive; meaning that the milk has not been heated before and the catalase enzyme remains in raw milk. When a color change is not observed, the test is considered negative; meaning that the milk is heated before.

31.2.8.6 Turbidity test on heated milk

Milk is commercially sterilized at a temperature over 100°C followed by rapid cooling and aseptic packaging for a long shelf-life. After this treatment, it must satisfy the turbidity test which is the official test for heated milk. The test examines the presence of soluble milk protein. The proteins are denatured when milk is boiled or heated over 100°C. The denatured protein cannot precipitate after the addition of ammonium sulfate. The formation of turbidity shows the presence of soluble milk protein that has not been denatured during heat treatment. This indicates inadequate commercial sterilization of milk or the contamination of commercially sterilized milk with raw milk.

Put 4 g of ammonium sulfate into a sterile Erlenmeyer flask and transfer 20 mL of milk into the flask. Shake the flask for 1 min to dissolve the ammonium sulfate. Filter through the filter paper into a sterile test tube and collect at least 5 mL of clean filtrate. Place the tube in a thermostatic shaking water bath at boiling for 5 min. Transfer the test tubes to cold water in a beaker. Cool and examine the contents of the tube for turbidity by moving the tube in front of an electric light shaded from the eyes of the observer. The milk is regarded as commercially sterilized when no turbidity is observed; proteins are denatured by heat treatment.

31.3 Milk powder analysis

In the production of milk powder, homogenized milk is heated to 65°C–85°C in the roller process or 68.8°C–93.3°C in the spray process. In the spray process, the milk is briefly subjected to dry air at high temperatures. Thermoduric bacteria may survive this treatment. During the concentration of the milk to a high solid content, surviving thermodurics, especially streptococci, may multiply if the temperature remains at a level favorable to their growth for a sufficiently long time.

Sample preparation. Aseptically weigh 10 g of milk powder and transfer to 90 mL of sterile distilled water at 50°C in an Erlenmeyer flask. Shake the flask 25 times in 12 s and place the flask in a thermostatic shaking water bath at 50°C for 15 min, then invert the flask several times. Prepare further dilutions in 9 mL of sterile 0.1% peptone water as required.

Microscopic analysis. Prepare a Breed's smear (Technique 3.2.3) and stain the smear by using Charlett's improved Newman's stain. Count the number of microorganisms from the stained smear.

Microbial counts. Aerobic mesophilic count. Refer to Technique 31.2.2 for APC by incubation at 35°C for 24 h.

Aerobic thermoduric count. Refer to Technique 12.3 for the aerobic thermoduric count.

Coliform count. Refer to Technique 13.3.2 for the coliform count by MPN.

Yeasts and molds count. Refer to Technique 6.3.2 for the yeasts and molds count.

Enterococcus. Refer to Technique 14.2.3.1 for the *Enterococcus* count.

For counting of individual pathogens, for example, *S. aureus*, *Y. enterocolitica*, *B. cereus*, *Salmonella*, and spores of *Bacillus* and *Clostridium*, refer to the corresponding practices of this book.

31.4 Canned and concentrated milk

31.4.1 Evaporated milk (unsweetened condensed milk)

Insufficient heat treatment during the canning process allows the survival of heat-resistant microorganisms, for example, *B. subtilis*, *Bacillus coagulans*, *B. cereus*, *G. stearothermophilus*, and *Clostridium*. The product should be commercially sterile. Test the product for commercial sterility after processing.

1. *Preincubation of cans*

   Incubate one can in duplicate at each of the following temperatures: 55°C for 7 days, 30°C for 14 days, and 25°C for 1 month. After incubation, analyze the cans and report their appearance. Cans containing gas-producing microorganisms may become swollen and are easily recognized.
2. *Bacteriological condition investigation*

Shake the can and its contents thoroughly. Wash the outside of the can with warm water. Wipe the can, dry with a clean paper towel, sponge with a suitable disinfectant, and finally dry. Swab the surface of the can with ethanol, remove excess ethanol and flame. Open the can with a sterile can opener (taking great care for blown cans) and cover the opened can with a sterile Petri dish lid. Weigh out 10 g of milk into a sterile Erlenmeyer flask containing 90 mL of 0.1% peptone water at

37°C and shake the flask 25 times. Prepare further dilutions in the usual way using 0.1% peptone water.

Microscopic count. To carry out microbial microscopic count (Technique 31.2.3). Prepare a milk smear on a slide and stain, defatting first with xylene, if necessary.

Growth in liquid media. To carry out plate counts, inoculate 1 mL milk into 7 mL CMP broth in duplicate, mix by gently shaking of tubes, and layer the broth by adding 2 mL of 2% agar onto the liquid for anaerobic bacteria. Add 1 mL of milk into 7 mL melted YEGL agar (at 45°C) in duplicate for LAB and mix by gently shaking the tubes. Incubate CMP broth and YEGL agar tubes at 30°C under anaerobic and microaerophilic conditions, respectively. After incubation examine the tubes for the presence or absence of respective microorganisms.

Plate counts. Spread plate 1 mL of diluted sample (Technique 2.3.4) to YESM agar (or PC agar) in duplicate and incubate at 35°C for up to 3 days for mesophilic aerobic count. Pour plate 1 mL of diluted sample (Technique 2.3.3) with YESM agar in duplicate and incubate at 55°C for up to 3 days for the thermophilic aerobic count. After incubation examine the colonies from the plates. Count the number of colonies from Petri plates containing 25–250 colonies. Calculate the number of respective bacteria by multiplying the average count by the dilution rate and dividing by the inoculation amount. Record the results as the mesophilic aerobic count and thermophilic aerobic count cfu per g of sample.

31.4.2 Sweetened condensed milk

Condensed milk has about 10%–12% fat and 36% total solids. The milk is initially heated at a low temperature, close to pasteurization temperature, and then subjected to evaporation under partial vacuum (at about 5°C). After treatment, thermoduric microorganisms can survive that subsequently can grow and cause spoilage. This product has a limited shelf life at refrigerated temperature. The average amount of sugar in the finished product is 55%–60% lactose plus added sugar. Microorganisms (e.g., yeasts, molds, micrococci, coliforms, and spore-forming aerobes) can become contaminants of condensed milk. The yeast tolerates sugar and can cause spoilage at refrigerated temperature. *Aspergillus* and *Penicillium* can survive in condensed milk.

Preparation of can. The product may be viscous and it is advisable to warm the contents of the can in a water bath at 45°C for not more than 15 min before opening, in order to reduce viscosity. Aseptically open the can as described above for evaporated milk (Technique 31.4.1). Weigh out 10 g of condensed milk into a sterile Erlenmeyer flask. Add 90 mL of 0.1% peptone water at 37°C and shake 25 times. Prepare further necessary dilutions in the usual way using 0.1% peptone water. Perform following plate counts.

Aerobic plate count. Refer to Technique 31.2.2 for APC. Use YESM agar (or PC agar) and incubate the inverted plates at 30°C for 3 days.

Lipolytic count. Refer to Technique 31.5.4 for lipolytic count but using tributyrin agar.

Coliform count. Refer to Technique 13.3.3 for coliform count. Use VRB agar and incubate the plates at 30°C for 24 h; alternatively, if small numbers are expected, use an MPN technique (Technique 13.3.2).

Yeasts and molds count. Refer to Technique 6.3.2 for yeasts and molds count. Contamination with low numbers of yeast may be detected using DYS broth by a MPN technique (Technique 4.2).

31.5 Butter and cream

Butter contains at least 80% milk fat. It can be salted or unsalted, and it may or may not contain added starter cultures consisting of *L. lactis*, *L. lactis* subsp. *cremoris* as an acid producer, and *Leuconostoc* species as a flavor producer. In ripened cream for butter making, cultures usually consist of citrate-fermenting bacteria producing diacetyl beside lactic acid from citrate, for example, *L. lactis* subsp. *lactis* biovar. diacetylactis and/or *Leuconostoc mesenteroides* subsp. *cremoris*. The bacteria that constitute the aroma and flavor producers are strains of *Leuconostoc* species (*L. mesenteroides* subsp. *cremoris*, *Leuconostoc mesenteroides* subsp. *dextranicum* and *L. mesenteroides*). The butter microflora depends on the cream quality, the equipment sanitary conditions, and the environmental and sanitary conditions during packaging and handling. The growth of bacteria on the butter surface can cause surface discoloration and flavor problems. These flavor and surface growth problems are usually caused by the species of *Pseudomonas*, *Streptococcus*, *Geotrichum*, and *Candida*. *S. aureus* food poisoning outbreaks can be associated with butter and cream.

31.5.1 Equipment, materials, and media

Equipment. Erlenmeyer flask, incubator, loop, Petri dishes, pipettes (1 mL) in pipette box, pipette discarding pad (containing 10% bleach solution), screw-capped test tube, thermometer, thermostatic shaking water bath, and waterproof pen.

Materials. Acetic acid (10%), bromocresol purple (BCP) solution (0.1%), copper sulfate solution (0.1%), Gram staining reagents, Gram staining stains, hydrochloric acid (1%, HCl), methylene blue indicator stain (0.1%), and quarter-strength Ringer's (QSR) solution with 0.1% agar.

Media. Brain heart infusion (BHI) broth, caseinate agar, nutrient broth containing NaCl (0.5%, 5.0%, 10%, 15%, and 20% NaCl), YESM agar, YEGL agar, and victoria blue butterfat (VBB) agar.

31.5.2 Sample preparation

Samples may consist either of individual cartons or packages. Melt the sample at a temperature not exceed 45°C by immersing the bottle for a short time in a water bath at 45°C and shaking. Mix the melted sample by shaking 50 times. Transfer 10 mL of melted sample into a sterile Erlenmeyer flask containing 90 mL of QSR solution with 0.1% agar to stabilize the emulsion. The pipettes and diluent should be at 45°C. Prepare the subsequent decimal dilution using QSR solution, as required.

31.5.3 Counting of proteolytic bacteria

Proteolytic bacteria are capable of hydrolyzing proteins, due to producing extracellular proteinases. Species in the genera *Alcaligenes*, *Bacillus*, *Micrococcus*, *Staphylococcus*, *Clostridium*, *Pseudomonas*, *Flavobacterium*, and *Brevibacterium* and some in the family Enterobacteriaceae are proteolytic.

Add 1 mL of the diluted sample to YESM agar (or caseinate agar) plate, and spread plate (Technique 2.2.4) and incubate the inverted plate at 30°C for up to 5 days. To count psychrotrophic proteolytic bacteria, incubate the inverted plates at 22°C for 3 days or at 7°C for up to 10 days. After incubation, count all colonies from Petri plates containing 25–250 colonies. Flood the YESM agar plates with an acid solution (1% HCl or 10% acetic acid) for 1 min. Pour off excess acid solution, then count colonies showing proteolytic activity by clear zone formation (or formation of white or off-white precipitate) around colonies on Petri plates. Strongly proteolytic bacteria can further break down the precipitate to soluble components with the formation of an inner transparent zone. Calculate the number of proteolytic bacteria by multiplying the average count by the dilution rate and dividing by the inoculation amount. Record the result as the number of proteolytic bacteria cfu per g of sample.

Select isolated colonies (at least three colonies per plate) showing proteolysis. Stain the smears by Gram staining, as explained in Appendix B under the topic "Gram Staining Reagents" and subculture BHI broth. Incubate at 30°C until growth occurs, then reexamine and streak on YESM agar plate to obtain pure cultures (Technique 13.4.1). Use for identification techniques, if necessary. Refer to identification practices in this book for isolated microorganisms.

31.5.4 Counting of lipolytic bacteria

Lipolytic bacteria are able to hydrolyze triglycerides due to the production of extracellular lipases. Species in the genera *Alteromonas*, *Flavobacterium*, *Micrococcus*, *Staphylococcus* and *Pseudomonas* are lipolytic.

Inoculate a predried plate of VBB agar by spread plating 1 mL of sample over agar (Technique 2.2.4). Incubate the inverted plates at 5°C for lipolytic *Pseudomonas* for up to 7 days and at 30°C for mesophilic lipolytic bacteria for up to 3 days. After incubation, flood the plates with 8–10 mL of saturated copper sulfate solution (0.1%) and allow them to stand for 15 min. Pour off the reagent from the plates and wash the plates gently with running water for 1 min to remove excess copper sulfate. Where lipolysis occurs, a bluish green color zone appears around the colonies, due to the formation of insoluble copper salts of the fatty acids and this indicates free fatty acids from the lipolysis of fat. Count colonies showing lipolytic activity from Petri plates containing 25–250 colonies. Calculate the number of lipolytic bacteria by multiplying the average count by the dilution rate and dividing by the inoculation amount. Record the result as the number of lipolytic bacteria cfu per g of butter (or cream).

Select least three isolated colonies per plate showing lipolysis. Stain smears by Gram staining as explained in Appendix B under the topic "Gram Staining Reagents" and subculture into YEGL agar. Incubate at 30°C until growth occurs, then reexamine and streak on VBB agar to obtain pure cultures (Technique 13.4.1). If necessary, refer to related practices in this book for the identification of isolated microorganisms.

31.5.5 Salt tolerance test

Inoculate one loopful of pure culture into each series of tubes of nutrient broth containing NaCl at a range of concentrations: 0.5%, 5%, 10%, 15%, and 20%. Incubate tubes at 30°C for 5 days. After incubation, record the relative amount of growth (turbidity) in each tube with respect to the concentration of salt.

31.5.6 Dye reduction test

Methylene blue reduction test. Transfer 7 mL of sterile QSR solution into a sterile test tube with a 10 mL mark and add 1 mL of methylene blue indicator stain. Using a wide-tipped pipette, add cream (butter) at 35°C up to the 10 mL mark. A control tube is prepared by transferring 8 mL of QSR solution into a sterile tube and adding cream (or butter) to the 10 mL mark. Insert a sterile rubber bung to the tube and invert the tube to mix. Incubate the tubes in a thermostatic shaking water bath (without shaking) at 20°C for 17 h. Transfer the test tubes to a water bath at 35°C. Examine and slowly invert once every 30 min during 4 h or until decolorization at 35°C is completed, compared with the control tube. Record the time taken for decolorization of the methylene blue at 35°C. If the methylene blue is decolorized at the time of removal from the 20°C water bath (within 17 h), the time is recorded as 0 h. Record the result by using the following suggestions for the dye reduction standard for butter (or cream):

Time taken to reduce methylene blue	Interpretation
Fails to decolorize in 4 h	Satisfactory
Decolorization in 1.5–4 h	Significance doubtful
Decolorization in 0 h	Unsatisfactory

Bromocresol purple test. Add 10 mL of cream (or butter) sample into each two sterile test tube. Add 1 mL of sterile BCP into one of the 10 mL cream and retain a second tube as a control. Insert a sterile rubber bung into the tube and invert the tubes to mix. Incubate the tubes at 30°C. Examine the test tubes, and look at the color changes in tubes after 16–17 h and again after 24 h. Record the decolorization time and indicate results by using the following:

Appearance of tubes	Results	Grades for cream (after 24 h)
No visible change	–	1
Slight acidity, faint yellow clot	+	2
Definite acidity, yellow color but no clot	++	3
Acid and clot	+++	4

31.5.7 Other counts

For APC (Technique 31.2.2), coliforms, yeasts and molds, psychrotrophic, and *S. aureus* counts refer to the corresponding practices of this book.

31.6 Ice cream

The types of spoilage microorganisms present in ice cream are streptococci, micrococci, sporeformers, coliforms, and *Pseudomonas*. The most common pathogenic microorganisms associated with ice cream are *E. coli*, *Salmonella*, and *S. aureus*. The other pathogens may be present in ice cream, for example, *L. monocytogenes*, *Y. enterocolitica*, *Mycobacterium*, *Campylobacter*, and *Brucella*.

31.6.1 Equipment, materials, and media

Equipment. Erlenmeyer flask, incubator, loop, Petri dishes, pipettes (1 mL) in pipette box, pipette discarding pad (containing 10% bleach solution), refrigerator, screw-capped test tube, spatula, thermometer, thermostatic shaking water bath, and waterproof pen.

Materials. Methylene blue indicator stain and 0.1% peptone water.

31.6.2 Sampling

Hard ice cream. From tubes and packages, take at least two unopened packages. From multilayered ice cream, remove a sample from each layer. For ice cream in bulk containers, first remove the surface layer with a sterile spatula or spoon and with a second sterile spatula take a sample of not less than 60 g into a sterile sample container.

Soft ice cream. Soft ice cream (freshly frozen ice cream) should be filled into the sample jars directly and take a sample of not less than 60 g.

Sample transport. Transport the sample to the laboratory in a frozen state and maintain at not more than −18°C until analyzed in the laboratory. Transport the sample to a laboratory within 2 h of the sampling time, but when this time limit is not practicable, samples should be packed in ice and should arrive at the laboratory within 6 h of sampling.

Sample preparation. Frozen samples should be left at room temperature for a maximum of 1 h until melting. Alternatively, liquefy the sample by holding the jar in a water bath at 43°C for no more than 15 min. When the sample is unfrozen, analyze immediately. The analysis should be carried out on a gravimetric basis since the weight of 10 mL of ice cream may range from 4.5 to 10.5 g. Invert the sample bottle at least three times to mix the sample. Weigh out 10 g of melted ice cream using a sterile 10 mL pipette into a sterile Erlenmeyer flask. Add 90 mL of sterile 0.1% peptone water and mix well by gently shaking. Prepare further required dilutions using 0.1% peptone water.

31.6.3 Microbiological analysis

Aerobic plate count. Refer to Technique 31.2.2 for APC and incubate inverted plates at 32°C for up to 3 days for mesophilic APC and at 5°C for up to 7 days for psychrotrophic APC.

For counting coliforms, yeasts and molds, thermoduric bacteria, and individual pathogenic, for example, *S. aureus*, *L. monocytogenes*, *Y. enterocolitica*, *E. coli*, *Salmonella* and *Campylobacter*, refer to the corresponding practices of this book.

31.6.4 Methylene blue reduction

Melt 5 mL of ice cream in a water bath at 37°C within 15 min and cool. Add 7 mL of QSR solution into a marked 10 mL tube and 1 mL of methylene blue indicator stain into a rubber-stoppered tube. Add the ice cream into the test tube up to the 10 mL mark. If the sample contains much air and the level falls as the air is freed, fill the sample up to the 10 mL mark of the tube. Close the tube with its own rubber stopper and mix well by using a vortex or by twice inverting the tube slowly. Incubate tubes in a thermostatic shaking water bath (without shaking) at 20°C for 17 h. After incubation, transfer the tubes to a water bath at 35°C, invert the tubes slowly once every 30 min until complete decolorization, compared with the control tubes. Record the decolorization time for methylene blue. If the methylene blue is decolorized at the time of removal from the 20°C water bath, the time is recorded as 0 h. Record the grade of ice cream in the methylene blue test as follows:

Time taken to reduce methylene blue	Grade
Fails to decolorization in 4 h	1
Decolorization in 2.5–4 h	2
Decolorization in 0.5–2 h	3
Decolorization in 0 h	4

31.7 Cheese

Microbial spoilage in cheeses is generally limited because of the combined effect of acid, salt, and low moisture. Fresh cheeses may be spoiled by Gram-negative psychrotrophic bacteria (*Flavobacterium*, *Pseudomonas*, and *Alcaligenes*), coliforms, yeasts, and molds. They can contaminate after the pasteurization of milk and during the processing of cheeses. The most common pathogens in cheeses are *S. aureus*, *L. monocytogenes*, *E. coli*, *Salmonella*, and *Brucella*. *M. tuberculosis* may survive for long periods in cheeses. Mycotoxin-producing molds can also be associated with cheeses.

31.7.1 Equipment, materials, and media (for cheese and yogurt)

Equipment. Microscope, microscope glass slide, pipette (1 and 10 mL) in pipette box, and pipette discarding pad (containing 10% bleach solution).

Materials. Agar solution (2%), quarter-strength Ringer's (QSR) solution, 0.1% peptone water, and sodium citrate (SC) solution (2%).

Media. Cheese agar containing 5% NaCl, Elliker agar, Lee's agar containing beta-glycerophosphate, M17 agar, nutrient agar containing 5% NaCl, sodium lactate (SL) agar, YEGL broth, and YEGL semi-solid medium (0.3% agar).

31.7.2 Sample preparation

Remove at least 60 g of cheese sample and transfer aseptically into a sterile sample container. Packet cheese is taken as a sample. From soft cheese and semihard cheese, the sample is removed with a sterile knife by making two cuts radiating from the center of the cheese. Any inedible surface layer should be removed before the removal of sampling. From hard cheeses, obtain the sample by a sterile cheese trier. The trier is inserted into the cheese's center on one of the flat surfaces. The outer 2 cm of the cheese's surface containing the rind is cut off.

Emulsify 50 g of the sample in either QSR solution or 2% SC solution in an Erlenmeyer flask at 45°C by shaking with hand (or homogenize by use of a stomacher). Prepare further dilutions in sterile 0.1% QSR solution in the usual way (Technique 2.2.2).

31.7.3 Microscopic analysis

Qualitative. Place a disinfected microscope glass slide firmly over a freshly cut cheese surface and slightly press on the slide. Press this slide surface with a second slide and press two slides, then separate the two slides and remove excess cheese from the slide surface with the edge of another slide. Place the slide into xylene and defat with xylene for 1 min, then pour xylene and allow it to dry in the air. Stain by simple staining as explained in Appendix B under the topic "Crystal Violet Stain" and examine using a microscope. Predict the relative number of microorganisms.

Quantitative. Examine the diluted cheese sample by DMC (Breed's smear) after staining by Charlett's Newman's stain (Technique 31.2.3).

31.7.4 Plate counts

Aerobic plate count. Refer to Technique 31.2.2 for APC. Use PC agar and YEGL agar. YEGL agar, as a nonselective medium, will support the growth of most LAB. *Streptococcus* in high numbers prevents the growth of *Lactobacillus*. Take 1 mL of sample from diluted sample for spread plating (Technique 2.2.4). Incubate the inverted plates at 30°C for 3 days.

Lactobacillus. Refer to Technique 33.2.3 and 33.2.7 for LAB and *Lactobacillus* count, respectively, and incubate the inverted plates at 30°C for up to 5 days.

For counting coliforms, yeasts and molds, and individual pathogens, for example, *S. aureus*, *L. monocytogenes*, *Y. enterocolitica*, *Salmonella*, *Campylobacter*, and *Clostridium*, refer to the corresponding practices of this book.

31.7.5 Counting of *Propionibacterium* from cheese

Propionibacterium is slow-growing, nonspore-forming, Gram-positive, mesophilic, and tends toward anaerobiosis. It can be rod-shaped or branched, pleomorphic, and can occur singularly, in pairs, or in groups (Fig. 31.2A). It generally produces lactic acid, propionic acid, acetic acid, and CO_2 from glucose. It may ferment lactic acid and other carbohydrates to produce propionic and acetic acid, and CO_2. These bacteria produce propionic acid and are used in dairy fermentation, for example, *Propionibacterium freudenreichii*. They can associate with raw milk, Swiss cheese, and silage.

MPN count. Melt and cool YEL semisolid medium (0.3% agar) to 45°C. Inoculate 1 mL of sample into 3-MPN tubes of YEL semisolid medium. Mix well by rotating the tubes between hands and allow to cool. Seal the surface of the medium in tubes with melted 2% agar (at 45°C). Incubate the tubes in an atmosphere enriched with 5% CO_2 at 30°C for 10 days (or aerobic incubation can also be used). After incubation, examine the tubes for growth and gas formation. Tubes of medium showing gas are recorded as a positive result and can be confirmed by streak plating to YES agar. Incubate in an atmosphere enriched with 5% CO_2 at 30°C for 7 days. Obtain a pure culture using YEL agar plates (Technique 13.4.1). After incubation, prepare a Gram stain as explained in Appendix B under the topic "Gram Staining Reagents." *Propionibacterium* frequently appear as small Gram-positive coccobacilli in 0.5 μm diameter. Pigment production is also a useful differentiation criterion. Indicate the number of

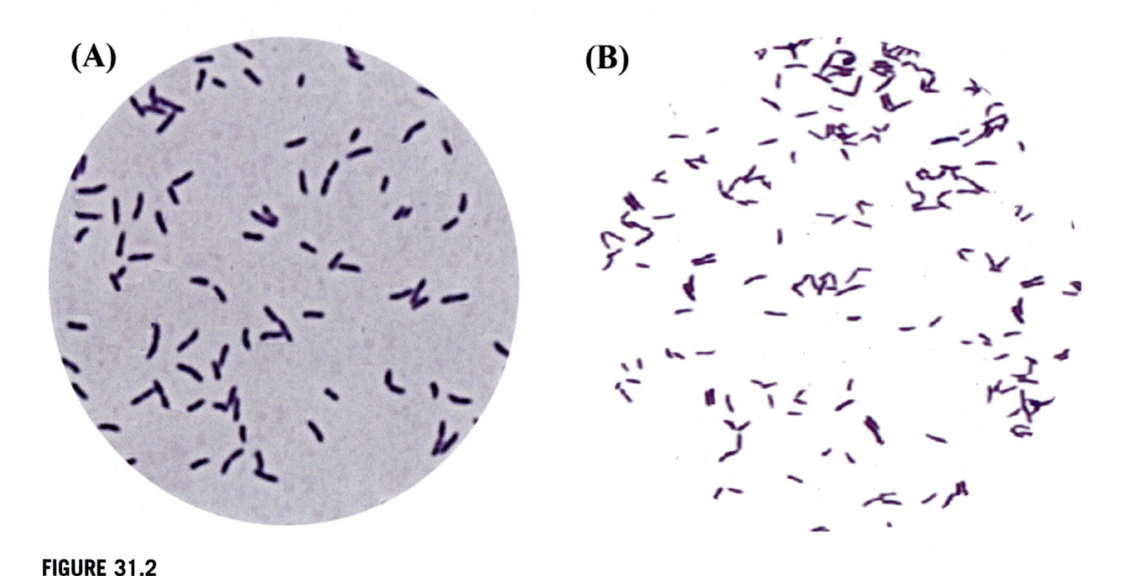

FIGURE 31.2

Staining appearance of *Propionibacterium* (A) and *Brevibacterium* (B).

positive tubes for each dilution, and calculate the MPN number as indicated in MPN Table 4.1. Report numbers of *Propionibacterium* MPN per g of cheese.

Plate count. Use SL agar for counting *Propionibacterium* on plates. This medium contains several salts, four vitamins, and cysteine ammonium sulfate as the nitrogen source, and SL as the carbon source. Several bacterial species cannot grow on this medium. Pour plate 1 mL of sample (Technique 2.2.3) and incubate at 32°C under anaerobic or microaerophilic conditions for up to 14 days. Colonies that appear within 2 days are probably not *Propionibacterium.* After incubation, analyze plates for the formation of characteristic colonies on SL agar. Confirm the individual colonies as *Propionibacterium* by Gram-staining and by detection of propionic acid production with gas or liquid chromatography. Count colonies from Petri plates containing 25–250 colonies. Calculate the number of *Propionibacterium* by multiplying the average count by the dilution rate and dividing by the inoculation amount. Record the result as the number of *Propionibacterium* cfu per g of cheese.

Storage of Propionibacterium. Inoculate one loopful of presumptive *Propionibacterium* colonies into YEL broth or YEGL broth and incubate at 30°C for 7 days. Maintain a pure culture by stab inoculation onto YEL agar slant, incubate until growth appears, and then cover with a layer of sterile liquid paraffin (or 2% agar) and store in the refrigerator.

31.7.6 Counting of *Brevibacterium linens*

Brevibacterium linens is a nonsporeforming, rod-shaped, nonmotile, mesophilic, and Gram-positive bacterium (Fig. 31.2B). It may be associated with the surface of bacterial ripened cheeses (e.g., Limburger cheese), soil, salt water, freshwater, and raw milk. *B. linens* and *Brevibacterium casei* cause surface smear or coat on the cheese, and cause a characteristic butyric texture and orange or orange brown color. Prepare a suspension from the cheese surface smear in 2 mL of QSR solution and streak onto the nutrient agar (or cheese agar) containing 5% NaCl. Incubate inverted plates at 25°C for 5 days. After incubation, analyze plates for the formation of characteristic colonies on nutrient agar. Count colonies from Petri plates containing 25–250 colonies. Calculate the number of *Brevibacterium* by multiplying the average count by the dilution rate and dividing by the inoculation amount. Record the result as the number of *Brevibacterium* cfu per g of cheese.

Storage of Brevibacterium. Subculture from isolated presumptive *B. linens* colonies onto sheep blood agar by streak plating. Incubate at 25°C for 5 days and streak onto nutrient agar slant. Maintain a pure culture on nutrient agar slants by subcultures at intervals of 3–6 months.

31.8 Yogurt

Yogurt fermentation involves a mixed culture of *L. bulgaricus* and *S. thermophilus*. Yogurt is prepared with the standardization of milk from 10.5% to 11.5% solids with heating to about 90°C for 30–60 min and cooling to 43°C. Then a mixed culture of *L. bulgaricus* and *S. thermophilus* in a 1:1 ratio is added to the milk. *S. thermophilus* and *L. bulgaricus* are the essential microbial species in yogurt and are active in a symbiotic relationship. The combined action of these two bacteria is needed to obtain the desired flavor and acid production. They ferment the milk sugar and convert it to lactic acid, reduce pH, and precipitate the casein.

Some species of *Bifidobacterium* (e.g., *B. bifidum*, *B. infants*, *B. adolescentis*) and *L. acidophilus* can be added into yogurt as probiotics after production. They are rods, single cells or in chains, arranged in V or star-like shape, nonmotile, mesophilic, and anaerobic. They metabolize carbohydrates to lactate and acetate, and are associated with the colon of humans, animals, and birds.

Yeasts and molds that tolerate low pH are more predominant microorganisms and they can be counted from yogurt. *B. subtilis* and *B. cereus* can cause bitter flavors in yogurt if large numbers survive after pasteurization. Coliforms decline rapidly during the production of yogurt. Yeasts and molds, and coliform counts may be used as indicators of process sanitation.

For yeasts and molds, coliforms, psychrotrophic microorganisms, and *Bacillus* counts refer to the corresponding practices of this book.

31.8.1 *S. thermophilus* and *L. bulgaricus* count

A morphological examination to determine the ratio of coccus to rod is inadequate because dead cells cannot be distinguished from viable ones. But adding methylene blue into the culture and microscopic counts can provide differentiation of living and dead cells (Technique 3.4).

Lee's agar can differentiate these two species. Carry out counts from a diluted sample or culture used for yogurt production. Weigh 25 g of yogurt into a sterile Erlenmeyer flask containing 225 mL of 0.1% peptone water and mix well by gently shaking. Prepare further dilutions using 0.1% peptone water. Spread plate 1 mL of diluted (or nondiluted) sample to Lee's agar surface (Technique 2.2.4). Incubate inverted plates at 35°C for up to 48 h in a 5% CO_2 incubator. After incubation, analyze plates for characteristic colony formations on the agar medium. *S. thermophilus* ferments sucrose and produces yellow colonies, and *Lactobacllus delbrueckii* subsp. *bulgaricus* does not ferment sucrose and produces white colonies on Lee's agar. For satisfactory differentiation, the total number of colonies on plates should not exceed 250. Count characteristic colonies of *L. bulgaricus* and *S. thermophilus* separately from Petri plates containing 25–250 colonies. Calculate number of *S. thermophilus* and *L. bulgaricus* separately by multiplying the average count by the dilution rate and dividing by the inoculation amount. Record the results as the number of *L. bulgaricus* and *S. thermophilus* cfu per g of yogurt. This medium is not a selective medium, and many other microorganisms can be expected to grow. Perform morphological (Fig. 31.3) and biochemical identification tests (Technique 33.4).

S. thermophilus and *L. bulgaricus* can also be counted in another way. The addition of beta-glycerophosphate into Lee's agar inhibits the growth of *L. bulgaricus*. Obtain a total LAB count on Lee's agar without beta-glycerophosphate and a differential *S. thermophilus* count on beta-glycerophosphate containing Lee's agar. Calculate the number of *S. thermophilus* by multiplying the average count by the dilution rate and dividing by the inoculation amount. Record the result as the number of *S. thermophilus* cfu per g of yogurt.

In another technique, count the total numbers of *S. thermophilus* and *L. bulgaricus* by pour plating using Elliker agar and *S. thermophilus* using M17 agar. Calculate the *L. bulgaricus* population by subtracting the *S. thermophilus* count from Elliker agar.

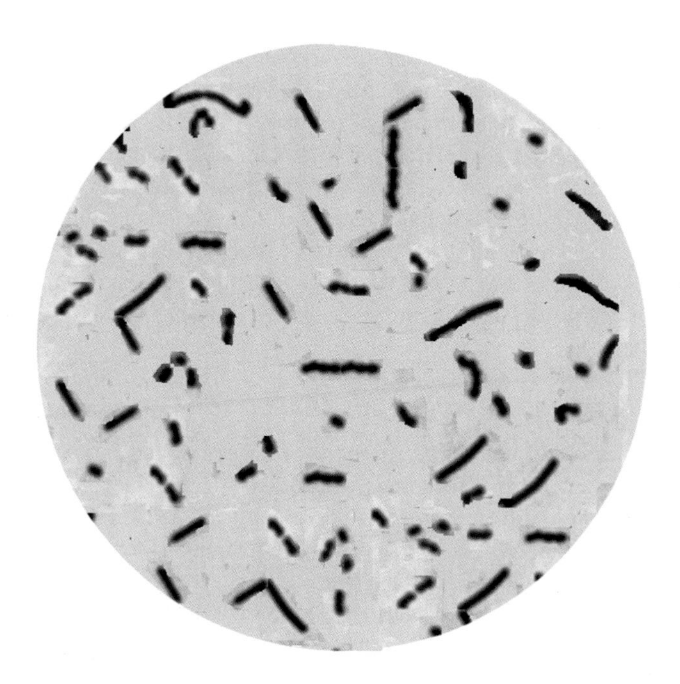

FIGURE 31.3

Staining appearance of *Streptococcus thermophilus* (coccus) and *Lactobacillus bulgaricus* (rod).

31.9 Dried dairy products

Dried dairy products include skim milk (nonfat dry milk), buttermilk, whey, cheese, and other fermented milk products. The microflora of dried milks is thermoduric *Micrococcus*, *Streptococcus*, *Corynobacterium*, and aerobic sporeformers. Dried milk must be considered as sensitive products from a public health aspect because they are often consumed after reconstitution without additional heating. They can contain pathogenic microorganisms, for example, *S. aureus* and *Salmonella*.

For counting of aerobic plate, coliforms, yeasts and molds, direct microscopic, psychrotrophic, and individual pathogens, for example, *S. aureus* and *Salmonella*, refer to the corresponding practices in this book.

Thermoduric count. Homogenize dairy dried products as explained in Technique 1.3.2. Aseptically place 5 mL of the homogenized sample into a sterile test tube in duplicate. Tightly close tubes and place in a water bath at 62.8°C by immersing the tubes completely in water that the water line is above the sample level. After the temperature of the sample reaches 62.8°C, allow it to remain for 30 min. After heating, cool the samples immediately in cold water. Perform the thermoduric count (Technique 12.3) by the pour plate technique (Technique 2.2.3) and incubate plates. Report the result as thermoduric bacteria cfu per g of sample.

31.10 Interpretation of results

It is very important that the quality of milk can be determined quickly before the milk is prepared for delivery to the consumer. Since bacterial contamination results from poor sanitation, it is possible to use the microbial content of milk as a measure of its sanitary quality. In a normal condition, no microorganisms in 1 mL of canned milk may be assumed to be commercially sterile.

Spray-dried milk powder should have a DMC of fewer than 10^6 cfu g^{-1}, APC of fewer than 10^4 cfu g^{-1}, and coliforms, yeasts and molds each have fewer than 10 g^{-1}. *S. aureus* should be lower than 10 cfu g^{-1} and *Salmonella* should be absent from 100 g. Rolled-dry powder milk should have APC lower than 10^3 g^{-1}, and coliforms, and yeasts and molds each have lower than 10 cfu g^{-1}.

Sweetened condensed milk should have APC of less than 100 cfu g^{-1}, and a lipolytic count of less than 10 cfu g^{-1}, and yeasts and molds, and coliforms count should each be less than 1 cfu g^{-1}.

The methylene blue reduction test can be used for routine screening purposes. The microbial content of frozen ice cream depends on the quality of the ingredients used for their production: milk, cream, nonfat, milk solids, sugar, chocolate, fruits, nuts, egg products, emulsifiers, and stabilizers. Contamination of the mix, temperature abuse of the mix, and inadequate cleaning and sterilization of the equipment can lead to APC in excess of 10^6 cfu g^{-1}. APC should be fewer than 10^3 cfu g^{-1} at 30°C and coliform counts should be less than 1 cfu g^{-1}.

Analysis of meat, poultry and their products 32

32.1 Introduction

The predominant flora on freshly slaughtered carcasses is mesophilic and cannot multiply at the chilling temperatures. *Clostridium perfringens*, *Escherichia coli*, *Listeria monocytogenes*, *Staphylococcus aureus*, *Campylobacter*, *Salmonella*, *Enterococcus*, and coliforms frequently associate with meat fresh tissues. The gut is the main bacterial source and contributes *C. perfringens*, *Staphylococcus*, *Salmonella*, and coliforms to the meat surfaces. Lymph nodes may contain any of the animal pathogens, for example, members of the genera *Staphylococcus*, *Clostridium*, *Streptococcus*, *Bordetella*, *Corynobacterium*, *Mycobacterium*, and *Salmonella*. During refrigerated storage, the carcass flora begins to change toward psychrotrophic species in the genera: *Pseudomonas*, *Acinetobacter*, and *Moraxella*. Spoilage of whole cuts of meat or poultry at refrigerator temperature mainly is associated with the formation of slime and off-odor on the surface area of the carcass. Raw chilled meats and poultry shelf life is prolonged by the control of factors affecting the psychrotrophs growth rate: dry surface, number of psychrotrophs, limitation of oxygen, and temperature. When the storage air contains fairly low relative humidity (e.g., cut surfaces are exposed to air) and there is forced air circulation in the chill room, the drying of the surface of the meat prevents psychrotrophic bacteria, molds and yeasts growth. The genera of molds growing under these conditions are *Sporotrichum* (causing "white spot"), *Cladosporium* (causing "black spot"), *Thamnidium*, *Geotrichum*, *Trichosporum*, *Mucor*, *Rhizopus*, and *Penicillium*. Wrapping meats in oxygen impermeable films retards surface growth of aerobic bacteria and provides selection for microaerophilic bacteria: *Lactobacillus* and *Brochothrix thermosphacta*.

Animal pathogens can cause human diseases with close animal–man contact, such as among farmers and persons working in slaughter plants. Some of these include *E. coli*, *S. aureus*, *L. monocytogenes*, *C. perfringens*, *Campylobacter jejuni*, *Yersinia enterocolitica*, *Coxiella burnetii*, *Clostridium tetani*, *Chlamydia psittaci*, *Salmonella*, *Brucella*, *Corynobacterium*, *Mycobacterium*, and *Leptospira*.

The microbiology of cooked meats begins with the cooking process. Cooking temperature destroys pathogenic and spoilage bacteria, but *Enterococcus* and bacterial spores may survive at cooking temperature. The presence of *E. coli* indicates unsanitary conditions. Cooked meat products offer an ideal environment for microbial growth due to high nutrients, a neutral pH, high water, and low salt content. Such foods will be spoiled by different contaminated microorganisms, including enterococci, pseudomonads, lactic acid bacteria (LAB), and yeasts. Heating at the cured meat process destroys most of the meat flora except thermoduric bacterial spores. Salt and nitrite in the cure inhibit the microbial survivors and contaminants. LAB, *Micrococcus*, *Enterococcus*, *Bacillus*, and yeasts may grow, and form slime during prolonged refrigeration. Pathogenic microorganisms associating with cooked cured meats are *E. coli*, *C. perfringens*, *S. aureus*, and *Salmonella*.

Microbiological Analysis of Foods and Food Processing Environments. DOI: https://doi.org/10.1016/B978-0-323-91651-6.00002-1

Canned cured meats are prepared in two forms: shelf stable and perishable. An underprocessing will permit the survival of *Enterococcus*, *Micrococcus*, and other nonsporulating thermoduric spoilage bacteria. During sausage storage, the organic acids, low pH, additives, and drying may destroy many bacteria. In a poorly controlled process, *S. aureus* may grow and produce toxin during the processing of sausage. *S. aureus* gradually dies out during drying and storage, but the toxin remains. Yeasts and molds may grow on the sausage surface during production and subsequent storage. The preparation of dried meat includes a cooking step that destroys the normal vegetative flora of the raw meat. An efficient and rapid drying process will reduce the water activity and inhibit microorganisms. Yeasts and molds may grow on the surface of dried meats including dried hams. *Micrococcus* is predominant, coliforms and *S. aureus* occasionally present, and *E. coli* rarely presents on dried meat.

Microbial contamination results from poor sanitation of processing environments and handling conditions. It is possible to use the microbial count of meat, meat products, and processing equipment as a measure of their sanitary quality. Microbiological quality and characteristics of meat and meat products can be indicated by microbiological analysis with respect to aerobic plate count (APC), coliforms count, lactic acid bacteria count, and *B. thermosphacta* count.

32.2 Microbiological analysis of meats

Sampling, sample preparation, and all experimental techniques should be performed under aseptic conditions, and in duplicate, unless otherwise specified. All microbiological media to be used in applications must be sterilized. Again, all solutions and equipment that may cause the contamination of microorganisms must be sterilized or disinfected.

32.2.1 Equipment, materials and media

Equipment. Blender (or Stomacher), filter paper, forceps (or scalpel), glass spreader, incubator, microscope, microscope glass slide, Petri dishes, pH meter, pipettes (1 and 10 mL) in pipette box, pipette discarding pad (containing 10% bleach solution), plastic bag, refrigerator, screw capped test tubes, spatula (or knife), swab, and vortex.

Materials. Alpha-naphthylamine reagent, 0.02% peptone water, arginine broth, di-tetramethyl-p-phenylene-diamine dihydrochloride, Gram-staining reagents, Hugh-Leifson's carbohydrate broth (oxidation–fermentation [o-f] broth), methylene blue indicator stain, phosphate buffer (PB) solution (0.1 M), sulfanilic acid, and xylene.

Media. 2% Agar, acetate differential (AD) agar, APT agar containing bromocresol purple, Baird-Parker (BP) agar, Brain heart infusion (BHI) broth, BHI agar, BHI agar slant, BHI blood agar, CPM agar, dichloran rose Bengal chloramphenicol (DRBC) agar, gelatin medium, glucose-peptone-meat extract (GPM) agar, de Man Rogosa Sharp (Mrs) agar, milk powder (MP) agar, yeast extract glucose (YEG) motility medium containing triphenyl tetrazolium chloride, plate count (PC) agar, PC agar containing 15% NaCl, reinforced clostridial (RC) agar, spirit blue milk fat (SBMF) agar, Streptomycin thallus acetate actidione (STAA) agar, and tryptone soy (TS) broth.

32.2.2 Sampling and sample preparation

Meat and meat product sampling. Refer to Technique 1.3.2 for solid sampling. Surface samples may be taken by scraping with a spatula or knife (not over 2 mm in thickness) or 60 g or more taken from various parts on the surface using a sterile scalpel and forceps. Interior samples from solid meat may be taken by breaking the sterile surface by searing and then cutting into the center area with the knife and removing samples with a sterile knife or scalpel. If a particular spot or area is obviously spoiled or discolored, a sample of spoiled and unspoiled portion should be taken from a similar location.

Refer to Technique 1.5 for sample preparation and dilutions. Weigh a representative 50 g of meat and meat products into a sterile blender (or stomacher bag), add 450 mL of sterile PB solution, and homogenize by blending (or stomaching) for 3 min. Prepare further dilutions using PB solution.

Poultry sampling. The microflora of poultry is largely confined to the skin surface or visceral cavity, and therefore the most appropriate sampling for the analysis of table poultry is a surface-sampling technique. Refrigerated samples should be kept cold up to the time that they are to be examined; frozen samples should remain frozen until they are ready for testing. Swab procedures can be used for sampling from a surface area of poultry or part of the carcass may be rinsed in a known volume of PB solution (Technique 1.3.7). Place poultry into a sterile plastic bag in a container. Add an equal volume of PB solution into a plastic bag to cover the poultry (or if necessary, add PB solution of twice the weight of the poultry). Loosely tie the open-end of the plastic bag (not airtight). Let it stand undisturbed for 60 min at room temperature. Lightly mix by gently swirling and do not adjust the pH. Prepare serial dilutions using PB solution (Technique 1.5.2).

32.2.3 Microbiological analysis

32.2.3.1 Aerobic plate count

Refer to Technique 2.2.4 for APC by the spread plate technique. Allow the surface of a PC agar plate to dry in an incubator or near Bunsen burner flame by slightly opening the lid of the plate (drying means the disappearance of moisture from the agar medium surface area). Spread plate 1 mL of diluted sample to two sets of PC agar in duplicate and allow the plate at room temperature to absorb liquid coming from inoculum. Incubate one set of inverted plates at 32°C for 1 day for mesophilic APC, and the other set of inverted plates at 20°C or 4°C for up to 3 days and up to 14 days, respectively, for psychrotrophic APC. After incubation, analyze colonies on the plates. Count characteristic bacterial colonies from Petri plates containing 25 to 250 colonies. Calculate the number of APC by multiplying the average count by the dilution rate and dividing by the inoculation amount. Record the results as the number of mesophilic and psychrophilic APC colony forming units (cfu) per g of meat.

After incubation identify bacteria at genus level depending on the characteristics in Fig. 32.1. Indicate the type of colonies and dominant colonies. Select typical colonies (according to pigmentation, size, appearance, etc.) from both incubations (at 32°C and 40°C) and label each colony by number. Perform a pure culture technique to purify the colonies (Technique 13.4.1). In the pure culture technique use BHI agar and BHI broth. Finally inoculate one loopful of pure culture to BHI agar slant and BHI broth, and use in the identification techniques (Technique 32.2.4).

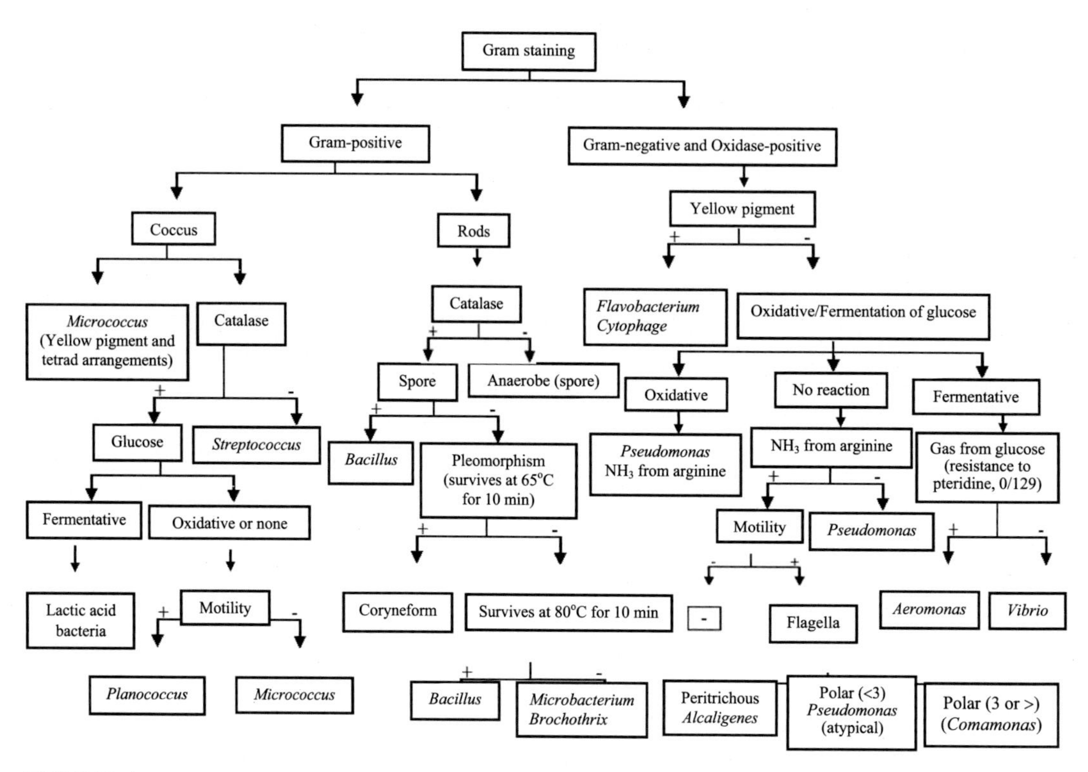

FIGURE 32.1

Identification flowchart of bacteria commonly associated with meats and meat products.

32.2.3.2 Counting of lactic acid bacteria

The total number of LAB is counted using a pour plate technique with Mrs agar. Refer to Technique 33.2.3 for details of the LAB count. APT agar is useful for counting of heterofermentative and homofermentative LAB, including *Lactobacillus*, *Leuconostoc*, and *Streptococcus* spp. (Technique 33.2.9). Because APT is nonselective, coliforms, and others grow well. The medium is suitable for the analysis of cured meats, pasteurized canned meats, poultry, and others. Count the colonies from Mrs agar and APT agar, and report the results as the total number of LAB and heterofermentative and homofermentative LAB cfu per g of sample.

32.2.3.3 Counting of Brochothrix thermosphacta

B. thermosphacta is a nonmotile, facultatively anaerobic or aerobic, rod-shaped, and Gram-positive bacterium. It appears singly, in short chains, or in long filamentous-like chains. It is nonpigmented and nonhemolytic, catalase-positive, oxidase-negative, H_2S-negative, indole-negative, methyl red-positive, and Vogues-Proskauer-positive. It reduces nitrate, hydrolyzes esculin, is gelatin-negative, citrate-negative, urease-negative, and produces acid but no gas from carbohydrates (e.g., glucose, fructose, lactose, maltose, mannose and xylose). It tolerates 8% NaCl, grows up to 20% CO_2

containing incubator, variably hydrolyzes casein, and does not hydrolyze arginine. Acetoin and acetate are the major end products of aerobic metabolism of glucose. The optimum growth temperature is 20°C–25°C; it grows from 0°C to 30°C; and rarely grows above 30°C. It is a significant proportion of the spoilage flora of meat and meat products stored aerobically or vacuum packed at chill temperatures. It is responsible for the off-odors with malodorous metabolic end products which can be used as a signal for the spoilage of vacuum-packed meat products. *B. thermosphacta* is not pathogenic, it is an economically important meat-spoilage bacterium.

Allow the surface of selective/differential STAA agar plate to dry in an incubator or near a Bunsen burner flame (drying means disappearance of moisture from the agar medium surface area). Spread plate 1 mL of diluted sample to STAA agar in duplicate (Technique 2.2.4) and leave the medium at room temperature to absorb the liquid originating from the inoculum. Incubate the inverted plates at 25°C for 48 h microaerobically (in 5% CO_2 incubator). After incubation, analyze the characteristic colonies on the plates. *B. thermosphacta* produces straw-colored colonies with 0.5–1.0 mm in diameter. Count the characteristic colonies from Petri plates containing 25 to 250 colonies. Calculate the number of *B. thermosphacta* by multiplying the average count by the dilution rate and dividing by the inoculation amount. Record the result as the number of *B. thermosphacta* cfu per g of meat. Streptomycin sulfate inhibits some Gram-positive bacteria and most Gram-negatives, while *B. thermosphacta* remains resistant. Thallous acetate inhibits many yeasts, and aerobic and facultative anaerobic bacteria. The incorporation of cycloheximide further inhibits yeasts and molds. *Pseudomonas* grows on STAA agar and they may be differentiated from *B. thermosphacta* by performing an oxidase test. *Pseudomonas* are oxidase positive and *B. thermosphacta* is oxidase negative.

B. thermosphacta will be further identified based on molecular analyses with MALDI-TOF Ms (Technique 13.6.3.1) polymerase chain reaction (PCR). PCR as a molecular biology technique is used in the detection of *Brochothrix* species or strains according to the genes encoding virulence factors or other characteristics. Some *Brochothrix*-specific primer sets that can be used in the identification are given in "Appendix A: Gene Primers Table A-1." Using *Brochothrix*-specific techniques and materials, *Brochothrix* can be identified by PCR analysis, as described in Technique 13.6.3.2.

Counting lipolytic B. thermosphacta. The lipolytic *B. thermosphacta* can be counted at two temperatures (4°C and 25°C) using SBMF agar containing tributyrin. It is a good substrate for the indication of lipolytic activity of microorganisms because some microorganisms hydrolyzing tributyrin cannot hydrolyze other triacylglycerols containing longer chain fatty acids in natural fats.

Spread plate 1 mL of diluted sample to SBMF agar in duplicate (Technique 2.2.4) and leave the medium at room temperature for a while to absorb the liquid originating from the inoculum. Incubate the inverted plates at 25°C for up to 72 h aerobically. After incubation, analyze the characteristic colonies on the plates. Lipolytic microorganisms metabolize the lipid in the medium and produce colonies with the surrounding blue zone (halos) on cloudy agar indicating lipase activity (lipid hydrolysis). *B. thermosphacta* hydrolyzes lipid tributyrin (lipolysis) in the medium and forms colonies with halos. Count characteristic colonies from Petri plates containing 25 to 250 colonies. Calculate the number of lipolytic *B. thermosphacta* by multiplying the average count by the dilution rate and dividing by the inoculation amount. Record the result as the number of lipolytic *B. thermosphacta* cfu per g of sample.

Counting of proteolytic and gelatin hydrolyzing B. thermosphacta. B. thermosphacta activity in terms of hydrolysis of casein and gelatin is assayed by culturing on CPM agar and GPM medium, respectively. Spread plate 1 mL of diluted sample to two sets of CPM agar and GPM medium in duplicate and leave the medium near the Bunsen burner flame for a while to absorb the liquid originating from inoculum. Incubate one set of inverted plates at 4°C for up to 7 days and the other set at 25°C for up to 72 h aerobically. After incubation, analyze the characteristic growth on the agar media. Some strains of *B. thermosphacta* can digest casein. *B. thermosphacta* cannot hydrolyze gelatin at either temperature. Indicate casein hydrolyzing and nongelatin hydrolyzing *B. thermosphacta* colonies. Calculate the number of lipolytic *B. thermosphacta* by multiplying the average count by the dilution rate and dividing by the inoculation amount. Record the result as the number of lipolytic *B. thermosphacta* cfu per g of sample.

32.2.3.4 Counting of greening bacteria

Add 1 mL of diluted sample into sterile Petri dishes in duplicate. Pour 17–19 mL of melted APT agar (at 47°C) into Petri dishes (Technique 2.2.3). Mix the agar and sample by thoroughly rotating the plates by drawing a figure of eight while lying on the table tap, and allow theagor to solidify in the plates (within 10 min). Incubate the inverted Petri plates at 21°C for up to 3 days. After incubation, analyze the plates for colony formation and record the characteristic colonies. Count the characteristic colonies from Petri plates containing 25 to 250 colonies. Calculate the number of greening bacteria by multiplying the average count by the dilution rate and dividing by the inoculation amount. Confirm the presumptive isolates for the formation of greening on meat and record the result as the number of greening bacteria cfu per g of sample.

Select at least three colonies per plate from each type of colony. Remove the colony by touching a loop onto the colony center and inoculate it into APT broth in tube in duplicate. Incubate at 25°C for 24 h. After incubation of APT broth, analyze broth for growth (turbidity) and color change. Place a frankfurter or other meat product into a sterile Petri dish in duplicate containing a sterile filter paper moistened with 2 mL of sterile distilled water to prevent drying. Remove a loopful of APT broth culture (or insert a cotton swab into APT broth culture) and spread onto the surface of a frankfurter or other meat product in the Petri dish. Incubate the inoculated meat product in dishes at 25°C for 24 h. During incubation, examine the meat product for the formation of greening on the surface. Uninoculated meat product slices should be similarly incubated as a control.

32.2.3.5 Indicating of nonbacterial greening

Nonbacterial greening in fermented sausages due to nitrite burn may result from excess bacterial nitrate reduction and will not be detected by the above technique. A qualitative test for nitrite may be performed by streaking the meat product surface (using pipettes) with sulfanilic acid reagent and then with alpha-naphthylamine reagent. Development of an intense red color on the discolored area indicates nitrite burn.

32.2.3.6 Methylene blue reduction test

After swabbing 10 cm of skin, break swabs into 10 mL of sterile 0.02% peptone water in screw cap tubes and shake until the swab is dispersed. Then add 1 mL of sterile TS broth and 1 mL of sterile skim milk (5%) to the tubes along with 1 mL of methylene blue indicator stain, and incubate the tubes at 30°C in dark and examine at 30 min intervals until the methylene blue is decolorized. The following decolorization time represents the quality of the poultry.

Quality of poultry	Decolorization time
Fresh	>8 h
Good	>5 h but <8 h
Fair	>3.1/2 h but <5 h

32.2.3.7 Direct microscopic count

Refer to Technique 3.2 for details of the direct microscopic count. Allow the homogenized solution to remain so that the meat particles settle in the liquid. Prepare a smear by transferring 0.01 mL of the sample to a microscope glass slide and spread over a 1 cm^2 area of the slide. Perform a simple stain as explained in Appendix B under the topic "Crystal Violet Stain." If the sample has a high fat content, rinse the fixed smear with xylene before staining and after staining, count 100 fields and report as the average number of DMC per g of sample.

32.2.3.8 Count from cured–pickled meat and other meat products

This group includes partially dried, smoked, and fermented products with a combination of brining. The brining solution may contain NaCl alone or be together with sodium or potassium nitrite (or nitrate). The pH of cured or pickled meats (pH 5.0–6.5) contributes flavor together with microorganisms. The bacterial flora of cured or pickled meat products will include LAB, halotolerant *Micrococcus*, and halotolerant yeasts (e.g., *Debaryomyces*). Molds may grow on the surfaces of dried products. Halophilic *Vibrio* may present in meat-curing brines. The diluent used in the preparation of the homogenization and dilution of the sample from the meat products (cured, pickled, and the others) should contain 15% NaCl.

Counting of halotolerant/halophilic bacteria. Perform viable counts on PC agar. Spread plate 1 mL of sample to PC agar containing 15% NaCl. Incubate the inverted plates at 7°C, 20°C, and 35°C for 14, 3, and 2 days, respectively. After incubation, analyze colony characteristics on plates and count the number of colonies for each incubation temperature. Calculate the number of bacteria by multiplying the average count by the dilution rate and dividing by the inoculation amount. Record the results for each temperature as follows: (1) record number of halotolerant/halophilic mesophilic aerobic bacteria cfu per g of meat product for 35°C; (2) record number of halotolerant/halophilic psychrotrophic aerobic bacteria cfu per g of meat product for 20°C; and (3) record number of halotolerant/halophilic psychrophilic aerobic bacteria cfu per g of meat product for 4°C.

Other counts. Perform the following counts as indicated in the respective practices of this book. Count *Lactobacillus* (on AD agar) at 30°C for 5 days in a microaerophilic atmosphere containing 5% CO_2 and *S. aureus* (on BP agar) at 35°C. Perform yeasts and molds count using DRBC agar and incubate at 25°C for 5 days.

32.2.3.9 Counting from cooked meat products

Cooking will destroy most of the meat microflora. Thermoduric bacteria (e.g., *C. perfringens* and *Bacillus cereus* spores) can survive after cooking and bacteria grow on sliced cooked meats. Perform the following counts as the complete procedure indicated in the respective practices of this book. Aerobic and anaerobic PCs at 25°C, 35°C, and 5°C. If cooked meat is stored at low temperature (e.g., refrigeration), perform low temperature count (7°C). For counting coliforms, fecal coliforms, and individual pathogens (e.g., *E. coli*, *B. cereus*, *S. aureus*, *C. perfringens*, *Salmonella* and *Enterococcus*), refer to the corresponding practices of this book.

32.2.3.10 Other counts

Viable counts of anaerobic bacteria can be performed using RC agar and BHI blood agar with incubation at 35°C for 2 days anaerobically. For counting coliforms, fecal coliforms, *Enterococcus*, yeasts and molds, and individual pathogens (*E. coli*, *S. aureus*, *B. cereus*, *C. perfringens*, *Y. enterocolitica*, *L. monocytogenes*, *Campylobacter* and *Salmonella*), refer to the corresponding practices in this book.

32.2.4 Identification of microorganisms

Gram-staining. Prepare a Gram stain from each pure culture as explained in Appendix B under the topic "Gram Staining Reagents." Indicate the following results for each selected colony from magnification under a microscope: Gram reaction type and size, shape and arrangement of the cells. If the Gram reaction is positive, perform the following tests with the isolate: motility test, catalase test, oxidase test, carbohydrate utilization, and NH_3 formation from arginine.

Motility. Refer to Technique 13.4.2 for the motility test. Indicate the motile and nonmotile characteristic for each isolate. The motility medium containing triphenyl tetrazolium chloride (TTC) is stab inoculated. The tube is incubated at 25°C and the medium is examined after 24 and 48 h. Motility is observed as a diffuse growth projecting from the stab line. Growth of bacteria capable of reducing TTC will appear as a red growth along the stab line as well as in the area into which the cells have migrated. Nonmotile cells grow only through the inoculation line. Motility can also be detected using the "hanging drop technique."

Catalase test. Perform the catalase test for Gram-positive bacteria. Refer to Technique 13.4.3.10 for the catalase test.

Oxidation–fermentation. Perform oxidation–fermentation characteristics as indicated in Technique 21.3.2.3. Incubate the test medium at 25°C for 3 days and indicate the type of carbohydrate utilization.

NH_3 from arginine. This test is very useful for identifying *Pseudomonas* species. Inoculate one loopful of bacterial culture to two sets of arginine broth in tubes in duplicate. Cover the surface of the broth in one tube with sterile liquid paraffin or sterile 2% melted agar (at 45°C) and use the second tube without covering as a control. Incubate the tubes at 25°C for 3 days. The production of NH_3 (color change from light pink to dark red) in the covered tube is a characteristic positive test reaction for most *Pseudomonas* species. This color change cannot be observed on the noncovered arginine tube. If the color change is not observed in the covered tube, record the test as negative for the tested culture.

Oxidase test. Refer to Technique 17.4.2.10 for the oxidase test. This enzyme hydrolyzes dimethyl-p-phenylenediamine and a-naphthol to form a blue color. Species of *Pseudomonas*, *Aeromonas*, and *Alcaligenes* are usually oxidase positive. The test is performed by moistening a piece of filter paper with a 1% aqueous solution of methyl-p-phenylene diamine, followed by rubbing a small amount of a bacterial colony onto the filter paper. A positive test is a deep blue color that develops in 30 s. No color change indicates the test is negative.

The results of these tests should be recorded, and with the aid of the identification schemes (Fig. 32.1), the isolate should be identified. Although all of the tests mentioned in the figure will not be performed, these tests provide preliminary information about the protocols necessary for identifying the Gram-positive and Gram-negative bacteria frequently encountered in meats. Perform MULDI-TOF Ms test for further identification of microorganisms (Technique 13.6.3.1).

32.3 Physical examination of meat and poultry products

Microorganisms associated with meat and poultry products can be divided into three groups: beneficial, spoilage, and pathogenic. Each product has a characteristic microflora called its "microbiota." Changes on the products can be observed by organoleptic characteristics of meat products. Organoleptic analyses are important in the investigations of meat product problems. Tasting products as part of a microbiological examination is a dangerous practice and should be avoided. Some of the organoleptic changes are off-odor, off-color, appearance, and texture.

A meat product sample can be analyzed for the following characteristics.

Appearance. Color changes due to fat hydrolysis and foreign materials (e.g., hair, feathers, charcoal, sand, metal, etc.) are unacceptable.

Texture. Changes in the characteristic of meat products are slime layer formation; consistency; structural organic compounds hydrolysis (e.g., proteolysis of proteins), etc.

Odor (taint). Changes in this characteristics of meat products are off-odors: fruity, sour (acidic), moldy, rancid, musty, fishy, yeasty (beer-like), and putrid.

pH. Homogenized sample pH is measured using a pH meter equipped with an electrode. If fat or oil causes fouling of the electrode, transfer a portion of the homogenate to a separatory funnel and separate the fat or oil from the aqueous phase of sample. For some types of product, a centrifuge can be used to separate particles from the sample. Certain low-fat product pH can be measured on the surface area using a surface electrode, but it must be ensured that the electrode has good contact with the meat surface. Record the measured pH result to the nearest 0.1 unit.

32.4 Interpretation of results

In case of spoilage, the results will give some idea as to the type of the predominant microorganisms. For example, the extensive growth of LAB will lower markedly the pH of products containing sugar, whereas the extensive growth of *Pseudomonas* and some *Micrococcus* may raise the pH. The combination of high nitrite and low pH can cause a greening of cured meats known as nitrite burn. The level of *S. aureus* or *C. perfringens* in ready-to-eat meat may have 10 cfu per g of sample, which may cause poisoning.

Frozen poultry when examined by rinsing should give a psychrotrophic APC at 20°C of less than 10^7 cfu per mL of the rinsing solution, and *Salmonella* should be detected in not more than one of five 25 g of sample. Mesophilic APC at 35°C (or at 20°C in the case of chilled meats for psychrotrophic APC) should be less than 10^7 cfu per g of meat, and *Salmonella* should be detected in not more than one of five 25 g of sample.

PRACTICE

Analysis of fermented foods

33

33.1 Introduction

Bacterial species from 11 genera are included in the group "lactic acid bacteria" (LAB) due to their ability to produce relatively higher amounts of lactic acids from carbohydrates. These genera are *Lactococcus*, *Leuconostoc*, *Streptococcus*, *Pediococcus*, *Lactobacillus*, *Bifidobacterium*, *Enterococcus*, *Vagococcus*, *Aerococcus*, *Tetragenecoccus*, and *Carnobacterium*. Species from the last five genera, except *Tetragenecoccus halophilus*, are not used as starter cultures.

The term "starter culture" refers to LAB that mostly induce lactic acid fermentation to produce fermented products. The type of LAB in any starter culture depends on the aim of their use. For example, in cheese production, active lactic acid production is an essential requirement and the LAB culture may consist of *Lactococcus lactis* subsp. *lactis* or *L. lactis* subsp. *cremoris*, *Lactobacillus helveticus*, *Lactobacillus delbrueckii* subsp. *bulgaricus*, *Streptococcus thermophiles*, or combination of these with or without aroma-forming LAB (e.g., *Leuconostoc*). In ripened cream for butter production, cultures usually consist of citrate-fermenting LAB able to produce diacetyl as well as lactic acid from citrate, for example, *L. lactis* subsp. *lactis* biovar diacetylacti*s* and/or *Leuconostoc mesenteroides* subsp. *cremoris*. LAB producing aroma and flavor are *Leuconostoc* species (*L. mesenteroides* subsp. *cremoris*, *L. mesenteroides* subsp. *Dextranicum*, and *L. mesenteroides*). *L. lactis* coagulates milk with an acidity of about 1.6% lactic acid.

Pasteurization and refrigeration are used to provide stability for these products. They have a pH of 4.6 or below. Pathogenic microorganisms, for example, *Clostridium botulinum* and *Listeria monocytogenes*, may associate with fermented vegetable products. The brine microflora is established early by coliforms and continued predominantly by the homofermentative *Pediococcus cerevisiae* and followed by *Lactobacillus plantarum*. Gas-forming heterofermentative LAB, for example, *Lactobacillus brevis* and *L. mesenteroides*, may also grow in certain vegetable brines. Nonpasteurized fermented vegetables contain many microorganisms, usually LAB, particularly *Lactobacillus*, *Leuconostoc*, and *Pediococcus*. Heterofermentative LAB can cause blotting by gas production in sweet and sour cucumbers. Yeasts, molds, members of Enterobacteriaceae, *Bacillus*, and *Clostridium* may cause spoilage on fermented vegetables. Some fermented vegetables are produced as the result of action by a mixture of LAB, yeasts, and molds. Coliforms, obligate halophiles, heterofermentative LAB, and fermentative yeasts may be associated with gaseous fermentation in brined fermented products. Growth of film yeasts (*Pichia*, *Hansenula*, *Debaryomyces*, *Candida*, and *Trichosporum*) may occur at different salt concentrations and will result in a loss of brine acidity.

The main lactic acid-producing LAB are homofermentative, such as *S. thermophilus*, *L. lactis* subsp. *Lactis*, and *L. lactis* subsp. *cremoris*. LAB vary in their acid production rate. The rate of acid production is influenced by pH, temperature, antimicrobials, bacteriophages, inhibitory compounds, milk composition, availability of nutrients, compatibility of strain, and microflora.

Microbiological Analysis of Foods and Food Processing Environments. DOI: https://doi.org/10.1016/B978-0-323-91651-6.00023-9

Lactobacillus species are microaerophilic and generally require pour plating for anaerobic growth. Heat injury or inactivation of heat-sensitive microbial cells may occur in the pour plate technique. This application describes microbiological techniques for determining the most common microbial contaminants, potential microbial spoilage and health hazards, and their hygienic significance in fermented foods. Some of the techniques to predict the shelf life of fermented foods involve counting of LAB, counting fermentative microbial groups, microscopic examination, and plate counting of bacteria and fungi.

33.2 Microbiological analysis techniques

Sampling, sample preparation, and all experimental techniques should be performed under aseptic conditions, and in duplicate, unless otherwise specified. All microbiological media to be used in applications must be sterilized. Again, all solutions and equipment that may cause the contamination of microorganisms must be sterilized or disinfected.

33.2.1 Equipment, materials, and media

Equipment. Blender (or Stomacher), capped jar, cover glass, Erlenmeyer flask, filter paper (0.45 μm), gasometer, hemocytometer counting chamber, incubator, microscope glass slide, microscope, Petri dishes, pH meter, pipettes (1 and 10 mL) in pipette box, pipette discarding pad (containing 10% bleach solution), platinum loop (3 mm diameter), test tubes, thermostatic shaking water bath, and vortex.

Materials. $BaCl_2$, calcium citrate solution, chloroform, creatin, erythrosine stain (0.02%), Gram staining reagents, HCl, methyl thymol blue (MTB) solution (0.1%), NaOH (0.1, 0.5 and 1.0N), NaOH (40%), 0.1% peptone water, phenolphthalein indicator (0.5%), potassium chromate solution (5%), quarter-strength Ringer's (QSR) solution, silver nitrate solution (0.171N; 29.063 $g\,L^{-1}$), sterile distilled water, trichloroacetic acid solution (4%), and xylen.

Media. 2.0% Agar, APT agar, bromocresol green ethanol (BGE) agar, chopped liver (CL) broth, dextrose sorbitol mannitol (DSM) agar containing bromocresol, Elliker agar, glucose yeast extract calcium carbonate (GEYC) broth supplemented with 100 mg pimaricin per L, HHD agar containing fructose, Lee's agar, litmus milk, M17 agar, de Man-Rogosa and Sharpe (MRS) agar, MRS broth, MRS broth containing 1%, 1.5%, and 2% bile salt, MRS broth (with pH 2.0, 3.0, 4.0, and 5.0), MRS supplemented with cysteine hydrochloride (mMRS) containing novobiocin and van comycin, neutral red chalk lactose (NRCL) agar, nitrate peptone water, Rogosa agar, skim milk (10%), yeast extract skim milk (YESM) agar, yeast extract glucose (YEG) broth, YEG broth containing acetic acid from 0.2% to 7.0%, YEG broth containing ethanol from 6% to 15%, yeast extract glucose lemco (YEGL) bromocresol purple agar, and yeast extract peptone (GYEP) agar.

33.2.2 Sample preparation

Refer to Technique 2.2.2 for details of sample preparation and dilutions. For a solid sample (e.g., sausage, vegetables, and fruits), add 25 g of the sample into 225 mL of sterile QSR solution in the blender (or stomacher bag). It is allowed to remain at 4°C for 10 min. Then the content is blended for 1 min (or stomached for 2 min) to homogenize the sample. For moist and liquid samples (e.g., yogurt, starter culture, and brine), directly add 25 mL of sample into 225 mL of sterile QSR solution in the Erlenmeyer flask and homogenize by gently shaking. Prepare further required dilutions using 0.1% peptone water and mix by vortexing.

33.2.3 Counting of lactic acid bacteria

MRS agar or broth supports the growth of LAB; *Lactobacillus*, *Streptococcus*, *Pediococcus* and *Leuconostoc*. Growth of LAB is enhanced by microaerophilic conditions. Generally, LAB shows delayed growth and smaller colony size than other microorganisms. Low pH (pH 5.7) may inhibit the main groups of competing microflorae. MRS agar pH can be reduced to 5.7 by using sorbic acid (or potassium sorbate). Generally, submerged or surface LAB colonies may be compact or feathery, small, opaque, and white on this medium. But *Lactobacillus* species produce large white-creamy, transparent, flat and irregular shaped, rough-surfaced, and shiny colonies on MRS agar.

Spread plate 1 mL of diluted sample onto YEGL bromocresol purple agar (Technique 2.2.4). Alternatively, add 1 mL of the diluted sample into a sterile Petri dish and pour 17–19 mL of MRS agar (at 45°C) into the Petri dish. Mix agar and 1 mL of the sample by thoroughly rotating the plates by drawing a figure of eight while lying on the table top, and allow the agar in the plates to solidify (within 10 min). Incubate the inverted Petri plates at 30°C for up to 4 days in a microaerophilic atmosphere in an incubator with 5% CO_2 (or in a candle jar). After incubation, Petri plates are analyzed for characteristic colony formation. Count all colonies from Petri plates containing 25 to 250 colonies. Calculate the number of LAB by multiplying the average count by the dilution rate and dividing by the inoculation amount. Record the result as the LAB colony forming units (cfu) per g or mL of sample.

Subculture the characteristic colonies to MRS broth and obtain a pure culture using MRS agar (Technique 13.4.1). Prepare a Gram stain from MRS broth culture, indicate the morphological characteristics, and test for catalase reaction. Catalase negative, Gram-positive, cocci or rods may be considered as LAB. Use pure cultures in identification tests (Technique 33.4).

33.2.4 Counting of total acid producers

Spread plate 1 mL of diluted sample to YEGL bromocresol purple agar (Technique 2.2.4). Alternatively, add 1 mL of diluted sample into a sterile Petri dish. Pour 17–19 mL of melted YEGL bromocresol purple agar (at 47°C) into Petri dishes (Technique 2.2.3). Mix agar and 1 mL of sample by thoroughly rotating the plates by drawing a figure of eight while lying on the table top, and allow the agar in the plates to solidify (within 10 min). Incubate the inverted Petri plates at 30°C for 3 days in a microaerophilic atmosphere in an incubator with 5% CO_2 (or in a candle jar). After incubation, the Petri plates are analyzed for yellow halo-forming colonies (acid producer). Count all the yellow colonies from plates containing 25 to 250 colonies. Calculate the number of total acid producers by multiplying the average count by the dilution rate and dividing by the inoculation amount. Record the result as the total acid producers cfu per g or mL of sample.

For coliforms and yeast and mold counts refer to the corresponding practices in this book.

33.2.5 Counting of lactic *Streptococcus* and *Lactobacillus*

Elliker agar supports the growth of *S. thermophilus* and *Lactobacillus*. Spread plate 1 mL of the diluted sample to Elliker agar (Technique 2.2.4). Alternatively, add 1 mL of the sample into a sterile Petri dish. Pour 13–15 mL of Elliker agar (at 45°C) into the Petri dishes. Mix the medium and 1 mL of the sample by thoroughly rotating the plates by drawing a figure of eight while lying on the table top, and allow the agar in the plates to solidify (within 10 min). Overlay an additional 4–6 mL Elliker agar and allow it to solidify. Incubate the inverted Petri plates at 30°C for up to 4 days in a microaerophilic atmosphere in an incubator with 5% CO_2 (or in a candle jar). After incubation, the Petri plates are analyzed for characteristic colony formation. Count all the colonies from plates containing 25 to 250 colonies. Calculate the number of lactic *S. thermophilus* and *Lactobacillus* by multiplying the average count by the dilution rate and dividing by the inoculation amount. Record the result as the lactic LAB cfu per g or mL of sample.

Select least five colonies per plate, subculture the selected colonies to MRS broth and agar, and obtain a pure culture (Technique 13.4.1). Follow further identification tests (Technique 33.4).

33.2.6 Counting of *Streptococcus thermophilus*

S. thermophilus is a Gram-positive, facultative anaerobic and has an optimum growth temperature of 40°C–45°C, a minimum of 20°C–25°C, and a maximum at 47°C–50°C. It cannot hydrolyze proteins and arginine. It ferments lactose, glucose, fructose, and sucrose. It does not ferment galactose. It is sensitive to antibiotics and sanitizers.

Add 1 mL of diluted sample into a sterile Petri dish. Pour 13–15 mL of NRCL agar (at 45°C) into Petri dishes. Mix the medium and 1 mL of the sample by thoroughly rotating the plates by drawing a figure of eight while lying on the table top, and allow the agar in the plates to solidify (within 10 min). Overlay an additional 4–6 mL NRCL agar and allow it to solidify. Incubate the inverted Petri plates at 30°C for up to 4 days in a microaerophilic atmosphere in an incubator with 5% CO_2 (or in a candle jar). After incubation, the Petri plates are examined for characteristic colony formation. On NRCL agar, acid-producing *S. thermophilus* produces small, deep red colonies that are surrounded by a clear zone. Count all the characteristic colonies from Petri plates containing 25 to 250 colonies. Calculate the number of *S. thermophilus* by multiplying the average count by the dilution rate and dividing by the inoculation amount. Record the result as the number of *S. thermophilus* cfu per g or mL of sample.

Select at least five characteristic colonies per plate. Subculture the characteristic colonies to MRS broth and agar, and obtain pure cultures (Technique 13.4.1). Subculture with a wire needle into litmus milk (10%) from MRS broth pure culture. Incubate the litmus milk at 30°C for acid production and observe the coagulation of proteins in milk. Follow further identification tests as indicated in Technique 33.4. Confirm the identity including the catalase test and prepare Gram stains from MRS broth pure culture that are to be examined morphologically. Gram-positive, catalase negative, spherical to ovoid nonmotile cocci, about 0.9 μm in diameter, appearing in pairs and in chains may be considered to be *S. thermophilus* (Fig. 33.1).

M17 agar is a suitable medium for the isolation and counting of *S. thermophilus* from cheese and yogurt. A high concentration of disodium-glycerophosphate in this medium results in the suppression of *L. bulgaricus*. M17 can also be used for the counting of *S. thermophilus*.

FIGURE 33.1

Staining appearance of *S. hermophilus*.

33.2.7 Counting of *Lactobacillus*

Lactobacillus is Gram-positive, rod-shaped, microaerophilic or facultatively anaerobic, catalase-negative, acid-tolerant, indole-negative, H_2S-negative, does not hydrolyze gelatin, and does not reduce nitrates. They hydrolyze proteins. Spread plate 1 mL of the diluted sample onto Rogosa agar. Alternatively, add 1 mL of the diluted sample into a sterile Petri dish. Pour 13–15 mL of Rogosa agar (at 45°C) into Petri dishes. Mix the medium and 1 mL of the sample by thoroughly rotating the plates by drawing a figure of eight while lying on the table top, and allow the agar in the plates to solidify (within 10 min). Overlay an additional 4–6 mL of Rogosa agar and allow it to solidify. The counting and incubation temperature depends on the particular species of *Lactobacillus* to be isolated. For example, thermobacteria *Lactobacillus* (e.g., *Lactobacillus acidophilus* and *L. bulgaricus*) require a higher optimum incubation temperature (at 42°C for 48 h) and counting by pour plating; psychrotrophic *Lactobacillus* require 28°C for 2 h and 22°C for up to 48 h in a microaerophilic atmosphere in an incubator with 5% CO_2 (or in a candle jar) and counting by spread plating. After incubation, Petri plates are examined for characteristic colony formation. On Rogosa agar, acid-producing *Lactobacillus* grow well and produce small grayish-white, raised or flat, smooth, rough, or intermediate colonies. Count all characteristic *Lactobacillus* colonies from Petri plates containing 25 to 250 colonies. Calculate the number of *Lactobacillus* by multiplying the average count by the dilution rate and dividing by the inoculation amount. Record the result as the number of *Lactobacillus* cfu per g or mL of sample.

Select at least five characteristic colonies per plate. Subculture the characteristic colonies to MRS broth and agar slant, and obtain pure cultures (Technique 13.4.1). Inoculate one loopful of pure culture into litmus milk. Incubate the litmus milk at 30°C for up to 4 days for acid production and observe the coagulation of proteins. Follow further identification tests as per Technique 33.4. Confirm the identity including catalase test and prepare a Gram stain from MRS broth pure culture, which is to be examined morphologically. Catalase negative, Gram-positive, rod-shaped, nonmotile, pairs and long chains are considered to be *Lactobacillus* (Fig. 33.2).

33.2.8 Counting of *Lactococcus*

Lactococcus has spherical or ovoid-shape cells and occurs singly, in pairs, or in chains (Fig. 33.3A). They are catalase-negative, Gram-positive, nonmotile, and facultatively anaerobic. Use Lee's agar medium for differential counting of *Lactococcus*. This medium can differentiate *L. lactis* and *Lac. lactis* subsp. *cremoris* and *Lac. lactis* subsp. lactis biovar. diacetylactis. Spread plate 1 mL of the diluted sample onto Lee's agar (Technique 2.2.4). Alternatively, add 1 mL of the diluted sample into a sterile Petri dish. Pour 13–15 mL of Lee's agar (at 47°C) into Petri dishes. Mix the medium and 1 mL of the sample by thoroughly rotating the plates by drawing a figure of eight while lying on the table top, and allow the agar in the plates to solidify (within 10 min). Overlay an additional 4–6 mL Lee's agar and allow it to solidify. Incubate the inverted plates in a microaerophilic atmosphere in an incubator with 5% CO_2 (or in a candle jar) at 32°C for up to 6 days. After inoculation, examine the colonies from plates. The production of acid from lactose causes yellow colony formation. Subsequent arginine utilization and liberation of NH_3 by *L. lactis* subsp. *lactis* biovar diacetylactis results in a localized pH change back to neutrality, with a return of the purple indicator. *L. lactis* subsp. *lactis* biovar. diacetylactis utilizes calcium citrate, and after 6 days of incubation, the citrate-degrading colonies exhibit clear zones against a turbid background. The buffering capacity of $CaCO_3$ limits the effects of acid and NH_3 production of individual colonies.

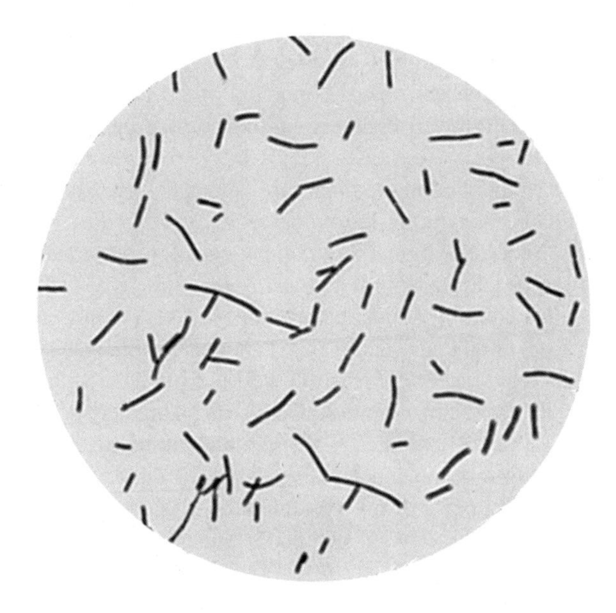

FIGURE 33.2

Staining appearance of *L. bulgaricus*.

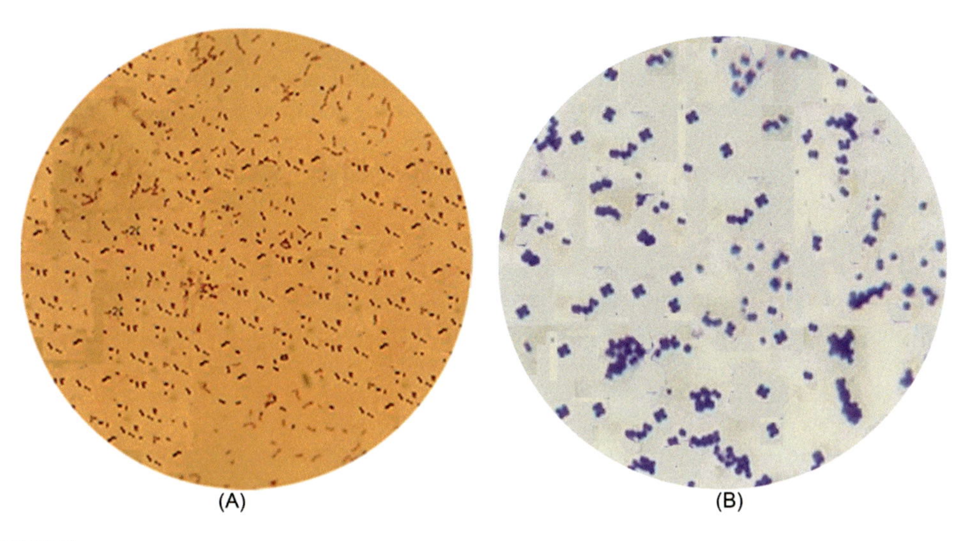

FIGURE 33.3

Staining appearance of *Lactococcus* (A) and *Pediococcus* (B).

After 36–40 h, count all colonies and then count the yellow *L. lactis* subsp. *cremoris* colonies from Petri plates containing 25 to 250 colonies. Record the results as the total *Lactococcus* and subsp. *cremoris* cfu per g or mL of the sample. Return the plates for an additional 4 days (total 6 days) incubation. After this period, count all the colonies and expose the plates to the air for 1 h. Count all colonies showing zones of clearing of the turbid suspension of calcium citrate (*L. lactis* subsp. *lactis* biovar. diacetylacti*s*). Subtract the species *L. lactis* subsp. *cremoris* (after 36–40 h) and *L. lactis* subsp. *lactis* biovar. diacetylacti*s* counts from the total count to obtain the remaining *L. lactis* population in the mixture. Slow arginine hydrolyzing or nonhydrolyzing strains of *L. lactis* subsp. *lactis* biovar. diacetylactis sometimes produce yellow colonies similar to subspecies *cremoris* after 36–40 h. In such cases, mark the yellow colonies (after 36–40 h) with a waterproof pen. In the final count, count the marked colonies that show clearing as *L. lactis* subsp. *lactis* biovar. diacetylactis and subtract their number from the original yellow colony count to obtain the accurate value for subsp. *cremoris*. Calculate the number of *Lactococcus* and *Lactococcus* subspecies by multiplying the average count by the dilution rate and dividing by the inoculation amount. Record the results as the number of subspecies and *Lactococcus* cfu per g or mL of sample.

Select at least five characteristic colonies per plate. Subculture the characteristic colonies to MRS broth and agar, and obtain a pure culture (Technique 13.4.1). Subculture with a wire needle into litmus milk from MRS broth pure culture. Incubate the litmus milk at 30°C for acid production and observe the coagulation of proteins. Follow further identification tests (Technique 33.4). Confirm the identity using the catalase test (Technique 13.4.3.10), Gram stain from MRS broth culture (Technique 13.4.2), and morphologically. Gram-positive, catalase negative, spherical to ovoid, nonmotile cocci, occurring in pairs and chains may be considered to be *Lactococcus*.

33.2.9 Counting of *Pediococcus*

Samples (10 g) are shaken well by hand in an Erlenmeyer flask containing 90 mL of sterilized distilled water and the required serial dilutions are prepared using serum physiological solution. From each dilution, spread plate 1 mL on mMRS agar (or GYEP agar) plates or pour plate. Incubate at 37°C under anaerobic conditions for 24 h. Count all characteristic (a yellowish colonies on mMRS agar or yellow colonies with a clear zone due to dissolving $CaCO_3$). Other LAB also produce the same colonies. Count the *Pediococcus* colonies from Petri plates containing 25 to 250 colonies. Calculate the number of *Pediococcus* by multiplying the average count by the dilution rate and dividing by the inoculation amount. Record the result as the number of *Pediococcus* cfu per g or mL of sample. Select at least five characteristic colonies per plate. Subculture the characteristic colonies to MRS broth and agar slant, and obtain pure cultures (Technique 13.4.1). Inoculate one loopful of pure culture into litmus milk. Incubate the litmus milk at 30°C for up to 4 days for acid production and observe the coagulation of proteins. Follow further identification tests as per Technique 33.4. Confirm the identity using the catalase test and prepare a Gram stain from MRS broth pure culture, examine morphologically. Detect growth at different temperatures in MRSs broth (10°C, 35°C, and 45°C within 7 days), bile tolerance in GYEP broth containing bile at 10%, 20%, 30%, and 40%, salt tolerance in MRS broth containing NaCl at 3.0% and 6.5%. Carbohydrate fermentation patterns tested in GYEP basal medium containing 1% (w/v) carbohydrate, grow from 10°C to 50°C, grow in 6.5% NaCl, nonmotile, not hydrolyze esculin (except *Pediococcus acidilactici*). *Pediococcus* is Gram-positive and catalase negative, is found in pairs to tetrad cocci (Fig. 33.3B), and cannot produce gas from glucose.

33.2.10 Counting of homofermentative and heterofermentative lactic acid bacteria

Use HHD agar for differential enumeration of homofermentative and heterofermentative LAB (Table 33.1). This medium contains fructose, which is reduced to mannitol by heterofermentative LAB but not by homofermentative LAB. In the agar medium, homofermentative colonies of LAB are blue to green, while heterofermentative LAB are white. HHD agar is not a selective medium for LAB. Spread plate 1 mL of a diluted sample to HHD agar (Technique 2.2.4). Alternatively, add 1 mL of the diluted sample into a sterile Petri dish. Pour 13–15 mL of HHD agar (at 47°C) into Petri dishes. Mix the medium and 1 mL of the sample by thoroughly rotating the plates by drawing a figure of eight while lying on the table top, and allow the agar in the plates to solidify (within 10 min). Overlay an additional 4–6 mL HHD agar and allow it to solidify. Incubate the inverted Petri plates aerobically at 32°C for up to 4 days. After incubation, the Petri plates are examined for characteristic colony formation. On HHD agar, count homofermentative and heterofermentative LAB characteristic colonies separately from Petri plates containing 25 to 250 colonies. Calculate the number of homofermentative and heterofermentative LAB separately by multiplying the average count by the dilution rate and dividing by the inoculation amount. Record the results as the homofermentative and heterofermentative LAB cfu per g or mL of sample.

Select at least five characteristic colonies per plate. Subculture the characteristic colonies to MRS broth and agar, and obtain pure cultures (Technique 13.4.1). Use pure cultures in identification tests. Subculture with a wire needle into litmus milk from MRS broth pure culture. Incubate

Table 33.1 Groups of LAB based on metabolic end products.

Homofermentative	Facultative heterofermentative	Heterofermentative
***Lactococcus* spp.**	***Lactobacillus* spp.**	***Lactobacillus* spp.**
Lac. lactic subsp. *lactis*	*Lb. animalis*	*Lb. brevis*
Lac. lactic subsp. *cremoris*	*L. bifermentants*	*Lb. buchneri*
***Lactobacillus* spp.**	*Lb. casei*	*Lb. cellobiosus*
Lb. acidophilus	*Lb. curvatus*	*Lb. confusus*
Lb. lactis	*Lb. homuhiochii*	*Lb. coprophilus*
Lb. bulgaricus	*Lb. murinus*	*Lb. fermentatum*
Lb. leichmannii	*Lb. plantarum*	*Lb. kefir*
Lb. salivarius	*Lb. pentosus*	*Lb. reuteri*
***Pediococcus* spp.**	*Lb. rhamnosus*	*Lb. sanfrancisco*
Ped. acidilactici	*Lb. sake*	*Lb. viridescens*
Ped. damnosus		***Leuconostoc* spp.**
Ped. pentosaceus		*Leu. dextranicum*
***Streptococcus* spp.**		*Leu. mesenteroides*
Str. bovis		*Leu. paramesenteroides*
Str. thermophilus		*Leu. carnosum*
***Enterococcus* spp.**		*Leu. gelidium*
Ent. faecium		
Ent. faecalis		

the litmus milk at 30°C for acid production. Confirm the identity by using the catalase test and prepare a Gram stain from MRS broth culture, which is examined microscopically. Homofermentative and heterofermentative LAB may be considered to be Gram-positive and catalase negative.

Use APT agar for the differential enumeration of heterofermentative LAB. Perform counts as indicated for HHD agar. Prepare pour plates using APT agar and incubate at 32°C for up to 48 h. Heterofermentative LAB produce yellow colonies on APT agar. Subtract the APT agar count from the HHD agar to find the homofermentative LAB. Record the number of heterofermentative and homofermentative LAB cfu per g or mL of sample separately.

33.2.11 Counting of butyric acid-producing bacteria

Butyric acid-producing bacteria are *Clostridium butyricum* and *Clostridium pasteurianum*. If a food product is acidic, it should be neutralized with sterile $CaCO_3$. Homogenize 50 g of the diluted sample in 450 mL of 0.1% peptone water using a blender (or stomacher bag). Pour the homogenized sample into an Erlenmeyer flask. For a liquid sample, homogenize 50 mL of the sample in 450 mL of 0.1% peptone water in an Erlenmeyer flask. Place the flask into a thermostatic shaking water bath at 80°C. Place an additional uninoculated Erlenmeyer flask containing 500 mL of distilled water and place a thermometer within for controlling the temperature. Hold the water level above the sample level in the flasks. The water bath should be shaken at low speed to assist with heat distribution in the flasks during heating. After the temperature reaches 80°C in the control flask, hold the flask for 20 min at that temperature to kill vegetative cells and to activate spores. After heating, subsequently cool the flasks in cold water and prepare decimal dilutions using 0.1% peptone water. Inoculate 1 mL of the diluted sample into previously heated and cooled 3-MPN tubes of CL broth. Seal the surface area of broth in the tube without disturbing the liquid with melted 2% agar (at 45°C). Incubate the tubes anaerobically at 32°C for up to 7 days. After incubation, analyze the tubes for growth (turbidity), gas formation (bubbles through medium), and strong butyric acid odor in tubes. Indicate the number of positive tubes for each dilution and refer to MPN Table 4.1. Calculate the number of butyric acid-producing bacteria and record the result as the number of butyric acid-producing bacteria MPN per g or mL of sample.

Caution: Spores of *C. botulinum* may also germinate and grow in this medium.

33.2.12 Counting of acetic acid-producing bacteria

Acetic acid bacteria (AAB) bacteria are *Acetobacter* (e.g., *Acetobacter aceti*) and *Gluconobacter* (e.g., *Gluconobacter oxydans*). They are used in making vinegar. AAB can be found in high-sugar fruits, from which yeast produces ethanol, then acetic acid bacteria oxidize ethanol to acetic acid. They can spoil beers, wines, and ciders. They are catalase positive; do not liquefy gelatin; produce acetic acid and form soft curd in litmus milk; ferment glucose, lactose, sucrose, and ethyl alcohol; and do not ferment maltose, fructose, and glycerol. They are cylindrical to ellipsoidal rods, occurring singly, in pairs, or in irregular clumps (Figs. 33.4 and 33.5). They are capsulated, nonmotile, nonsporulating, Gram-negative, and do not give the cellulose reaction with iodine solution. The optimum growth temperature range is 25°C–30°C. The optimum pH for their growth ranges from 5.0 to 6.5. They tolerate 5% of ethanol and some of them will tolerate 10% of ethanol.

Use DSM agar for counting acetic acid-producing bacteria. This medium contains calcium lactate as a main carbon source as well as smaller amounts of other carbon sources. When *Acetobacter* grows on DSM agar, it can utilize (oxidize) lactate to increase pH and the colony color changes from yellow to

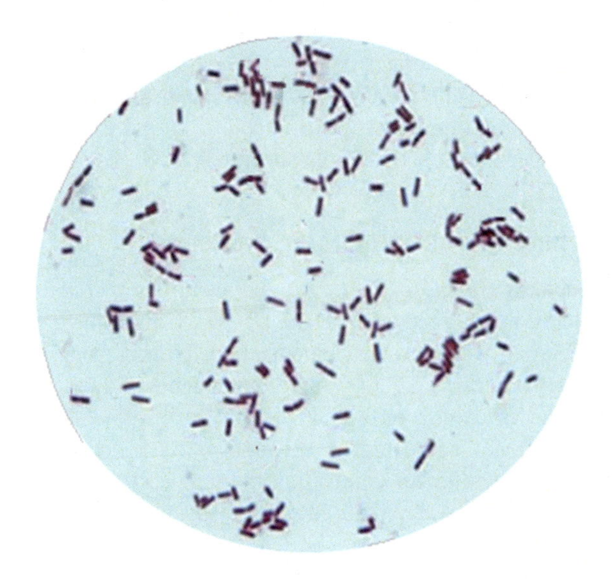

FIGURE 33.4

Staining appearance of *Acetobacter*.

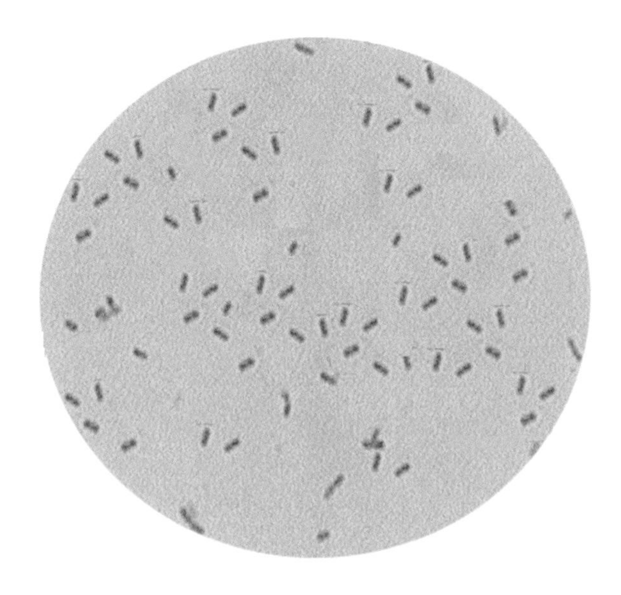

FIGURE 33.5

Staining appearance of *Gluconobacter*.

purple. *Gluconobacter* cannot oxidize lactate and oxidizes the minor carbohydrates, producing acid and yellow colonies.

Spread plate 1 mL of diluted sample to the surface of dried DSM agar (Technique 2.2.4). Allow the plate to stand for the medium to absorb the liquid. Incubate the inverted Petri plates aerobically at 25°C for up to 4 days. During incubation, examine the plates for characteristic colony formation, since the color of colonies changes depending on the utilization of the carbon source. After incubation, examine the plates for characteristic colony formation. *Acetobacter* produces purple colonies and *Gluconobacter* produces yellow colonies on this medium within 48 h. Label these colonies and count the colonies again after 4 days. Count *Acetobacter* and *Gluconobacter* colonies separately from Petri plates containing 25 to 250 colonies. Calculate the number of *Acetobacter* and *Gluconobacter* separately by multiplying the average count by the dilution rate and dividing by the inoculation amount. Record the results as the number of *Acetobacter* and *Gluconobacter* cfu per g or mL of sample.

Select at least three characteristic colonies per plate. Subculture the characteristic colonies to DSM broth and agar, and obtain pure cultures (Technique 13.4.1). Use pure cultures in the identification tests (Technique 33.4.2).

33.2.13 Counting of yeasts

Use a microscopic technique to count the yeasts from fermenting brines and finished pickle products undergoing gaseous spoilage by yeasts, where populations are in excess of 10^4 cells per mL of sample. Add 1 mL of a liquor sample to 1 mL of erythrosine stain in a sterile small beaker. Shake the stain mixture to obtain an even suspension. Using a pipette, transfer enough mixture to the area under the cover glass of a counting chamber to fill the chamber in one operation (Technique 3.4). Allow the cells to settle for approximately 5 min and count the yeast cells. Record the cells stained pink as dead yeast cells and unstained cells as live yeast cells. The number of live yeast cells per mL of sample may be calculated from:

$$\text{Numbers per mL} = (n \times \text{dilution rate})/10^{-4}\text{mL}$$

where n is average number of yeast cells per small square, and the Neubauer improved counting slide volume is 10^{-4} mL.

33.3 Quality test on fermented products

Incubate the finished fermented products in their original equipment at 7°C for up to 7 days. After incubation perform the following standard plate counts. The results of any test to predict shelf life should be correlated with actual product shelf life. Some of the techniques to predict the shelf life of fermented products are counting of lactic acid bacteria (Technique 33.2.4), counting psychrotrophic bacteria at 21°C for 25 h (Technique 9.2), aerobic plate count (Technique 2.2.4), coliforms count (Technique 13.3.2), yeasts and molds count (Technique 6.3.3), proteolysis (Technique 34.2.3.2), titratable acidity (Technique 33.3.1), and pH (Technique 33.3.1).

33.4 Chemical tests

33.4.1 Titratable acidity and pH

Titratable acidity and pH of the samples are extremely useful in providing information supplementary to microbiological analysis. Determine the titratable acidity of 10 mL of the brine or liquor by diluting the sample with 20–50 mL of distilled water and titrate with 0.1 N NaOH, using phenolphthalein. Report the values for a brined sample as g of lactic acid per 100 mL of sample and for finished liquor samples as g of acetic acid per 100 mL of sample. Use the following calculations:

$$\text{mL of 0.1 N NaOH use} \times 0.09 = \text{g of lactic acid per 100 mL.}$$

$$\text{mL of 0.1 N NaOH use} \times 0.06 = \text{g of acetic acid per 100 mL.}$$

Carry out pH determinations with a pH meter using an electrode.

33.4.2 Determination of chlorine and calcium contents of brine

Transfer 1 mL of the sample to an Erlenmeyer flask and dilute with 15–20 mL of distilled water. Titrate with a silver nitrate solution (0.171N) using three drops of potassium chromate solution (5%) as an indicator. Agitate to keep the precipitate broken up until a light salmon pink color develops. Calculate the amount of NaCl in the sample by using the following formula and report the result as g of NaCl per 100 mL of the sample.

$$\text{NaCl content (\%)} = (V \times N \times F \times 0.0585 \times 100)/\text{sample amount}$$

where V is Consumption of 0.1N $AgNO_3$ in titration (mL), N is Normality of $AgNO_3$, F is factor of 0.1N $AgNO_3$, and m is the amount of sample (g or mL)

A rapid calorimetric procedure based on calcium binding by methyl thymol blue stain can be used to measure the calcium content of brines or blended tissue. Mix the samples with an equal volume of trichloroacetic acid solution. An immediate absorbance change at 612 nm occurs. The relationship between calcium concentration and absorbance is hyperbolic. A standard curve can be constructed to find the concentration of calcium in the sample.

33.4.3 Determining dissolved carbon dioxide

Take 100 mL of the sample in a beaker and add 10 drops of phenolphthalein indicator. If a pink color develops, no carbon dioxide is present in the water sample. Add NaOH solution (1/44N) from a buret to the sample and stir gently until a slight permanent pink color appears, compared with distilled water. Record the number of milliliters of NaOH used. Since excess CO_2, if present, easily escapes to the atmosphere, tests should be performed immediately after the collection of the water sample. If this is not possible the sample bottle should be completely filled and stoppered and be kept at a temperature lower than that at which it was collected.

$$\text{Amount of carbon dioxide}(\text{mg/L}) = (\text{Normality of NAOH} \times \text{equivalent weight of}$$
$$CO_2 \times 1000 \times \text{milliliter of used NaOH})/(\text{amount of sample})$$

33.5 Identification of lactic acid bacteria

33.5.1 Phenotypic identification techniques

33.5.1.1 Morphological analysis

Prepare a Gram stain as explained in Appendix B under the topic "Gram Staining Reagents" from a broth culture. Examine it microscopically and test for catalase reaction. Perform a motility test as explained in Technique 13.4.2. Gram-positive, catalase negative, cocci or rods may be considered to be LAB.

33.5.1.2 Starter culture activity test

Add 1 mL of starter pure culture to 100 mL of sterile 10% skim milk; incubate at 30°C in a thermostatic shaking water bath for 6 h. After incubation, determine the acidity that is produced by adding 1 mL of phenolphthalein solution as an indicator into 10 mL of the incubated skim milk and titrate with NaOH (0.1N) until a faint pink color is obtained. Using the formula, calculate the percentage of lactic acid production by culture:

$$\%\ \text{Lactic acid} = [(S \times N \times F \times M)/(\text{amount of culture})] \times 100$$

where M is lactic acid milliequivalent weight (0.09), S is amount of used NaOH, F is factors of NaOH, and N is normality of NaOH (0.1N).

Normally, the acidity of the culture will be above 0.5% in 6 h if there is no any inhibitory agent in the milk.

33.5.1.3 Carbon dioxide production rate

In general, aroma-producing LAB produce CO_2 (e.g., *Lac. lactis* subsp. *lactis* biovar. diacetylactis, *L. mesenteroides* subsp. *cremoris*, *Lactobacillus casei*, and *L. acidophilus*). In the CO_2 production test, add 0.5% pure culture and 9 mL of pasteurized milk into a fermentation tube (Fig. 31.1). Adjust a CO_2 gasometer over the inoculated tube. Incubate the tubes at 30°C for 4 h. Measure the amount of CO_2 after the incubation period. When the starter culture produces more than 200 μL CO_2 in 4 h, this culture is not available as a starter culture, for example, in cheese production.

33.5.1.4 Bitterness

A pure culture producing a bitter taste is not used in cheese production, for example, Cheddar cheese. Add 0.5% of starter culture into YESM broth and incubate at 22°C until the pH drops to 6.0. Then cool quickly to 4°C. Place the medium at 37°C and adjust the pH to 6.3 using 1.0 N NaOH. In this condition, when the cells are not alive, they are not able to produce a bitter taste, and if the cells are live, they are able to produce bitterness. Detect whether the cells are live or not in the treated samples using a microscope (Technique 3.4) or suitable agar media (Technique 33.2).

33.5.1.5 Bacteriocin production test

Many strains of LAB from the genera *Lactococcus*, *Streptococcus*, *Leuconostoc*, *Pediococcus*, *Bifidobacterium*, and *Propionibacterium* produce bacteriocins. Bacteriocins are bactericidal agents that act against other bacteria. Some are relatively large molecules and heat labile, whereas others are small molecules and heat stable. *L. lactis* subsp. *lactis* produces nisin, *L. lactis* subsp. *cremoris* produces diplococcin, *P. acidilactici* produces sakacin A, *Lactobacillus sake* produces LB 706 and leucocin A, *L. plantarum* produces plantaricin, and *L. acidophilus* produces 11988 and 55 lactococcin A. Bacteriocins of LAB have the potential of being used as food preservatives. For use in food, a bacteriocin must be produced economically.

In the bacteriocin test, use a suitable medium depending on the bacterium, for example, Elliker agar or M17 agar. Here Elliker agar is used as an example. Label the Petri plate with a small circular point at the bottom. Place one loopful of overnight bacterial culture onto the labeled part of Elliker agar by point inoculation. Allow the plates to stand to allow the medium to absorb the liquid at points. Incubate the inverted plates at 32°C for 24 h. After incubation, expose the colonies to chloroform vapor for 30 min. Remove the chloroform from the plate and pour indicator bacteria (sensitive to bacteriocin) onto the point of plate exposed to chloroform. Incubate the plates at 32°C for 24 h. After incubation, analyze the inactivation zone around the test bacterium. Zone formations around the colonies of test bacterium indicate the production of bacteriocin by bacterium.

33.5.1.6 Bile salt and acid tolerance tests

LAB bile salt tolerance is performed in MRS broth containing 1.0%, 1.5%, and 2.0% bile salts. Inoculate 1 mL of overnight MRS broth cultures (in the early stage of stationary phase) into each MRS broth containing bile salt.

LAB acid tolerance is performed in MRS broth with pH adjusted to 5.0, 4.5, 4.0, 3.5, 3.0, and 2.0 by 5 N HCl. Inoculate 1 mL of overnight MRS broth cultures (in the early stage of stationary phase) into each acidified MRS broth.

Incubate all tubes at 32°C for up to 48 h. After incubation, analyze the broth for growth (turbidity) and color change. Indicate bile salt and pH tolerance of isolates.

33.5.1.7 Other tests

Carbohydrate fermentation tests. Refer to Technique 21.3.2.3 for carbohydrate fermentation tests using MRS broth with citrate omittted and using an inverted Durham tube. Perform a fermentation test using carbohydrates.

Catalase test. LAB are catalase negative. A positive reaction indicates cross-contamination. Refer to Technique 13.4.3.10 for catalase test.

Diacetyl test (Voges-Proskauer test). Refer to Technique 13.4.3.3 for the Voges-Proskauer test.

Indole test. Refer to Technique 13.4.3.2 for the indole test.

H_2 S test. Refer to Technique 13.4.3.1 for the H_2S production test on TSı agar slant.

Gelatin hydrolysis test. Refer to Technique 23.3.2.3 for the gelatin hydrolysis test.

Nitrate reduction. Refer to Technique 19.3.2.11 for the nitrate reduction test.

Oxidase test. Refer to Technique 21.3.2.7 for the oxidase test.

Esculin hydrolysis test. Refer to Technique 14.3.2.1 for the esculin hydrolysis test.

Perform other psychochemical characteristics of LAB as indicated in Table 33.2, if required.

33.5.1.8 Characterization of lactic acid bacteria isolates from the fermented products

LAB species are characterized according to the phenotypic results. For phenotypic identification, all isolates should be initially subjected to Gram staining, catalase test, growth at 10°C and 45°C in MRS and M17 broth, gas production from glucose, arginine hydrolysis, and growth in the presence of 2%, 3%, 4%, 6.5%, and 10% of NaCl. The fermentation of carbohydrates (starch, amygdalin, arabinose, cellobiose, fructose, galactose, glucose, lactose, maltose, mannitol, mannose, melizitose, melibiose, raffinose, rhamnose, ribose, sucrose, salicin, sorbitol, trehalose, and xylose) can be performed in MRS broth (without glucose in original medium) containing 1% solution of carbohydrate and 0.025% bromocresol purple as a pH indicator. Carbohydrate fermentation results will be recorded after 48 h of incubation at 30°C. The isolates can be subdivided into eight groups depending on these results:

- Thermoduric and homofermentative; Gram-positive, motile (but *Enterococcus casseliflavus* and *Enterococcus gallinarum motile*), catalase negative, cocci, grow at 10°C and at 45°C, do not produce gas from glucose, do not ferment pentosans (ribose, xylose, and arabinose), hydrolyze bile esculin, grow in the presence of 6.5% of NaCl and at pH 9.6—considered to be presumptive *Enterococcus*.
- Mesophilic and homofermentative; Gram-positive, nonmotile, nonmotile, catalase negative, spherical to ovoid cocci, grow at 10°C but not at 45°C, do not ferment pentosans, do not produce gas from glucose, grow on Lee's agar with yellow and later purple colonies with or without zone, and grow in 4% NaCl but not in 6.5% NaCl—considered to be presumptive *Lactococcus*.

Table 33.2 Physicochemical characteristics of isolated genera of LAB.

Characteristic	*Enterococcus*	*Lactobacillus*	*Leuconostoc*	*Streptococcus*	*Lactococcus*	*Pediococcus*
Cell morphology	Cocci/round	Rod	Cocci/ovoid	Cocci/chain	Cocci/round	Cocci/round
Catalase	−	−	−	−	−	−
Oxidase	−	−	−	−	−	−
CO_2 from glucose	+	+	+	−	+	−
Esculin hydrolyze	+	−	−	−	−	+
Nitrate reduction	+	−	−	−	+	−
Arginine hydrolysis	±	−	−	+	+	+
Fermentation of						
Glucose	−	+	+	+	+	−
Lactose	+	+	+	+	±	+
Starch						
Xylose	+	±	±	±	±	+
Sucrose	+	+	±	±	−	+
Mannose	+	−	+	+	+	±
Melibiose	±	±	−	−	−	−
Raffinose	−	−	−	−	−	±
Sorbitol		+	±	−	−	±
Pentosans	−	±	+	−	−	−
Growth at						
10°C	−	−	−	−	±	−
15°C	+	+	+	−	+	+
45°C	±	±	−	±	±	±
Growth in % NaCl						
2	+	−	+	+	±	+
4	+	±	−	+	+	+
6.5	±	−	−	−	−	−
Growth at pH 9.6	+	−	−	−	−	−

+, positive; −, negative; ±, various between isolates.

- Mesophilic and homofermentative; Gram-positive, catalase negative, nonmotile, pairs to tetrad cocci with short or long chains, grow well at 35°C (also at 45°C), do not produce gas from glucose, do not ferment pentosans, starch, mannose, raffinose, and sorbitol, do not hydrolyze esculin (except *P. acidilactici*), and grow in 6.5% NaCl—considered to be presumptive *Pediococcus*.
- Thermoduric and homofermentative; Gram-positive, catalase negative, cocci, exist in diplococci to long chain cells, do not produce gas from glucose, do not grow at 10°C and in 6.5% NaCl, do not ferment sucrose and pentosans, do not produce small, deep red colonies with a clear zone on NRCL agar, grow on M17 agar, and grow at 45°C—considered to be presumptive *Streptococcus salivarius* subsp. *thermophilus*.
- Mesophilic and heterofermentative; Gram-positive, catalase negative, diplococci in oval chain, ferment pentosans (ribose, xylose and arabinose), produce gas from glucose, grow in 6.5% NaCl and at 35°C, and do not hydrolyze arginine—considered to be presumptive *Leuconostoc*.
- Thermoduric and homofermentative, Gram-positive, catalase negative, rods in pairs or long chains, grow at 45°C but not at 15°C, do not ferment pentoses and gluconate, do not produce gas from glucose, and produce small grayish-white colonies on Rogosa agar—considered to be presumptive homofermentative *Lactobacillus*.
- Mesophilic and facultative heterofermentative; Gram-positive, catalase negative, rods in pairs or long chains, grow at 15°C, produce gas from glucose and pentoses, and produce small grayish-white colonies on Rogosa agar—considered to be presumptive facultative heterofermentative *Lactobacillus*.
- Mesophilic and Heterofermentative; Gram-positive, catalase negative, rods in pairs or long chains, produce gas from glucose and pentoses, grow at 15°C, and produce small grayish-white colonies on Rogosa agar—considered to be presumptive heterofermentative *Lactobacillus*.

33.5.2 Identification tests for acetic acid bacteria

The characteristics of *Acetobacter* and *Gluconobacter* are given in Table 33.3.

33.5.2.1 Morphological analysis

Purified colonies are inoculated into BHI broth and incubated at 30°C for 48 h. After incubation, one loopful of BHI culture is streak plated to GEYC agar. Incubate the inverted plates at 30°C for 48 h. The colonies are analyzed on the basis of microscopic examination. AAB colonies are small, white, spherical, pinpoint, raise, off-white, and show a clear halo on GEYC agar. A clear halo indicates the ability of AAB to dissolve calcium carbonate with the production of acid. Perform morphological tests as indicated in Technique 13.4.2.

33.5.2.2 Acetic acid production

Purified colonies are inoculated into BHI broth and incubated at 30°C for 48 h.

On solid medium. Streak plate one loopful of BHI broth culture onto BGE agar. Incubate the inverted plates at 30°C for 24 h. After incubation, analyze the colonies from plates. Acid production is indicated by colony formation with a yellow zone appearance on the agar plate.

In liquid medium. Inoculate from the BHI broth culture into YEG broth. The YEG broth is aerated and the broth is incubated at 30°C. Samples from YEG broth are taken at 48 h intervals.

Table 33.3 Characteristics of *Acetobacter* and *Gluconobacter*.

Characteristic	*Acetobacter*	*Gluconobacter*
Motility	Peritrichous flagella or nonmotile	Polar flagella or nonmotile
Gram reaction	−	−
Capsule	+	+
Ethanol oxidation to acetic acid	+	+
Acetic acid oxidation to CO_2 and H_2O	+	−
Growth in the presence of 0.35% acetic acid	+	+
Nitrate reduction	+	+
Catalase/oxidase test	+/−	+/−
Urea/starch hydrolysis	−/−	−/−
Fermentation of glucose, saccharose, mannitol, and melibiose	+	+
Fermentation of lactose, maltose, arabinose, and citrate	−	−
Gelatin liquefaction	−	−

Remove 5 mL of YEG broth culture sample after 48 h incubation and mix with 20 mL of distilled water in a sterile Erlenmeyer flask and add 3−5 drops of phenolphthalein. Titrate the mixture against NaOH (0.1 N). Calculate the amount (g) of acetic acid production in 100 mL of culture using the following formula:

$$\text{Amount of acetic acid g per 100 mL} = \text{Volume of used NaOH (mL) used} \times 0.06$$

33.5.2.3 Acetic acid and ethanol tolerance

Use YEG broth containing different ethanol concentrations for the ethanol tolerance test and YEG broth containing different acetic acid concentrations for the acetic acid tolerance test. Inoculate 0.1 mL of BHI broth culture into both of the media and incubate broths at 30°C for up to 10 days. After incubation, examine the growth (turbidity) in tubes. Indicate the ethanol and acetic acid tolerance level of AAB.

33.5.2.4 Other tests

The following tests can identify AAB: catalase (Technique 13.4.3.10), production of acid from D-glucose (Technique 15.3.2.2), and nitrate reduction (Technique 19.3.2.11). Nitrate reduction is tested from nitrate peptone water.

33.5.3 Matrix-mediated laser desorption ionization-time-of-flight-mass spectrometry identification of fermentative bacteria

Fermentative bacteria can be rapidly identified with the matrix-mediated laser desorption ionization-time-of-flight-mass spectrometry (MALDI-TOF Ms) technique, which is based on the principle of ionizing a specific protein profile of microbial cells. Fermentative bacteria can be easily identified by comparing these profiles with a reference spectrum. MALDI-TOF Ms identification has been explained in Technique 13.6.3.1.

33.5.4 Molecular identification of fermentative bacteria

One of the genotypic identification techniques for the identification of fermentative bacteria is the polymerase chain reaction (PCR) assay. PCR is a molecular biology technique that is used for the detection of various fermentative microbial species according to the genes encoding specific characteristics. Some fermentative microorganism-specific primer sets that can be used in the identification are given in "Appendix A: Gene Primers (Table A-1)." Using fermentative microorganism-specific techniques and materials, they can be identified by PCR analysis as described in Technique 13.6.3.2.

33.6 Interpretation of results

Fermentation is a process that helps break down large organic compounds into simpler ones by the action of fermentative microorganisms. Fermentative microorganisms play an important role in the fermentation of foods by causing changes in the raw foods' chemical and physical properties. Fermented foods have several advantages: (1) fermented foods have a longer shelf life than their raw foods; (2) the enhancement of organoleptic properties of fermented products; (3) the harmful/unwanted ingredients can be reduced from raw materials, for example, removal of poisonous cyanide content of cassava during garri preparation; (4) fermenting microorganisms increase the nutritional properties of fermented products; (5) the fermentation process reduces the cooking time of food; (6) the fermented products consist of higher in vitro antioxidant capacity. For example, fermented milk and yogurt consist of higher antioxidant properties compared to milk, as there is a release of biopeptides that follows the proteolysis of milk proteins, particularly α-casein, α-lactalbumin, and β-lactoglobulin.

Some microorganisms that can degrade pectic substances from fruits and vegetables are used in fermentation. Salt, titratable acidity, pH, fermentable sugars, and dissolved CO_2 are very useful for detecting the quality of brined vegetables. High salt concentrations will prevent the softening of fruits and vegetables. Properly acidified, packaged, and pasteurized pickle products are not subject to microbial spoilage. When spoilage occurs in pasteurized pickle products, it is usually due to underpasteurization or postprocess contamination. The shelf life of fermented products mainly depends on the contamination of microorganisms after pasteurization. Gram-negative psychrotrophic bacteria (e.g., *Pseudomonas*) cause most of the important shelf-life problems in fermented dairy products. Another major problem is associated with yeasts and molds. Fermented products are routinely tested by organoleptic characteristics, level of acidity, coliform count, and yeasts and molds count. For example, yoghurt should contain specific LAB of not less than 10^6 cfu per g or mL of sample, aerobic plate count of not more than 10^5 cfu per g or mL, coliforms of not more than 10 cfu per g or mL, yeasts and molds of not more than 100 cfu per g or mL, and pathogens should be absent.

PRACTICE

34 Analysis of fruits, vegetables and precooked frozen foods

34.1 Introduction

The natural bacterial microflora of foods of plant origin are *Achrobacter*, *Flavobacterium*, *Micrococcus*, coliforms, and lactic acid bacteria (LAB). Microflora of a soil-contaminated fruit or vegetable is primarily sporeformers, coryneform, molds, and others. The high acidity and sugar content of fruits often permits a predomination of yeasts and molds, while the high carbohydrates and low acidity of vegetables permits LAB to predominate. The fruits have high sugar content (especially fructose, glucose, and sucrose) and a high acidity with low pH (usually pH 4.5 or lower). The pH of vegetables ranges from 5 to 7.

Nonfecal coliforms and *Enterococcus* are part of the natural microflora of many foods. *Streptococcus* and *Leuconostoc* are the most numerous on vegetables. The use of polluted water in irrigation can increase the incidence of enteric pathogens. Vegetables (e.g., lettuce and cabbage) may contain enteric pathogens (e.g., *Salmonella*, *Listeria*, *Shigella*, *Aeromonas*, viruses, and parasites). *Enterococcus* is a common contaminant of frozen vegetables. *Escherichia coli* and fecal coliforms are relatively rare contaminants of many frozen vegetables. Canned fruits may contain some gas-producing butyric anaerobic (e.g., *Clostridium pasteuranum*) and aciduric flat sour bacteria (e.g., *Bacillus coagulans*). Most vegetables and a few fruits are blanched as one of the early processing steps. The application of temperature (at 86°C–98°C) to vegetables destroys most vegetative microorganisms but the heat-resistant microbial spores usually survive. Vegetables may be contaminated with *Clostridium perfringens* and *Clostridium botulinum* from soil. Leaves of lettuce and cabbage may be contaminated with *Salmonella* and *E. coli*.

Aciduric microorganisms are the predominant contaminants on low-pH fruits. Yeasts and molds are often the most numerous microorganisms in fruits. *Acetobacter* and other acid-tolerant bacteria are most common microorganisms in the acid environment of fruit processing. Coliforms can present on various fruits. Fruits are generally too acid for the growth of most foodborne pathogens. Many microorganisms cannot survive in acidic environment, for example, *Staphylococcus, E. coli*, and *Shigella* die off rapidly in citrus juices. Toxigenic molds can cause potential health problems since they can easily grow on fruits and vegetables. For example, *Penicillium patulum*, *Penicillium expansum*, and other mold species can grow and produce mycotoxins.

Decay caused by molds and certain bacteria can cause much of the decay of fresh fruits and vegetables. Many of these decaying microorganisms are true plant pathogens and they can invade healthy plant tissues. Market or storage diseases are rot formations on the fresh fruits and vegetables. Many of the diseases are named for their appearance, for example, gray, brown, cottony, and rots. Molds generally responsible for the spoilage of fruits and vegetables are *Alternaria*, *Penicillium*, and *Phytophthora*. The bacterium that commonly causes spoilages of fruits and vegetables is *Erwinia carotovora*. Some members of *Pseudomonas*, *Bacillus*, and *Clostridium* are

Microbiological Analysis of Foods and Food Processing Environments. DOI: https://doi.org/10.1016/B978-0-323-91651-6.00021-5

also important spoilage bacteria in fruits. *Leuconostoc mesenteroides* produces pectic enzymes. Vegetables can be affected by bacterial soft rots, particularly by Gram-negative microorganisms of the genera *Erwinia* (e.g., *Erwinia chrysanthemi* and *E. carotovora*) and *Pseudomonas*. Probable pathogens able to associate with vegetables and salads are *Leuconostoc monocytogenes*, *Salmonella*, *E. coli* O157:H7, *Staphylococcus aureus*, and hepatitis A virus.

This practice describes microbiological techniques for determining the most common microbial contaminants, potential microbial spoilage and health hazards, and their hygienic significance in fruits and vegetables.

34.2 Microbiological analysis techniques

Sampling, sample preparation, and all experimental techniques should be performed under aseptic conditions, and in duplicate unless otherwise specified. All microbiological media to be used in applications must be sterilized. Again, all the solutions and equipment that may cause contamination of microorganisms must be sterilized or disinfected.

34.2.1 Equipment, materials, and media

Equipment. Blender (or stomacher), microscope, microscope glass slide, pipettes (1 mL) in pipette box, pipette discarding pad (containing 10% bleach solution), scalpel, vortex, and waterproof pen.

Materials. Citrate buffer (0.18 M), Gray's double-dye stain, immersion oil, North's aniline methylene blue stain, 0.1% peptone water, sterile distilled water, and toluene.

Media. Fluorescent pectolytic pseudomonads (FPP) agar, plate count (PC) agar, and polypectate gel (PG) agar.

34.2.2 Sampling

Cut fruits or vegetables into small pieces. Packaged frozen fruits and vegetables should be thawed at room temperature for 2 h before opening. Open the fruit and vegetable package, and note the opening conditions of the product. Record observations, for example, presence or absence of liquid, moisture condition, unnatural color or odor or microbial growth (e.g., pink colonies of *Torulae*), and note any other abnormalities. The interior of the product should be checked for discoloration and off-odor. Sour, putrid, or other off-odors will indicate fermentation during storage.

Weigh a 100 g of sample into a sterile blender (or stomacher bag). Add 900 mL of 0.1% peptone water and blend for 1 min (or stomach for 2 min). Allow the sample to stand for 3 min to exclude foam. Prepare further dilutions using 0.1% peptone water.

34.2.3 Microbial counts from fruits and vegetables

34.2.3.1 Microscopic counting of microorganisms

Refer to Technique 3.2.2 for details of direct microscopic count (DMC). Transfer 0.01 mL of the sample onto a microscope glass slide. Spread the sample over a 1 cm^2 area of the slide (labeled with a waterproof pen). Allow the slide to dry in air, heat fix the slide. Stain with Gray's double-dye stain (or North's aniline oil methylene blue stain), rinse, dry, and examine under a microscope using the oil-immersion objective. Calculate the number of microorganisms and record the result as the number of DMC per g of sample.

34.2.3.2 Counting of pectolytic bacteria

Genera of microorganisms producing pectolytic enzymes are *Achromobacter*, *Aeromonas*, *Arthrobacter*, *Bacillus*, *Clostridium*, *Enterobacter*, *Erwinia*, *Flavobacterium*, *Pseudomonas*, *Xanthomonas*, many yeasts, molds (e.g., *Botrytis*, *Fusarium*, and *Rhizopus*), and protozoa. Some pectolytic bacteria grow at 37°C, but do not produce pectate lyase until the temperature is 32°C or below.

Spread plate 1 mL of diluted sample to PG agar (Technique 2.2.4). Allow the medium in the plates to absorb the water of the inoculum. Incubate the inverted Petri plates aerobically at 30°C for 48 h. After incubation, analyze the plates for characteristic colony formation. Pour pectin precipitating reagent (0.1% polygalacturonic acid) onto colonies on the PG agar and the appearance of a clear zone around the colonies indicates the hydrolysis of pectin. Detect fluorescent *Pseudomonas* from FPP agar with long-wavelength UV light before adding the pectin precipitant. After pouring pectin precipitant onto colonies on FPP agar, the appearance of a clear zone around colonies indicates the hydrolysis of pectin. When clear zones are not formed around colonies on the agar medium, this indicates that bacteria cannot produce pectinase enzymes and cannot hydrolyze pectin. Count all the characteristic colonies from Petri plates containing 25–250 colonies. Calculate the number of pectolytic bacteria and *Pseudomonas* by multiplying the average count by the dilution rate and dividing by the inoculation amount. Record the results as the pectolytic bacteria and *Pseudomonas* cfu per g of sample.

34.2.3.3 Other microbial counts

Aerobic plate count. Refer to Technique 2.2.4 for aerobic plate count (APC) by spread plate technique. Incubate at 32°C for up to 3 days, and report the result as the number of APC cfu per g of fruits or vegetables.

For counting coliforms, fecal coliforms, *Enterococcus*, yeasts and molds, and individual pathogens, for example, *E. coli*, *S. aureus*, *C. perfringens*, *L. monocytogenes*, *Campylobacter*, *Salmonella*, *Shigella*, and *Aeromonas*, refer to the corresponding practices in this book.

Howard mold count. Refer to Technique 3.5 for the Howard mold count.

34.2.4 Counting of microorganisms from precooked frozen products

Some frozen precooked products contain ingredients which readily promote bacterial growth, including that of food-poisoning pathogens. They can be analyzed for the following microorganisms:

Aerobic plate count. Refer to Technique 2.2.4 for APC by spread plate technique. Incubate at 32°C for 1–3 days, and report the result as APC per g of sample.

Thermoduric aerobic and anaerobic plate count. Refer to Technique 12.3 for thermoduric APC and for thermoduric anaerobic plate count by the pour plate technique using PC agar. Incubate the inverted plates at 35°C for 1–3 days under aerobic and anaerobic conditions, respectively. After incubation, report the results as the number of thermoduric APC and thermoduric anaerobic PC per g of sample.

For counting coliforms, fecal coliforms, yeasts and molds, and individual pathogens, for example, *E. coli*, *S. aureus*, *C. perfringens*, *L. monocytogenes*, *Campylobacter*, *Salmonella*, *Shigella*, and *Enterococcus*, refer to the corresponding practices in this book.

34.3 Interpretation of results

Microbial growth on packaged products is indicative of improper handling conditions. Color change or a bleached appearance is especially visible on frozen fruits and vegetables by reason of a slow freeze or thaw. A dry bleaching surface is indicative of improper packaging, excessive storage, or widely varying temperatures during frozen storage. APC of more than 4×10^5 cfu per g of vegetables or DMC of over 10^6 cfu per g of vegetables may be considered as poor sanitary handling conditions. The bacterial counts of precooked frozen food should be low. Any carelessness in handling or lack of good sanitary practices may allow bacterial populations before they were frozen. Coliforms and *Enterococcus* are the most common normal flora of vegetables during processing, but populations of 100–1000 cfu per g of processed product are not acceptable. *S. aureus* may be present on vegetables, but usually low in numbers, fewer than 10 cfu per g. The routine microbiological analysis for *S. aureus* is not recommended for fruits and vegetables. Microbiological analysis should be made with respect to human pathogens, for example, *L. monocytogenes*, *C. perfringens*, *Salmonella*, and *Shigella*. APC and DMC in excess of 10^5 cfu and 5×10^6 cfu per g, respectively, are indicative of the poor handling of precooked frozen foods and indicate that a check of plant operations should be performed. Fruit should not show more than 20% positive fields for hyphae from the Howard mold count. The maximum Howard mold count limit for tomato paste is 40% positive fields for the appearance of mold hyphae and for tomato juice 20%.

PRACTICE

35 Analysis of fruit juices and concentrates

35.1 Introduction

Fruit received at a processing plant will be contaminated with high numbers of yeasts from $10^3-10^7\ g^{-1}$. Fruit juices are obtained directly from fruits, for example, grapes, cherries, berries, apples, etc. Fruit juice drinks may contain 5%–20% or more juice, and 40%–60% juices in concentrates. The fruit juice concentrates can have about 0.90 water activity (a_w). Different ingredients can be added into fruit juices before concentration, for example, colors, flavors, and others (e.g., benzoic and sorbic acid). The low pH of juices and concentrates limits the microorganisms that can survive and grow. For example, due to the pH range of lemon or lime juice of 2.2–2.6, the spoilage bacteria cannot survive. Orange juice at pH 3.4–4.0 can allow the survival of *Acetobacter*, *Lactobacillus*, *Leuconostoc*, yeasts, and molds, and they can cause spoilage. But molds and *Acetobacter* can grow in fruit juices if enough dissolved oxygen is available. Yeasts can cause both oxidation with the production of CO_2 and H_2O, and fermentation with the production of alcohol and CO_2 in fruit juices. *Acetobacter* can oxidize alcohol to acetic acid. *Lactobacillus fermentum* and *Leuconostoc mesenteroides* can ferment carbohydrates in fruit juice to lactate. *L. mesenteroides* and *Lactobacillus plantarum* can form a slime layer due to dextrin production. Coliforms and *Enterococcus* can contaminate citrus and other fruit products. They may become part of the normal processing plant microflora. Most bacteria cannot grow in concentrates over 30°Brix (% total soluble solids) or higher. Yeasts are usually the most important group in apple and grape juices. *L. mesenteroides* and *Lactobacillus brevis* can produce off-flavors in concentrates. Refrigeration temperature supports fungal growth in concentrate. The yeasts most frequently associated with commercial orange concentrates are *Candida*, *Saccharomyces*, *Torulapsis*, and *Rhodotorula* (e.g., *Rallina rubra*). Machinery mold, *Geotrichum candidum*, may contaminate fruit products from unsanitary equipment. Low numbers of heat-resistant molds (e.g., *Byssocchlamys* spp. and *Neosartorya fischeri*) often are associated with raw fruit and may contaminate products and survive during the processing steps. Other molds frequently associated with fruits and contaminating products are *Alternaria*, *Botrytis*, *Colletotrichum*, *Diplodia*, *Fusarium* (e.g., *Fusarium oxysporum* associates with potato, sugar cane, and garden bean), *Penicillium*, and *Phonopsis*.

Yeasts are primarily responsible for the spoilage of raw chilled juices. Coliform and *Erwinia* are frequently present on or in oranges before they are harvested and therefore they contaminate orange juice. Tomato juice has a pH of about 4.3; bacterial spores (e.g., *Bacillus coagulans*) can survive at this low pH. Most other vegetable juices with low pH (up to 5.0) and microbial growth factors support the growth of lactic acid bacteria (LAB).

The microflora of juice will modify from the original microflora of the raw fruit and vegetables. Hot-filling of fruit juice at a fill temperature of around 88°C can produce a shelf life of many months in a glass bottle. Ascospores of *Bysocchlamys*, *Talaromyces*, and *Neusartorya* may survive

Microbiological Analysis of Foods and Food Processing Environments. DOI: https://doi.org/10.1016/B978-0-323-91651-6.00045-8

this process, and produce mycelial masses and a haze during the storage of juice. Ready-to-drink concentrates, preservative-containing cold-filled fruits, and fruit drinks have a limited shelf life with high bacterial counts. The growth of acetic acid bacteria in juice can cause off-flavors and the reduction of oxygen from the head space can result in the partial collapse of plastic containers.

Some groups of microorganisms (e.g., *Escherichia coli*, aerobic bacteria, heat-resistant molds, and osmophilic yeasts) can be used as indices of unsanitary handling, improper processing conditions, or microbiological qualities of products. Microbiological quality can be detected by plate count and direct microscopic count techniques.

35.2 Microbiological analysis techniques

Sampling, sample preparation, and all experimental techniques should be performed under aseptic conditions, and in duplicate, unless otherwise specified. All microbiological media to be used in applications must be sterilized. Again, all solutions and equipment that may cause contamination of microorganisms must be sterilized or disinfected.

35.2.1 Equipment, materials and media

Equipment. Automatic pipette, beaker, blender (or stomacher), brown bottle, distillation equipment, Erlenmeyer flask, Howard mold count slide, incubator, microscope, microscope glass slide (in bottle containing alcohol), Petri dishes, pipettes (0.5, 1 and 10 mL) in pipette box, pipette discarding pad (containing 10% bleach solution), refrigerator, test tubes, thermometer, thermostatic shaking water bath, vortex, waterproof pen, and Whatman no 1 filter paper.

Materials. Alpha-naphthol reagent, antifoam, cold water, crystal violet solution (0.075% aq.), isopropyl alcohol (99%), KOH solution (40%), 0.1% peptone water, stabilizer solution, and sterile distilled water.

Media. Dichloran rose Bengal chloramphenicol (DRBC) agar, EC broth containing CC disk (ColiComplete is a substrate containing both XGAL and MUG), osmophilic agar (with pH 5.4 and 3.5), universal preenrichment (UP) broth, and yeast extract glucose lemco (YEGL) bromocresol purple agar.

35.2.2 Sample preparation

Refer to Technique 2.2.2 for details of sample preparation and dilutions. Add 25 mL or g of sample into 225 mL of sterile 0.1% peptone water in an Erlenmeyer flask. Homogenize by gently shaking and allow it to remain at 4°C for 10 min. Then the content is mixed well again by gently shaking. Prepare further required dilutions using 0.1% peptone water.

35.2.3 Microbiological analysis of fruit juices and concentrates

Aerobic plate count. Refer to Technique 2.2.4 for details of aerobic plate count (APC) by the spread plate technique.

Lactic acid bacteria count. Refer to Technique 33.2.3 for details of the LAB count.

35.2.3.1 Yeasts and molds count

Spread plate 1 mL of diluted (or nondiluted) sample to the surface area of well-dried DRBC agar (Technique 6.3.3) for yeasts and molds counts. Allow plates to stand for the medium to absorb the liquid. Incubate the Petri plates in the incubator in the dark at 25°C for up to 5 days; do not stack Petri plates higher than three and do not invert the Petri plates. Do not disturb plates until the colonies are to be counted. After incubation, count the characteristic yeasts and molds colonies from plates containing 15–150 colonies. If excessive molds grow on the plates, count starting from 1–5 days. Counting plates from the cover of plates is sometimes helpful if they are overgrown with molds (usually not a problem with DRBC agar). Inverting and reversing will spread mold spores and produces noncountable plates. Calculate the number of yeasts and molds by multiplying the average count by the dilution rate and dividing by the inoculation amount. Report the results as the number of yeasts and molds colony forming units (cfu) per g or mL of sample.

35.2.3.2 Counting of osmophilic microorganisms

Refer to Technique 11.2.1.3 for osmophilic microbial count. Spread plate 1 mL of diluted (or nondiluted) sample to the surface area of well-dried osmophilic agar. Allow the plates to stand for the medium to absorb liquid. Incubate the plates in the incubator in the dark at 25°C for up to 5 days; do not stack plates higher than three and do not invert the plates. Do not disturb plates until colonies are to be counted. After incubation, count the colonies from plates containing 15–150 colonies. If excessive mold growth develops on the plates, count starting from 1–5 days. Report results as the number of osmophilic microorganisms cfu per g or mL of sample.

35.2.3.3 Counting of heat-resistant molds

Refer to Technique 12.2 for heat-resistant mold count. There are low incidences of heat-resistant mold ascospores on many foods, so 100 g or mL or more sample is cultured for counting. Add 100 g or mL of the sample into a sterile Erlenmeyer flask containing 100 mL of distilled water. Homogenize the sample by gently shaking. Place the flask into a thermostatic shaking water bath at 80°C. Place an additional uninoculated control Erlenmeyer flask containing 200 mL of water and place a thermometer within. Adjust the temperature of the water bath to 80°C and hold the water level above the sample level in the flasks. After the temperature reaches 80°C, hold the flasks for 30 min. Shake the water bath at a slow speed to assist heat distribution during heating. After heating, immediately cool the mixture to 45°C under running water or in cold water. Prepare further decimal dilutions using 0.1% peptone water. Add 1 mL of the heated sample into sterile Petri dishes in duplicate and pour 17–19 mL of DRBC agar (at 45°C) into the plates. Mix agar and 1 mL of the sample by thoroughly rotating the plate by drawing a figure of eight while it is lying on the table tap, and allow the agar to solidify in the places (within 10 min). Incubate the Petri plates at 21°C for up to 14 days and do not stack plates higher than three and do not invert the plates. After incubation, count the colonies from the plates containing 15–150 colonies. The heat-resistant species can be distinguished from each other by the color of colonies on acidified agar after 14 days of incubation: *B. nivea* colonies are predominantly white, and *B. fulva* colonies are dull yellow in color. The colonies on agar start as small mycelia zones and spread rapidly to cover the entire plate. Therefore, count colonies starting from 3 days up to 14 days. Calculate the number of heat-resistant molds by multiplying the average count by the dilution rate and dividing by the inoculation amount. Report the results as the number of heat-resistant molds cfu per mL or g of sample.

35.2.3.4 Direct microscopic count

Refer to Technique 3.2.2 for details of the direct microscopic count (DMC). Take a microscope glass slide immersed in alcohol from a bottle and pass over a flame. Draw a 1 cm^2 area under the slide by a waterproof pen. For citrus products, stain 5 mL of 12°Brix juice with equal volume of crystal violet solution (0.075% aq.) in a test tube. After mixing thoroughly, distribute 0.01 mL of stained sample over 1 cm^2 on the slide. With a sterile 0.1 mL pipette, withdraw a sample slightly above the graduation mark. Wipe the exterior of the pipette tip with clean dry paper tissue. Then absorb liquid at the tip to give the 0.1 mL label. Place the tip of pipette to the center of slide and expel the test portion up to the 0.09 mL label (0.01 mL). Dry the smear in a dust-free area, such as in an incubator. After drying, do not heat fix the slide. Examine the slide under the oil-immersion objective of microscope, count at least 25 fields, and find the average counts per field. Calculate the number of microorganisms in a 1 mL of sample by using the microscope factor (Technique 3.2.2.1). Report the results as the estimated number of microorganisms per mL or g of sample.

35.2.3.5 Counting of total acid producers

Refer to Technique 33.2.4 for details of the total acid producers count. Spread plate 1 mL of diluted sample to YEGL bromocresol purple agar (Technique 2.2.4). Alternatively, add 1 mL of diluted sample into a sterile Petri dish. Pour 17–19 mL of melted YEGL bromocresol purple agar (at 45°C) into Petri dishes. Mix agar and 1 mL of sample by thoroughly rotating the plates in the route by drawing a figure of eight while it is lying on the table tap, and allow the plates to solidify the agar (within 10 min). Incubate the inverted Petri plates at 32°C for up to 3 days. After incubation, count the colonies with a yellow halo from the Petri plates containing 25–250 colonies. Calculate the number of total acid producers by multiplying the average count by the dilution rate and dividing by the inoculation amount. Record the result as the number of total acid producers cfu per g or mL of sample.

35.2.3.6 Counting of aerobic thermoduric sporeforming

Transfer 10 mL of the sample into cotton wool plugged test tubes in duplicate. Place tubes in a thermostatic shaking water bath at 80°C. Place an additional uninoculated control tube containing 10 mL of the sample and place a thermometer. Hold the water level above the sample level in the tubes. After the temperature reaches 80°C, hold the tubes for 15 min. Shake the water bath at low speed to assist with heat distribution during heating. After heating, remove the tubes and cool quickly under running water or in cold water. Prepare the required dilution using sterile 0.1% peptone water. Spread plate 1 mL of the heated and diluted sample to DTBP agar in duplicate (Technique 2.2.4). Incubate the inverted plates at 42°C for up to 3 days. After incubation, count the colonies from Petri plates containing 25–250 colonies. Calculate the number of aerobic thermoduric spore-forming bacteria by multiplying the average count by the dilution rate and dividing by the inoculation amount. Record the result as the number of aerobic thermoduric spore-forming bacteria cfu per g or mL of sample.

35.2.3.7 Enrichment isolation of Escherichia coli

E. coli analysis can be performed to indicate potential contamination of juices or to indicate the effectiveness of sanitation during juice processing. The MPN count is not used for juice because the acidity (pH 3.6–4.3) of juices can inhibit bacteria.

Aseptically, add 10 mL of fruit juice into each of 10 tubes of 90 mL of UP broth and incubate the tubes at 35°C for 24 h. After enrichment, mix the contents of the tubes by gently shaking through hand or vortexing, and transfer 1 mL of each incubated UP broth into 9 mL of 10 EC broth in tubes containing a CC disk. Incubate the tubes at 41.5°C in a thermostatic shaking water bath for 24 h. After incubation, analyze the tubes in the dark and under long-wave UV light. The presence of blue fluorescence in either tube indicates the presence of *E. coli* in the sample. Note: The CC disks also contain X-gal; cleaving by β-galactosidase will yield a blue color on or around the disk. The presence of a blue color indicates the presence of coliforms.

35.2.3.8 Counting of Geotrichium

Remove the food from the container. Wash the container using 300 mL of sterile distilled water. Transfer the liquid portion to a 2-liter sterile beaker. Wash the residue from the container with 50 mL of sterile distilled water. Transfer 10 mL of residue wash water into a sterile test tube, add 1 drop of crystal violet stain, and add 1 mL of stabilizer. After mixing the stained residue in the stabilizer, transfer 0.5 mL to a Howard mold count slide (Technique 3.5). Examine the slide with a light microscope for the *Geotrichium* hyphae (Fig. 35.1). Careful examination includes varying the intensity of the light occasionally and continuous use of the fine adjustment on the microscope. In some cases, a 20 × magnification (20 × objective) may be necessary to identify mold hyphae. Then

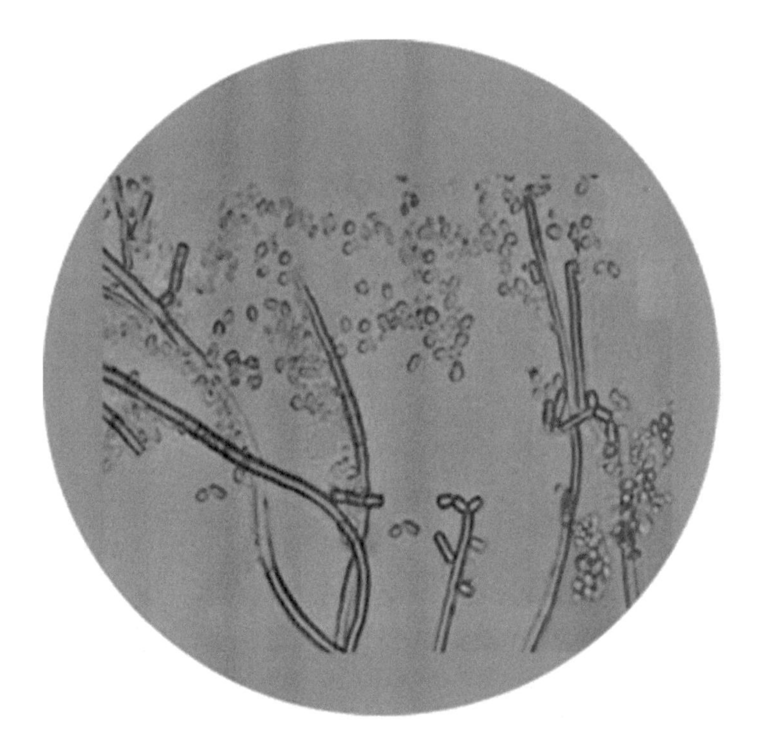

FIGURE 35.1

Appearance of *Geotrichium* hyphae and spores.

return to 10× magnification and magnify. Calculate the proportion of positive fields from the results of the analysis of all observed fields (F) and the number of positive fields (P), and report the result as the percentage of positive fields containing mold hyphae.

$$\text{Percent positive field} = (P/F) \times 100$$

35.2.4 Microbiological analysis of sugars and sugar syrups

Sugar syrups are used as an ingredient in the preparation of many food products. The genera of microorganisms associated with sugars and sugar syrups are *Bacillus*, *Lactobacillus*, *Leuconostoc*, *Hansenula*, *Pichia*, *Saccharomyces*, and *Torulapsis*. The sugar syrup samples can be examined for the presence of flat-sour bacteria (*Geobacillus stearothermophilus*), hydrogen sulfide-producing *Clostridium*, and nonhydrogen sulfide-producing *Clostridium*. Refer to the respective practices of this book for counting these microorganisms. Perform the yeasts and molds count as indicated in Technique 35.2.3.1. Mesophilic aerobic plate count, thermoduric count, flat sour sporeformers count, and thermophilic sporeformers count can also be performed according to the requirements as indicated in the related practices in this book.

35.3 Diacetyl test from fruit juices

This test is a calorimetric procedure. It detects within 30 min diacetyl and acetylmethylcarbinol that are produced by *Lactobacillus* and *Leuconostoc* in the orange and apple juices.

Add 300 mL of juice (12°Brix) into a 1-liter Erlenmeyer flask connected to a condenser for a rapid distillation. Add two drops of antifoam solution (10% v/v). Place a 25 mL graduated cylinder at the effluent end of condenser. Heat the flask to boil the juice. Separate the oil by filtering the condenser effluent through Whatman No. 1 filter paper and collect the filtrate in a graduated cylinder. Collect 25 mL of distillate into the cylinder. Remove the first graduated cylinder and place a second graduated cylinder and collect the next 25 mL fraction. Repeat to collect a third 25 mL fraction in the first graduated cylinder. Discard the second fraction. Pipette 10 mL of distillate from the first fraction. Avoid disturbing the oil layer if the distillate is not filtered. Transfer to a clean 50 mL Erlenmeyer flask. Add 5 mL of alpha-naphthol solution (5% w/v) and 2 mL of KOH solution with creatine. Mix by gently shaking the flask for 20 s, and allow 10 min for full color development. Prepare a blank at the same time using sterile distilled water in place of the juice distillate; add reagents in the same way. Transfer to a calorimetric tube and record the reading in 60 s at 530 nm optical density (OD). Prepare a standard curve from the known amounts of diacetyl. Read the ppm diacetyl from standard curve.

If a differentiation between diacetyl and acetylmethylcarbinol is desired, the test is run on the third fraction. Acetylmethylcarbinol will present in the third fraction at approximately the same concentration as in the first fraction. Since diacetyl comes off in the first fraction, color production in the third fraction will be due to acetylmethylcarbinol only. If the third fraction value is subtracted from the first, the difference represents diacetyl.

35.4 Interpretation of results

High plate count from juices in various stages of preparation prior to concentration indicates improperly cleaned equipment, operation without cleaning and sanitizing, poor fruit or vegetable quality, unsanitary equipment, or allowing the opportunity for microbial growth in the food at the same stage in the process. The diacetyl test is used for most citrus juice concentrates as a rapid technique to indicate spoilage. This technique can also detect microbial activity during processing of juice prior to juice concentration. Diacetyl-producing microorganisms exhibit little or no growth after the product concentration reaches 45°Brix.

High heat-processed juices can be free of viable aciduric microorganisms but yield low numbers of viable spore-forming bacteria when cultured on nonselective media. Frozen concentrated juices generally contain microbial populations from 10^2 to 10^5 cfu per mL. Heat-resistant mold spores may present in low numbers in fruit juice concentrates. Nonsterile fruit juices may contain low numbers of coliforms and *Enterococcus*. The low pH of most fruit juices prevents the growth of enteric pathogens. Products that receive a heat treatment at some stage of the process should be free of enteric pathogens. In rare instances, *Salmonella* and *E. coli* may be introduced into nonpasteurized apple juice or contaminate after pasteurization, and they survive for an extended storage period.

PRACTICE

Analysis of eggs and egg products 36

36.1 Introduction

Shell eggs can be used to produce different types of products: liquid eggs, frozen eggs, and dried eggs. Microorganisms cannot penetrate easily thorough the egg shell. Spoilage of eggs starts as a result of cracking the egg shell, improper washing, and inadequate storage conditions. The outsides of egg shells are not sterile and different microorganisms are associated with the surface of egg shells. Poultry frequently carries *Salmonella* in their intestines; it may become part of the bacterial flora of the egg shell during ovulation and content from the intestine in the nest environment. *Salmonella enterica* subsp. *enterica* var. Enteritidis, *Escherichia*, and fecal *Enterococcus* are important contaminants from fecal matter. Other bacteria associated with egg shells are *Acetobacter*, *Aeromonas*, *Arthrobacter*, *Bacillus*, *Flavobacterium*, *Micrococcus*, *Pseudomonas*, *Serratia*, *Streptococcus*, and coliforms. *Alcaligenes*, *Escherichia*, *Proteus*, and *Pseudomonas* are common spoilage bacteria in improperly cleaned eggs and eggs stored for extended periods.

The egg shell, and the outer and inner membranes are the main barriers against the penetration of bacteria. Certain environmental conditions may affect the easy penetration of bacteria. When moisture presents and the temperature is favorably high during storage, microbial penetration is favored. Eggs are very warm at the time they are laid (about 41°C) while air temperature is around 20°C. The eggs cools to the surrounding temperature. As cooling takes place, the contents of the egg contract more than does the shell of the egg. This creates a vacuum in the egg (negative pressure in the egg), and air and microbial cells are drown through the pores into the egg. When eggs are removed from refrigerated storage and placed at room temperature, they may "sweat" due to the condensation of water droplets on the egg surface. Sweat does not allow further penetration of microbial cells. The egg is most susceptible to penetration at the point of laying since where the newly laid eggs are deposited has a major influence of contaminant on eggs. Eggs laid into a heavily contaminated environment suffer more bacterial spoilage (due to contamination of 10^7-10^9 cfu per egg) than those laid in a clean environment (due to contamination around 10^3 cfu per egg). When the shell is penetrated by microorganisms, two membranes separate the egg white from the shell as a barrier. The inner membrane is rich in lysozyme, which is inhibitory to Gram-positive bacteria. However, membrane penetration may take place depending on storage conditions. At refrigeration temperature, penetration is retarded. The egg shell and the albumen provide physical and chemical barriers against microorganisms. The egg yolk is an excellent medium for microbial growth.

The most common bacterial genera in liquid eggs are Gram-negative (e.g., *Escherichia*, *Alcaligenes*, *Pseudomonas*, and *Proteus*) and Gram-positive (e.g., *Micrococcus*, *Bacillus*, *Micrococcus*, and *Enterococcus*), contaminating from the egg shell. Gram-positive bacteria survive at egg pasteurization temperature. Liquid whole eggs can be pasteurized at 62°C for 3 min. Pasteurized frozen or dried egg products may have total microbial counts ranging from 100 cfu to a few thousand cfu per g.

Microbiological Analysis of Foods and Food Processing Environments. DOI: https://doi.org/10.1016/B978-0-323-91651-6.00024-0

The major pathogens associated with eggs and egg products are *Salmonella* spp. (Salmonella enteritidis is a major one). Other pathogenic bacteria associated with eggs and egg products are *Listeria monocytogenes* and *Yersinia enterocolitica*. They are resistant to lysozyme enzyme and survive at low temperature under anaerobic and aerobic conditions.

The insides of undamaged shell eggs are essentially free of bacteria. However, shell eggs are susceptible to spoilage if microbial invasion occurs into the eggs. Microbial spoilage of eggs is caused primarily by Gram-negative bacteria of the genera *Achromobacter, Aeromonas, Alcaligenes, Enterobacter*, *Escherichia*, *Proteus, Pseudomonas*, and *Serratia*. These bacteria can cause different types of rot spoilage. Fecal material associated with the nest or cage material, hens, feed, dirt or dust, hands of the person, and equipment used for handling are the main microbial sources of contamination to eggs. The extent of the bacterial contamination can be controlled to a certain degree by sanitation practices. Before the spoilage of eggs by microorganisms, the responsible microorganism(s) must overcome a series of mechanical and biochemical barriers of eggs. Certain characteristics of the albumen provide undesirable conditions for bacterial growth due to the presence of lysozyme, conalbumin, ovomucoid, avidin, the low content of nonprotein nitrogen, and a high pH.

Ineffective cooling, sweating (condensation of water on shell), and improper washing procedures may allow the penetration of microorganisms through the shell and speed up bacterial spoilage of eggs. Improper washing or handling may allow damage to the cuticle. The cuticle of the shell generally provides at least four days of protection against bacterial penetration into eggs. Eggs without the cuticle spoil much faster than normal eggs.

The most important pathogenic hazard in fresh eggs is *S. enteritidis*. This pathogen is associated with the gastrointestinal tract of healthy hens and contaminates the egg during its formation. If eggs contaminated with this pathogen are not cooled appropriately, this bacterium grows in eggs and its number increases, therefore increasing the risk of infection in people if the egg is not properly cooked before eating.

In this practice, the microbiological analyses of eggs and egg products are performed by plate and microscope counting techniques. Their specific sampling techniques are also explained. *Escherichia coli*, coliforms, yeasts and molds, and *Pseudomonas* counts are performed as indicators of unsanitary handling and storage conditions. *Salmonella* and *E. coli* counts can also indicate the potential degree of food poisoning with eggs and egg products.

36.2 Microbiological analysis techniques

Sampling, sample preparation, and all experimental techniques should be performed under aseptic conditions, and in duplicate unless, otherwise specified. All microbiological media to be used in applications must be sterilized. Again, all solutions and equipment that may cause contamination of microorganisms must be sterilized or disinfected.

36.2.1 Equipment, materials and media

Equipment. Beaker, container, Erlenmeyer flask, glass beads, gloves, heater, incubator, microscope, microscope glass slide, pipettes (0.1 and 1 mL) in pipette box, pipette discarding pad (containing 10% bleach solution), refrigerator, spoon, vortex, and waterproof pen.

Materials. Alcohol (70%), lithium hydroxide solution, North's aniline oil methylene blue stain, and saline solution (0.85% NaCl).

Media. Holmon's alkaline cooked meat (HACM) medium, liver broth, and trypticase soy (TS) broth.

36.2.2 Sampling

36.2.2.1 Shell eggs

Select a number of shell eggs representing the lot. Transfer the eggs to a laboratory in clean cartons at 10°C and store at that temperature until analysis is made. Before analysis, wash each egg with a brush using soap and warm water, drain and immerse in 70% alcohol for 10 min. Remove from alcohol, drain, and flame. Handle the alcohol flamed eggs with sterile gloves and aseptically break the egg shells with a flame-disinfected knife and place the content of the eggs into a sterile container. Homogenize the eggs by gently mixing with a sterile spoon. For the separate analysis of egg white and yolk, use a sterile commercial separator or a sterile spoon to separate them. They can be individually homogenized as well as whole eggs. Weigh 11 g of the sample into 99 mL of sterile saline solution in an Erlenmeyer flask containing glass beads. Gently shake 25 times by hand thoroughly for 7 s. Prepare serial dilutions by transferring 11 mL of 1:10 dilution to 99 mL of saline solution. Prepare further required dilutions using saline solution.

36.2.2.2 Liquid eggs

Obtain liquid egg samples from vats or tanks at the plant or from containers. Make sure that the product has been mixed thoroughly by mechanical agitation or by hand mixing in the container. Transfer samples to the laboratory at 4°C and store the sample at 4°C for no more than 4 h. Avoid freezing, as this will destroy many of the bacteria in the liquid eggs. Homogenize the sample and prepare further serial dilutions as described in Technique 36.2.2.1.

36.2.2.3 Frozen eggs

Select a number of cans containing frozen liquid eggs representative of the lot. Thaw the contents of the eggs in the can as rapidly as possible. Use running tap water on the can and submerge up to two-thirds of the can in water. Rotate the can in the container frequently until the content of can is thawed. An alternative technique is to thaw overnight in a refrigerator, but the growth of psychrotrophic bacteria should be considered. Disinfect the surface area of the opening side of the cans by pouring and burning alcohol. Open the cans, remove any ice or frost on the frozen egg with a sterile spoon. Homogenize the sample and prepare further serial dilutions.

36.2.2.4 Dried eggs

Take samples of powdered or spray dried albumen using a sterile spoon. If the product is in small packages, select several unopened packages. Transfer samples to the laboratory at 4°C and store at that temperature. Disinfect the open side of packages by pouring alcohol for 10 min and then cleaning. Open them in the laboratory under aseptic conditions and thoroughly mix with a sterile spoon to obtain a homogenous mixture. Remove the required quantity of the sample into a sterile Erlenmeyer flask. Weigh 11 g of dried egg product sample into 99 mL of sterile saline solution in the Erlenmeyer flask and gently shake until no evidence of lumps. Place the flask in a refrigerator. Shake once or twice every few min until the sample has dissolved, but the sample should not be stored more than 10 min. Homogenize the sample and prepare further dilutions as described in Technique 36.2.2.1.

36.2.3 Plate counting techniques

36.2.3.1 Counting of putrefactive anaerobes

Perform putrefactive anaerobic counts from frozen or dried eggs. Add 10 mL of homogenized sample into tubes in duplicate. Place tubes in a thermostatic shaking water bath at 75°C. Place an additional uninoculated control tube containing 10 mL of the same sample and place a thermometer within. Hold the water level above the sample level in the tubes. After the temperature reaches 75°C, allow the tubes to remain for 20 min at that temperature. Agitate the tubes gently to assist heat distribution during heating. This heating exhausts the oxygen, inactivates vegetative bacterial cells, and inactivates endospores. After heating, quickly cool below 35°C in cold water. Add 1 mL of heated sample into 3-MPN tubes of liver broth. Incubate the tubes anaerobically at 35°C for up to 3 days. Examine the tubes starting from 24 h for growth (turbidity), digestion of meat, and gas formation (evidencing by bubbles through medium). Indicate positive tube numbers for each dilution and refer to MPN Table 4.1. Calculate the number of putrefactive anaerobes MPN per g of sample.

36.2.3.2 Direct microscopic count

Liquid and frozen eggs. Refer to Technique 3.2.3 for details of the direct microscopic count. Label a disinfected microscope glass slide at the bottom with a 2 cm^2 area (a circular area with a diameter of 1.6 cm is preferred) with a waterproof pen. Place 0.01 mL of homogenized egg sample on a slide using a sterile 0.1 mL pipette. In this transfer, withdraw a sample slightly above the graduation mark. Wipe the exterior of the pipette tip with a clean dry paper tissue. Then absorb the liquid at the tip to give exactly 0.1 mL. Place the tip of the pipette to the center of labeled area on the slide and expel the test portion from 0.1 to 0.09 label (0.01 mL). Spread the sample over an area of 2 cm^2. Dry the smear in a dust-free area such as in an incubator (at 35°C). Immerse the slide in xylene for a minimum of 1 min, remove and dry. Then immerse in 95% alcohol for 1 min using forceps. Remove the slide from an alcohol immersed slide two or three times in distilled water in a beaker to remove alcohol and dry in the air. Pour North's aniline oil methylene blue stain for 20 min. Was the stained smear by repeated immersions in a beaker of distilled water. Dry thoroughly before analysis, but do not use blotting techniques because this may remove some of the smear material from the surface of the slide. Examine under the microscope using the oil-immersion objective. Count the number of microorganisms in least 12 fields and multiply the average count per field by two due to the use of a 2 cm^2 area and microscope factor (Technique 3.2.2.1) and multiply by dilution rate. Report results as the estimated number of bacteria per g of sample.

Dried egg products. Perform a microscope count as indicated in the part "liquid and frozen eggs" of this technique. Lithium hydroxide solution (at 22°C) may be used for whole egg and yolk products if that is more suitable for the homogenization and dilution of sample. If there is a difficulty in spreading the smear, the addition of a drop of water will be helpful to allow homogenization. Report the results as the estimated number of bacteria per g of sample.

36.2.3.3 Other counts

Aerobic plate count. Refer to Technique 2.2.4 for details of aerobic plate count (APC) by spread plate. Prepare two sets of plates in duplicate and incubate one set of inverted plates at 22°C for up to 5 days for psychrotrophic APC and the other set at 35°C for up to 3 days for mesophilic APC. Report the results as the number of respective APC cfu per g of sample.

Yeasts and molds count. Refer to Technique 6.3.2 for details of the yeasts and molds count. Incubate at 25°C for up to 5 days and report the result as the number of yeasts and molds cfu per g of sample.

Coliform count. Refer to Technique 13.3.2 for details of the MPN coliforms count. Incubate at 32°C for up to 48 h and report the result as the number of MPN coliforms per g of egg products.

E. coli. Refer to Technique 13.3.6.2 for counting *E. coli*. Incubate at 32°C for up to 3 days and report the results as the number of *E. coli* cfu per g of sample.

Salmonella. Refer to Technique 15.2.3 for counting *Salmonella*. Incubate at 32°C for up to 3 days. Report the results as the number of *Salmonella* cfu per g of sample.

For counting other individual pathogens, for example, *L. monocytogenes*, *C. jejuni*, *Y. enterocolitica*, and *Bacillus cereus*, refer to the corresponding practices in this book.

36.2.4 Microbiology of shell eggs

Use five fresh eggs in duplicate at a temperature of 30°C in each of the following analyses. One of the eggs will also be marked as the control for each analysis. The other eggs (experimental egg) are used in analyses. Bring the temperature of fresh *Pseudomonas fluorescens* TS broth culture in the beakers to 30°C and 4°C, respectively.

1. Using the steel egg holder or disposable gloves, immerse the experimental egg-1 for 15 min into *P. fluorescens* culture in a beaker at 30°C. Remove the egg from the culture and place this egg together with the control egg in the egg carton and store at 30°C.
2. Using the steel egg holder or disposable gloves, immerse the experimental egg-2 for 15 min into *P. fluorescens* culture in a beaker at 4°C. Remove the egg from the culture and place this egg together with the control egg in the egg carton and store at 4°C.
3. Wash the experimental egg-3 with tap water and a brush for approximately 5 min. Then use the steel egg holder or disposable gloves to immerse the experimental egg for 15 min into *P. fluorescens* culture in a beaker at 30°C. Remove the egg from the culture and place this egg together with the control egg in the egg carton and stored at 30°C.
4. Wash the experimental egg-4 with tap water and a brush for approximately 5 min. Then use the steel egg holder or disposable gloves to immerse the experimental egg for 15 min into *P. fluorescens* culture in a beaker at 4°C. Remove the egg from the culture and place this egg together with the control egg in the egg carton to be stored at 4°C.

Note: *P. fluorescens* is used to simulate dirty wash water. This bacterium can cause typical rots and produces a bright fluorescence in the albumen when viewed under ultraviolet light.

Eggs should be examined every week during storage for 3–4 weeks with the standard candling light and under ultraviolet light. Any abnormal conditions of the interior contents of the eggs (e.g., increasing transparence of light) should be noted and recorded.

At the end of storage, immerse the eggs into a beaker containing an equal volume of water and eggs. Prepare further serial dilutions using 0.1% peptone water. Refer to Technique 43.2.5.2 for counting *Pseudomonas aeruginosa*. Carry out the count by spread plating 1 mL of sample onto cetrimide agar and incubate at 35°C for up to 48 h. After incubation, analyze colony formation on the agar medium. *P. aeruginosa* produces yellow-green or yellow-brown colonies (Figure 42.2A) and fluoresces under UV light. Presumptive identification by colonial morphology should be confirmed by pure culturing (Technique 13.4.2). Perform further identification tests and the oxidase test (Technique 43.2.5.2). Inoculate onto P agar (Seller's differential agar) for confirmation (incubate at 35°C for 48 h). *P. aeruginosa* produces straw-colored colonies with green pigmentation on P agar.

36.3 Interpretation of results

APC of the interior contents of shell eggs should be less than 10 cfu per g. A count of more than 100 cfu per g usually indicates bacterial invasion through the shell. This is possibly due to improper washing and sanitizing of the shell's surface, followed by a period of storage or excessive handling of "sweated" eggs. The storing of eggs in a cooler at a relative humidity above 85% encourages mold growth on the surface of shell eggs. APC exceeding 10^7 cfu per g in liquid egg usually indicates the use of poor-quality eggs for breaking but may also be due to poor sanitation or improper storage conditions of the liquid eggs. Yeasts and molds must be less than 10 cfu per g of liquid eggs, and *Salmonella* less than 1 cfu per g of liquid eggs. Coliforms as well as yeasts and molds counts are generally used as indicators of unsanitary conditions. The *E. coli* count may be more important than coliforms count. The amount of lactic acid formation in eggs during bacterial growth can be used to show improper handling and poor sanitation.

PRACTICE

Analysis of cereals and cereal products

37

37.1 Introduction

Cereal grains include wheat, oats, corns, rye, barley, millet, and rice. Soybeans are not grain. Cereal products are flour, breakfast cereals, snack foods, corn meal, dough, pasta, cakes, and breads. Many cereal products are used in the formulation and manufacturing of other products (e.g., sausage, confectioneries, and baby food). The majority of spoilage microorganisms are yeasts, molds (e.g., *Cladosporium*, *Aspergillus*, *Fusarium*, and *Alternaria*) and bacteria (psychrotrophic, thermophilic, thermoduric, lactic acid, and rope forming). Pathogenic microorganisms may be present if cereal grains and products are improperly stored, processed, or handled. They are *Staphylococcus aureus*, *Clostridium botulinum*, *Bacillus cereus*, *Clostridium perfringens*, *Listeria monocytogenes*, *Escherichia coli*, *Aeromonas hydrophilia*, *Yersinia enterocolitica*, *Salmonella*, and toxigenic molds.

The bacterial count from cereal grains is not usual case and the main concern is the association of mold growth due to major spoilage microorganisms and the production of potentially hazard mycotoxins. A large number of mycotoxins are produced by a wide range of molds, for example, *Penicillium*, *Aspergillus*, *Fusarium*, and *Alternaria*. Certain mycotoxins have greater public health concern.

When flour is used as an ingredient in food production, the coliforms count is normally performed. The flour products are usually subjected to a heat treatment but the most significant microorganisms surviving that treatment are the spores of *Bacillus* and *Clostridium*. Certain *Bacillus* species (mostly *B. subtilis*) can cause ropiness in bakery products and *C. perfringens* can be present in the flour used as a thickener in meat-containing products. The microbiological examinations for cereal and cereal products may involve coliforms, molds, and mesophilic and thermophilic bacterial counts. Coliforms, *Bacillus*, and *Clostridium* in the cereal and their products can be used as sanitary index conditions of processes. A yeasts and molds count can be performed to indicate sanitary conditions as well as for a prediction of potential spoilage during storage.

37.2 Microbiological analysis of cereals and cereal products

Sampling, sample preparation, and all experimental techniques should be performed under aseptic conditions, and in duplicate, unless otherwise specified. All microbiological media to be used in applications must be sterilized. Again, all solutions and equipment that may cause the contamination of microorganisms must be sterilized or disinfected.

Microbiological Analysis of Foods and Food Processing Environments. DOI: https://doi.org/10.1016/B978-0-323-91651-6.00007-0

37.2.1 Equipment, materials, and media

Equipment. Blender (or Stomacher), Erlenmeyer flask, incubator, Petri dishes, pipettes (1 mL) in pipette box, pipette discarding pad (containing 10% bleach solution), tubes, thermostatic shaking water bath, and vortex.

Materials. 0.1% Peptone water.

Media. Amos and Kent Jones (AKJ) broth (in 75 mL tube), Davis's yeast salt (DYS) agar, DYS containing 10 ppm cycloheximide (DYSC18) agar, dichloran glycerol 18% (DG18) agar, and plate count (PC) agar containing 10 ppm cycloheximide (PCC18).

37.2.2 Sample preparation

If the cereal sample consists of whole grain, it may be tested without grinding. There is no difference in count from whole grain and ground flour. Weigh 50 g of the sample into an Erlenmeyer flask containing 450 mL of 0.1% peptone water and add approximately 10 g of sea sand. Shake vigorously for at least 2 min and allow to stand for 2 or 3 min. Prepare further dilutions if necessary, using 0.1% peptone water.

37.2.3 Counting of mesophilic aerobic rope sporeformers

Refer to Technique 10.2.3 for the mesophilic aerobic rope sporeformers (MARS) count.

Alternative technique for rope spore count. Weigh 20 g of the sample into 180 mL of 0.1% peptone water in the Erlenmeyer flask to give a 1:10 dilution and shake well. Place the flask into a thermostatic shaking water bath. Place an additional uninoculated control flask containing 200 mL of 0.1% peptone water and place a thermometer. Adjust the temperature of the water bath to 95°C and hold the water level 2 cm above the sample level in the flasks. After the temperature reaches 95°C, hold the flask for 15 min at that temperature. Shake the water bath at low speed to assist with heat distribution during heating. After heating, agitate the tubes thoroughly and immediately cool to 35°C in cold water. Prepare further dilutions using 0.1% peptone water if necessary. After shaking, inoculate 5-MPN tubes of AKJ broth with 1 mL of the heated sample. Incubate test tubes at 32°C for up to 48 h. After incubation, indicate the number of positive tubes with pellicle growth for each dilution. Calculate the number of mesophilic aerobic rope sporeformers colony forming units (cfu) per g of sample by referring to MPN Table 4.2.

When required, obtain a pure culture from each tube (Technique 13.4.1) and use in the identification of *B. cereus*. Conduct the following tests: catalase (Technique 13.4.3.10), acetoin (acetylmethylcarbinol) production (Technique 13.4.3.3), nitrate reduction (Technique 19.3.2.11), utilization of citrate (Technique 13.4.3.4), hydrolysis of starch (Technique 19.3.2.14), hydrolysis of casein (Technique 31.5.3), and hydrolysis of gelatin (Technique 23.3.2.3). Positive results should be obtained for each test for *B. cereus*. Additionally, the isolate should be examined microscopically for morphology (Technique 13.4.2) by Gram stain reaction and the presence of ellipsoidal or cylindrical spores with a distinct swelling of the sporangium (Technique 19.3.1).

37.2.4 Other counts

Yeasts and molds count. Refer to Technique 6.3.2 for the yeasts and molds count.

Coliforms and E. coli. Refer to Technique 13.3.2 for coliforms and *E. coli* counts.

S. aureus. Refer to Technique 21.2.3.1 for *S. aureus*.

Salmonella. Refer to Technique 15.2.3 for enumeration of *Salmonella*.

B. cereus. Refer to Technique 19.2.3.1 for *B. cereus*.

Clostridium perfringens. Refer to Technique 20.2.4.1 for the *C. perfringens* count.

Mesophilic aerobic plate count. Spread plate 1 mL of sample (Technique 2.2.4) onto PCC18 agar and incubate the inverted plates at 32°C for up to 3 days. Count the colonies and report the result as the number of mesophilic aerobic plate count cfu per g of sample.

Mesophilic anaerobic plate count. Pour plate 1 mL of sample onto thioglycolate agar (Technique 2.2.3), and incubate inverted plates under anaerobic conditions at 32°C up to 3 days. After incubation, count colonies and report result as the number of mesophilic anaerobic plate count cfu per g of sample.

37.3 Confectionery products

Confectionery products usually are low moisture foods. Water activity values (a_w, 0.50–0.80) prevent the growth of different microorganisms. Values less than 0.95 prevent the most pathogenic bacteria, whereas 0.90 prevents the growth of most spoilage bacteria. *S. aureus* may be an exception, some growing as low as 0.86 but these cannot produce enterotoxin. Growth of most yeasts is prevented at 0.88 and most mold growth is stopped at a_w values of less than 0.80. Halophilic bacteria will grow at a_w values as low as 0.75, xerophilic molds will tolerate a_w values down to 0.65, and the osmophilic yeasts will tolerate a_w as low as 0.61. In general, only xerophilic molds and osmophilic yeasts can spoil adequately processed confectionary products held in proper storage conditions. Pathogenic and spoilage bacteria cannot grow in properly produced confectionery products.

Bacteria that produce lactic acid (e.g., *Leuconostoc mesenteroides*, *Lactobacillus* and *Streptococcus*) and acetic acid (e.g., *Aerobacter*, *Bacillus* and *Clostridium thermoaceticum*) are commonly present in confectionery along with common yeasts (e.g., *Rhodotorula* and *Saccharomyces*) and molds (e.g., *Aspergillus* and *Penicillium*). Pathogenic microorganisms (e.g., *Salmonella*, *L. monocytogenes*, *S. aureus*, *B. cereus*, and *C. perfringens*) can associate with confectionary products. Typical spoilage yeast species are *Zygosaccharomyces rouxii*, *Saccharomyces heterogenicus*, *Saccharomyces mellis*, *Torulopsis colliculosa*, *Hansenula anomala*, and *Pichia membranefaciens*. Typical spoilage mold species are *Aspergillus glaucus*, *Aspergillus niger*, *Aspergillus soydowi*, and *Penicillium expansum*.

For counting osmophilic, aerobic plate count, coliforms, yeasts and molds, and individual pathogens, for example, *L. monocytogenes*, *B. cereus*, *S. aureus*, *E. coli*, *C. perfringens*, *Enterococcus*, and *Salmonella*, refer to the corresponding practices in this book.

37.4 Bread, cakes, and bakery goods

These products do not support the growth of bacteria, but cream and cakes are highly favorable for bacterial growth. *Bacillus* may cause a defect in bread known called "ropiness." The most important hazard for cream or other fillings is *S. aureus* contamination from a food-handler. *Shigella* and hepatitis A virus rarely contaminate these products. Microbiological control mainly involves the counting of molds. *Wallemia sebi* is the spoilage mold on wholemeal or brown bakery products and appears as light to brown color. It grows up to a limit of a_w 0.65 on foods.

For counting osmophilic count, aerobic plate count, coliforms, yeasts and molds, and individual pathogens, for example, *B. cereus*, *E. coli*, *S. aureus*, *Enterococcus*, and *Salmonella*, refer to the corresponding practices in this book.

37.4.1 Counting of xerophilic fungi and *Wallemia sebi*

Spread plate 1 mL of diluted sample to DG18 agar (Technique 2.2.4) and allow the medium in the plates to absorb the liquid. Incubate the inverted Petri plates at 25°C for up to 2 weeks. During incubation, examine plates for mold colony formations. If enough colonies are formed, end the incubation and count xerophilic mold colonies from plates containing 15–150 colonies. Identify *W. sebi* colonies and count. *W. sebi* produces brown colonies on DG18 agar. It is easily differentiated from yeasts since *Saccharomyces cerevisiae* produces whitish, dry colonies. Calculate the number of xerophilic and *W. sebi* cfu per g of sample. Follow further identification tests for morphological (Fig. 37.1) and biochemical analysis.

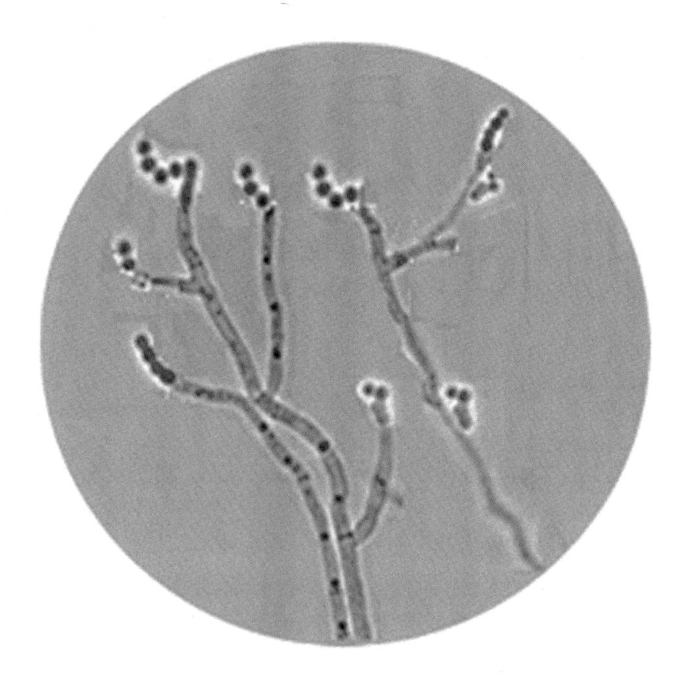

FIGURE 37.1

Appearance of hyphae and spores of *Wallemia sebi*.

37.5 Compressed bakers' yeast

The microbiological analysis for bakers' yeast can involve the detection of bacterial contamination and wild yeasts. In microbiological analysis, homogenize 10 g of sample in 90 mL of 0.1% peptone water and prepare further serial dilutions using 0.1% peptone water. Perform a microscopic count of yeasts either by counting the slide. A microscopic morphological analysis can be used to determine the ratio of dead and living yeasts by adding Liefson's methylene blue stain into yeast culture (Technique 3.4). From the microscopic count, calculate the percent viability in bakers' yeast.

Plate count. Spread plate 1 mL of sample to DYS agar in duplicate (Technique 2.2.4) and incubate inverted plates at 30°C for up to 5 days. After incubation, analyze yeast colonies on the agar medium and count the number of characteristic yeast colonies. Calculate the number of yeasts by multiplying the average count by the dilution rate and dividing by the inoculation amount. Record the result as the number of viable bakers' yeast cfu per g of sample.

Predict bacterial contaminants using PCC10 agar by aerobic plate count (Technique 2.2.4) and incubate the plates at 30°C for 5 days. Count *Lactobacillus* using RC10 agar (or AC10 agar) (Technique 33.2.7) and incubate the plates at 28°C for 5 days. Count wild yeasts using DYSC10 agar by the spread plate technique (Technique 2.2.4) and incubate the plates at 30°C for up to 5 days.

37.6 Interpretation of results

Aerobic plate counts (APC) in confectionery products immediately after processing will range from 10^3 to 10^6 cfu per g. Coliforms may range from 10 to 100 cfu per g. *E. coli* should be absent in 1 g of samples. *Enterococcus* may be present at levels ranging from 100 to 10^4 cfu per g. Yeasts and molds may vary between 10 to 10^3 cfu per g. Yeast counts usually do not exceed 100 cfu per g. *S. aureus*, *C. perfringens*, *B. cereus*, and *Salmonella* are rarely present when ingredients and processing conditions are properly controlled. The finding of coliforms in cereal grains and flour is common and does not necessarily imply mishandling. *E. coli* in a finished ready-to-eat product may be of public health concern. Enterotoxin-producing *S. aureus* represents a potential hazard. Generally, *S. aureus* contaminates the product during processing rather than entering with the flour. Therefore testing of the raw product or the finished product is a better determination of a potential *S. aureus* problem. *B. cereus* commonly is associated with cereal product foodborne disease. In food poisoning where rice is the suspected vehicle, analyses should be done for *B. cereus*. Soy proteins added to products should be analyzed for *C. perfringens*. The yeasts and molds count is an indication of the sanitary history of the product as well as a prediction of potential spoilage during storage.

PRACTICE

Analysis of seafoods

38

38.1 Introduction

Living fish carry predominantly Gram-negative psychrotrophic bacteria on their external surfaces. The internal tissues of healthy fish are usually sterile. Naturally occurring potential pathogenic bacteria on seafoods are *Clostridium botulinum* especially type A and *Vibrio parahaemolyticus.* Fresh fish carry populations of 10^2–10^3 bacteria per cm^2 of skin surface or per g of gill tissue. Bacterial genera *Pseudomonas*, *Acinetobacter*, *Flavobacterium*, and *Vibrio* are normally present in the gut system of fish. *Micrococcus*, *Bacillus*, and *Clostridium* may be present at a low number in the intestine. *Pseudomonas* species rapidly become dominant in the microflora of chilled fish; *Acinetobacter* and *Flavobacterium* (e.g., *Flavobacterium columnare*, *Flavobacterium psychrophilum* and *Flavobacterium johnsoniae*) make up the remaining microflora of chilled fish. Insanitary processing conditions can lead to the contamination of seafoods with potential pathogenic bacteria (e.g., *Staphylococcus* and *Salmonella*). *Salmonella enterica* subsp. *enterica* var. Typhi is the most significant pathogenic microbial contaminant with regard to fish. *Listeria monocytogenes*, *Campylobacter jejuni*, *Vibrio cholerae*, *Vibrio vulnificus*, and *V. parahaemolyticus* can also contaminate seafoods. Microbiological analysis of seafoods and their products can be performed by plate and microscopic count techniques. Coliforms and fecal coliforms are the most common sanitary indicators of seafoods. Inadequate storage and unsanitary handling conditions can also be indicated by these indicator microorganisms. The microbiological analysis of seafoods can also be performed with regard to *Staphylococcus aureus*, *Escherichia coli*, *C. botulinum*, *L. monocytogenes*, *C. jejuni*, *Vibrio*, and *Salmonella*.

38.2 Microbiological analysis of seafoods

Sampling, sample preparation, and all experimental techniques should be performed under aseptic conditions, and in duplicate, unless otherwise specified. All microbiological media to be used in applications must be sterilized. Again, all solutions and equipment that may cause the contamination of microorganisms must be sterilized or disinfected.

38.2.1 Equipment and materials

Equipment. Blender (or stomacher), bottle, container, Erlenmeyer flask, forceps, glass bead (or sea sand), incubator, microscope glass slide, microscope, pipettes (1 mL) in pipette box, plastic bag, pipettes (1 mL) in pipette box, pipette discarding pad (containing 10% bleach solution), refrigerator, scalpel, swab, and vortex.

Materials. Trypticase soy (TS) agar, Baird-Parker (BP) agar, 0.85% NaCl solution, and 70% alcohol.

Microbiological Analysis of Foods and Food Processing Environments. DOI: https://doi.org/10.1016/B978-0-323-91651-6.00052-5

38.2.2 Sampling and sample preparation

The hands of the examiner must be scrubbed thoroughly with soap and water, and rinsed with 70% alcohol to disinfect before sampling and analysis. Collect samples into sterile containers and transport to the laboratory with sufficient speed. Frozen seafoods can be packaged with dry ice in an insulated container for no more than 24 h without significant change by the extreme cold conditions (−30°C within 1 h); prolonged frozen storage should be avoided. Wet ice can be used for 24 h. A frozen product may be shipped in an insulated container without additional cooling, but the temperature will start to increase gradually after 6 h. Bacterial counts are relatively stable after 12 h, but they show a dramatic increase after 24 h. Frozen samples must be kept frozen and transported under wet ice or in freezer. Unfrozen samples should not be frozen, especially if *V. parahaemolyticus* and other bacteria are to be counted or isolated. Refrigerated samples (except shellfish and shell stock) are transported with refrigeration at 0°C–4°C until arrival at the laboratory. Do not freeze refrigerated samples. Pack samples of shucked shellfish immediately in crushed ice until analysis; keep shellfish above freezing but below 10°C. Examine refrigerated fish, shellfish, and shell samples within 6 h of collection but in no case more than 24 h after collection. When freezing a fresh product is unavoidable because of delays in shipping or analysis, destruction or injury of microorganisms can be expected. Special protections should be applied on products to be examined for *V. parahaemolyticus* and *Clostridium perfringens* because both are sensitive to refrigeration.

Skin samples. A corner of skin is lifted with sterile forceps or scalpel, and adhering muscle or a known area of skin is removed by cutting. Weigh the skin sample to the nearest gram. Place the sample quickly into a sterile Erlenmeyer flask and add 0.85% NaCl solution to provide 1:10 dilution. Add sterile sand or glass beads to the flask to ensure removal of the microorganisms from the sample. Shake the flask by using a mechanical shaker, but shaking by hand for 3 min is recommended. Prepare further dilutions using 0.85% NaCl solution if necessary.

Surface sample. (1) Swab sample. Remove a sterile swab from a test tube, moisten in the 0.1% peptone water, and swab the entire skin surface of seafood (2 cm^2). The swab should be applied firmly and slowly twirled on the area of the tissue (Technique 1.2.7). Remove the closure from a dilution bottle and insert the swab about 5 cm within the neck of the bottle. Break the tip off below where the analyzer has touched the swab stick, thereby dropping the cotton tip into 0.85% NaCl solution. Prepare further dilutions using 0.85% NaCl solution if necessary. (2) Rinse sample. Contaminant microorganisms are related to the surface area of seafoods. Place the sample into a sterile plastic bag and weigh. Add an appropriate equal amount of sterile 0.85% NaCl solution, and massage the bag by hand for 2 min. Prepare further dilutions using 0.85% NaCl solution if necessary.

Tissue samples. Scrape off all growth and loose material from the shell. Open the shellfish, collect the appropriate quantities of shell meat, and place into suitable sterile container. Weigh the sample to the nearest gram. A representative 50 g sample of the fish meat should be blended (or stomached) with 450 mL of cold 0.85% NaCl solution for 2 min. Analyze the homogenized sample within 2 min after blending. Prepare further dilutions using cold 0.85% NaCl solution if necessary. In the case of shellfish, transfer the weighed sample into a sterile blender (stomacher bag), add an equal amount (by weight) of sterile cold 0.85% NaCl solution, and blend (stomach) for 2 min at approximately 14,000 rpm. Two milliliters of this homogenate contains 1 g of shellfish meat.

38.2.3 Microbiological analysis

Aerobic plate count. Refer to Technique 2.2.4 for aerobic plate count (APC) by spread plate. The medium should contain 0.85% NaCl (nonselective media already contain 0.85% NaCl, e.g., TS agar). For the routine assessment of quality, incubate the plates at 25°C for fresh and frozen seafoods, and 35°C for cooked products. Occasionally, both temperatures can be used to assess the quality as well as the safety of a product.

Coliforms count and fecal coliform counts. Use five tubes of most probable number (MPN) technique, as described in Technique 13.3.2.

S. aureus. The MPN technique has limited value in the analysis of fresh and fresh frozen seafood products because of the high number of halotolerant microorganisms that grow in media containing 10% NaCl. Use spread plate counting on BP agar (Technique 21.2.3.1) for fresh and fresh frozen seafoods.

Direct microscopic count (DMC). Refer to Technique 3.4 for detailed information about the DMC using a counting chamber. This is not a suitable substitute for an APC but may be used as a rapid, simple means of assessing the quality of seafoods and their products that are subject to surface contamination. DMC normally will be 1–2 logs higher than the plate count techniques.

PSP and DSP analysis. Paralytic shellfish poisoning (PSP) and diarrhetic shellfish poisoning (DSP) are other hazards of microbial origin. They result from the growth of shellfish in waters containing dinoflagellates (e.g., *Gonyaulax* and *Dinophysis*). These microorganisms multiply in coastal waters in large numbers, resulting in the appearance of "redblooms." Analysis of samples for the presence of PSP and DSP can be achieved by chemical analytical techniques. These techniques are not microbiological and therefore are not considered further here.

For counting coliforms, fecal coliforms, halophilic microorganisms, and individual pathogens, for example, *E. coli*, *L. monocytogenes*, *V. parahaemolyticus*, *V. cholerae*, *V. vulnificus*, and *Salmonella*, refer to the corresponding practices in this book.

38.3 Interpretation of results

APC at 35°C may be useful to provide an index of unsanitary handling or high-temperature storage. APC of 10^6 cfu per g at 20°C indicates spoilage of seafoods. Cooked picked crabmeat normally should be expected to have APC less than 10^5 cfu per g. *Staphylococcus* and fecal coliforms should not exceed 100 cfu per g and 40 cfu per g in good quality products, respectively. *V. parahaemolyticus* should be absent or at most present to less than 100 cfu per g. Fish and shellfish of good quality will have APC of less than 10^5 cfu per cm^2 (or per g of seafood) at 20°C. The presence of significant numbers of fecal coliforms in excess of 10 cfu per g and *Staphylococcus* in excess of 100 cfu per g should be taken as an indication of contamination from human sources. Fish harvested from water polluted by human sewage should be tested for *E. coli* and *Salmonella*. *V. parahaemolyticus* can be expected to occur in water during warm periods of the year. The counting of *V. parahaemolyticus* in excess of 100 cfu per g should be considered an alarm and require further analysis.

Precooked seafood products may be associated with hazards from *S. aureus* and this bacterium should be lower than 100 cfu per g to achieve Good Manufacturing Practice. The potential hazard of *S. aureus* to the consumer from contaminated precooked products is much higher than from raw seafoods. In the absence of indicator bacteria (fecal coliforms), high APC ($>5 \times 10^5$ cfu per g) at 20°C indicates faulty processing, contamination, or poor temperature control during cooking or storage. Good quality cooked fish products should have APC of less than 5×10^5 cfu per g, *E. coli* fever than 10 cfu per g, *V. parahaemolyticus* fever than 10 cfu per g, and *Salmonella* not detected in five 25 g samples.

Cold smoked fish should have APC less than 10^5 cfu per g, *Staphylococcus* less than 100 cfu per g, and fecal coliforms less than 40 cfu per g. The presence of *C. botulinum* types A, B, E, and F in smoked fish indicates a potentially hazardous product.

Fresh or fresh frozen shellfish generally are considered to be satisfactory at the wholesale market if the fecal coliform does not exceed 230 cfu per 100 g, and APC at 35°C is no more than 5×10^5 cfu per 100 g. *E. coli* MPN should not exceed more than 1 cfu per mL of oysters and mollusks ordinarily eaten raw. *Salmonella* and *Shigella* should not be present in shellfish. The other acceptable microbiological criteria for shellfish (e.g., crustacea) are as follows: fecal coliforms lower than 300 cfu per 100 g, *E. coli* lower than 230 cfu per 100 g, *Salmonella* absent in 25 g, PSP not more than 80 μg per 100 g, and DSP not detected.

PRACTICE

Analysis of canned foods

39

39.1 Introduction

Canned foods are preserved by heat treatment in hermetically sealed containers. Hermetically sealed usually means a container that excludes the gas and microbial passages. Under proper handling and storage conditions, the high heat-resistant and nontoxic thermophilic sporeformers survive in canned foods. Commercially sterile food is defined as: (1) the absence of growing and spoiling microorganisms under normal storage conditions; and (2) the absence of pathogenic microorganisms. Viable microorganisms surviving in commercially sterile foods can be included in three groups: (1) thermophilic spore-forming bacteria cannot grow under normal storage temperature; (2) acid-tolerant microorganisms survive the high-acid (<3.7) to acid (4.6) canned foods due to underprocessing and leakage; and (3) mesophilic spore-forming microorganisms survive in low-acid canned foods due to underprocessing and leakage.

When spoilage occurs in a canned food, it involves gas production, swelling the lid of the container or flat sour spoilage without swelling of container; changes in the consistency, odor, and pH of the product; and increasing microbial numbers. Pure or mixed cultures of bacteria, yeasts, and molds contaminations occur with canned foods due to the leakage of the can.

Commercial sterility in canned foods can be achieved by using test microorganisms. Test microorganisms providing the commercial sterility can be selected depending on the food pH. (1) Foods with pH higher than 4.6; *Geobacillus stearothermophilus* or *Clostridium sporogenes* can be used as the test microorganism. (2) Foods with pH between 4.0 and 4.6; *Bacillus coagulans* or *G. stearothermophilus* can be used as the test microorganism. (3) Foods with pH lower than 4.0; *Byssochlamys fulva* can be used as the test microorganism.

39.2 Appearance of can

Swelling can occur in metal containers, plastic pouches, or plastic trays. Swelling in glass containers may appear on the lid. Swelling of cans results from gas production with the loss of vacuum on the threaded or twisted lids of the container, losing the hermetic seal. Plastic pouches tend to balloon. Plastic tray covers swell. Can spoilage can show different types of appearance: hard swells, soft swells, springers, and flippers.

1. *Flat (normal) can*. Flat can has concave shape at both ends due to partial vacuum in headspace.
2. *Flipper can*. One end of the can will be flat but other will be swollen. When finger pressure is applied to this end, it returns to normal appearance and both ends of the can appear flat. Flipper type of appearance may result from a lack of vacuum. Flipper cans have equal pressure to atmospheric pressure or inside pressure is slightly higher than atmospheric pressure. Flipper spoilage may result from underprocessing, brought down sharply on a place, hydrogen gas formation, and start of microbial growth.

Microbiological Analysis of Foods and Food Processing Environments. DOI: https://doi.org/10.1016/B978-0-323-91651-6.00053-7

3. *Springer can.* One end of the can will be permanently bulged and the other end of the can remains flat (convex). When finger pressure is applied on the bulging end of can, it will return to its normal position, but the other end will bulge out. The reasons for this appearance are the same as in flipper, but it results from a higher amount of gas formation.
4. *Soft swell can.* Both ends of the can will bulge. When finger pressure is applied, it gets depressed but when the finger pressure is removed, the ends return to the original bulge.
5. *Hard swell can.* Both ends of the can will bulge. When the finger pressure is applied, ends cannot return to the normal position. Soft and hard swellings are due to high pressure gases, more frequent formation of hydrogen gas or gases from the growth of bacteria.

Gaseous swelling in cans can be caused by a chemical reaction or the generation of gas with microbial growth. Improper cooling can cause cans to warp and hard bulges. Also, hitting the can against a hard surface can cause a lid of the can to "swell." Appropriate pressure control should be done to prevent buckling during cooling. Swelling on the can due to overfilling can cause flipper or springer. Another cause of swelling in the cans can be the chemical reaction between the product acidity and the metal part of the cans, which usually results in the formation of hydrogen gas.

39.3 Reasons for microbial spoilage in canned foods

1. Low-acid canned foods (pH above 4.6)
   a. Spoilage in commercial sterile canned foods

      When low-acid canned foods are processed sufficiently (commercial sterile), only thermophilic spore-forming bacteria survive and they cannot cause spoilage under normal storage conditions. The minimum growth temperature of the thermophilic bacteria is 35°C and the optimum growth temperature ranges from 55°C to 60°C. Therefore these bacteria cannot cause problems in low-acid canned foods. With underprocessing and unfavorable storage, these bacteria can grow in low-acid canned foods and cause the following types of spoilage:

      Flat sour spoilage is indicated by lowering of the pH of the product without swelling the can. The spore-forming thermophilic facultative aerobic bacteria (e.g., *G. stearothermophilus*) cause this type of spoilage.

      Thermophilic anaerobic (TA) spoilage results in the swelling of the container and bursting. TA spore-forming bacterium (*Thermoanaerobacterium thermosaccharolyticum*) causes this type of spoilage and it produces large quantities of hydrogen gas. The spoiled product usually has a cheesy odor.

   b. Spoilage due to insufficient processing

      Insufficient processing allows the survival of different microorganisms, mainly sporeformers. They are chiefly the thermoduric *Clostridium* and *Bacillus* spp., which subsequently spoil the low-acid canned foods at mesophilic storage temperature. From the public health standpoint, this is a most serious problems because of the potential growth of *Clostridium botulinum* and the presence of its toxin. Generally, putrefactive anaerobic spoilage, sulfide spoilage, and aerobic spore-forming spoilage can occur in insufficiently processed low-acid canned foods.

Sulfide spoilage is characterized by a flat container and darkening products with a rotten egg odor due to H_2S production. The spore-forming, anaerobic and thermoduric bacteria (e.g., *Clostridium nigrificans* and *Clostridium fermentum*) can cause sulfide spoilage. Swelling cannot appear since H_2S is very soluble in the food; sulfide and H_2S react with iron of the can to form black iron sulfide (black color).

Putrefactive anaerobic spoilage is indicated by the swelling and bursting of the container. *Clostridium putrefaciens* and *C. botulinum* cause this type of spoilage with the production of gases in the can.

Aerobic sporeformers spoilage is indicated by a higher amount of gas production in the can. *Bacillus polymyxa* and *Bacillus macerans* can cause this type of spoilage.

c. Spoilage due to leakages

When a can leaks during transport or storage of low-acid canned foods, environmental sources (e.g., water or dirty surfaces) contribute microbial contamination to canned foods. Different types of microorganisms may contaminate leaking cans, including bacteria, yeasts, molds, and bacterial and mold spores. They can cause spoilage and this type of product is not suitable for consumption.

2. Acidic canned foods (pH 4.6 or lower)

C. botulinum does not grow in these acidic conditions of foods. The surviving microorganisms are usually vegetative members of thermoduric *Lactobacillus*, *Streptococcus*, and *Pediococcus* due to insufficient processing. They may be differentiated from aerobic sporeformers by the catalase test. Aerobic spore-forming bacteria (e.g., *B. coagulans*) can cause spoilage due to insufficient processing. Some anaerobic spore-forming bacteria can also survive and grow in acidic canned foods due to insufficient processing, and cause butyric anaerobic spoilage. Aciduric molds, if they survive, easily grow in these foods with characteristic spoilage. These microorganisms can also contaminate foods due to can leakage. The types of microbial spoilage associated with acidic canned foods due to insufficient processing or leakage can be indicated as follows:

Butyric anaerobes produce butyric acid as well as CO_2 and hydrogen. This type of spoilage is caused by *Clostridium butyricum* and *Clostridium pasteurianum*.

Aciduric flat sour spoilage occurs on canned foods without gas production. This type of spoilage is caused by *B. coagulans* and the bacterium grows at both 30°C and 55°C.

Heat-resistant mold can cause spoilage characterized by a moldy taste and odor, color change with the presence of mold mycelia in the canned product. Sometimes slight swelling may occur of the can lid. Heat-resistant mold spoilages (e.g., *Byssochlamys fulva*) mostly is associated with fruits and juice concentrates.

Nonsporeformer spoilage is indicated by swelling of the can and bursting. Thermoduric lactic acid bacteria (LAB; for example, *Lactobacillus*, *Leuconostoc*, *Streptococcus*, *Micrococcus* and *Enterococcus*), thermoduric spore-forming bacteria and yeasts cause spoilage in acidic canned foods.

39.4 Physical and microbiological analysis of canned foods

Sampling, sample preparation, and all experimental techniques should be performed under aseptic conditions, and in duplicate, unless otherwise specified. All microbiological media to be used in applications must be sterilized. Again, all solutions and equipment that may cause contamination of microorganisms must be sterilized or disinfected.

39.4.1 Equipment, materials, and media

Equipment. Blender (or stomacher bag), cover slip, Erlenmeyer flak, incubator, loop, microscope glass slide, microscope, Petri dishes, pH meter, pipettes (1 and 10 mL) in pipette box, pipette discarding pad (containing 10% bleach solution), spoon, test tubes, thermostatic shaking water bath, and vortex.

Materials. Alcohol, sterile distilled water, spore stains, Gram staining reagents, waterproof pen, xylene, 0.1% peptone water, crystal violet stain (0.5%–1%), and hydrogen peroxide (H_2O_2, 3%).

Media. 2.0% Agar, APT agar, cooked meat (CM) medium, dextrose tryptone bromocresol purple (DTBP) agar, DTBP broth, liver veal (LV) agar, nutrient agar containing manganese (NAMn agar), NAMn broth, orange serum (OS) broth, OS agar, nutrient agar slant, nutrient broth, PE-2 agar, peptone yeast extract (PYE) broth, potato dextrose (PD) agar, and thermoacidurans (TA) agar.

39.4.2 Sampling

Select a sufficient number of cans at random; at least 12 cans from a level of production. The cans are examined for physical defects including faulty side or end seams, perforations, rust or other corrosion, dents, and swollen ends. Swelling at one or both ends of the can may be due to the growth of microorganisms, chemical reactions, or denting. If any of the cans show such defects, examine these cans separately from the normal appearance of cans.

Opening can. A swollen can should be refrigerated before opening. After cooking, scrub the can end with soap and water. Rinse and dry the end of the can with a clean disinfected towel. Swab the top of the can with alcohol. Expose the surface area of the can to alcohol for 30 min and wipe off the alcohol with disinfected towel and then flame. Do not flame the swollen can. Disinfect the can opener by immersing into alcohol and flame or use separate sterile can openers for each can. Aseptically, open the can. A vertical laminar flow hood can be used during opening of the can.

39.4.3 Physical analysis of can and products

1. Physical examination of can

   Before examining the cans for classification, cans should be at room temperature. With a waterproof pen, label all information of can for the side of can for a code, size of can, weight of can, name of product, pinholes or rusting, dents, buckling or other abnormality if present. Classify each can according to the descriptive terms. Measure the net weight of the can, determine the drained weight, vacuum, and headspace of the can. Examine the integrity of the metal container. Indicate the appearance of the can: springers, that is, swollen, flat, or flipper. Retain a normal can as a control. Place the swollen and a control cans in an incubator at 35°C. Examine at frequent intervals during 14 days. Bring cans back to room temperature before analysis. Clean, disinfect, and open the cans as indicated in Technique 39.4.2.

2. Removal of sample for analysis

   Remove at least 60 g of solid sample from the center of can. If the can content is liquid, remove 60 mL a liquid sample with a sterile pipette into a sterile container. Use a sterile pipette for liquid products, either regular or wide-mouthed. Transfer a solid sample with a sterile spoon or another sterile device. Always use safety devices for pipetting. Refrigerate samples at about 4°C until analysis.

3. Physical examination

Analyze the normal can food sample physically, including pH determination. Determine the pH of sample using a pH meter electrode. Do not use pH paper. Examine the can lining for blackening, denting, and pitting. In describing the product in the can, indicate the liquid level and any other unwanted characteristics, leakage, etching, corrosion, etc. Pour an amount of the can onto examination pans. Examine for odor, color, consistency, texture, and overall quality. Do not taste the product.

39.4.4 Microscopic analysis

Examine the contents of each food microscopically. Refer to Technique 3.2.3 with regard to the microscopic analysis of foods. With a flamed loop, make a smear from each sample on a microscope glass slide. Let the smear air dry, flame fix, stain with crystal violet solution (Appendix B under the topic "Crystal Violet Stain"), and dry. Never Gram stain a sample because a Gram stain result depends on the age of the culture. A Gram stain should be made from 18 h cultures only. If a smear will not adhere to the slide during staining, rinse a second smear with xylene and allow it to dry before simple staining with crystal violet. Examine the slides using the oil-immersion objective and record the morphological characteristics of microorganisms.

Prepare a wet-mount by transferring a small drop of food slurry to a slide and overlaying it with a cover slip (Technique 13.4.2). Examine the slide using a high-power objective and record observations.

A normal product may show a few microbial cells in every microscopic field analysis. Mixtures of rods, coccobacilli, yeasts, and molds usually indicate leakage. Pure cultures of medium to long or large rods with or without spores may indicate insufficient processing or leakage.

In order to demonstrate spore formation, subculture the aerobic nutrient broth culture on to two nutrient agar slants in duplicate and incubate one slant at 32°C and the other at 55°C for 18 h or up to 3 days if growth is not observed. Make smears from the agar slant and stain with spore stains as explained in Appendix B under the topic "Spore Staining." Flood with aqueous malachite green stain for 60 s and then heat to steaming four times. During heating, add dye to prevent drying of the smear. After staining, wash excess stain in running water. Immerse in safranin for about 30 s. Wash, blot, dry, and analyze under the oil-immersion objective of a microscope. The spores will stain green and the vegetative cells red.

39.4.5 Microbiological analysis of low-acid foods (pH greater than 4.6)

Weigh 25 g or mL of sample and add to a sterile Erlenmeyer flask containing 225 mL of sterile 0.1% peptone water. Shake vigorously for at least 2 min and allow it to stand for 2 or 3 min. Prepare further dilutions if necessary, using 0.1% peptone water. Undiluted liquid sample can also be used in the culturing of microorganisms from the sample.

1. Anaerobic culturing

Plate count. Label two sets of Petri dishes in duplicate for PE-2 agar with sample number, dilution and your initials. Label one Petri dish with 32°C and the other with 55°C. Add 1 mL of the sample (diluted or undiluted) into two sterile Petri dishes in duplicate and add 13–15 mL of melted PE-2 agar (at 46°C). Mix agar and 1 mL of sample by thoroughly rotating the plates in the route while drawing a figure of eight while it is lying on the table top. Allow the agar in the plates to solidify (within 10 min). Layer the surface of the plates with 3–5 mL of melted PE-2 agar. Incubate one set of inverted Petri plates in duplicate in the following incubation conditions: (1) in an anaerobic incubator at 32°C for up to 4 days for anaerobic mesophilic count and (2) the other set in an anaerobic incubator at 55°C for up to 10 days for anaerobic thermophilic count. After incubation, analyze the plates for colony formation and record characteristic colonies. Count colonies from Petri plates containing 25–250 colonies. Calculate the number of respective microorganisms by multiplying the average count by dilution rate and dividing by the inoculation amount. Record results as the number of respective bacteria cfu per g or mL of sample. If necessary, perform identification tests on isolates (Technique 39.4.2).

Isolation. CM medium (or PE-2 broth) must be exhausted by heating at 94°C for 20 min and cooled immediately before use. In the isolation of anaerobic microorganisms from canned foods, carry out two sets of 1 mL inoculation from the sample (homogenized or not homogenized) to CM medium (or PE-2 broth) in duplicate. Incubate each set at the following incubation conditions in duplicate: (1) in an anaerobic incubator at 32°C for up to 4 days for anaerobic mesophilic bacteria; and (2) in an anaerobic incubator at 55°C for up to 10 days for anaerobic thermophilic bacteria. After incubation, analyze the medium for growth (turbidity) and color change. Remove the cap from each liquid medium and note the odor; replace cap. Record the results from the liquid medium as the presence or absence of anaerobic mesophilic or thermophilic bacteria. Use cultures in the identification of isolates (Technique 39.3.5c).

2. Aerobic culturing

Plate count. Label two sets of Petri dishes in duplicate for DTBP agar with sample number, dilution, and your initials. Pour plate 1 mL of sample into two sets of DTBP agar (Technique 2.2.4) in duplicate. Incubate each set at the following incubation conditions: (1) in an aerobic incubator at 32°C for up to 4 days for aerobic mesophilic count; and (2) in an aerobic incubator at 55°C for up to 10 days for aerobic thermophilic count. After incubation, analyze the plates for colony formation and record characteristic colonies. Count the colonies from Petri plates containing 25–250 colonies. Calculate the number of respective microorganisms by multiplying the average count by the dilution rate and dividing by the inoculation amount. Record the results as the number of respective bacteria cfu per g or mL of sample. If necessary, perform identification tests on isolates (Technique 39.3.5c).

Isolation. In the isolation of aerobic microorganisms, carry out 1 mL of sample inoculation into two sets of DTBP broth in duplicate using a canned food sample (homogenized or not homogenized). Spread plate can also be performed using DTBP agar. Incubate each set at the following incubation conditions: (1) in an aerobic incubator at 32°C for up to 4 days for mesophilic aerobic bacteria and (2) in an aerobic incubator at 55°C for up to 10 days for thermophilic aerobic bacteria. After incubation, analyze tubes for growth (turbidity) and color change. Record the results from tubes for each incubation temperature. Streak plate to PE-2 agar plates from DTBP broth. Use cultures for the identification of isolates (Technique 39.3.5c).

3. Identification of isolates

From anaerobic (Technique 39.4.2) or aerobic (Technique 39.3.5b) culture, select least three colonies per plates to obtain pure cultures using broth or agar media (Technique 13.4.1). Examine the pure cultures microscopically by simple staining using crystal violet stain (Appendix B under the topic

"Crystal Violet Stain"), Gram staining (Appendix B under the topic "Gram Staining Reagents"), and spore staining (Appendix B under the topic "Spore Stain"). During microscopic examination, observe the morphology, shape of cells, presence or absence of spores within cells, and/or free spores. If growth consists of mixed or pure cultures of cocci, yeasts, or molds, this indicates leakage or inadequate processing. Perform a gas formation test by inoculation of each isolate into PE-2 broth and layer the surface area of broth by melted 2% agar (at 45°C) with up to 1 cm without disturbing the broth. After incubation at their respective temperature, analyze the broth for the formation of gas bubbles through the medium. Record the characteristics of isolates and perform identification techniques for nonsporeformers, as outlined in this book for the relevant microorganism. If isolates are sporeformers, obtain a pure culture of sporeformers, as explained in Technique 39.3.5d.

4. Isolation of sporeformers

a. Aerobic spore-forming bacteria (*Bacillus* spp.)

Inoculate one loopful of aerobic mesophilic culture (single colony, from Technique 39.3.5c) into NAMn broth in duplicate, and incubate tubes at 32°C for up to 4 days. Inoculate one loopful of aerobic thermophilic culture (single colony from Technique 39.3.5c) into NAMn broth in duplicate, and incubate tubes at 55°C for up to 10 days. After incubation, analyze the tubes for growth (turbidity).

Place the NAMn broth culture from each incubation separately in the thermostatic shaking water bath at 80°C. Use a control medium containing a thermometer to measure the temperature. Shake the water bath at low speed to assist heat distribution during heating. This heat treatment activates spores and destroy vegetative cells. During heating, the level of the culture (sample) in the tube should be 2.5 cm below the level of water in the water bath. After the temperature reaches 80°C, hold the tubes (or flasks) for 20 min at that temperature. Following heat treatment of the culture (sample), immediately cool to 35°C in cold water. Streak plate one loopful of heated culture onto two sets of NAMn agar plates in duplicate. Incubate one set of inverted plates at 32°C for up to 4 days from aerobic mesophilic culture and the other at 55°C for up to 10 days from aerobic thermophilic culture. After incubation, analyze the plates for the formation of colonies on the agar medium to indicate spore-forming bacteria. Record the characteristics of isolates and perform identification techniques for sporeformers as outlined in Technique 39.3.5c.

Aerobic spore-forming bacteria can also be isolated directly from canned food samples. Place 10 g or mL of the sample into 90 mL of 0.1% peptone water in duplicate and homogenize by gently shaking. Heat the homogenate at 80°C for 20 min as indicated above. After heating, streak plate one loopful of heated culture onto NAMn agar plates in duplicate. Incubate the inverted plates at 32°C for up to 4 days for aerobic mesophilic sporeformers and at 55°C for up to 10 days for aerobic thermophilic sporeformers. After incubation, analyze the plates for the formation of colonies on the agar medium. Record the characteristics of isolates and identify aerobic sporeformers, as outlined in Technique 39.3.5c.

b. Anaerobic spore-forming bacteria (*Clostridium* spp.)

Inoculate one loopful of mesophilic anaerobic culture (single colony, Technique 39.3.5c) into CM medium in duplicate and incubate tubes at 32°C for up to 4 days under anaerobic conditions. Inoculate one loopful of thermophilic anaerobic culture (Technique 39.3.5a) into CM medium in duplicate and incubate tubes at 55°C for up to 10 days under anaerobic conditions. After incubation, analyze the tubes for growth (turbidity).

Place the CM medium culture from each incubation separately in the thermostatic shaking water bath at 80°C. Use a control medium containing a thermometer to measure the temperature. After the temperature reaches 80°C, hold the tubes for 20 min at that temperature. Shake the water bath at low speed to assist heat distribution during heating. This heat treatment activates spores and destroy vegetative cells. During heating, the level of the culture (or sample) in the tube should be 2.5 cm below the level of water bath. Following the heat shock of the culture (or sample), immediately cool to 35°C in cold water. Streak plate a loopful of heated culture onto each of two sets of LV agar plates in duplicate. Incubate one set of inverted Petri plates (from heated mesophilic culture) at 32°C for up to 4 days and the other (from heated thermophilic culture) at 55°C up to 10 days under anaerobic conditions. After incubation, analyze the plates for the formation of colonies to indicate anaerobic spore-forming bacteria.

Anaerobic spore-forming bacteria can also be isolated directly from canned food samples. Place 10 g or mL of the sample into 90 mL of CM medium in duplicate and homogenize by gently shaking. Heat the homogenate at 80°C for 20 min as indicated above. After heating, streak plate one loopful of heated culture onto two sets of LV agar plates in duplicate. Incubate one set of inverted plates anaerobically at 32°C for up to 4 days for anaerobic mesophilic sporeformers and the other set anaerobically at 55°C for up to 10 days for anaerobic thermophilic sporeformers. After incubation, analyze the plates for the formation of colonies on the agar medium. Record the characteristics of isolates and identify anaerobic sporeformers as outlined in Technique 39.3.5c.

39.4.6 Microbiological analysis of acid foods (pH of 4.6 or lower)

Weigh 25 g or mL of the sample and add to a sterile Erlenmeyer flask containing 225 mL of sterile 0.1% peptone water. Shake vigorously for at least 2 min and allow it to stand for 3 min. Prepare further dilutions if necessary, using 0.1% peptone water. Undiluted liquid sample can also be used in the culturing of microorganisms from canned foods.

Plate count. From an homogenized sample, carry out the following inoculation into media: two TA agar plates (or OS broth) for butyric anaerobic bacteria by pour plate technique (Technique 2.2.3) with layering medium (Technique 39.3.5a), PD agar plates for yeasts and molds count by spread plating (Technique 2.2.4), two DTBP agar plates for aerobic acid formers by spread plating, and Mrs agar plates for LAB by spread plating. After the inoculation of media, incubate the inverted plates in duplicate at the following incubation conditions for their respective microbial counts: (1) DTBP agar plates; in an aerobic incubator at 32°C for up to 4 days and in an aerobic incubator at 55°C for up to 10 days; (2) TA agar plates; in an anaerobic incubator at 32°C for up to 4 days and in an anaerobic incubator at 55°C for up to 10 days; (3) PD agar plates at 25°C for up to 3 days; and (4) Mrs agar plates microaerobically (with 5% CO_2) at 30°C for up to 3 days.

After incubation, analyze the plates for colony formation and record the characteristic colonies. Indicate the presence of acid formers on TA agar and DTBP agar with color change (from purple to yellow). Count the bacterial colonies from Petri plates (TA, DTBP, and Mrs agars) containing 25–250 colonies. Count yeasts and molds colonies from PDA agar plates containing 15–150 colonies. Calculate the number of counts by multiplying the average count by the dilution rate and dividing by the inoculation amount. Confirm the presumptive isolates and record the results as the number of respective microorganisms cfu per g or mL of sample.

Isolation. In the isolation of microorganisms from canned foods, inoculate 1 mL of the sample into four sets of OS broth in duplicate from a canned food sample with or without homogenization. Incubate one set of broths in duplicate at the following incubation conditions: (1) in an aerobic incubator at 32°C for up to 4 days for mesophilic aerobic bacteria; (2) in an aerobic incubator at 55°C for up to 10 days for thermophilic aerobic bacteria; (3) in an anaerobic incubator at 32°C for up to 4 days for mesophilic anaerobic bacteria; and (4) in an anaerobic incubator at 55°C for up to 10 days for thermophilic anaerobic bacteria. After incubation, analyze tubes (or plates) for growth (turbidity) and color changes.

After growth is obtained in media, perform a pure culture technique (Technique 13.4.1) using the respective media to obtain pure cultures from isolates. Prepare a crystal violet simple stain (Appendix B under the topic "Crystal Violet Stain"), a wet-mount (Technique 13.4.2), Gram stain (Appendix B under the topic "Gram Staining Reagents"), and spore stain (Appendix B under the topic "Spore Stain") from pure cultures. Look for the presence of refractile spores in sporangia and/or free spores, as well as other morphological characteristics of spore-forming and nonspore-forming vegetative microorganisms. From pure cultures, perform a catalase test (Technique 13.4.3.10) using 3% H_2O_2. Gas bubbles from colonies indicate a positive catalase test; a negative catalase reaction is indicated by the absence of bubbles. LAB are indicated by a negative catalase test. For spore-forming isolates, continue to isolate spore-forming bacteria (Technique 39.3.5c).

With or without spores, identification techniques can be continued for each isolate as outlined in this book for the relevant bacteria.

39.4.7 Sterility tests on can

Unspoiled cans may be subjected to incubation tests to determine the keeping quality of canned food. It is often desirable to incubate cans as an acceleration test to determine the possibilities of spoilage during storage. The can sample should include at least two sets of 12 cans. Use one can for each 12 cans as a control.

Low-acid canned foods (with pH > 4.6): Incubate one set of cans at 35°C for up to 30 days and the other set at 55°C for up to 10 days. Retain control cans at room temperature. *Exception*: no incubation at 55°C is necessary for meat and fish products. Meat products should be incubated at 55°C if cereals are present.

Acid canned foods (with pH ≤ 4.6): Incubate one set of cans for up to 14 days at 35°C. Retain control cans at room temperature. No incubation is carried out at 55°C. *Exception*: Incubate whole tomatoes at 55°C.

Examine the cans at frequent intervals during the incubation period to indicate defects and swells. With or without defects or swells, perform physical and microbiological analysis on products. Open all cans at the end of the incubation period as indicated in Technique 39.3.2. Observe the condition of contents, test for flat sours with a glass electrode pH meter. Certain indicators (such as bromocresol purple) may be used to test for flat sour acidic spoilage (in corn and pear) in culture media if desired. Indicate the following results from sterility tests.

1. Results of underprocessing

   Low-acid products. Refer to Technique 39.3.5a for the isolation of microorganisms from low-acid canned foods. From culturing, gas in the anaerobic tubes at 55°C and a cheesy odor in the tube indicates spoilage by thermophilic anaerobes. Spoilage by putrefactive anaerobes is indicated by gas and putrid odor in the anaerobic tubes at 35°C when rods and spores are observed in microscopic examination. Spoilage by mesophilic and/or thermophilic flat sour types is indicated by acid production in the aerobic broth tubes with yellow color or yellow colonies on plates incubated at 35°C and/or 55°C (Technique 39.3.5b).

Acid products. Refer to Technique 39.3.6 for the isolation of microorganisms from acid canned foods. Nonspore-forming LAB and yeasts generally cause spoilage in acid products due to underprocessing or leakage. Spoilage in acid products may have a butyric acid odor. Thermophilic spoilage may appear in tomato juice, tomatoes, pears, pineapples, and nectarines due to spore-forming anaerobes, for example, *C. pasteuranum.*

2. Results of leakage

Perform simple staining (Appendix B under the topic "Crystal Violet Stain"), wet-mount (Technique 13.4.2), and isolation of microorganisms (Technique 39.3.5 and 39.3.6) from products to indicate the types and level of microorganisms. The presence of mixed flora of rods and cocci on microscopic examination and the isolation of these microorganisms from culturing are indicative of leakage.

3. Results of precanning spoilage

In some instance, spoilage may result from the growth of microorganisms in the food before canning. In this type of spoilage, a mixture of bacterial types is usually observed microscopically from normal-appearing cans, but none of the microorganisms are viable. Perform simple staining (Appendix B under the topic "Crystal Violet Stain"), wet-mount (Technique 13.4.2), and isolation of microorganisms (Technique 39.3.5 and 39.3.6) from products to indicate types and level of microorganisms. The viability of cells can also be indicated microscopically (Technique 3.4). The pH, odor, and appearance may be abnormal with the precanning spoilage.

4. Results of hydrogen swells

Perform simple staining (Appendix B under the topic "Crystal Violet Stain"), wet-mount (Technique 13.4.2), and isolation of microorganisms (Techniques 39.3.5 and 39.3.6) from products to indicate the types and level of microorganisms. If no microorganisms are detected on smears or cultures of swollen cans and the product appears normal, there is a possibility that the swelling of the cans is due to the formation of hydrogen gas by chemical action of the acidic foods on the iron portion of the cans.

39.5 Interpretation of results

The presence of only spore-forming bacteria growing at 35°C in cans produced with heat equal to commercial sterility without leakage indicates underprocessing. Thermophilic anaerobic *T. thermosaccharolyticum* spoilage can be indicated by gas and a cheesy odor in CM medium at 55°C. Spoilage by thermoduric *Clostridium* spp. (e.g., *C. botulinum*, *C. sporogenes* or *C. perfringens*) may be indicated in CM medium at 35°C by gas and a putrid odor; and rods and spores from microscopic examination. Spoilage by mesophilic bacteria and/or thermophilic bacteria (e.g., *Bacillus stearothermophilus*), which are flat-sour types, may be indicated by acid production in medium at 35°C and/or 55°C in high-acid or low-acid canned foods. Spoilage in acid canned foods is usually caused by nonspore-forming *Lactobacillus* and yeasts. Canned tomatoes and tomato juice spoilages may remain flat but the products have an off-odor with or without lowered pH due to aerobic, mesophilic, and thermophilic sporeformers. Many canned foods contain thermophiles that do not grow under normal storage conditions, but which grow and cause spoilage when the product is subjected to elevated temperatures (over 40°C). *B. coagulans* and *G. stearothermophilus* are thermophiles responsible for flat sour spoilage in acid and low-acid foods, respectively. In flat sour spoilage, incubation at 55°C will not cause a change in the appearance of the can, but the product has an off-odor with or without lowering the pH. A mixed microflora of viable bacterial rods and cocci usually indicates leakage. A mixed microflora in the product, as shown by direct smear, in which there are large numbers of bacteria visible but no growth in the cultures, may indicate precanning spoilage. The product may be abnormal in pH, odor, and appearance. If no evidence of microbial growth can be found in swollen cans, the swelling may be due to the development of hydrogen by the chemical action of contents on container interiors. Chemical breakdown of the product may result in the evolution of carbon dioxide. This results from concentrated products containing sugar and some acid, such as tomato paste, molasses, mincemeats, and highly sugared fruits. The chemical reaction is accelerated at elevated temperatures. Canned pasteurized foods requiring refrigeration (whole pies, sausage, chicken portions; canned ham, fruit juice, and soups) are satisfactory with APC $<10^4$ cfu per g or mL of sample, need precautions between 10^4 and $<10^7$ cfu per g or mL of sample, and are unsatisfactory with $\geq 10^7$ cfu per g or mL of sample. Canned ultrahigh-temperature (UHT) products (e.g., tuna, salmon, corned beef, soups, desserts, and fruit) are satisfactory with APC <10 cfu. Ten grams of sugar should not contain more than 150 thermophilic spores. Flat sour spores for five samples examined, there shall be a maximum of not more than 75 spores and an average of not more than 50 spores per 10 g of sugar. Thermophilic anaerobic spores shall be present in not more than three of the five samples of starch. Sulfide spoilage spores shall be present in not more than two of the five samples and in any one sample of starch to the extent of not more than five colonies per 10 g.

PRACTICE

40 Analysis of salad dressings and spices

40.1 Introduction

Mayonnaise, cooked starch-based dressings, and pourable dressings are types of salad dressings. Excess cooking would destroy the physical integrity of these products; preservation usually depends on the amount of vinegar (acetic acid) or lemon juice. Mayonnaise is a creamy pale-yellow product. The pH value usually ranges from 3.6 to 4.0 and the products contain 0.5%–1.2% acetic acid. Sodium benzoate and potassium sorbate are also included as preservatives. The oil contents range from 65% to 80%, salt is about 9%–11%, and sugars are usually 7%–10%. Microbial spoilage of salad dressing can be associated with *Bacillus subtilis*, *Bacillus mesentericus*, *Candida krusei*, *Candida pseudotropicalis*, *Micrococcus*, *Lactobacillus*, *Saccharomyces*, *Hansenula*, *Zygosaccharomyces*, *Pichia*, and molds (e.g., *Geotrichum*). *Salmonella*, *Staphylococcus*, and *Escherichia coli* are the pathogens in salad dressing. Three microbial groups can be used as microbial indicators for salad dressings: yeasts and molds, *Lactobacillus*, and *Bacillus*. High numbers of these groups are indicative of poor sanitation and potential spoilage problems. Only bacterial spores survive in salad dressing indefinitely. Aerobic plate count (APC), yeasts and molds count, homofermentative lactic acid bacteria (LAB) and heterofermentative counts, *Lactobacillus* count, *Bacillus* count, and osmophilic count can be used to indicate the sanitary quality of process conditions and product quality.

Spices are plant products (seed, flower, leaf, bark, roots, or bulb) used whole or ground, singly or mixed. They are used in relatively small amounts for aroma and color. Some spices, unless given antimicrobial treatments (e.g., irradiation and ethylene oxide), may be contaminated by microorganisms by as much as 10^6–10^7 cfu per g. The most important microorganisms are the spores of molds, *Bacillus*, and *Clostridium*. Also, *Micrococcus*, *Enterococcus*, yeasts, and several pathogens (e.g., *Salmonella*, *Escherichia coli*, *Staphylococcus aureus*, and *Bacillus cereus*) may be associated with spices. They may also harbor mold toxins. Therefore, spices can be the source of spoilage and pathogenic microorganisms. Some spices (e.g., cloves, allspices, and garlic) have antimicrobial properties. Certain species of microorganisms may give very high viable counts in pepper. Black pepper may contain a high number of aerobic spore-forming bacteria. The use of pepper in the production of food products may introduce pathogenic and spoilage microorganisms (e.g., *Bacillus*). Spices can be examined with a plate count technique for APC, yeasts and molds, and coliforms. If coliforms are present, spices should be analyzed for *E. coli*. When unsanitary storage of spices is indicated, they should also be analyzed for *S. aureus*, *B. cereus*, *Salmonella*, and *Shigella*. Coliforms, *Lactobacillus*, *Enterococcus*, and spore counts will indicate public health hazards, quality, and sanitary conditions of the production and handling of spices.

Microbiological Analysis of Foods and Food Processing Environments. DOI: https://doi.org/10.1016/B978-0-323-91651-6.00003-3

40.2 Microbiological analysis of salad dressings

Sampling, sample preparation, and all experimental techniques should be performed under aseptic conditions, and in duplicate, unless otherwise specified. All microbiological media to be used in applications must be sterilized. Again, all solutions and equipment that may cause the contamination of microorganisms must be sterilized or disinfected.

40.2.1 Equipment, materials and media

Equipment. Blender (or stomacher), incubator, microscope, microscope glass slide, Petri dishes, pipettes (1 mL) in pipette box, pipette discarding pad (containing 10% bleach solution), and vortex.

Materials. Crystal violet stain (0.5%) and 0.1% peptone water.

Media. Lactobacillus heterofermentative screen (LHS) broth, de Man-Rogosa and Sharpe (Mrs) agar, plate count (PC) agar containing 100 μg per mL cycloheximide, potato dextrose (PD) agar containing 100 mg per L chlortetracycline, and chloramphenicol.

40.2.2 Sample preparation

Refer to Technique 2.2.2 for sample preparation and dilution. Place 50 g of sample into a sterile blender (or stomacher bag), add 450 mL of sterile 0.1% peptone water, and blend (or stomach) for 2 min. Make subsequent dilutions with 0.1% peptone water if necessary.

40.2.3 Microbiological counts

40.2.3.1 Homofermentative and heterofermentative counts

Refer to Technique 33.2.9 for homofermentative and heterofermentative LAB counts. Use HHD agar for differential enumeration of homofermentative and heterofermentative LAB. Incubate the inverted Petri plates aerobically at 32°C for up to 4 days. This medium contains fructose, which is reduced to mannitol by heterofermentative LAB but not homofermentative. In agar medium, homofermentative colonies of LAB are blue to green, while heterofermentative colonies are white. On HHD agar, count homofermentative and heterofermentative LAB characteristic colonies separately from Petri plates containing 25 to 250 colonies. Calculate the number of homofermentative and heterofermentative LAB separately by multiplying the average count by the dilution rate and dividing by the inoculation amount. Record the results as the homofermentative and heterofermentative LAB cfu per g or mL of sample. Perform identification as indicated in Technique 33.2.9.

40.2.3.2 Other counts

Microscopic analysis. Yeasts and bacteria are readily simple stained for analysis. Refer to Technique 3.2.3 for details of direct microscopic count using crystal violet stain (0.5%, aq.).

Yeasts and molds count. Refer to Technique 6.3.3 for details of yeasts and molds count. Use acidified PD agar to inhibit bacteria. Incubate the inverted plates at 25°C for up to 5 days.

Aerobic plate count. Refer to Technique 2.2.4 for APC by spread plating. Use PC agar containing cycloheximide to inhibit fungi. Some yeasts can be resistant to a low level of cycloheximide. Incubate plates at 32°C for up to 5 days.

Bacillus count. Refer to Technique 19.2.3.1 for *Bacillus* count by spread plating. Incubate plates at 32°C for 24 h.

Lactobacillus count. *Lactobacillus* (e.g., *Lactobacillus fructivorans* and *Lactobacillus brevis*) cannot be detected easily on PC agar, but they grow on Mrs agar. Refer to Technique 33.2.7 for details of the *Lactobacillus* count. Incubate plates at 25°C for up to 14 days in a 5%–7% CO_2 incubator. Alternatively, overlay Mrs agar and incubate the inverted Mrs agar plates at 25°C for up to 14 days.

Osmophilic count. Refer to Technique 11.2.1.3 for details of osmophilic count.

Others. For counting coliforms, fecal coliforms, and individual pathogens, e.g., *E. coli*, *Staphylococcus aureus*, *Listeria monocytogenes*, *Yersinia enterocolitica*, *Salmonella*, and *Bacillus*, refer to the corresponding practices in this book.

40.3 Microbiological analysis of spices

Sampling, sample preparation, and all experimental techniques should be performed under aseptic conditions, and in duplicate, unless otherwise specified. All microbiological media to be used in applications must be sterilized. Again, all solutions and equipment that may cause the contamination of microorganisms must be sterilized or disinfected.

40.3.1 Equipment, materials and media

Equipment. Bottle, Erlenmeyer flask, incubator, pipette (1 mL) in pipette box, pipette discarding pad (containing 10% bleach solution), Screw-cap jar, spatula, thermostatic shaking water, and vortex.

Materials. Phosphate buffer (PB) solution.

Media. Cooked meat agar, dextrose tryptone bromocresol purple (DTBP) agar, FT agar, Jensen's pork sucrose nitrite (JPSN) medium, liver agar, PE-2 agar, Sulfide iron (SI) agar, tomato juice (TJ) agar with pH 5.0, and tryptone glucose extract (TGE) agar.

40.3.2 Sampling

Aseptically, obtain at least 100 g of spices from each spice into sterile bottles. If spices are in small packages, send the packages to the laboratory. In the homogenization of the sample, weigh 25 g of spices into a Erlenmeyer flask containing 225 mL of PB solution using a flamed spatula. Shake the flask for 5 min with a mechanical shaker. Allow it to stand for 5 min to settle out the coarse particles and follow further serial dilutions using PB solution if necessary.

40.3.3 Counting of aerobic spores

Mesophilic aerobic acid-producing sporeformers count. Refer to Technique 10.2.4 for the mesophilic aerobic sporeformers count for a heated sample. Pour plate 1 mL of heated samples in duplicate to TGE agar without layering. Incubate the inverted plates at 32°C for 3 days. Count the yellow (acid producing) colonies and report the results as the number of mesophilic aerobic spoilage acid-producing colony forming units (cfu) per g of spice.

Mesophilic aerobic gas-producing sporeformers count. Refer to Technique 10.2.5 for mesophilic aerobic gas-producing sporeformers (MAGS) count from a heated sample. For counting MAGS, inoculate 1 mL of heated sample into each of 5-MPN tubes of JPSN broth for each serial dilution and incubate the tubes at 35°C for 3 days. After incubation, look for gas production in the Durham tube. Refer to MPN Table 4.2 and calculate the number of MAGS. Report the results as the number of MAGS MPN per g of spice.

Thermophilic aerobic flat sour sporeformers count. Refer to Technique 10.5.3 for the thermophilic aerobic flat sour sporeformers (TAFS) count. After heating, cool to about 45°C, distribute agar among five sterile Petri dishes, and allow it to solidify. Incubate the inverted plates at 55°C for 48 h and count the typical colonies. Flat sour colonies are characteristic in appearance, being round, 2–3 mm in diameter with an opaque central spot, and usually surrounded by a yellow halo. Combine counts from five plates and calculate the number of TAFS by multiplying by the dilution rate and dividing by 10 mL and report the result as the number of TAFS cfu per g of spice.

40.3.4 Counting of anaerobic spores

Mesophilic anaerobic sporeformers count. Refer to Technique 10.3.3 for the mesophilic anaerobic sporeformers count. Add 1 mL of heated sample in duplicate from the dilutions in to sterile Petri dishes, pour with melted FT agar (at 45°C) with layering. Incubate the inverted plates in an anaerobic incubator at 32°C for 3 days. After incubation, analyze the colony formation and count the number of colonies on the agar medium. Calculate the number of anaerobic spores by multiplying the average count by the dilution rate and dividing by the inoculation amount. Report the results as the number of mesophilic anaerobic sporeformers cfu per g of spice.

Mesophilic anaerobic putrefactive sporeformers count. Refer to Technique 10.3.4 for the heating of the sample. Transfer 1 mL of boiled spice suspension to each of 5-MPN tubes of freshly steamed liver broth and layer with nutrient agar. Incubate tubes at 32°C for 5 days. After incubation, analyze the liver broth for turbidity and decomposition of livers. Refer to MPN Table 4.2 for calculation of mesophilic anaerobic putrefactive sporeformers. Report the results as the number of mesophilic anaerobic putrefactive sporeformers MPN per g of spice.

Thermophilic anaerobic sulfide spoilage sporeformers count. Refer to Technique 10.7.3 for details of thermophilic anaerobic sulfide spoilage sporeformers with H_2S production. Use SI agar. Report results as the number of thermophilic anaerobic sulfide spoilage sporeformers cfu per g of spice.

40.3.5 Counting of acid-tolerant bacteria

Spread plate 1 mL of diluted sample to TJ agar (Technique 2.2.4) and incubate the inverted plates at 32°C for up to 3 days. After incubation, analyze the colony formation and count the number of colonies on the agar medium. Calculate the number of acid-tolerant bacteria by multiplying the average count by the dilution rate and dividing by the inoculation amount. Report the results as the number of acid-tolerant bacteria cfu per g of spice.

40.3.6 Other counts

Aerobic plate count. Refer to Technique 2.2.4 for APC by spread plating. Incubate plates at 32°C for 2 days. Report the results as the number of APC cfu per g of spice.

Yeasts and molds count. Refer to Technique 6.3.3 for the yeasts and molds count. Incubate plates at 30°C for up to 5 days. Report the results as the number of yeasts and molds cfu per g of spice.

Bacillus count. Refer to Technique 19.2.3.1 for the *Bacillus* count by spread plating. Incubate plates at 32°C for 24 h. Report the results as the number of *Bacillus* cfu per g of spice

For counting coliforms, *Lactobacillus* and pathogenic microorganisms, for example, *E. coli*, *S. aureus*, *Clostridium perfringens*, *Salmonella*, *Shigella*, and *Enterococcus*, refer to the corresponding practices in this book.

40.4 Interpretation of results

Sometimes spoiled dressings contain few APC (fewer than 100 cfu per g) or none at all. In these instances, the microorganisms have probably died after the nutrients were exhausted or after the accumulation of metabolic by-products. A direct microscopic examination will usually reveal dead cells. Gaseous fermentation may not occur in spoiled dressings until after several weeks. *L. fructivorans* and other *Lactobacillus* spp. grow slowly, and considerable time is needed to increase their population and visible gas production. Some spoilage may be observed only by an increase in acid or a flavor change. These may result from spoilage due to the homofermentative and heterofermentative *Lactobacillus* (e.g., *L. casei* and *L. plantarum* respectively). *Zygosaccharomyces bailii* and certain other species of *Zygosaccharomyces* ferment glucose and sucrose quickly. Salad dressing with low pH due to LAB accounts for the absence of food-poisoning microorganisms. Dressings at a significantly higher pH should be examined for the presence of food-poisoning microorganisms.

The thermophilic flat sour sporeformers and thermophilic anaerobic nonsulfide spoilage sporeformers should be low if the spices are to be used in canned food production. Spices using in canned cured meat products should be free of microorganisms producing gas in JPSN broth. Spices showing very high counts are likely to be of low sanitary quality and indicative of the conditions of production and handling. Microorganisms of public health significance should not be present in spices. Large numbers of aerobic spore-forming microorganisms in spices may be potential spoilage agents or may contribute to quality loss in the foods in which they are used. Generally, the microbial loads of *Bacillus* spp. are high and some common molds may be present.

PRACTICE

Analysis of bottled soft drinks

41

41.1 Introduction

Bottled soft drinks are a class of nonalcoholic beverages that contain water, nutritive or nonnutritive sweeteners, acids, flavors, colors, emulsifiers, preservatives, and various other compounds that are added for their functional properties. Some fruit juices contain citrus, apple, pear, and grape. Soft drinks are carbonated with 1.5–4 volumes of CO_2. Carbonated and noncarbonated soft drinks are low-pH products (pH ranges from 2.5 to 4.0). Citric acid is the most widely used acidulant. Colas have a pH of 2.5–2.8, and generally are acidified with phosphoric acid. The process of bottled soft drink preparation may be divided into three parts: (1) the returnable glass bottle washing, (2) the flavoring syrup preparation together with water processing, and (3) the syrup filling into bottle and hermetically closing the bottle.

Bacterial spores, yeasts, and molds would be expected to survive in soft drinks. Important spoilage microorganisms in bottled soft drinks are acid-tolerant yeasts and some yeasts can survive and grow under anaerobic conditions. Certain lactic acid bacteria (LAB) can also grow in acidic conditions. Acetic acid bacteria and molds are also aciduric, and they grow only in the presence of dissolved oxygen. Microbial growth produces haze, sediment, off-flavors, and gas. When yeasts are responsible for the spoilage of bottled soft drinks, excess CO_2 is produced, which can cause swelling and bursting of the bottle. Some yeast species from the genera *Torulapsis*, *Candida*, *Pichia*, *Hansenula*, and *Saccharomyces* can grow and produce carbonated turbid products. Some *Lactobacillus* and *Leuconostoc* spp. can also grow in bottled (carbonated) soft drinks to cause cloudiness and ropiness (due to the production of dextran from carbohydrates). Yeasts, *Lactobacillus*, and *Leuconostoc* spp. also can cause similar spoilage in noncarbonated beverages. If there is enough dissolved oxygen, *Acetobacter* and molds (*Penicillium*, *Aspergillus*, *Mucor*, and *Fusarium*) can cause spoilage of bottled soft drinks. Morphological analysis indicates the sanitary quality of production conditions and the main spoilage microorganisms in products, via aerobic bacteria counting, yeasts and molds count, and LAB count.

41.2 Microbiological analysis of bottled soft drinks

Sampling, sample preparation, and all experimental techniques should be performed under aseptic conditions, and in duplicate, unless otherwise specified. All microbiological media to be used in applications must be sterilized. Again, all solutions and equipment that may cause the contamination of microorganisms must be sterilized or disinfected.

Microbiological Analysis of Foods and Food Processing Environments. DOI: https://doi.org/10.1016/B978-0-323-91651-6.00015-X

41.2.1 Equipment, materials and media

Equipment. Bottle with closure, Erlenmeyer flask, incubator, membrane filter (0.22 μg pore size), membrane filter system, Petri dishes, pipette (1 and 10 mL) in pipette box, pipette discarding pad (containing 10% bleach solution), refrigerator, spoon, stopper, tongs (or slide forceps), vortex, and wide-mounted jar.

Materials. 70% Alcohol, 0.1% peptone water, phosphate buffer (PB) solution, and sterile distilled water.

Media. Plate count (PC) agar and malt extract (ME) agar.

41.2.2 Water analysis

All water used for the preparation of bottled soft drinks, rinse water, and cooling water should meet the requirements of drinking water. The sample should be collected in a sterile bottle with a closure. Collect water entering the plant, after plant processing (filtering, chemical treatment, etc.), and as delivered at the filling machine head. If another source of water is used for a general plant service or bottle rinsing, it should be similarly sampled. Prepare dilutions with the use of 0.1% peptone water (Technique 1.5.1.1). For aerobic plate count (APC) refer to Technique 42.2.2.1, and for coliforms and fecal coliforms refer to Technique 42.2.2.3.

41.2.3 Sugar analysis

Sugar should be relatively free of potential spoilage microorganisms. Obtain a sugar sample representative of the lot being used. Place it in a sterile wide-mounted jar using a sterile spoon and taking precautions to prevent secondary contamination. Weigh 10 g of sugar aseptically into a sterile Erlenmeyer flask and add 90 mL sterile distilled water and homogenize by gently shaking. Prepare further dilutions using 0.1% peptone water. Spread plate 1 mL sample onto PC agar in duplicate for APC and ME agar in duplicate for yeasts and molds count. Incubate the inverted PC agar plates at 32°C up to 3 days and the ME agar plates at 25°C for up to 5 days. After incubation, analyze colony formation on agar medium. Count colonies from PC agar containing 25 to 250 colonies. Count yeasts and molds colonies from ME agar containing 15 to 150 colonies. Calculate the number of APC, yeasts, and molds by multiplying the average count by the dilution rate and dividing by the inoculation amount. Record the results as the number of APC and yeasts and molds cfu per g of sugar.

For coliforms and *Escherichia coli* counts refer to Technique 13.3.2. Record the results as the number of coliforms and *E. coli* MPN per g of sugar.

41.2.4 Syrup (liquid sugar) analysis

The syrups prepared in the plant should contain relatively low numbers of microorganisms that can cause spoilage. A sample of at least 100 mL of syrup should be taken into a sterile colored bottle under aseptic conditions. In cases where the syrup has undergone any processing or is mixed with other ingredients, the sample should be taken before and after each step.

Adjust liquid sugar °Brix to 51.37 by using sterile distilled water, which will contain 0.6367 g of sugar per mL. Prepare further dilutions using 0.1% peptone water if necessary. Follow procedures as given in Technique 42.2.2.1 for APC, and Technique 6.33 for yeasts and molds count. Spread plate 1 mL of diluted sample into Petri plates. Calculate the number of APC, and yeasts and molds by multiplying by the dilution rate and dividing by the inoculation amount and 0.6367. Record the results as the number of APC, and yeasts and molds cfu per g of sample.

Refer to Technique 33.2.3 for LAB count.

41.2.5 Flavoring and coloring ingredients analysis

Remove at least 50 mL of sample into a sterile Erlenmeyer flask and add 450 mL of 0.1% peptone water. Mix well by gently shaking. Spread plate 1 mL of sample to PC agar in duplicate for APC (Technique 2.2.4) and to ME agar in duplicate for yeasts and molds count (Technique 6.3.4). Incubate the inverted PC agar plates at 32°C for up to 3 days and the inverted ME agar plates at 25°C for up to 5 days. After incubation, analyze the colony formation on the agar media. Count the aerobic bacteria colonies from Petri plates containing 25 to 250 colonies. Count yeasts and molds colonies from plates containing 15 to 150 colonies. Calculate the number of APC, and yeasts and molds by multiplying the average count by the dilution rate and dividing by the inoculation amount. Record the results as the number of APC, and yeasts and molds cfu per g of sample.

41.2.6 Washed bottles analysis

Before being filled with soft drink, the washed bottles should be free from spoilage and pathogenic microorganisms. Take at least five bottles at random from the washer discharge end over a 1 min period. In more extensive production, take at least one bottle from each discharge during a 5 min period. If the bottles are not to be analyzed immediately, close with sterile stoppers, store at 4°C, and analyze within 4 h after collection.

For routine analysis of bottles. When bottles (250 mL) permit observation of the insides, pour 20 mL of melted PC agar (at 45°C) in duplicate for APC and ME agar, separately, into the bottles for the yeasts and molds count, and roll the bottles thoroughly to cover all portions of the inside. When cool, incubate the bottles at 32°C for up to 5 days. Preferably, sterile cotton plugs should be used to allow the exchange of gases. After incubation, analyze the colony formation on the agar media. Count the aerobic bacterial colonies from Petri plates containing 25 to 250 colonies. Count yeasts and molds colonies from plates containing 15 to 150 colonies. Calculate the number of APC, yeasts and molds by multiplying the average count by the dilution rate and dividing by the inoculation amount. Record the results as the number of APC, and yeasts and molds cfu per g of sample.

For normal analysis. Pour 10 mL of sterile distilled water into bottles (250 mL) in duplicate and swirl the water thoroughly over the inside surfaces. Spread plate 1 mL of the water sample into each PC agar for APC and ME agar for the yeasts and molds count. Incubate inverted PC agar plates at 32°C for up to 3 days for APC and the ME agar plates at 25°C for up to 5 days for the yeasts and molds count. After incubation, analyze the colony formation on agar media. Count the aerobic bacterial colonies from Petri plates containing 25 to 250 colonies. Count the yeasts and molds colonies from plates containing 15 to 150 colonies. Calculate the number of APC, yeasts and molds by multiplying the average count by the dilution rate and dividing by the inoculation amount. Record the results as the number of APC, and yeasts and molds cfu per g of sample.

Coliforms. Refer to Technique 13.3.2 for coliforms count.

41.2.7 Closure analysis

The inner surface of the closures should always be free from microorganisms. Select at least 10 crown closures randomly from each unopened manufacturer's package. Select closures with disinfected stainless-steel tongs and place in a container. Immerse tongs into alcohol and flame for disinfection.

Remove the crowns from the container with disinfected tongs. Flame very lightly on the backside of the crown, take care not to expose the inner surface to the flame, or wipe the back with 70% alcohol. Using disinfected tongs, place the crowns inner side up in sterile Petri dishes and pour agar with a pipette on the linear surface to fill or cover the open crown. Separately use PC agar for APC and ME agar for yeasts and molds count. Incubate Petri dishes containing closure together with PC agar at 32°C for up to 3 days for APC and ME agar at 25°C for up to 5 days for yeasts and molds count. After incubation, count the colonies on each closure. Calculate the numbers of colonies per closure. Report the results as the number of APC, and yeasts and molds cfu per closure.

Alternate technique 1. Place five crown closures into 20 mL of sterile distilled water in a bottle and agitate vigorously for at least 5 min. Spread plate 1 mL of water sample to each of PC agar for APC and ME agar for yeasts and molds count (Technique 2.2.4). Incubate the inverted PC agar plates at 32°C for up to 2 days for APC and ME agar plates at 25°C for up to 5 days for the yeasts and molds count. After incubation, analyze the colonies on the agar media. Count the aerobic bacterial colonies from Petri plates containing 25 to 250 colonies. Count yeasts and molds colonies from plates containing 15 to 150 colonies. Calculate the number of APC, and yeasts and molds by multiplying the average count by the dilution rate and dividing by the inoculation amount and the number of closures. Record the results as the number of APC, and the yeasts and molds cfu per closure.

Alternate technique 2. Wet a swab in a sterile PB solution, and swab the entire inside surface of a closure twice. Insert the swab in a tube of 10 mL sterile PB solution and break off the stick below the area touched by the fingers. Mix well. Use 1 mL of sample, perform APC (Technique 2.2.4), coliforms count (Technique 13.2.2), and yeasts and molds count (6.2.3.1) by spread plating (Technique 2.2.4). Calculate the number of respective microorganisms in cfu per closure.

41.2.8 Finished beverage analysis

The finished beverage should meet all requirements as drinking water. It should be free of potential spoilage and pathogenic microorganisms. Take at least two representative bottles. Open the bottle aseptically and flame the top lightly. Flat-top metal cans should be washed, the end wiped with alcohol, flamed, and pierced with a disinfected opener. Using a 10 mL pipette, withdraw 10 mL of beverage sample (if a smaller pipette is used, difficulty will be experienced with gassing of the beverage) and add into a sterile Erlenmeyer flask containing 90 mL of 0.1% peptone water. Prepare further dilutions using sterile 0.1% peptone water.

Spread plate 1 mL of sample (nondiluted and diluted) to each PC agar in duplicate for APC (Technique 2.2.4), and ME agar for yeasts and molds count (Technique 2.2.4). Incubate the inverted PC agar plates at 32°C for up to 2 days and ME agar plates at 25°C for up to 5 days. After incubation, analyze the colonies on agar media. Count the aerobic bacteria colonies from Petri plates containing 25 to 250 colonies. Count the yeasts and molds colonies from plates containing 15 to 150 colonies. Calculate the number of APC, and yeasts and molds by multiplying the average count by the dilution rate and dividing by the inoculation amount. Record the results as the number of APC, and yeasts and molds cfu per mL of beverage.

41.2.9 Membrane filtration

Refer to Technique 5.2 for membrane filtration (MF). MF is used when the culturing of microorganisms from soft drinks or rinse water supplies is necessary to detect low numbers of contaminant. Filter 100 mL of water or beverages through a membrane filter. Place the filter onto an agar plate. Use PC agar for APC and ME agar for yeasts and molds count. Incubate the inverted PC agar plates at 32°C for up to 3 days and ME agar plates at 25°C for up to 5 days. After incubation, analyze the colonies on the agar media. Count the aerobic bacterial colonies from Petri plates containing 25 to 250 colonies. Count yeast and mold colonies from plates containing 15 to 150 colonies. Calculate the number of APC, and yeasts and molds count by multiplying the average count by the dilution rate and dividing by the inoculation amount. Record the results as the number of APC, and yeasts and molds cfu per 100 mL of sample.

For counting of LAB, pathogenic, and spoilage microorganisms refer to the corresponding practices in this book.

41.3 Interpretation of results

Results from bottled soft drink and plants are useful in assessing either sanitation or spoilage risks. When general media are used and fresh (less than 1 h old) samples are tested, APC, and yeasts and molds count results can indicate sanitation efficiency. Sanitation cannot be assessed by testing old rinse water samples, because of the frequent growth of microorganisms, or by testing old beverage, because of the frequent death of microorganisms. The interpretation of results depends on the type of beverage and formulation. An excessive number of colonies (yeasts, molds, or bacteria) in the rolled bottles indicate the lack of proper cleaning. All solutions of flavoring, acidulants, colors, etc., should be relatively free of spoilage microorganisms. Table 41.1 indicates the interpretation of microbiological results for bottled soft drinks.

Table 41.1 Interpretation of results for bottled soft drink.

Sample	Count	Result	Interpretation
Beverage or rinse water	Spoilage yeast[a]	≥ 50 cfu[b] per 100 mL	Extreme risk; shut line down and sanitize until counts are zero.
		10 to 50 cfu per 100 mL	High risk; improve sanitation. Hold product 21 days and retest.
		1 to 10 cfu per 100 mL	Moderate risk; improve sanitation.
Finished beverage	LAB	<10 cfu per 100 mL	Low risk.
		>10 cfu per 100 mL	Moderate risk; improve sanitation.
Swab from sugar tank	Spoilage yeast	>10 cfu per 100 mL	High risk of contamination spread to line; sanitize tank.

(Continued)

Table 41.1 Interpretation of results for bottled soft drink. ***Continued***

Sample	Count	Result	Interpretation
Rinse water	APC[c]	≥200 cfu per 100 mL	High risk; improve sanitation. Hold product for 21 days and retest.
		50 to 200 cfu per 100 mL	Moderate risk; improve sanitation.
		<50 cfu per 100 mL	Low risk.
Beverages or rinse water	Yeasts	≥200 cfu per 100 mL	Extreme risk; shut line down and sanitize until counts are below 10 cfu per mL. Hold product for 21 days and retest.
		50 to 200 cfu per 100 mL	High risk; improve sanitation. Hold product for 21 days and retest.
		10 to 50 cfu per 100 mL	Moderate risk; improve sanitation.
		<10 cfu per 100 mL	Low risk.

[a]Zygosaccharomyces bailii, Zygosaccharomyces rouxii, Zygosaccharomyces bisporus, *and* Saccharomyces cerevisiae, *which are preservative resistant.*
[b]*Colony forming units.*
[c]*Aerobic plate count.*

PRACTICE

Analysis of bottled and process water

42

42.1 Introduction

The drinking of bottled water is increasing rapidly worldwide. The FDA defines bottled water as "water that is intended for human consumption and that is sealed in bottles or other containers with no added ingredients except that it may contain safe and suitable antimicrobial agents" and, within limitations, some added fluoride. There are four basic types of water offered to the consumer: (1) Spring or well water, taken directly from spring or well and bottled with minimum treatment. (2) Specifically prepared drinking water, where the mineral content has been adjusted and controlled. (3) Purified water removal by minerals to less than 10 mg per L. Distillation, ion exchange, or reverse osmosis can purify water. Preparation techniques must be indicated on the bottle. Only water prepared by distillation is called distilled water. (4) Fluorinated water, fluoride is added into water at the optimum concentration.

Ozone is typically used as a disinfectant for the disinfection of the bottles. Disinfection is applied just before bottling water. The predominant microorganisms in bottled water are in the genera *Pseudomonas*, *Flavobacterium*, and *Moraxella*. Coliforms, which are found naturally in uncontaminated soil, water, and the environment, are transmitted to water. Coliforms can be used as indicators of bottled water quality. Fecal coliforms indicate a possibility of sewage contamination with the associated possibility of pathogenic contaminant.

Legionella is a motile, Gram-negative, and oxidase and catalase positive bacterium. Human inhalation of large numbers of *Legionella* species results in *Legionella* infection and causes Legionnaires' disease. *Legionella pneumophila* disease can quickly become fatal without antibiotic treatment. The spread of bacteria takes place through the air-conditioning system, which is cooled by blowing air through water, after circulating a large amount of water. Within warm circulated water in the system, large quantities of *Legionella* multiply rapidly and spread via aerosols, through the air-cooling system. Large numbers of *Legionella* inhalation cause atypical pneumonia. Cooling systems should be disinfected with antimicrobial disinfectants at regular and frequent intervals in order to prevent *Legionella* proliferation in air-cooled water systems and to prevent the contamination of water systems of large amounts of *Legionella*.

Different microorganisms, especially saprophyte bacteria, may multiply as a result of storing water at around 22°C. Most microorganisms can be reduced as a result of water filtering and chlorination before distribution to the public. Bacterial contamination of water may be from peoples with disease or carriers of enteric pathogens, for example, *Salmonella enterica* subsp. *enterica* var. Typhi. The detection of pathogens in a water supply is difficult because the numbers of pathogens are often low and their incidence sporadic. From a public health point of view, coliforms and *Escherichia coli* tests are the most important for the indication of public health hazards. The quantitative microbiological tests are the most probable number (MPN) or membrane filter (MF)

Microbiological Analysis of Foods and Food Processing Environments. DOI: https://doi.org/10.1016/B978-0-323-91651-6.00020-3

techniques. *Clostridium perfringens* and *Enterococcus faecalis* can survive and grow for a long time in water sources. In the absence of coliforms in water, the presence of these bacteria can be considered to be an indicator of fecal contamination. *Legionella* can be isolated from water using its selective medium. Proper cleaning and sanitation of equipment can be indicated by yeasts and molds, and coliforms counts. Microbiological quality of water can be identified by coliforms, fecal coliforms, yeasts and molds counts.

42.2 Microbiological analysis techniques

Sampling, sample preparation, and all experimental techniques should be performed under aseptic conditions, and in duplicate, unless otherwise specified. All microbiological media to be used in applications must be sterilized. Again, all solutions and equipment that may cause the contamination of microorganisms must be sterilized or disinfected.

42.2.1 Equipment, materials, and media

Equipment. Centrifuge, centrifuge tube, container (containing sodium thiosulfate), container, glass spreader, incubator (at 25°C and 32°C), incubator, membrane filtration system, MF (0.22 μm pore size), microscope, microscope glass slide, Petri dishes, pipette (1 and 10 mL) in pipette box, pipette discarding pad (containing 10% bleach solution), spectrophotometer, tongs, and vortex.

Reagents. Alcohol, 70% alcohol, hypochlorite (5%), cetrimide, Gram staining reagents (carbol-fuchsin instead of safranin), phosphate buffer (PB) solution, 0.1% peptone water, and HCl-KCl buffer (mix 3.9 mL of 0.2 M HCl with 25 mL of 0.2 KCl, adjust pH to 2.2 using 1.0 M KOH).

Media. Double strength lauryl sulfate tryptose (DSLST) broth, Legionella CYE (LCYE) agar (with and without selective agent), LES endo agar, lauryl sulfate tryptose (LST) broth, malachite broth, membrane lauryl sulfate (MLS) agar, M-endo agar, mFC agar, m-HPC agar, P agar (Seller's differential agar), plate count (PC) agar, Pseudomonas cetrimide agar, RA-2 agar, tryptone glucose yeast extract (TGYE) agar, and tryptone soy (TS) agar containing 5% sheep blood agar.

42.2.2 Bottled water analysis

42.2.2.1 Aerobic plate count

Label PC agar plates with dilution rate, sample name, and your initials. Shake the water sample vigorously. Add 1 mL of water sample into 9 mL of sterile 0.1% peptone water in the test tube to prepare a 1:10 dilution by a fresh pipette. Transfer 1 mL of 1:10 dilution into a second 9 mL diluent with the second pipette. Prepare further serial dilutions in the same manner, if necessary. Transfer 1 mL of sample onto labeled PC agar and spread plate (Technique 2.2.4). When the agar absorbs water from the sample, incubate the inverted Petri plates at 32°C for up to 48 h for aerobic plate count (APC). After incubation, count the colonies on plates containing colonies between 25 and 250 colonies. Calculate the number of APC by multiplying the average count by the dilution rate and dividing by the inoculation amount. Report the results as the number of APC cfu per mL of water.

42.2.2.2 Counting of heterotrophic bacteria

Obtain a water sample, shake the water sample vigorously, and dilute using sterile 0.1% peptone water. Spread plate 1 mL of sample (nondiluted or diluted) to m-HPC agar (Technique 2.2.4). Spread over the agar medium and allow the medium to absorb the inoculum liquid. Incubate the inverted plates at 22°C for up to 7 days. After incubation, analyze the colonies on the agar medium. Count the colonies from Petri plates containing 25 to 250 colonies. Calculate the heterotrophic plate count (HPC) by multiplying the average count by the dilution rate and dividing by the inoculation amount. Record the result as the number of HPC cfu per mL of water.

42.2.2.3 Counting of coliforms and fecal coliforms

MPN. Refer to Technique 13.3.2 for the coliform count by MPN with the following exception. Inoculate 10 mL of water sample into 3-MPN tubes of 10 mL double strength LST broth, 1 mL into 3-MPN tubes of 10 mL LST broth, and 0.1 mL into 3-MPN tubes of 10 mL LST broth. Alternatively, inoculate 10 mL of water sample into 10 mL of each of 10 double strength LST broth. Incubate tubes at 35°C for up to 5 days. If growth is negative at 24 h, reincubate tubes for an additional 24 h and examine again for gas formation. After incubation, analyze the tubes for growth (turbidity), yellow color formation (acid production), and gas (and acid) formation in the Durham tube. Indicate the number of presumptive positive tubes for each inoculum and refer to MPN Table 4.1 for three tubes inoculation (or MPN Table 42.1). Calculate the number of coliforms MPN per mL of water.

Refer to Technique 13.2.3 for the fecal coliforms from coliforms presumptive test.

Membrane filter technique. Filter two 100 mL of water through separate MFs. Place one of the MFs onto MLS agar (or M-endo agar or LES endo agar) for coliforms and the other onto mFC agar for *E. coli*. Incubate the inverted MLS agar plates at 30°C for 4 h followed by incubation at 37°C for 14 h and incubate the inverted mFC agar plates at 41.5°C for 24 h. After incubation, analyze the colonies on the agar medium. Count the colonies from plates containing 15 to 150 colonies. Calculate the number of coliforms and *E. coli* by multiplying the average count by the

Table 42.1 Ten MPN tubes estimates for positive results per 10 mL of sample.

Positive tubes	MPN per mL
0	<1.1
1	1.1
2	2.2
3	3.6
4	5.1
5	6.9
6	9.2
7	12
8	16
9	23
10	>23

dilution rate and dividing by the inoculation amount. For the confirmation of coliforms and *E. coli*, refer to Technique 13.3.2.2. Record the results as the number of coliforms and *E. coli* cfu per mL of water.

For counting fecal coliforms, *Salmonella*, *Shigella*, *E. faecalis*, and *C. perfringens* from bottled water refer to the corresponding practices in this book.

42.2.3 Potable water (household water) analysis

A household water sample (500 mL) should be collected in a sterile container containing sodium thiosulfate. Water should be taken from the consumption points of the houses. Samples should be transported to the laboratory at room temperature within 4 h of collection. It must be analyzed within 24 h of reaching the laboratory. The turbidity of the water is measured by the spectrophotometric technique.

Perform a heterotrophic count as indicated in Technique 43.2.2.2, a *Pseudomonas* count as per Technique 43.2.5.2, and a *Legionella* count as per Technique 43.2.5.1. For counting fecal coliforms, *E. coli*, *E. faecalis*, *Salmonella*, *Shigella*, and *Aeromonas* from potable water refer to the corresponding practices in this book.

42.2.4 Containers and bottles analysis

Container. Refer to Technique 41.2.6 for the counting of microorganisms from containers or bottles. Make a heterotrophic count (Technique 42.2.2.2), APC (Technique 2.2.4), and coliforms count (Technique 13.3.2) per container or bottle.

Closure. Refer to Technique 42.2.7 for APC from closures. Use PB solution in the dilution of water. Perform a coliform count (for sample preparation refer to Technique 42.2.7 and count 13.3.2). Calculate the number of APC and coliforms cfu per closure.

42.2.5 Water supply or water distribution system analysis

42.2.5.1 Counting of Legionella

For the isolation of *Legionella* species from a water supply or water distribution system, take two sets of 10 mL of water sample from the water supply or water distribution system. Centrifuge the water sample at 25,000 rpm for 20 min under aseptic conditions. Resuspend the pellet from one of the centrifuged sets in 2 mL of 0.1% peptone water. Perform dilutions if necessary. (1) Spread plate 1 mL of suspension onto LCYE agar in duplicate with and without selective agents using a sterile glass spreader. (2) Add other pellet into 9 mL of HCl-KCl buffer into the remaining suspension; shake gently, and leave for 5 min. Spread plate this suspension onto other LCYE agar plates in duplicate. Incubate all inverted plates at 35°C and examine daily for up to 7 days. Count the white-green colonies on LCYE agar plates. Calculate the number of *Legionella* by multiplying the average count by two and the dilution rate, and dividing by the inoculation amount. Report the results as the number of *Legionella* cfu per 10 mL of water from the water supply.

Subculture presumptive *Legionella* colonies to TS blood agar and LCYE agar plates by streak plating (Technique 13.4.1). Isolates that grow on LCYE agar but fail to grow on TS blood agar and have characteristic morphology may be presumed to be *Legionella*. Perform a confirmation with biochemical and serological tests (Table 42.2).

Morphological characteristics of Legionella. L. pneumophila produces 1–2 mm colonies on LCYE agar are white, shiny, circular, smooth, raised with entire margin, buff-white or cream and slightly raised mucoid. It is a Gram-negative, short, pleomorphic rod with 1 μm diameter and 1–4 μm long (Fig. 42.1). Carbol-fuchsin should be used as a counter stain in the Gram staining of *Legionella* (Appendix B under the topic "Gram Staining Reagents").

Table 42.2 Differentiation of *Legionella* species.

Characteristics	*L. pneumophila*	*L. bozemanii*	*L. umoffii*	*L. micdadei*	*L. gormanii*	*L. longbeachae*	*L. jordanis*
Primary isolation on							
LCYE agar	+	+	+	+	+	+	+
TS blood agar	−	−	−	−	−	−	−
Colony color on medium	W/G	G	G	B/Gr	G	W/G	W/G
Gram reaction	−	−	−	−	−	−	−
Flagella	+	+	+	+	+	+	+
Oxidase	+	−	−	+	−	+	+
Catalase	+	+	+	+	+	+	+
β-Lactamase	+	+	+	−	+	+	+

+, positive; −, negative; W, white; G, green; B, blue; Gr, gray.

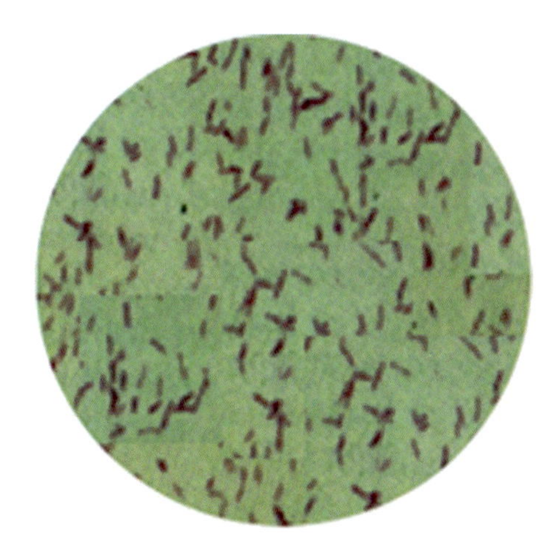

FIGURE 42.1

Staining appearance of *Legionella pneumophila*.

42.2.5.2 Counting of Pseudomonas aeruginosa

A *P. aeruginosa* count can be performed using the membrane filtration technique. Filter 250 mL of water through a MF. Isolation of *P. aeruginosa* is carried out by MF enrichment in malachite broth. The food sample should be incubated in malachite broth at 35°C for long enough to allow sublethally injured *Pseudomonas* to repair, but not to multiply (usually 2–5 h). After the incubation of the enrichment broth, carry out spread plate onto Pseudomonas cetrimide (PC) agar and incubate at 35°C for up to 48 h. After incubation, analyze the colony formation on agar medium. *P. aeruginosa* produces green color (Fig. 42.2A) and fluoresces under UV light. Presumptive identification by colonial morphology should be confirmed by pure culturing (Technique 13.4.2). Inoculate onto P agar (Seller's differential agar) for confirmation (incubate at 35°C for 48 h). *P. aeruginosa* produces straw-colored colonies with green pigmentation on P agar.

Characteristics of P. aeruginosa. Gram-negative, unicellular, aerobic rods 1.5 to 3 μm long (Fig. 42.2B), motile by single polar flagellum, nonfermenting, oxidative, and noncapsulated. Refer to Table 42.3 for other characteristics of *P. aeruginosa.*

For counting fecal coliforms, *E. coli*, *Enterococcus*, *Salmonella*, *Shigella*, *Aeromonas*, *Legionella*, *Mycobacterium*, and *Pseudomonas* from a water supply refer to the corresponding practices in this book.

42.3 Interpretation of results

An excess number of colonies (yeast, mold, or bacteria) from bottled water indicates the lack of proper cleaning and sanitation. Standards for bottled water for microbiological quality are based on coliforms detection levels. No more than one coliform is allowed per 100 mL of water when the water is analyzed with the MF technique; and no more than 10 coliforms are allowed per 100 mL of water when the water is analyzed with the MPN technique. Bottled water and public water require weekly coliform analyses in plant. Microbiological tests on containers and closures should be made. At least four containers and closures are selected just before filling for microbiological

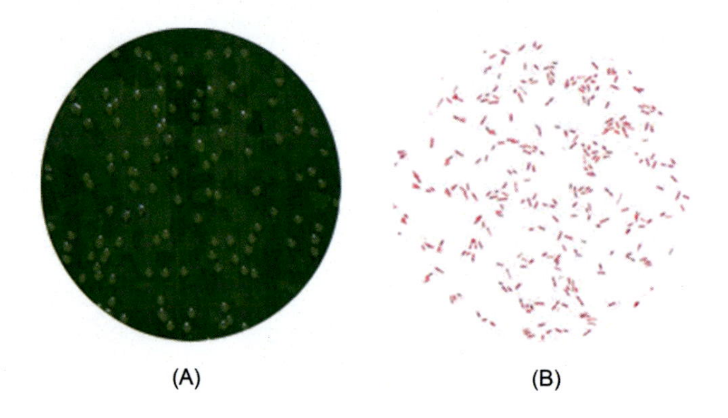

FIGURE 42.2

Colony appearance of *P. aeruginosa* on PC agar (A) and staining appearance of *P. aeruginosa* (B).

Table 42.3 Biochemical characteristics of *P. aeruginosa*.

Characteristic	Result
Catalase/citrate	+/+
Gas/H_2S	−/−
Gelatin hydrolysis	−
Indole/ urease	−/−
Nitrate reduction	+
OF (oxidative−fermentative)	Oxidative
Glucose, fructose, malonate, mannitol and oxidase.	+
Pigment	Pigmented (blue/green)
TSI agar reactions (slant/butt)	Red/Red
Mr/VP	−/
Enzyme reactions	
Acetate utilization	+
Arginine dehydrolase	+
Esculin hydrolysis	−
Lecithinase/lysine	−/−
Lipase	+
Ornithine decarboxylase	−
Phenylalanine deaminase	−

+, positive; −, negative.

analysis. All samples from bottles and closures should be free of coliforms. No more than one colony per milliliter capacity of the container or one colony per cm^2 of surface area of the closures are permitted.

Low APC (less than 100 cfu per mL) at the time of bottling of water serves as an indicator of Good Manufacturing Practice (GMP). The presence of coliforms serves as an indicator of a lack of GMP and potential health problems.

Legionella species are highly pathogenic bacteria if inhaled. Avoid creating aerosols and handle liquid cultures or suspensions of bacteria in a protective cabinet. Disinfect working surfaces with 5% hypochlorite.

The usual microorganisms that grow in bottled water are heterotrophic bacteria. The microbiological quality of water can be considered acceptable if *E. coli* and coliforms are not isolated from 100 mL of water sample. Since a HPC count >500 cfu per mL will inhibit coliforms and *E. coli* on lactose medium, HPC counts >500 cfu per mL can also be unacceptable for water. The microbiological limits for HPC in water are that no increase in the number of microorganisms above normal levels should be detected after incubating water at 22°C and 37°C for 12 h, and after incubating bottled water at 22°C for 72 h it should not contain more than 100 cfu per mL, and at 37°C for 48 h it should not contain more than 20 cfu per mL.

Gene primers

Table A10.1 PCR gene primers for bacterial species

Bacteria	Target gene	Primer sequence (5'-3')	
		Forward gene	**Reverse gene**
Bacteria (lactic acid bacteria)	*16S rRNA*	- AAGGAGGTGATCCAGCC-	AGAGTTTGATCCTGGCTCAG-
	LAB primers	-AGAGTTTGATCCTGGCTCAG	CTACGGCTACCTTGTTACGA-
ETEC	*elt*	-GAACAGGAGGTTTCTGCGTTAGGTG-	-CTTTCAATGGCTTTTTTTGGGAGTC-
	estla	-CCTCTTTTAGYCAGACARCTGAATCASTTG-	-CAGGCAGGATTACAACAAAGTTCACAG-
EPEC	*eae*	-TCAATGCAGTTCCGTTATCAGTT-	-GTAAAGTCCGTTACCCCAACCTG-
	bfp	-GACACCTCATTGCTGAAGTCG-	-CCAGAACACCTCCGTTATGC-
EAEC	*aa*	-CTGGCGAAAGACTGTATCAT-	-CAATGTATAGAAATCCGCTGTT-
	pic	-AGCCGTTTCCGCAGAAGCC-	-AAATGTCAGTGAACCGACGATTGG-
EHEC	stx_1	-CGATGTTACGGTTTGTTACTGTGACAGC-	-AATGCCACGCTTCCCAGAATTG-
	stx_2	-GTTTTGACCATCTTCGTCTGATTATTGAG-	-AGCGTAAGGCTTCTGCTGTGAC-
EIEC	*invE*	-CGATAGATGGCGAGAAATTATATCCCG-	-CGATCAAGAATCCCTAACAGAAGAATCAC-
	ipaH	-GAACTCAAATCTTGCACCATTCA-	-CGTCCGGAGAACAATTAAG-
DAEC	*daaD*	-TGAACGGGAGTATAAGGAAGATG-	-GTCCGCCATCACATCAAAA-
	alr	-CTGGAAGAGGCTAGCCTGGACGAG-	-AAAATCGCCACCGGTGGAGCGATC-
Enterococcus	*16S rRNA*[1]	-AGAGTTTGATCCTGGCTCAG-	-CGACTACCAGGGTATCT-
Enterococcus	*ent*	-TACTGACAACCATTCATGATG-	-AACTTCGTCACCAACGCGAAC-
Enterococcus faecalis	*faecalis2*	-ACACTTGGAAACAGGTGC-	-AGTTACTAACGTCCTTGTTC-
Enterococcus casseli	*casseli2*	-GGAAGAAAGTTGAAAGGC-	-TCGGTCAGACTTKCGTCC-
Enterococcus faecium	*faecium2*	-TGCTCCACCGGAAAAAGA-	-CACCAACTAGCTAATGCA-
Salmonella	*invA*	-GTGAAATTATCGCCACGTTCGGGCAA-	-TCATCGCACACGTCAAAGGAACC-
Salmonella	*sdiA*	-AATATCGCTTCGTACCAC-	-GTAGGTAAACGAGGAGCAG-
Salmonella	*IE-1*	-AGTGCCATACTTTTAATGAC-	-ACTATGTCGATACGGTGGG-
Salmonella	*Flic-C*	-CCCGCTTACAGGTGGACTAC-	-AGCGGGTTTTCGGTGGTTGT-
Salmonella	*16S rRNA*[1]	-CCTACGGGAGGCAGCAG-	-CCGTCAATTCCTTTRAGTTT-
Listeria monocytogenes	*hly*	-CATTAGTGGAAAGATGGAATG-	-GTATCCTCCAGAGTGATCGA-
L. monocytogenes	*iap*	-GGGCTTTATCCATAAAATA-	- TTGGAAGAACCTTGATTA-
L. monocytogenes	*hly A*	-CATGGCACCACCAGCATCT-	- ATCCGCGTGTTTCTTTTCGA-
Listeria	*prs*	-GCTGAAGAGATTGCGAAAGAAG-	-CAAAGAAACCTTGGATTTGCGG-
Listeria	*16S rRNA*	-CACGTGCTACAATGGATAG-	-AGAATAGTTTTATGGGATTAG-
Campylobacter jejuni	*mapA*	-CTATTTTATTTTTGAGTGCTTGTG-	-GCTTTATTTGCCATTTGTTTTATTA-
C. jejuni	*ceuE*	-AATTGAAAATTGCTCCAACTATG-	-TGATTTTATTATTTGTAGCAGCA-
Campylobacter	*cadF*	-TTGAAGGTAATTTAGATATG-	-CTAATACCTAAAGTTGAAAC-

(*Continued*)

Table A10.1 PCR gene primers for bacterial species *Continued*

Bacteria	Target gene	Primer sequence (5'-3')	
		Forward gene	**Reverse gene**
Yersinia enterocolitica	*inv*	CGGTACGGCTCAAGTTAATCTG	-CCGTTCTCCAATGTACGTATCC-
Y. enterocolitica	*virF*	GGCAGAACAGCAGTCAGACATA	-GGTGAGCATAGAGAATACGTCG-
Yersinia	*16S rRNA*	-AGAGTTTGATCCTGGCTCAG-	-GGTTACCTTGTTACGACTT-
Bacillus cereus	*hblA*	-GTGCAGATGTTGATGCCGAT-	-ATGCCACTGCGTGGACATAT-
B. cereus	*nheB*	-CTATCAGCACTTATGGCAG-	-ACTCCTAGCGGTGTTCC-
B. cereus	*bceT*	-CGTATCGGTCGTTCACACTCGG-	-GTTGATTTTCCGTAGCCTGGG-
Bacillus	*16S rRNA*	-AGAGTTTGATCCTGGCTCAG-	-AAGGAGGTGATCCAGCCGCA-
Clostridium perfringens	*plc*	-CCGTTGATAGCGCAGGACA-	-CCCAACTATGACTCATGCTAGCA-
C. perfringens	*cpe*	-GGTTCATTAATTGAAACTGGTG-	-AACGCCAATCATATAAATTACAGC-
Clostridium botulinum	*type A*	-GTGATACAACCAGATGGTAGTTATAG-	-AAAAAACAAGTCCCAATTATTAACTTT-
C. botulinum	*type B*	-GAGATGTTTGTGAATATTATGATCCAG -	-GTTCATGCATTAATATCAAGGCTGG-
C. botulinum	*type E*	-CCAGGCGGTTGTCAAGAATTTTAT-	-TCAAATAAATCAGGCTCTGCTCCC-
C. botulinum	*type F*	-GCTTCATTAAAGAACGGAAGCAGTGCT-	-GTGGCGCCTTTGTACCTTTTCTAGG-
Clostridium	*16S rRNA*	-AAAGGAAGATTAATACCGCATAA-	-ATCTTGCGACCGTACTCCCC-
Staphylococcus aureus	*seh*	-CAATCACATCATATGCGAAAGCAG-	-CATCTACCCAAACATTAGCACC-
S. aureus	*nuc*	-GCGATTGATGGTGATACGGTI-	-AGCCAAGCCTTGACGAACTAAAGC-
S. aureus	*sei*	-CTCAAGGTGATATTGGTGTAGG-	-AAAAAACTTACAGGCAGTCCATCTC-
Staphylococcus	*16S rRNA*	-GTAGGTGGCAAGCGTTACC-	-CGCACATCAGCGTCAG-
Vibrio cholerae	hlyA	-TGCGTTAAACACGAAGCGAT-	-TCAACCGATGCGATTGCCCAAGA-
Vibrio parahaemolyticus	toxR	-CCACTACCACTCTCATATGC-	-GGTCTAAATGGCTGACATC-
Vibrio vulnificus	*vvhA*	-CCGTTAACCGAACCACCCGCAA-	-TTCTTTATCTAGGCCCCAAACTTG-
Vibrio	*dnaJ*	-TTTTAYGAAGTDYTDGGYGT-	-GACAVGTWGGACAGGYYTGYTG-
Vibrio	*16S rRNA*	-AGAGTTTGATCMTGGCTCAG-	-TACGGYTACCTTGTTACGACTT-
Shigella dysenteriae	*rfpB*	-TCTCAATAATAGGGAACACAGC-	-CATAAATCACCAGCAAGGTT-
Shigella sonne	*wbgZ*	-TCTGAATATGCCCTCTACGCT-	-GACAGAGCCCGAAGAACCG-
Shigella flexneri	*rfc*	-TTTATGGCTTCTTTGTCGGC-	-CTGCGTGATCCGACCATG-
Shigella	*16S rDNA*	-AGACFGCTACGGGAGGCAGCAGT-	-GTTGCGCTCGTTGCGGGACTTAA-
Brucella spp.	*bcsp31*	-GCTCGGTTGCCAATATCAATGC-	-GGGTAAAGCGTCGCCAGAAG-
Brucella abortus	*alkB*	-GCGGCTTTTCTATCACGGTATTC-	-CATGCGCTATGATCTGGTTACG-
Brucella melitensis	*BMEI1162*	-AACAAGCGGCACCCCTAAAA-	-CATGCGCTATGATCTGGTTACG-
Aeromonas hydrophila	*aerA*	-CAAGGCTGATATCTCCTATCCCTATG-	-GCCACTCAGGGTCAGGTCAT-
A. hydrophila	*alt*	-TGCTGGGCCTGCGTCTGGCGG-	-AGGAACTCGTTGACGAAGCAGG-
A. hydrophila	*ast*	-GACTTCAATCGCTTCCTCAAC-	-GCATCGAAGTCACTGGTGAAGC-
Aeromonas	*AHCYTOEN*	-GAGAAGGTGACCACCAAGAACA-	-TGACATCGGCCTTGAACTC-
Aeromonas	*16S rRNA*	-GGCCTTGCGCGATTGTATAT-	-GTGGCGGATCATCTTCTCAGA-
Pleisiomonas shigelloides	*hugA*	-GCGAGCGGGAAGGGAAGAACC-	-GTCGCCCCAAACGCTAACTCATCA-
Pleisiomonas	*23S rRNA*	-CTCCGAATACCGTAGAGTGCTATCC-	-CTCCCCTAGCCCAATAACACCTAAA-
Alicyclobacillus acidoterrestris	16S rRNA	-AGAGTTTGATCCTGGCTCAG-	-GGYTACCTTGTTACGACTT-
Brochothrix thermosphacta	16S rRNA	-GAATTTGATCCTGGCTCAGGA-	-GGAGGTGATCCAGCCGC-

Table A10.1 PCR gene primers for bacterial species *Continued*

Bacteria	Target gene	Primer sequence (5'-3')	
		Forward gene	Reverse gene
Acetic acid bacteria	16S rRNA	-AGAGTTTGATCCTGGCTCAG-	-ACGGCTACCTTGTTACGACCT-
Fungi	18S rRNA	-CGAGCGTCATTWCACCAC-	-AATGAACGCTCGRACAGG-
Fungi	Actin (act1)	-GAGGCYCCCRTCAAC-	-GGCCAGCCAKRTCRAB-
Fungi	B-tubulin (TUBB5)	-GAGGGYGCYGARCTBRT-	-GGACCRSMNYKGAYR-
Fungi	Redesigned-Actin (act1)	-CCACCATHTWCCCHGGTATT-	-TTTCTYTCKGGAGGAGCRATR-
Fungi	Elongation Factor 1-α (ef1-α)	-GWGGTAAYGTGYGGTGACTC-	-CCWGGRTGGTTSAAGAYRATRA-
Byssochlamys	*idh*	-CGCCGATGCATATGGAAGGCGAGAC-	-CTGCGCTGCCTTGCAGGGCCC-

alt=heat labile enterotoxin; ast=heat stable enterotoxin.

APPENDIX

Media, stains, and reagents

B

A2.1 Alphabetical listing of culture media

The media are arranged in alphabetical order.

A2.1.1 Abeyta-Hunt-Bark agar

Heart infusion agar	40.0 g
Yeast extract	2.0 g

Dissolve ingredients in 950 mL of distilled water and adjust pH to 7.4 and sterilize at 121°C for 15 min. Cool to 50°C and add sodium cefoperazone (6.4 mL for broth or 4 mL for agar preparation), 4 mL rifampicin, 4 mL amphotericin B, and 50 mL lysed horse blood. After pouring plates, dry plates overnight on bench. If plates must be used the same day, place them in 42°C incubator for several hours. Do not dry in a hood with lids open. Even very brief surface drying will inhibit *Campylobacter* growth.

Sodium cefoperazone. Prepare with concentration of 32 mg L^{-1}.

Rifampicin. Dissolve 0.25 g slowly into 60–80 mL alcohol, swirling repeatedly. When powder is dissolved completely, bring to 100 mL with distilled water.

Amphotericin B. Dissolve 0.05 g in water in a 100 mL volumetric flask and bring to the line. Filter sterilize.

A2.1.2 Acetate differential agar (acetate agar) (a selective media for lactobacilli)

Sodium acetate	10.0 g
NaCl	5.0 g
Monoammonium phosphate	2.0 g
K_2HPO_4	1.0 mL
Bromothymol blue	0.08 g
Agar	15.0 g
Magnesium sulfate ($MgSO_4.7H_2O$)	0.1 g

Dissolve ingredients except agar in 1.0 L of distilled water. Heat to dissolve and adjust pH to 5.0. Add agar, heat to dissolve the agar. Sterilize at 121°C for 15 min. Cool to 45°C and pour into sterile Petri dishes.

A2.1.3 Agar solidifying agent

Dissolve 0.75 g agar in 100 mL of distilled water and heat to boiling to completely dissolve the agar. Sterilize at 121°C for 15 min.

A2.1.4 Alkaline peptone water

Dissolve ingredients in 1.0 L of distilled water, and adjust pH to 8.4–8.6. Dispense into tubes or bottles, and sterilize at 121°C for 10 min.

A2.1.5 Amos and Kent Jones Broth (use for rope spore count)

Peptone	10.0 g
Beef extract	5.4 g
NaCl	9.0 g
Bromocresol purple	0.04 g

Dissolve ingredients in 100 mL distilled water by heating. Adjust pH to 7.2. Dispense 5.0 mL into tubes and sterilize at 121°C for 15 min.

A2.1.6 Anaerobic agar

A2.1.6.1 Base medium

Trypticase soy agar	40.0 g
Yeast extract	5.0 g
Agar	5.0 g
L-Cysteine (dissolved in 5 mL 1 N NaOH)	0.4 g

Dissolve ingredients except agar in 1.0 L of distilled water by steaming. Adjust pH to 7.5. Add agar, bring to the boil and continue to heat until melting the agar. Sterilize at 121°C for 15 min. Cool to about 50°C and add following ingredients to prepare final medium.

Hemin solution. Dissolvel g hemin in 100 mL distilled water. Sterilize at 121°C for 15 min. Refrigerate at 4°C.

Vitamin K_1 solution. Dissolve 1 g vitamin K_1 in 100 mL 95% ethanol. Solution may require 2–3 days with intermittent shaking to dissolve. Refrigerate at 4°C.

Final medium. To 1 L of sterilized base medium, add 0.5 mL hemin solution and 1 mL vitamin K_1 solution. Mix and pour 20 mL portions into 15 × 100 mm petri dishes. Medium must be reduced before inoculation by 24 h anaerobic incubation in anaerobic glove box or GasPak jar.

To prepare anaerobic blood agar: Add 5% of defibrinated horse blood to cooled anaerobic agar (at 50°C). Mix and pour 20 mL into sterile Petri dishes.

A.2.1.7 Anaerobic egg yolk agar (use for anaerobic microorganisms)

Fresh eggs	3.0 g
Yeast extract	5.0 g
Tryptone	5.0 g
Proteose peptone	20.0 g
NaCl	5.0 g
Agar	20.0 g

Wash eggs with a stiff brush and drain. Soak in 70% alcohol for 15 min. Remove eggs and allow to air-dry. Crack eggs aseptically, separate egg yolk, and discard the whites. Add the yolks to equal volume of sterile saline and mix thoroughly.

Combine the remaining ingredients, dissolve by heating, and adjust pH to 7.0. Sterilize at 121°C for 15 min. Cool to 50°C, and add 80 mL of the egg yolk emulsion. Mix thoroughly and pour into sterile Petri dishes immediately.

A2.1.8 APT agar/broth (use for heterofermentative lactobacilli)

Trypton	12.5 g
Yeast extract	7.5 g
K_2HPO_4	5.0 g
NaCl	5.0 g
Sodium citrate	5.0 g
Sodium carbonate	1.25 g
Glucose	10.0 g
Tweeen 80	0.2 g
Magnesium sulfite	0.8 g
Magnesium chloride	0.14 g
Ferrous sulfate	0.04 g
Bromocresol purple	0.02 g
Agar	15.0 g

Dissolve ingredients except agar in 1.0 L of distilled water, adjust pH to 6.7, and heat to boiling to completely dissolve the agar. Sterilize at 121°C for 12 min. Cool to 45°C and pour into sterile Petri dishes.

A2.1.9 Arginine-glucose agar

Peptone	5.0 g
Yeast extract	3.0 g
Tryptone	10.0 g
NaCl	20.0 g
Glucose	1.0 g
L-Arginine (hydrochloride)	5.0 g
Ferric ammonium citrate	0.5 g
Sodium thiosulfate	0.3 g
Bromocresol purple	0.02 g
Agar	13.5 g

Dissolve ingredients except agar in 1.0 L of distilled water by steaming. Adjust pH to 6.9. Add agar, bring to the boil and continue to heat until the agar dissolves. Dispense into test tubes (use 5 mL). Sterilize at 121°C for 12 min.

A2.1.10 Aspergillus flavus-parasiticus agar (a selective medium for *Aspergillus flavus* and *Aspergillus parasiticus*)

Yeast extract	20.0 g
Peptone	10.0 g
Ferric ammonium citrate	0.5 g
Chloramphenicol	0.1 g
Agar	15.0 g
Dichloran (2,6-dichloro-4-nitroanaline) solution (0.2% in ethanol)	1.0 mL

Dissolve ingredients except agar in 1.0 L of distilled water by steaming. Adjust pH to 6.0–6.5. Add agar, bring to the boil and continue to heat until melting the agar. Sterilize at 121°C for 15 min. Cool to about 45°C and pour into sterile Petri dishes.

A2.1.11 *Bacillus cereus* agar (for *Bacillus cereus*)

Yeast extract	4.0 g
Peptone	10.0 g
N_2HPO_2	2.52 g
PH_2PO_2	0.28

Sodyum pyruvate	10.1 g
Chromogenic mix	1.2 g
Agar	13.0 g

Dissolve ingredients except agar in 1.0 L of distilled water, adjust pH to 7.2, and heat to boiling to completely dissolve the agar. Sterilize at 121°C for 12 min. Cool to 50°C and add sterile selective supplement into medium, mix thoroughly and pour into sterile Petri dishes.

Selective supplement: Polymyxin B 106000 IU and Trimethoprim 10.0 g. Add ingredients into 1.0 L of distilled water. Mix well and bring to the boil to dissolve completely. Sterilize by autoclaving at 121°C for 15 mins.

A.2.1.12 Baird–Parker agar (a selective an differential medium for *Staphylococcus aureus*)

A2.1.12.1 Base

Tryptone	10.0 g
Meat extract	5.0 g
Yeast extract	1.0 g
Glycine	12.0 g
Sodium pyruvate	10.0 g
Lithium chloride	5.0 g
Agar	20.0 g

Dissolve ingredients except agar in 1.0 L of distilled water by steaming. Adjust pH to 7.0. Add agar, bring to the boil and continue to heat until the agar is dissolved. Dispense 90.0 mL portions in screw-capped bottles, and sterilize at 121°C for 15 min.

A2.1.12.2 Rabbit plasma fibrinogen solution

Mix 50.0 mL of rehydrated rabbit plasma-EDTA with 50.0 mL of 15% bovine fibrinogen. Add 20 mg of salt-free soybean trypsin inhibitor, and dissolve by mixing. Filter sterilize the mixture using sterile 1-μm millipore filters.

A2.1.12.3 Preparation of complete medium

Prepare filter sterilized solutions of 1% potassium tellurite and 20% sodium pyruvate.

Mix 10.0 mL of prewarmed rabbit plasma-fibrinogen (45°C to 50°C), 90.0 mL of melted and cooled (45°C–50°C) base (Baird–Parker) medium, and 1.0 mL of 1% potassium telluride. Avoid entrapping air bubbles. Pour into sterile Petri dishes. After solidification store plates in plastic bags at 4°C. Just prior to use, spread 0.5 mL of sterile 20% sodium pyruvate solution on each agar plate, and allow to dry at 50°C or 35°C (2 or 4 h, respectively) with agar surface upwards.

A2.1.13 Bacteriological analytic medium (for *Alicyclobacillus acidoterrestris*)

A2.1.13.1 Base medium

$CaCl_2 \cdot 7H_2O$	0.25 g
$MgSO_4 \cdot 7H2O$	0.50 g
$(NH_4)_2SO_4$	0.20 g
Yeast extract	2.00 g
Glucose	5.00 g
KH_2PO_4	3.00 g

Dissolve ingredients in 500 mL distilled water and adjust pH to 4.0 with 1 N H_2SO_4. Sterilize at 121°C for 10 min.

A2.1.13.2 Trace elements

$CaCl_2 \cdot H_2O$	0.66 mg
$ZnSO_4 \cdot 7H_2O$	0.18 mg
$CuSO_4 \cdot 5H_2O$	0.16 mg
$MnSO_4 \cdot H_2O$	0.15 mg
$CoCl_2 \cdot 6H_2O$	0.18 mg
H_3BO_3	0.10 mg
$Na_2MoO_4 \cdot 2H_2O$	0.30 mg

Dissolve ingredients except agar in 1.0 L distilled water. The solution was sterilized at 121°C for 15 min and kept in the refrigerator upon use.

A2.1.13.3 Agar

Add agar into 500 mL distilled water and bring to the boil and continue to heat until agar is dissolved. Sterilize at 121°C for 15 min.

After sterilization of the solution, cool to 47°C. Add 1 mL of trace element and pour 500 mL of agar into 500 mL base medium in the flask. Gently shake the flask to mix the solutions completely.

To prepare **BAM broth**; omit agar.

A2.1.15 Bile esculin agar (A selective and differential media for group D-streptococci)

Peptone	8.0 g
Oxgall (bile salts)	20.0 g
Esculin	1.0 g
Ferric citrate	0.5 g
Agar	15.0 g

Dissolve ingredients except agar in 1.0 L of distilled water by steaming. Adjust pH to 7.1. Add agar bring to the boil and continue to heat until agar is dissolved. Dispense into test tubes and sterilize at 121°C for 15 min.

A2.1.15 Bile peptone broth

Na_2HPO_4	8.23 g
$NaH_2PO_4 \cdot H_2O$	1.2 g
Bile salts No.3	1.3 g
NaCl	5.0 g
Sorbitol	10.0 g
Peptone	5.0 g

Dissolve ingredients in distilled water by steaming. Adjust pH to 7.6. Dispense 100 mL into test tubes. Sterilize at 121°C for 15 min. Cool to 45°C and pour into sterile Petri dishes.

A2.1.16 Bismuth sulfite agar (A selective and differential medium for *Salmonella*)

Meat extract	5.0 g
Peptone	10.0 g
Glucose	5.0 g
Na_2HPO_4	4.0 g
Ferrous sulfate	0.3 g
Bismuth sulfide indicator	8.0 g
Brilliant green	0.025 g
Agar	20.0 g

Dissolve ingredients except agar in distilled water by steaming. Adjust pH to 7.7. Add agar, bring to the boil and continue to heat until the agar is dissolved. Avoid overheating and do not sterilize. Cool to 45°C and pour into sterile Petri dishes.

A2.1.17 Blood agar

Tryptone	15.0 g
Phytone or soytone	5.0 g
NaCl	5.0 g
Agar	15.0 g

Dissolve ingredients except agar in 1.0 L of distilled water by steaming. Adjust pH to 7.3. Add agar, bring to the boil and continue to heat until the agar dissolves. Sterilize at 121°C for 15 min.

Cool to 50°C. Add 5% of defibrinated sheep blood to melted agar (at 50°C). Mix and pour 20 mL into sterile Petri dishes.

For *Vibrio hollisae*, add NaCl to a final concentration of 1%.

A2.1.18 Bolton selective enrichment broth (campylobacter enrichment broth)

Peptone	10.0 g
Lactalbumin hydrolysate	5.0 g
Yeast Extract	5.0 g
Sodium chloride	5.0 g
a-Ketoglutaric acid	1.0 g
Sodium pyruvate	0.5 g
Sodium metabisulfite	0.5 g
Sodium carbonate	0.6 g
Hemin	0.01 g

Dissolve ingredients and adjust pH to 7.4, sterilize at 121°C for 15 min and cool to 45°C–50°C. Aseptically add 25 mL defibrinated horse blood and the content of one vial of Bolton selective supplement. Mix well and distribute the broth into sterile screw top containers. After the addition of the sample the space between screw top and broth should be approximately 2 cm.

Microaerophilic incubation is not required with this medium. Bolton selective supplement inhibits Gram-positive and Gram-negative bacteria, yeasts and molds.

A2.1.19 Bolton elective supplement

Cefoperazone	20.0 g
Vancomycin	20.0 g
Trimethoprim	20.0 g
Cycloheximide	50.0 g

Add these antibiotics into 2 mL of distilled water and mix well.

A2.1.20 Brain heart infusion agar/broth (a highly nutritious medium for the cultivation of a wide variety of microorganisms)

Calf brains, infusion from	200.0 g
Beef hearts, infusion from	250.0 g
Proteose peptone	10.0 g
Glucose	2.0 g
NaCl	5.0 g

Na_2HPO_4	2.5 g
Agar	15.0 g

Dissolve ingredients except agar in 1.0 L of distilled water by steaming. Adjust pH to 7.4. Add agar, bring to the boil and continue to heat until the agar is dissolved. Sterilize at 121°C for 15 min. Cool to about 45°C and pour into sterile Petri dishes.

To prepare **BHI broth**, omit the agar.

To prepare **BHI blood agar**, cool sterilized BHI agar to about 45°C, add aseptically 5% of sterile sheep blood, and pour into Petri dishes in a thick layer.

A2.1.21 Brilliant green agar/broth (a selective and differential medium for *Salmonella*, other than *Salmonella typhi*)

Proteose peptone	10.0 g
Yeast extract	3.0 g
Lactose	10.0 g
Saccharose	10.0 g
NaCl	5.0 g
Brilliant green	0.0125 g
Phenol red	0.08 g
Agar	20.0 g

Dissolve ingredients except agar in 1.0 L of distilled water by steaming. Adjust pH to 7.4. Add agar, bring to the boil and continue to heat until the agar is dissolved. Sterilize at 121°C for 15 min. Cool to about 45°C and pour into sterile Petri dishes.

To prepare **brilliant green broth**, omit the agar.

A2.1.22 Brilliant green lactose bile broth (selective medium for coliforms and *Escherichia coli*)

Peptone	10.0 g
Lactose	10.0 g
Ox-bile	20.0 g
Brilliant green (1% aq. solution)	13.3 mL

Dissolve peptone and lactose in 500 mL of distilled water. Dissolve ox-bile in 200 mL of distilled water. Mix two solutions, complete to 987 mL using distilled water, and adjust pH to 7.4. Add 13.3 mL of brilliant green. Dispense into test tubes containing inverted Durham tubes, plug and sterilize at 121°C for 15 min.

A2.1.23 Bromocresol green ethanol agar

Ethanol	20.0 g
Yeast extract	3.0 g
Bromocresol green	0.1 g
Agar	15.0 g

Dissolve ingredients except agar in 1.0 L of distilled water. Heat to dissolve and adjust pH to 6.5. Add agar, heat to dissolve the agar. Sterilize at 121°C for 15 min. Cool to 45°C and pour into sterile Petri dishes.

A2.1.24 Brucella agar/broth (a medium for *brucella*)

Tryptone	10.0 g
Peptamin	10.0 g
Glucose	1.0 g
Yeast extract	2.0 g
NaCl	5.0 g
Sodium bisulfite	0.1 g
Sodium citrate	1.5 g
Agar	15.0 g

A2.1.25 Brucella selective supplement

Polymyxin B	2500.0 IU
Bacitracin	12,500.0 IU
Cycloheximide (or *Natamycin)	50.0 mg (25 mg)
Nalidixic acid	2.5 mg
Nystatin	50,000.0 IU
Vancomycin	10 mg

Dissolve ingredients except agar in 1.0 L of distilled water by steaming. Adjust pH to 7.0. Add agar, bring to the boil and continue to heat until the agar is dissolved. Sterilize at 121°C for 15 min. Cool to 50°C, add sterile Brucella selective supplement into medium, and mix thoroughly. Add 5%–10% inactivated horse serum, mix well, and pour into sterile Petri dishes.

To prepare **Brucella broth**, omit the agar and add 1%–2% sodium citrate during dissolving ingredients.

To prepare **Brucella blood agar**, add 10% blood into cooled Brucella agar (at 50°C).

*for modified supplement.

A2.1.26 Biphasic brucella blood media (for the isolation of *brucella*)

Use ingredients from Brucella Agar/Broth. Dissolve ingredients except agar in 1.0 L of distilled water by heating to 45°C. Adjust pH to 7.0. Add agar (increased agar at a final concentration of 2.5%), bring to the boil and continue to heat until the agar is dissolved. Dispense 30 mL of melted agar into screw capped square bottle (125 mL) and loosely close cap. Sterilize at 121°C for 15 min. Cool to 50°C, add sterile Brucella selective supplement into medium, mix thoroughly, add 5%–10% defibrinated horse blood, and mix thoroughly. The bottle is tilted to allow the agar to solidify on a surface. Aseptically, add at least 50 mL of Brucella broth into bottle to prepare biphasic medium.

Prepare **Brucella broth** separately (with supplements antibiotic, blood) with 40 mL amount in 150 mL Erlenmeyer flask. Add 1%–2% sodium citrate in the liquid phase. After sterilization cool the broth and add into Brucella agar in the bottle to prepare biphasic medium.

A2.1.27 Brucella-FBP agar/broth (a medium for *Brucella* spp.)

Brucella broth	28.0 g
$FeSO_4$	0.25 g
Sodium metabisulfite (anhydrous)	0.25 g
Sodium pyruvate (anhydrous)	0.25 g

Suspend Brucella broth in 970.0 mL of distilled water, and boil for approximately 1 min to dissolve completely. Sterilize at 121°C for 15 min, and cool to 50°C. Prepare the FBP solution by dissolving $FeSO_4$, sodium metabisulfite, and sodium pyruvate together in 30 mL of distilled water, and sterilize by filtering through a 0.22 μm filter. Add the filter-sterilized FBP solution (30 mL) to the cooled medium, and distribute into tubes.

To prepare **Brucella-FBP agar** and **semisolid medium** in the same manner, but add 15 g or 1.6 g of agar, respectively, to the Brucella broth before sterilization.

A2.1.28 Buffered motility-nitrate medium (for *Clostridium perfringens*)

Beef extract	3.0 g
Peptone	5.0 g
KNO	1.0 g
Na HPO	2.5 g
Galactose	5.0 g
Agar	3.0 g
Glycerin (reagent grade)	5.0 mL

Dissolve ingredients except agar in 1.0 L of distilled water. Adjust pH to 7.3. Add agar and boil to dissolve agar completely. Dispense 11 mL portions into tubes. Sterilize at 121°C for 15 min. If not used within 4 h, heat 10 min in boiling water or flowing steam. Chill in cold water.

A2.1.29 Buffered listeria enrichment broth (a selective medium for *Listeria* spp.)

Tryptic (trypticase) soy broth	30.0 g
Yeast extract	6.0 g
Monopotassium phosphate (anhydrous)	1.35 g
Disodium phosphate (anhydrous)	9.6 g
Sodium pyruvate (sodium salt)	1.11 g

Dissolve ingredients in 1.0 L distilled water and adjust pH to 7.3. Sterilize at 121°C for 15 min. Cool medium to 47°C and aseptically add the three selective supplements to medium after 4 h incubation at 30°C.

Not: Optionally a filter-sterilized 10% (wt./vol.) sodium pyruvate solution may be added after autoclaving (11.1 mL L^{-1}).

A2.1.30 Selective supplements

Acriflavin HCl	10 mg L^{-1}
Nalidixic acid (sodium salt)	40 mg L^{-1}
Cycloheximide	50 mg L^{-1}

Prepare acriflavin and nalidixic acid supplements as 0.5% (wt./vol.) stock solutions in distilled water. Prepare cycloheximide supplement as 1.0% (wt./vol.) stock solution in 40% (vol./vol.) solution of ethanol in distilled water. Filter-sterilize stock solutions and store at 4°C, protect acriflavin from light.

A2.1.31 Buffered peptone broth (buffered peptone water)

Peptone	10.0 g
NaCl	5.0 g
Na_2HPO_4	3.5 g
KH_2PO_4	1.5 g

Dissolve ingredients in 1.0 L of distilled water and adjust pH to 7.2. Sterilize at 121°C for 15 min.

A2.1.31 Campy-BAP agar/broth (a selective medium for *Campylobacter* spp.)

Brucella agar	43.0 g
Lysed horse blood	50.0 mL
Vancomycin	10.0 mg
Polymyxin B	2500.0 IU

Trimethoprim lactate	5.0 mg
Amphotericin B	2.0 mg
Cephalothin	15.0 g

Dissolve Brucella agar in 1.0 L of distilled water by steaming. Adjust pH to 7.0. Add agar, bring to the boil and continue to heat until the agar is dissolved. Sterilize at 121°C for 15 min. Cool the medium to 50°C, and add the blood and filter sterilized solutions of antibiotics. Prepare solutions of each antimicrobial compound for 1.0 L of medium in 5 mL of *Brucella* broth. Pour into sterile Petri dishes.

A2.1.32 Campylobacter charcoal differential agar (preston blood-free medium) (a selective medium for *Campylobacter* spp.)

Nutrient broth	25.0 g
Charcoal	4.0 g
Casein hydrolysate	3.0 g
Sodium deoxycholate	1.0 g
Ferrous sulfate	0.25 g
Sodium pyruvate	0.25 g
Agar	12.0 g
Sodium cefoperazone	32.0 g

Dissolve ingredients except agar and cefoperazone in 1.0 L of distilled water. Adjust pH to 7.4. Add agar and boil to dissolve agar completely. Sterilize at 121°C for 15 min. Cool to 50°C, and add filter-sterilized (through a 0.22 μm filter) cefoperazone (16 mg/500 mL medium). Mix well, and pour into sterile Petri dishes.

A2.1.33 Campylobacter enrichment broth

Meat peptone	10.0 g
Lactalbumin hydrolysate	5.0 g
Yeast extract	5.0 g
NaCl	5.0 g
Hemin	0.01 g
Sodium pyruvate	0.5 g
α-Ketoglutaric acid	1.0 g
Sodium metabisulfite	0.5 g
Sodium carbonate	0.6 g

Dissolve ingredients except agar and cefoperazone in 1.0 L of distilled water. Prepare the broth base in bottles. Once the powder is dissolved, adjust to pH 7.4 and autoclave 15 min at 121°C. Tighten the caps after the broth has cooled. Before use, add 50 mL lysed horse blood and two rehydrated (5 mL per vial 50:50 sterile filtered $H_2$0-ethanol solution) vials of Campylobacter enrichment broth (Bolton formula) supplement.

A2.1.34 Campylobacter modified agar/broth

Campylobacter enrichment broth	27.6 g
Agar	15.0 g
Neutral red	2.5 mL (0.2%)
Antibiotic supplement	1.8 g

Use Campylobacter enrichment broth without (Bolton enrichment broth without blood and supplements)

Mix ingredients in 1 L distilled water (except neutral red), then boil. After cooling, divide into four 250 mL portions. Add or not add 2.5 mL neutral red 250 mL medium.

Campy semisolid broth–nitrate: Add 2.5 g potassium nitrate into 250 mL broth (1%, wt./vol.) without neutral red.

Campy semisolid broth–glycine: Add 2.5 g glycine into 250 mL broth (1%, wt./vol.) with neutral red.

Campy semisolid broth–NaCl: Add 7.5 g NaCl into 250 mL broth (wt./vol.) with neutral red.

Campy semisolid broth–cysteine: Add 0.05 g Cysteine-HCl into 250 mL broth (0.02%, wt./vol.) with neutral red.

Adjust the pH of each portion to 7.4. Dispense 10 mL per 16 × 125 mm screw-cap tube. Sterilize at 121°C for 15 min.

Campy modified agar. Mix ingredients in 1 L distilled water (without neutral red). Adjust the pH of each portion to 7.4, then boil. Add agar and boil to dissolve agar completely. Sterilize at 121°C for 15 min. Cool to 50°C and pour into sterile Petri dishes.

A2.1.35 Carbohydrate fermentation broth (refer to oxidation-fermentation broth)

A2.1.35.1 Cary-blair transport medium

Sodium thioglycolate	1.5 g
Na_2HPO_4	1.1 g
NaCl	5.0 g
Agar	5.0 g
$CaCl_2$ (1% solution)	9.0 mL

Heat with agitation to dissolve dry ingredients in 991 mL distilled water. Cool to 50°C and add $CaCl_2$ solution. Adjust pH to 8.4. Dispense 7 mL portions into 9 mL screw-cap tubes and sterilize at 121°C for 15 min. Cool and tighten caps.

A2.1.36 Cefsulodin-irgasan-novabiocin Agar (a differential and selective medium for *Y. enterocolitica*)

Peptone	17.0 g
Yeast extract	2.0 g
Mannitol	3.0 g
Sodium pyruvate	2.0 g
NaCl	1.0 g
Sodium cholate	0.5 g
$MgSO_4.7H_2O$	10.0 mg
Sodium deoxycholate	0.5 g
Neutral red	30.0 mg
Crystal violet	1.0 mg
Agar	15.0 g

Dissolve ingredients except agar in 1.0 L of distilled water. Adjust pH to 7.0. Add agar, heat to boil and continue to heat until the agar dissolves. Sterilize at 121°C for 15 min. Cool to 50°C and add the following filter-sterilized (0.22 μm pore size) antibiotics dissolved in 2 mL of distilled water and 1 mL of ethanol.

A2.1.36.1 Antibiotic solution

Cefsulodin	7.5 mg
Novabiocin	1.25 mg
Irgasan	2.0 mg

A2.1.37 Cellobiose polymyxin colistin agar (a selective medium for *Vibrio* spp.)

A2.1.37.1 Base medium

Peptone	10.0 g
Beef extract	5.0 g
NaCl	20.0 g
Bromothymol blue	0.04 g
Cresol red	0.04 g
Agar	15.0 g

Dissolve ingredients except agar in 900 mL of distilled water by steaming. Adjust pH to 7.6. Add agar, bring to the boil and continue to heat until the agar dissolves. Sterilize at 121°C for 15 min.

A2.1.37.2 Dye solution

Bromothymol blue	4.0 g
Cresol red	4.0 g
Ethanol, 95%	100 mL

For consistent medium color, use dye solution rather than repeatedly weighing out dry dyes. Dissolve dyes in ethanol for 4% (wt./vol.) stock solution. Using 1 mL of this solution per liter of mCPC agar gives 40 mg bromothymol blue and 40 mg cresol red per liter.

A2.1.37.3 Antibiotic supplement

Cellobiose	10.0 g
Colistin	400,000 IU
Polymyxin	100,000 IU

Dissolve cellobiose in distilled water by heating gently. Cool.

Add dye solution, antibiotic supplement into bases medium at 45°C and mix gently and pour into sterile Petri dishes.

A2.1.38 **Cheese agar (a selective medium for *Brevibacterium linens*)**

Ripened cheese	100.0 g
Potassium citrate	10.0 g
Peptone	10.0 g
NaCl	50.0 g
Sodium oxalate	2.0 g
Agar	15.0 g

Dissolve potassium citrate in 300 mL distilled water and add the cheese. Warm the suspension to 50°C, then transfer to a funnel with rubber tubing and spring clip and allow standing for 30 min to separate fat. Run off the aqueous portion and add to the remainder of the ingredients dissolved in 700 mL of distilled water. Adjust pH to 7.4 and sterilize at 121°C for 25 min. Cool to about 45°C and pour into sterile Petri dishes.

A2.1.39 **CHROMagar listeria**

Peptone	23.0 g
Yeast extract	5.0 g
NaCl	5.0 g
Chromogenic selective mix	0.8 g
Selective supplement	9.0 g
Growth factors	1.7 g
Agar	15.0 g

Dissolve ingredients except agar in 1.0 L of distilled water by steaming. Adjust pH to 7.4 and sterilize at 121°C for 25 min. Cool in water bath to about 47°C, add chromogenic selective supplement and pour into sterile Petri dishes.

A2.1.40 Chopped liver broth/agar

Ground beef liver	500.0 g
Soluble starch	1.0 g
Peptone	10.0 g
K_2HPO_4	1.0 g

Add ground beef liver into 1.0 L of distilled water, and boil for 1 h. Adjust pH to 7.0, and boil for another 10 min. Press through cheesecloth, and bring the broth to 1.0 L with distilled water. Add the peptone and K_2HPO_4, and adjust pH to 7.0. Place liver particles from the pressed cake in the bottom of culture tubes (about 1 cm deep), cover with 8 to 10 mL of broth, sterilize at 121°C for 20 min. Before use, exhaust the medium for 20 min in flowing steam.

CL agar. Dissolve ingredients except agar and in 1.0 L of distilled water by heating. Adjust pH to 7.3. Add agar (1.5%), bring to the boil and continue to heat until the agar dissolves. Sterilize at 121°C for 15 min. Cool to 50°C and pour into sterile Petri dishes.

A2.1.41 Christensen citrate agar

Sodium citrate	3.0 g
Glucose	0.2 g
Yeast extract	0.5 g
Cysteine monohydrochloride	0.1 g
Ferric ammonium citrate	0.4 g
$KH_2 PO_4$	1.0 g
NaCl	5.0 g
Sodium thiosulfate	0.08 g
Phenol red	0.012 g
Agar	15.0 g

Dissolve ingredients except agar in 1.0 L of distilled water by steaming. Adjust pH to 6.9. Add agar, bring to the boil and continue to heat until the agar dissolves. Fill tubes 1/3 full and cap or plug to maintain aerobic conditions and sterilize at 121°C for 15 min. Before media solidify, incline tubes to obtain 4–5 cm slant and 2–3 cm butt.

A2.1.42 Chromocult coliform agar (counting of coliforms on solid medium)

Peptones	3.0 g
Sodium chloride	5.0 g
Sodium dihydrogen phosphate	2.2 g
Disodium hydrogen phosphate	2.7 g
Sodium pyruvate	1.0 g

Tryptophane	1.0 g
Agar	10.0 g
Sorbitol	1.0 g
Tergitol 7	0.15 g
Chromogenic mixture	0.4 g

Dissolve ingredients in 1.0 L of distilled water. Adjust pH to 6.8. Sterilize at 121°C for 15 min. Cool to 47°C and add chromogenic mixture *E. coli*/coliform selective supplement (X-glucuronide/ Salmon-GAL respectively). Pour into sterile Petri dishes.

A2.1.43 Chromocult enterococci agar (selective medium for *enterococcus*)

Peptone	6.0 g
Sodium chloride	5.0 g
Bile salt mixture	1.5 g
5-Bromo-4-chloro-3-indolyl-αD-glucopyranoside 1	0.1 g
Agar	12.0 g

Dissolve ingredients except agar in 1.0 L of distilled water by steaming. Adjust pH to 7.0 and sterilize at 121°C for 15 min. Cool in water bath to about 47°C, add chromogenic selective supplement and pour into sterile Petri dishes.

A2.1.44 Colombia blood agar base (for culturing fastidious organisms)

Peptone	23.0 g
Starch	1.0 g
Sodium chlorine	5.0 g
Agar	15.0 g

Dissolve ingredients except agar and in 1.0 L of distilled water by heating. Adjust pH to 7.3. Add agar, bring to the boil and continue to heat until the agar dissolves. Sterilize at 121°C for 15 min. Cool to 47°C and add 5% sterile defibrinated sheep blood. Pour into sterile Petri dishes.

A2.1.45 Congo red–brain heart infusion agar

Brain heart infusion	37.0 g
$MgCl_2$	1.0 g
Agar	12.0 g
Congo red dye solution	20.0 mL

Dissolve ingredients except agar and in 980 mL of distilled water by heating. Adjust pH to 7.4. Add agar, bring to the boil and continue to heat until the agar dissolves. Sterilize at 121°C for 15 min. Cool to 47°C and pour into sterile Petri dishes.

Congo red dye solution: Add 375 mg of Congo red into 100 mL of distilled water. Mix to dissolve Congo red in water.

A2.1.46 Cooked meat agar/broth (for aerobic and anaerobic microorganisms)

Fresh ground beef	454.0 g
Proteose peptone	20.0 g
Glucose	2.0 g
NaCl	5.0 g
Agar	15.0 g

Add fresh ground beef to 1.0 L of distilled water, remove fat, and boil for 1 h. Adjust pH to 7.2, and boil for another 10 min. Press through cheesecloth, and bring to 1.0 L with distilled water. Add the proteose peptone, glucose and NaCl, and adjust pH to 7.0. Dispense into test tubes in 10 mL amounts, and sterilize at 121°C for 15 min. Before use, exhaust the medium for 20 min in flowing steam.

For agar medium: add agar indu medium during preparation.

A2.1.47 Cystine agar/broth (for detecting the ability to produce H_2S)

Add cystine to a final concentration of 0.1% to a base medium of peptone water or nutrient broth. Add 15 g agar in the preparation of agar medium and bring to the boil and continue to heat until the agar dissolves. Distribute 10 mL into tubes, plug, and sterilize at 121°C for 15 min.

A2.1.48 Czapek yeast autolysate agar

Yeast extract	5.0 g
Sucrose	30.0 g
$NaNO_3$	3.0 g
KCl	0.5 g
K_2PO_4	1.0 g
$MgSO_4 \cdot 7H_2O$	0.5 g
$FeSO_4 \cdot 5H_2O$	0.01 g
Agar	15.0 g

Dissolve ingredients except agar in 1.0 L of distilled water by heating. Add agar, bring to the boil and continue to heat until the agar dissolves. Sterilize at 121°C for 15 min. Cool to 50°C and pour into sterile Petri dishes.

A2.1.49 Davis's yeast salt agar/broth (a selective medium for yeasts and molds)

Ammonium nitrate	1.0 g
Ammonium sulfate	1.0 g
NaH_2PO_4	4.0 g
KH_2PO_4	2.0 g

Magnesium sulfate	0.5 g
Glucose (filter-sterilized)	10.0 g
Sodium citrate · $3H_2O$	0.5 g
NaCl	1.0 g
Yeast extract	2.0 g
Agar	20.0 g

Dissolve ingredients except agar and glucose in 1.0 L of distilled water by steaming. Adjust pH to 6.6. Add agar, bring to the boil, and continue to heat until the agar is dissolved. Sterilize at 121°C for 15 min. Filter-sterilize glucose as 10% solution. Cool slowly to about 50°C, aseptically add the glucose for a final concentration of 1.0%. Adjust pH to 3.5 if required, by the addition sterile 10% citric acid solution. Pour into sterile Petri dishes.

To prepare **DYS broth**, omit the agar.

To prepare **antibiotic containing DYS agar**: Prepare an antibiotic solution containing 500 mg each of chlortetracycline HCl and chloramphenicol in 100 mL of sterile phosphate buffered distilled water and mix. (Not all material dissolves; therefore, the suspension must be evenly dispersed before pipetting into the medium.) Add 2 mL of this antibiotic solution per 100 mL of sterilized and cooled (about 50°C) DYS agar or broth. After being swirled to mix, and pour the medium into sterile Petri dishes.

A2.1.50 Decarboxylase broth (use for differentiation of bacteria on the basis of their ability to decarboxylate amino acids)

Peptone	5.0 g
Yeast extract	3.0 g
Glucose	1.0 g
Amino acid	5.0 g
Bromocresol purple	0.02 g

Dissolve ingredients except agar in 1.0 L of distilled water by steaming. Adjust pH to 6.8, and dispense 5 mL into screw-capped tubes. Sterilize at 121°C for 15 min.

A2.1.51 Dey–Engley neutralizing broth

Tryptone	5.0 g
Yeast extract	2.5 g
Glucose	10.0 g
Sodium thioglycolate	1.0 g
Sodium thiosulfate	6.0 g
Sodium bisulfite	2.5 g
Polysorbate-80 (Tween 80)	5.0 g
Lecithin (soy bean)	7.0 g
Brom cresol purple	0.02 g

Dissolve ingredients in 1.0 L of distilled water by heating. Adjust pH to 7.6 and dispense 5 mL into screw-capped tubes. Sterilize at 121°C for 15 min.

A2.1.52 Dextrose sorbitol mannitol agar

Dextrose (glucose)	1.0 g
Sorbitol	1.0 g
Mannitol	2.0 g
Yeast extract	3.3 g
Proteose-peptone	10.0 g
Calcium lactate	15.0 g
KH_2PO_4	1.0 g
$MnSO_2 \cdot H_2O$	0.02 g
Cycloheximide	0.004 g
Bromocresol purple	0.03 g
Brilliant green	0.0295 g
Agar	15.0 g

Dissolve ingredients except agar in 1.0 L of distilled water by heating. Adjust pH to 7.2. Add agar, bring to the boil and continue to heat until the agar dissolves. Sterilize at 115°C for 15 min. Cool slowly to about 50°C and aseptically add 25 mL of sterile 40% urea solution. Mix thoroughly and pour into sterile Petri dishes.

A2.1.53 Dextrose tryptone bromocresol purple agar/broth (use to isolate thermophilic flat sour bacteria)

Tryptone or trypticase	12.0 g
Glucose	5.0 g
Bromocresol purple	0.04 g
Agar	18.0 g

Dissolve ingredients except agar in 1.0 L of distilled water by steaming. Adjust pH to 6.8. Add agar, bring to the boil and continue to heat until the agar dissolves. Sterilize at 121°C for 15 min. Cool to about 45°C and pour into sterile Petri dishes.

To prepare **DTBP broth**, omit the agar.

(The dilution to be created by distributing 20 mL of sample to six Petri dishes was taken into account in the calculation.)

A2.1.54 Dicloran rose bengal chloramphenicol agar (a selective medium for yeasts and molds associated with food spoilage)

Peptone	5.0 g
Glucose	10.0 g

KH_2PO_4	1.0 g
Magnesium sulfate	0.5 g
Rose Bengal	0.025 g
Dicloran (2,6-dichloro-4-nitroaniline)	0.002 g
Chloramphenicol	0.1 g
Agar	15.0 g

Dissolve ingredients except agar and chloramphenicol in 1.0 L of distilled water by steaming. Adjust pH to 6.5. Add agar, bring to the boil and continue to heat until the agar dissolves. Sterilize at 121°C for 15 min. Cool slowly to about 50°C and add 2 mL of filter sterilized chloramphenicol into medium. Mix thoroughly and pour into sterile Petri dishes.

Notes: DRBC agar is especially useful for analyzing sample containing "spreader" molds (e.g., *Mucor*, *Rhizopus*, etc.), since the added dicloran and rose Bengal effectively slow down the growth of fast-growing fungi, thus readily allowing detection of other yeast and mold, which have lower growth rates. Media containing rose Bengal are light-sensitive; relatively short exposure to light will result in the formation of inhibitory compounds. Keep these media in a dark, cool place until used. DRBC agar should be used for spread plates only.

The reduced pH of DRBC Agar increases the inhibition of yeasts by Rose-Bengal. pH is lowered to 5.6, the Rose-Bengal content is reduced by 50% and Dicloran is added.

A2.1.55 Dicloran glycerol (G18) agar (for selective isolation of xerophilic molds)

Dextrose	$10.0\ g\ L^{-1}$
Magnesium sulfate	$0.5\ g\ L^{-1}$
Dicloran	$0.002\ g\ L^{-1}$
Monopotassium phosphate	$1.0\ g\ L^{-1}$
Peptic digest of animal tissue	$5.0\ g\ L^{-1}$
Glycerol	110 mL
Chloramphenicol	$0.1\ g\ L^{-1}$
Agar	$15.0\ g\ L^{-1}$

Dissolve ingredients except agar in 1.0 L of distilled water by steaming. Adjust pH to 5.6. Add agar, bring to the boil and continue to heat until the agar is dissolved. Do not sterilize. Cool to 45°C and pour into sterile Petri dishes.

A2.1.56 *Escherichia coli* broth (a selective medium for fecal and nonfecal coliforms, and *Escherichia coli*)

Tryptose (or trypticase)	20.0 g
Lactose	5.0 g
Bile salts	1.5 g
K_2HPO_4	4.0 g

KH_2PO_4	1.5 g
NaCl	5.0 g

Dissolve ingredients except agar in 1.0 L of distilled water by steaming. Adjust pH to 6.8, and distribute in required amounts into test tubes containing inverted Durham tubes. Sterilize at 121°C for 15 min.

Modified *Escherichia coli* **broth**: Add novobiocin (20 mg L^{-1}) into sterilized and cooled (at 50°C) *Escherichia coli* (EC) broth.

EC-methylumbelliferyl-β-D-glucuronide (**MUG) medium:** add 50 mg 4-MUG per liter before sterilization.

A2.1.57 Edward's esculin crystal violet blood agar (a selective medium for streptococci)

Nutrient agar (pH 7.4)	100.0 mL
Crystal violet (0.05% aq. solution)	0.4 mL
Defibrinated blood	5.0 mL
Esculin	0.1 g

Add esculin to the melted nutrient agar and sterilize by steaming for 30 min on each of three successive days. Add 0.4 mL of sterile 0.05% crystal violet solution (sterilized by autoclaving) into 100 mL of sterile esculin nutrient agar cooled to 50°C, and mix well. Next add 5 mL of sterile blood, mix thoroughly, and pour into sterile Petri dishes.

A2.1.58 Egg-yolk emulsion

Separate aseptically the yolk from the white of the eggs by pipette into a sterile measuring cylinder and add 4 parts of 0.85% sterile saline solution to 1 part of egg yolk. Mix thoroughly and heat in a water bath at 45°C for 2 h. Centrifuge to remove the precipitate (alternatively, stand the mixture overnight in refrigerator). Decant the supernatant liquid. Sterilize by filtration (0.45 μm). Filtration can be employed with a number of filters with progressively smaller pore size; for example, fiber glass prefilter followed by a filter of pore size 1.2 μm, pore size 0.65 μm, pore size 0.45 μm, and finally the sterilizing grade of filter with a pore size of 0.22 μm.

A2.1.59 EHEC enrichment broth

Add 0.2 mL novobiocin solution to trypticase soy broth modify.

Dissolve 100 mg novobiocin (sodium salt) novobiocin in 1.0 mL water and filter-sterilize, using 0.2 μm filter and syringe.

A2.1.60 Eijkman lactose broth (a medium for differentiate *Escherichia coli* from other coliforms)

Tryptose	15.0 g
Lactose	3.0 g
K_2HPO_4	4.0 g
KH_2PO_4	1.5 g
NaCl	5.0 g

Dissolve ingredients in 1.0 L of distilled water by steaming. Adjust pH to 6.8, and dispense 5 mL into test tubes. Sterilize at 121°C for 15 min.

A2.1.61 L-Eosin methylene blue agar/broth (Levine) (a differential medium for Gram-negative enteric bacteria)

Peptone	10.0 g
Lactose	5.0 g
Sucrose	5.0 g
K_2HPO_4	2.0 g
Eosin-Y	0.4 g
Methylene blue	0.065 g
Agar	3.5 g

Dissolve ingredients except agar in 1.0 L of distilled water by steaming. Adjust pH to 7.2. Add agar, bring to the boil and continue to heat until the agar dissolves. Sterilizes at 121°C for 15 min. Avoid overheating. Sterilization reduces the methylene blue, leaving the medium orange color. Gentle shaking may restore the normal purple color of the medium. If the reduced medium is not shaken to oxidize the methylene blue, a dark zone beginning at the top and extending downward through the medium will gradually appear. The sterilized medium normally contains a flocculent precipitate that should not be removed. Cooling to 50°C and gently agitating the medium before pouring into sterile Petri dishes will finally disperse this flocculation.

To prepare **EMB broth**, omit the agar.

A2.1.62 Enterococcosel (ECA) broth/agar (selective differential medium for *Enterococcus*)

Pancreatic digest of casein	17.0 g
Sodium citrate	1.0 g
Peptic digest of animal tissue	3.0 g
Esculin	1.0 g
Yeast extract	5.0 g

Ferric ammonium citrate	0.5 g
Oxgall	10.0 g
Sodium azide	0.25 g
Sodium chloride	5.0 g

Dissolve ingredients except agar in 1.0 L of distilled water, and adjust pH to 7.1. Add agar, and heat while stirring until the agar dissolves. Sterilize at 121°C for 15 min. Cool to 45°C, pour into sterile Petri dishes to a depth of 4 to 5 mm, and allow to solidify.

To prepare **ECC broth**, omit the agar.

A2.1.63 Esculin iron agar (use to verify colonies as *enterococcus*)

Esculin	1.0 g
Ferric ammonium citrate	0.5 g
Agar	15.0 g

Dissolve ingredients except agar in 1.0 L of distilled water, and adjust pH to 7.1. Add agar, and heat while stirring until the agar dissolves. Sterilize at 121°C for 15 min. Cool to 45°C, pour into sterile Petri dishes to a depth of 4 to 5 mm, and allow solidifying.

A2.1.64 Fluid thioglycolate medium (a medium for anaerobes, aerobes, and microaerobic)

Yeast extract	2.0 g
Lab-lemco powder	1.0 g
Peptone	5.0 g
NaCl	5.0 g
Glucose	5.0 g
Sodium thioglycolate	1.1 g
Agar	1.0 g

Dissolve ingredients except agar in 1.0 L of distilled water by steaming. Adjust pH to 7.2. Add agar, bring to the boil, and continue to heat until the agar is dissolved. Dispense into test tubes, filling them to at least two-thirds of tube capacity and sterilize at 121°C for 15 min. Cool to 25°C and store at room temperature, not in the refrigerator. After storage of prepared medium, if more than 30% of the medium demonstrates oxidation (color change of indicator), reheat medium once in a boiling water or steam bath to drive off absorbed oxygen.

To prepare fluid thioglycolate (FT) **medium containing cycloserine**; Dissolve 1.0 g D-cycloserine (white crystalline powder) in 200 mL of distilled water. Sterilize by filtration and store at 4°C until use. add 20 mL of sterile D-cycloserine solution to 250 mL base.

A2.1.65 Fluorogenic gentamicin-thallous-carbonate agar (a selective medium for *enterococcus*)

Trypticase soy agar	1.0 L
KH_2PO_4	5.0 g
Amylase azure	3.0 g
Galactose	1.0 g
Thallous acetate (acetic acid, thallous salt)	0.5 g
Tween 80 (polyoxyethylene sorbitol monooleate)	0.75 mL
4-methylumbelliferyl-alpha-D-galactoside	100.0 mg
Gentamycin sulfate	2.5 mg

Dissolve ingredients except agar in 1.0 L of distilled water by steaming. Adjust pH to 7.1. Add agar, bring to the boil and continue to heat until the agar dissolves. Sterilize at 121°C for 15 min. Cool to about 45°C and pour into sterile Petri dishes.

A2.1.66 Gelatin agar (use for gelatin utilization test)

Gelatin	30.0 g
NaCl	10.0 g
Trypticase	10.0 g
Agar	15.0 g

Dissolve ingredients except agar in 1.0 L of distilled water by steaming. Adjust pH to 7.2. Add the agar, bring to the boil and continue to heat until the agar dissolves. Dispense 8 mL into test tubes, and sterilize at 121°C for 15 min.

A2.1.67 Glucose azide agar/broth (a selective medium for *Enterococcus*)

Peptone	10.0 g
NaCl	5.0 g
K_2HPO_4	5.0 g
KH_2PO_4	2.0 g
Glucose	5.0 g
Yeast extract	3.0 g
Sodium azide	0.25 g
Bromocresol purple (1% aq. sol)	3.0 mL

Dissolve ingredients in 1.0 L of distilled water. Adjust pH to 7.0. Add the agar, bring to the boil and continue to heat until the agar dissolves. Distribute in 5 mL into test tubes. Sterilize at 121°C for 15 min.

Double strength medium can be prepared in a half quantity of distilled water.

To prepare **GPA agar**, add 1.5% agar into broth.

A2.1.68 Gelatin medium

Peptone	5.0 g
NaCl	5 0.0 g
Meat extract	3.0 g
Gelatin	150.0 g

Dissolve ingredients in 1.0 L of distilled water. Adjust pH to 6.8. Add the agar, bring to the boil and continue to heat until the agar dissolves. Sterilize by autoclaving at 121°C for 15 min. Mix well before pouring into Petri dishes.

A2.1.69 Gelatin phosphate salt agar/broth (use for isolation of *Vibrio* spp.)

Gelatin	10.0 g
NaCl	10.0 g
K_2HPO_4	5.0 g
Agar	15.0 g

Dissolve all ingredients in distilled water by heating. Adjust pH to 7.2. Dispense in tubes or bottles, and sterilize at 121°C for 15 min.

To prepare **GPS broth**, omit the agar.

A2.1.70 Glucose yeast extract agar (for counting of acetic acid producing bacteria)

Glucose	50.0 g
Yeast extract	10.0 g
Calcium carbonate	5.0 g
Agar	2.0 g

Dissolve ingredients except agar in 1.0 L of distilled water. Adjust pH to 7.3. Add agar, bring to the boil and continue to heat until the agar dissolves. Sterilize at 121°C for 15 min. Cool to 50°C and add 100 mg pimaricin per L and pour into sterile Petri dishes.

A2.1.71 Glucose salt teepol broth (an enrichment medium for *Vibrio parahaemolyticus*)

Beef extract	3.0 g
Peptone	10.0 g
NaCl	30.0 g

Glucose	5.0 g
Methyl violet	0.002 g
Teepol	4.0 mL

Dissolve ingredients in 1.0 L of distilled water. Adjust pH to 8.8, dispense in tubes, and sterilize at 121°C for 15 min.

A2.1.72 Glucose–peptone–meat extract agar

Glucose	10.0 g
Peptone	5.0 g
Meat extract	5.0 g
NaCl	3.0 g
Agar	12.0 g

Dissolve ingredients except agar in 1.0 L of distilled water by steaming. Adjust pH to 6.9. Add agar, bring to the boil and continue to heat until the agar dissolves. Sterilize at 121°C for 15 min.

A2.1.73 Glycerol agar/broth (for xerophilic molds)

Peptone	5.0 g
Dextrose	10.0 g
KH_2PO_4	1.0 g
$MgSO_4 \cdot 7H_2O$	0.5 g
Agar	15.0 g
Glycerol	110.0 g

Dissolve ingredients in 1.0 L of distilled water. Add the agar, bring to the boil and continue to heat until the agar dissolves. Sterilize at 121°C for 15 min. Cool to 50oC and adjust pH to 5.5. Add 250 mg L^{-1} chloramphenicol and mix well. Pour into sterile petri dishes.

To prepare glycerol broth; omit agar and distribute in 5 mL into test tubes.

A2.1.74 Hektoen enteric agar (a selective medium for *Shigella* and *Salmonella* spp.)

Proteose peptone	12.0 g
Yeast extract	3.0 g
Bile salts No. 3	39.0 g
Lactose	12.0 g
Sucrose	12.0 g
Salicin	2.0 g

NaCl	5.0 g
Sodium thiosulfate	5.0 g
Ferric ammonium citrate	1.5 g
Bromothymol blue	0.065 g
Acid fuchsine	0.1 g
Agar	14.0 g

Dissolve ingredients except agar in 1.0 L of distilled water by steaming. Adjust pH to 7.5. Add agar, bring to the boil, and continue to heat until the agar is dissolved. Sterilize at 121°C for 15 min. Cool to 45°C and pour into sterile Petri dishes.

A2.1.75 Heterofermentative broth (for heterofermentative *Lactobacillus*)

Proteose peptone No. 3	10.0 g
Yeast extract	5.0 g
Tween 80	1.0 mL
K_2HPO_4	2.0 g
Sodium acetate	5.0 g
Ammonium citrate	2.0 g
Magnesium sulfate	0.1 g
Magnesium sulfite	0.05 g
Glucose	20.0 g
Bromocresol green	0.04 g
2-Phenylethyl alcohol (to inhibit Gram-negatives)	3.0
Actidione (to inhibit yeasts)	4.0 mg

Dissolve ingredients in 1.0 L of distilled water by steaming. Adjust pH to 4.3 with concentrated HCl, and then dispense the medium into test tubes containing inverted Durham tubes. Sterilize at 121°C for 15 min.

A2.1.76 HHD agar (a selective medium for homofermentative and heterofermentative lactic acid bacteria)

Fructose	2.5 g
KH_2PO_4	2.5 g
Typticase peptone	10.0 g
Phytone peptone	1.5 g
Casamino acids	3.0 g
Yeast extract	1.0 g
Tween 80	1.0 g

Bromocresol green (1% aq. solution)	20.0 mL
Agar	20.0

Dissolve ingredients except agar in 1.0 L of distilled water by steaming. Adjust pH to 7.0. Add agar, bring to the boil, and continue to heat until the agar is dissolved. Sterilize at 121°C for 15 min.

A2.1.77 HL agar

(A nutrient medium for the cultivation of fastidious organisms)

A2.1.77.1 Base layer

Prepare Colombia blood agar base, and sterilize at 121°C for 15 min. Place 10 mL into sterile Petri dishes.

A2.1.77.2 Top layer

Prepare Colombia blood agar base, and sterilize at 121°C for 15 min. Cool to 46°C, and add 5% horse blood and mix thoroughly. Pour 5.0 mL of blood agar over the base layer while the base is still warm. HL agar may be stored under refrigeration in plastic bags; discard any plates that become discolored or show hemolysis.

A2.1.78 Holmon's alkaline cooked meat medium (a medium for putrefactive anaerobes)

Infuse 500 g of ground fresh beef in 1.0 L of distilled water overnight in a refrigerator. Skim off fat. Filter infusion through several layers of cheese cloth and press out broth. Retain the pressed beef cake. Make broth to original volume with water and add 5.0 g of peptone. Heat in steam for 10 min. Filter and add 5.0 g of NaCl. Add normal NaOH until alkaline to phenolphthalein. Heat in steam for 15 min and filter. Transfer about 2.0 g of pressed out beef to each test tube (20 × 150 mm), add 10 mL of clear alkaline broth to tubes, and sterilize at 121°C for 20 min. Final pH should be 7.2–7.4. Store media at 4°C. Immediately before use, heat tube medium in boiling water bath for at least 15 min to expel adsorbed oxygen and cool promptly to 37°C before inoculation.

A2.1.79 Inositol brilliant green bile salt agar (a selective medium for *Plesiomonas shigelloides*)

Peptone	10.0 g
Meat extract	5.0 g
NaCl	5.0 g
Bile salt No. 3	8.5 g
Brillant green	0.33 mg
Neutral red	25.0 mg

Inositol	10.0 g
Agar	15.0 g

Dissolve ingredients except agar in 1.0 L of distilled water by steaming. Adjust pH to 7.1. Add agar, bring to the boil and continue to heat until the agar dissolves. Sterilize at 115°C for 15 min. Cool to 50°C and pour into sterile Petri dishes.

A2.1.80 Inositol gelatin deeps (a medium to test gelatin utilization by *Plesiomonas shigelloides*)

Gelatin	120.0 g
Na_2HPO_4	5.0 g
Yeast extract	5.0 g
Inositol	10.0 g
Phenol red	0.05 g

Heat to dissolve ingredients in 1.0 L of distilled water, and adjust pH to 7.4. Distribute the medium into tubes (5 mL per tube), and sterilize at 115°C for 15 min.

A2.1.81 Irgasan–Ticarcillin–Chlorate Broth

Tryptone	10.0 g
Yeast extract	1.0 g
MgCl2 · 6H2O	60.0 g
NaCl	5.0 g
Potassium chlorate	1.0 g
Malachite green, 0.2%	5.0 mL

Dissolve ingredients in 1.0 L distilled water, mix thoroughly and heat with occasional agitation. Boil about 1 min to dissolve ingredients. Adjust pH to pH 7.6 and Sterilize at 121°C for 15 min.

A2.1.81.1 Add the following

Ticarcillin (1 mg per mL)	1.0 mL
Irgasan (1 mg per mL)	1.0 mL

A2.1.82 Jensen's pork sucrose nitrite medium (use for isolation of mesophilic aerobic sporeforming bacteria)

Fill culture tubes approximately 1/2 full with finely ground or chopped spiced ham and wet down with a solution composed of 0.4% sodium nitrate and 2% sucrose. The meat should be covered at least 0.7 cm by solution. No trapped bubbles of air should appear in medium. Tubes should not be more than 1/2 full of meat and broth. Sterilize at 121°C for 15 min. When used inoculate at the bottom of medium, taking care not to introduce air bubbles.

A2.1.83 KCN broth (a test medium for growth *Enterobacteriaceae* in the presence of potassium cyanide)

Proteose peptone	3.0 g
Na_2HPO_4	5.64 g
KH_2PO_4	0.225 g
NaCl	5.0 g

Dissolve ingredients in 1.0 L of distilled water by steaming. Adjust pH to 7.6. Dispense in 10 mL amount into test tubes and sterilize at 121°C for 15 min. Cool to 50°C, add 0.15 mL of the potassium cyanide solution to each 10 mL medium, and stopper immediately with paraffin stoppers. (Prepare a 0.5% potassium cyanide solution in distilled water, sterilize by heat and cool to room temperature.)

A2.1.84 KF streptococcal agar/broth (a selective medium for *Enterococcus*)

Proteose peptone	10.0 g
Yeast extract	10.0 g
NaCl	5.0 g
Sodium glycerophosphate	10.0 g
Maltose	20.0 g
Lactose	1.0 g
Sodium azide	0.4 g
bromocresol purple	0.015 g
Agar	20.0 g

Dissolve ingredients except agar in 1.0 L of distilled water by steaming. Adjust pH to 7.2. Add agar, bring to the boil, and continue to heat until the agar is dissolved. Sterilize at 121°C for 15 min. Cool to 50°C and pour into sterile Petri dishes.

To prepare KF broth, omit agar.

A2.1.85 Kim–Goeppert agar (a selective and differential medium for *Bacillus*)

A2.1.85.1 Base

Peptone	1.0 g
Yeast extract	0.5 g
Phenol red	0.025 g
Agar	18.0 g

Dissolve the ingredients except agar in 900 mL of distilled water by steaming. Adjust pH to 6.8. Add agar, bring to the boil and continue to heat until the agar dissolves. Sterilize at 121°C for 20 min. Cool to 50°C, and add 100 mL of egg yolk and 1.0 mL of polymyxin B sulfate solution. Mix well and pour into sterile Petri dishes.

A2.1.85.2 Egg yolk

Look at to egg yolk emulsion.

A2.1.85.3 Polymyxin B

Add 5 mL of sterile distilled water into sterile 50.0 mg (500,000 units) polymyxin B sulfate with a sterile syringe. Mix to dissolve the power.

A2.1.86 Kligler iron agar

Polypeptone peptone	20.0 g
Lactose	20.0 g
Dextrose	1.0 g
NaCl	5.0 g
Ferric ammonium citrate	0.5 g
Sodium thiosulfate	0.5 g
Phenol red	0.025 g
Agar	15.0 g

Dissolve ingredients except agar in 1.0 L distilled water, mix thoroughly. Adjust pH 7.3. Add agar, bring to the boil and continue to heat until the agar dissolves. Dispense into tubes and sterilize at 121°C for 15 min. Cool and slant to form deep butts.

For halophilic *Vibrio* spp., add NaCl to a final concentration of 2%–3%.

A2.1.87 Lactic (Elliker) agar/broth (a medium for streptococci and lactobacilli)

Tryptone	20.0 g
Yeast extract	5.0 g
Gelatin	2.5 g
Glucose	5.0 g
Lactose	5.0 g
Saccharose	5.0 g
NaCl	4.0 g
Sodium acetate	1.5 g
Ascorbic acid	0.5 g
Agar	15.0 g

Dissolve ingredients except agar in 1.0 L of distilled water by steaming. Adjust pH to 6.8. Add agar, bring to the boil, and continue to heat until the agar is dissolved. Add skim milk and mix thoroughly. Sterilize at 121°C for 15 min. Cool to about 50°C and pour into sterile Petri dishes.

A2.1.88 Lactose broth (use in lactose utilization tests and humanizing solution)

Peptone	5.0 g
Yeast extract	3.0 g
Lactose	5.0 g
Phenol red	0.05 g

Dissolve ingredients in 1.0 L of distilled water by steaming. Adjust pH to 6.9. Dispense into tubes containing inverted Durham tubes. Sterilize at 121°C for 15 min. Before opening autoclave, allow temperature to drop below 75°C to avoid entrapped air bubbles in inverted Durham tubes.

A2.1.89 Lactose egg-yolk milk agar (a medium for the differentiation of *Clostridium* spp.)

A2.1.89.1 Base medium

Nutrient broth	800.0 g
Lactose	9.6 g
Neutral red (1% solution)	2.6 g
Agar	12.0 g

Dissolve agar in 800.0 mL of nutrient broth by heating, then add the lactose and neutral red solution. Mix well and distribute in 80 mL amounts in screw-capped bottles. Sterilize at 121°C for 20 min.

A2.1.89.2 Egg-yolk emulsion

Separate aseptically the yolk from the white of the eggs by pipette into a sterile measuring cylinder. Mix the yolk with an equal volume (about 20.0 mL) of 0.85% sterile saline solution, and transfer aseptically to a sterile screw-capped bottle. This emulsion can be used for routine purposes; it can be tested for sterility by plating out 1 mL quantities.

A2.1.89.3 Complete medium

Cool the base medium to 50°C. Add 3 mL of egg-yolk emulsion and 12 mL of sterile reconstituted skim milk (10%) to 80 mL of melted base medium. Mix well and pour into sterile Petri dishes.

A2.1.90 Lactose gelatin medium (use in gelatin test for *Clostridium perfringens*)

Tryptose	15.0 g
Yeast extract	10.0 g

Na_2HPO_4	5.0 g
Lactose	10.0 g
Gelatin	120.0 g
Phenol red	0.05 g

Dissolve ingredients except gelatin and phenol red in 400 mL of distilled water by heating gently while stirring. Suspend the gelatin in 600 mL of cold distilled water, dissolve by heating in a water bath at 50 to 60°C with frequent stirring. When the gelatin dissolved, combine with the other dissolved ingredients and adjust pH to 7.2. Add the phenol red, mix well, and dispense 10 mL into test tubes. Sterilize at 121°C for 15 min. Just before use, heat to boiling, or expose to flowing steam for 10 min to remove dissolved oxygen. Cool rapidly to incubation temperature.

A2.1.91 Lauryl sulfate tryptose broth (a selective medium for coliforms and *Escherichia coli*)

Tryptose (or pancreatic digest of casein)	20.0 g
Lactose	5.0 g
K_2HPO_4	2.75 g
KH_2PO_4	2.75 g
NaCl	5.0 g
Sodium lauryl sulfate	0.1 g

Dissolve ingredients in 1.0 L of distilled water by steaming. Adjust pH to 7.2. Dispense into tubes containing inverted Durham tubes in desired amounts. Sterilize at 121°C for 12 min, but not exceeding 15 min. After sterilization, cool broth as quickly as possible.

For **double strength** Lauryl sulfate tryptose (**LST**) **broth**; multiply ingredients by two.

For **LST Broth-MUG**; add 4-methylumbelliferyl-β-D-glucuronide 0.1 g into LST broth and perform as LST broth preparation.

A2.1.92 Lee's agar (a selective and differential medium for *Streptococcus thermophilus* and *Lactobacillus bulgaricus*)

Tryptose	10.0 g
Yeast extract	10.0 g
Lactose	5.0 g
Sucrose	5.0 g
Calcium carbonate	3.0 g
K_2HPO_4	0.5 g
Bromocresol purple (0.2% aq. sol.)	10.0 mL
Agar	18.0 g

Dissolve ingredients except agar and bromocresol purple in 1.0 L of distilled water. Adjust pH to 7.0. Add agar, bring to the boil and continue to heat until the agar dissolves. Sterilize at 121°C for 20 min. Carefully mix the melted medium to suspend the calcium carbonate evenly. Cool to 50°C and add 10 mL of sterile (121°C for 15 min) 0.2% bromocresol purple solution. Pour the medium into sterile Petri dishes to obtain a layer of medium 4–5 mm thick. After solidification, dry the plates in a 30°C incubator for 18 h.

Lee's agar containing disodium-β-glycerophosphate: Add 10 g to Lee' agar.

A2.1.93 Legionella CYE agar (a selective medium for *Legionella*)

Activated charcoal	2.0 g
Yeast extract	10.0 g
Agar	13.0 g

A2.1.94 Selective supplement

ACES buffer/KOH	1.0 g
Ferric pyrophosphate	0.025 g
L-cysteine HCl	0.04 g
Alpha-ketoglutarate	0.1 g

Dissolve ingredients except agar and bromocresol purple in 1.0 L of distilled water. Adjust pH to 6.9. Add agar, bring to the boil and continue to heat until the agar is dissolved. Sterilize at 121°C for 20 min. Cool to 50°C and add selective supplement (dissolved in 10.0 mL of sterile distilled water). Mix gently and pour into sterile Petri dishes.

A2.1.95 LES endo agar (for coliforms by membrane filter technique)

Yeast extract	1.2 g
Casitone or trypticase	3.7 g
Thiopeptone or thiotone	3.7 g
Tryptose	7.5 g
Lactose	9.4 g
Dipotassium hydrogen phosphate	3.3 g
Potassium dihydrogen phosphate	1.0 g
Sodium chloride, NaCl	3.7 g
Sodium deoxycholate	0.1 g
Sodium lauryl sulfate	0.05 g
Sodium sulfite, Na_2SO_3	1.6 g
Basic fuchsin	0.8 g
Agar	15.0 g

Dissolve ingredients in 1.0 L distilled water containing 20 mL 95% ethanol by gently heating. Adjust pH to 7.2. Bring to a near boil to dissolve agar, then remove from heat and cool to 47°C. Do not sterilize by autoclaving. Dispense into sterile petri dishes. Do not expose poured plates to direct sunlight; refrigerate in the dark, preferably in sealed plastic bags to reduce moisture loss.

A2.1.96 Litmus milk (use for maintenance of lactic acid bacteria)

Skim milk powder	100.0 g
Litmus (or azolitmin)	1.0 g

Dissolve litmus in 10 mL of water. Add a few drops of 1 N NaOH. Dissolve skim milk in 900 mL of distilled water. Adjust pH to 6.5. Add indicator solution and sufficient water to make 1.0 L. Mix thoroughly. Dispense into test tubes and sterilize at 121°C for 15 min.

A2.1.97 Liver agar/broth (a liquid medium, containing liver particles for saccharolytic or putrefactive mesophilic and thermophilic anaerobes)

Fresh ground beef liver	500.0 g
Soluble starch	1.0 g
Peptone	10.0 g
K_2HPO_4	1.0 g

Remove the fat from 500 g beef liver, mix with 1.0 L of distilled water, and boil slowly for 1 h. Adjust pH to 7.0, and boil for another 10 min. Remove the liver particles by pressing through cheesecloth, and complete broth to 1.0 L with distilled water. Add the peptone, starch, and K_2HPO_4. Refilter broth and adjust pH to 7.0. Place liver particles from the pressed cake in the bottom of culture tube (about 2 cm deep), cover with 8–10 mL of broth, and sterilize at 121°C for 20 min. Before use, exhaust the medium for 20 min in flowing steam.

A2.1.98 Liver egg-yolk agar (a differential medium for anaerobic bacteria)

Fresh eggs	3.0
Liver veal agar	1.0 L

Wash eggs with a stiff brush, and drain. Soaks eggs in 0.1% mercuric chloride solution for 1 h. Pour off the mercuric chloride solution, and replace with 70% ethyl alcohol. Soak in 70% ethyl alcohol for 30 min. Crack the eggs aseptically, and discard the whites. Remove the yolk with a sterile 50 mL syringe. Place in a sterile container, and an equal volume of sterile sale. Mix thoroughly. To each 500 mL of melted liver veal agar tempered to 50°C, add 40.0 mL of the egg yolk–saline solution. Mix thoroughly and pour plates. Dry plates at room temperature for 2 days or at 35°C for 24 h. Discard contaminated plates, and store sterile plates under refrigeration.

A2.1.99 Lithium chloride phenylethanol moxalactam agar (for *listeria monocytogenes*)

Glycine anhydride	10.0 g
Lithium chloride	5.0 g
Sodium chloride	5.0 g
Casein peptone (pancreatic)	5.0 g
Peptone (animal)	5.0 g
Beef extract	3.0 g
Phenylethyl alcohol	2.5 g
Agar	15.0 g

Dissolve ingredients except agar in 1.0 L of distilled water by steaming. Adjust pH to 7.3. Add agar, bring to the boil and continue to heat until the agar dissolves. Sterilize at 115°C for 15 min. Cool to about 45°C and pour into sterile Petri dishes.

A2.1.100 Luria bertani broth (for *Escherichia coli*)

Tryptone	10.0 g
Yeast extract	5.0 g
Sodium chlorine	5.0 g

Dissolve ingredients except agar in 1.0 L of distilled water by steaming. Adjust pH to 6.9. Sterilize at 121°C for 15 min.

A2.1.101 Lysine arginine iron agar

Peptone	5.0 g
Yeast extract	3.0 g
Glucose	1.0 g
L-Lysine	10.0 g
L-Arginine	10.0 g
Ferric ammonium citrate	0.5 g
Sodium thiosulfate	0.04 g
Bromcresol purple	0.02 g
Agar	15.0 g

Dissolve ingredients except agar in 1.0 L of distilled water by steaming. Add agar, bring to the boil and continue to heat until the agar dissolves. Adjust pH to 6.8 and dispense 5 mL into each 13 × 100 mm screw-cap tube. Sterilize at 121°C for 12 min. Cool tubes in slanted position.

A2.1.102 Lysozyme broth

A2.1.102.1 Nutrient broth

Prepare nutrient broth, and dispense 99 mL into bottles or flasks. Sterilize at 121°C for 15 min.

A2.1.102.2 Lysozyme solution

Dissolve 0.1 g of lysozyme in 65 mL of sterile 0.01 N HCl acid. Heat to boiling for 20 min, and dilute to 100 mL with sterile 0.01 N HCl acid. Alternatively, dissolve 0.1 g lysozyme chloride in 100 mL of distilled water, and sterilize by filtration. Test solution for sterility before use.

Add 1 mL of sterile 0.1% lysozyme solution to each 99 mL of nutrient broth. Mix thoroughly, and aseptically dispense 2.5 mL of the complete medium into sterile tubes.

A2.1.103 Lysed horse blood

Use fresh blood and freeze to lyse upon receipt. To freeze, resuspend blood cells gently and pour ~40 mL portions into sterile 50 mL disposable centrifuge tubes. Freeze at −20°C. Thaw and refreeze once more to complete lysis. Store blood up to 6 months. Unused portions

A2.1.104 Lysine iron agar (a differential medium for *Salmonella arizonae* and *Arizona arizonea*)

Peptone	5.0 g
Yeast extract	3.0 g
Glucose	1.0 g
L-Lysine hydrochloride	10.0 g
Ferric ammonium citrate	0.5 g
Sodium thiosulfate	0.04 g
bromocresol purple	0.02 g
Agar	15.0 g

Dissolve ingredients in 1.0 L of distilled water by steaming. Adjust pH to 7.2. Dispense into tubes containing inverted Durham tubes in desired amounts. Sterilize at 121°C for 12 min, but not exceeding 15 min. After sterilization, cool the broth as quickly as possible.

To prepare **LI agar**, add 1.5% agar into LI broth.

A2.1.105 M17 agar/broth (a selective and differential medium for lactic streptococci and their bacteriophages, and *Streptococcus thermophilus*)

Phytone peptone (soya peptone)	5.0 g
Tryptone (or ploypeptone)	5.0 g
Yeast extract	2.5 g
Ascorbic acid	0.5 g

Magnesium sulfate	0.25 g
Beta-disodium glycerophosphate	19.0 g
Agar	11.0 g

Dissolve ingredients except agar in 1.0 L of distilled water, and adjust pH to 6.9. Add agar and heat until the agar dissolves. Cool to 50°C and add 50 mL of sterile lactose solution (10% wt./wt.).

A2.1.105.1 Lactose solution

Dissolve 10 g of lactose in 100 mL of distilled water. Sterilize at 121°C for 15 min or by membrane filtration through a 0.2 μm membrane.

To prepare **M 17 broth**, omit the agar.

A2.1.106 MacConkey agar (a differential medium for lactose fermenting organisms)

Peptone	20.0 g
Lactose	10.0 g
Bile salts No. 3	5.0 g
NaCl	5.0 g
Neutral red	0.075 g
Agar	12.0 g

Dissolve ingredients except agar in 1.0 L of distilled water by steaming. Adjust pH to 7.4. Add agar, bring to the boil and continue to heat until the agar dissolves. Sterilize at 121°C for 15 min. Cool to about 45°C and pour into sterile Petri dishes.

A2.1.107 Malachite green broth (for the selective enrichment of *Pseudomonas aeruginosa*)

Meat peptone	15.0 g
Meat extract	9.0 g
Malachite green oxalate	0.03 g
Dipotassium hydrogen phosphate	1.1 g

Dissolve ingredients except agar in 1.0 L of distilled water by steaming. Adjust pH to 7.6. Add agar, bring to the boil and continue to heat until the agar dissolves. Sterilize at 121°C for 15 min. Cool to about 45°C and pour into sterile Petri dishes. Protected from direct light.

A2.1.108 Malonate broth/agar (differentiate enterobacteriaceae on the basis of malonate utilization)

Yeast extract	1.0 g
Ammonium sulfate	2.0 g

K_2HPO_4	0.6 g
KH_2PO_4	0.4 g
NaCl	2.0 g
Sodium malonate	3.0 g
Bromothymol blue (1% aq. solution)	2.5 mL

Dissolve ingredients in 1.0 L of distilled water, distribute 5 mL into test tubes and sterilize at 121°C for 15 min.

To prepare **malonate agar**, add 1.5% agar into broth.

A2.1.109 Malt extract agar/broth (a selective medium for yeasts and molds)

Malt extract	6.0 g
Maltose	1.8 g
Glucose	6.0 g
Yeast extract	1.2 g
Agar	15.0 g

Dissolve ingredients in 1.0 L of distilled water by steaming. Add agar, bring to the boil and continue to heat until the agar dissolves. Sterilize at 121°C for 15 min. Cool to 50°C and acidify to pH 4.5 or 3.5. Acidification may be achieved by adding aseptically sterile 10% lactic acid (or citric acid) solution to the medium. Mix well and pour into sterile Petri dishes. Never heat medium after the addition of acid, as heating in the acid state will hydrolyze the agar, reducing its solidifying properties so that the resulting medium will be soft.

To prepare **ME broth**, omit the agar.

To prepare **ME agar with antibiotic**: Prepare an antibiotic solution containing 500 mg each of chlortetracycline HCl and chloramphenicol in 100 mL of sterile phosphate buffered distilled water and mix. (Not all material dissolves; therefore the suspension must be evenly dispersed before pipetting into the medium.) Add 2 mL of this antibiotic solution per 100 mL of sterilized and cooled (about 50°C) ME agar or broth. After being swirled to mix, pour the medium into sterile Petri dishes.

A2.1.110 macconkey sorbitol agar (a selective and differential medium *Escherichia coli* O157:H7)

Peptone	20.0 g
Sorbitol	10.0 g
Bile salts No. 3	5.0 g
NaCl	5.0 g
Neutral red	0.075 g
Crystal violet	0.001 g
Agar	12.0 g

Dissolve ingredients except agar in 1.0 L of distilled water by steaming. Adjust pH to 7.1. Add agar, bring to the boil and continue to heat until the agar dissolves. Sterilize at 121°C for 15 min. Cool to about 45°C and pour into sterile Petri dishes.

A2.1.111 Malt extract yeast extract glucose (MY40GA) agar/broth (a medium for halophilic and osmophilic microorganisms)

Malt extract	12.0 g
Glucose	400.0 g
Yeast extract	3.0 g
Agar	12.0 g

Dissolve ingredients except glucose in 550 mL of distilled water, adjust pH to 5.5, and steam to dissolve the agar. Immediately complete to 1.0 L with distilled water. While the solution is still hot, add the glucose all at once, and stir rapidly to prevent the formation of hard lumps of glucose monohydrate. If lumps form, dissolve them by steaming for a few min. Steam the medium for 30 min. Do not sterilize. Cool to about 45°C and pour into sterile Petri dishes.

To prepare **MYG broth**, omit the agar.

A2.1.112 Mannitol egg yolk polymyxin agar (a selective medium for *Bacillus* spp.)

Beef extract	1.0 g
Peptone	10.0 g
Mannitol	10.0 g
NaCl	10.0 g
Phenol red	0.025 g
Agar	15.0 g

A2.1.112.1 Egg yolk

See egg yolk emulsion.

A2.1.112.2 Polymyxin B

Dissolve 500,000 units of sterile polymyxin B sulfate in 50.0 mL of sterile distilled water.

Dissolve ingredients except agar in 1.0 L of distilled water by steaming. Adjust pH to 7.2. Add agar, bring to the boil and continue to heat until the agar dissolves. Sterilize at 121°C for 15 min. Cool to 50°C in a water bath, and add 50 mL of egg yolk and 10 mL of polymyxin B solution. Mix well and pour into sterile Petri dishes.

A2.1.113 M-E agar (a recovery medium for *Enterococcus*)

Pancreatic digest of gelatin	10.0 g
Yeast extract	30.0 g

NaCl	15.0 g
Sodium azide	0.15 g
Esculin	1.0 g
Cycloheximide	0.05 g
Nalidixic acid	0.25 g
Agar	15.0 g

Dissolve ingredients except agar in 1.0 L of distilled water by steaming. Adjust pH to 7.1. Add agar, bring to the boil and continue to heat until the agar dissolves. Sterilize at 121°C for 15 min. Cool to 50°C and add 15 mL of sterile 1.0% triphenyl tetrazolium chloride solution. Pour into sterile Petri dishes to a depth of 4–5 mm.

A2.1.114 Membrane lauryl sulfate broth (for coliforms)

Peptone	39.0 g
Yeast extract	6.0 g
Lactose	30.0 g
Phenol red	0.2 g
Sodium lauryl sulfate	1.0 g

Dissolve ingredients in 1.0 L of distilled water by steaming. Adjust pH to 7.4 and dispense into tubes. Sterilize at 121°C for 20 min.

A2.1.115 M-HPC (a selective medium for heterotrophic plate counts)

Pancreatic digest of casein	20.0 g
Gelatin	25.0 g
Glycerol	10.0 mL
Agar	15.0 g

Dissolve ingredients except agar and glycerol in 1.0 L of distilled water by steaming. Adjust pH to 6.8. Add agar, bring to the boil and continue to heat until the agar dissolves. Add 10 mL of glycerol. Sterilize at 121°C for 20 min. Cool to about 45°C and pour into sterile Petri dishes.

A2.1.116 M-Endo agar/broth (for coliforms)

Tryptose or polypeptone	10.0 g
Thiopeptone or thiotone	5.0 g
Casitone or trypticase	5.0 g
Yeast extract	1.5 g
Lactose	12.5 g

Sodium chloride	15.0 g
Dipotassium hydrogen phosphate	4.375 g
Potassium dihydrogen phosphate	1.375 g
Sodium lauryl sulfate	0.05 g
Sodium deoxycholate	0.10 g
Sodium sulfite, Na_2SO_3	2.10 g
Basic fuchsin	1.05 g
Agar	15.0 g

Dissolve all ingredients except agar in 1.0 L distilled water containing 20 mL 95% ethanol. Lightly heat to dissolve ingredients. Adjust pH to 7.2. Add agar and heat to near boiling to dissolve agar, then promptly remove from heat and cool to 47°C. Dispense into sterile petri dishes. Do not sterilize by autoclaving.

To prepare **M-Endo broth**, omit the agar.

A2.1.117 Milk powder agar (or CPM agar)

Milk powder	100 g
Agar	15 g

Dissolve ingredients in 1 L of distilled water (except agar) and heat to dissolve ingredients completely. Adjust pH to 6.8. Add agar, heat until boil to dissolve the agar. Sterilize at 105°C for 10 min.

A2.1.118 Milk salt agar (for *Staphylococcus*)

Peptic digest of animal tissue	5.0 g
Beef extract	3.0 g
Sodium chloride	65.0 g
Agar	15.0 g

Dissolve ingredients except agar in 1 L of distilled water by steaming. Adjust pH to 7.4. Add agar, bring to the boil and continue to heat until the agar dissolves. Sterilize at 121°C for 15 min. Cool to about 45°C and pour into sterile Petri dishes.

To prepare **Milk salt broth**, omit the agar.

To prepare **Milk (Skim milk) Agar/Broth;** omit the salt.

A2.1.119 Modified AE sporulation medium (a sporulation medium)

Polypeptone	10.0 g
Yeast extract	10.0 g
K_2HPO_4	4.36 g

KH_2PO_4	0.25 g
Ammonium acetate	1.5 g
Magnesium sulfate	0.2 g

Dissolve ingredient in 1.0 L of distilled water, and adjust pH to 7.5 with 2 M sodium carbonate. Dispense the medium in 15 mL amount in test tubes, and sterilize at 121°C for 15 min. Cool to 50°C and add 0.66 mL of sterile 10% raffinose, and 0.2 mL from each of filter sterilized 0.66 M sodium carbonate and 0.32% cobalt chloride ($COCl_2 \cdot 6H_2O$) drop wise to each 15 mL of base medium. Check pH of one or two tubes. The pH should be near 7.8. Just before use, steam the medium for 10 min, and after cooling, add 0.2 mL of filter sterilized 1.5% sodium ascorbate/prepared daily to each 15 mL medium.

A2.1.120 Modified buffered peptone water with pyruvate broth (selective enrichment broth for pathogenic *Escherichia coli*)

Peptone	10.0 g
NaCl	5.0 g
Na_2HPO_4	3.6 g
KH_2PO_4	1.5 g
Casamino acids	5.0 g
Yeast extract	6.0 g
Lactose	10.0 g
Sodium pyruvate	1.0 g

Dissolve ingredients in 225 L distilled water, mix thoroughly. Adjust pH 7.2. Add agar, bring to the boil and continue to heat until the agar dissolves. Sterilize at 121°C for 15 min. Cool to 47°C and add selective supplement by mixing gently.

Selective supplement: Mix 1.125 g acriflavine, 1.125 g cefclidine, and 0.90 g vancomycin in 500 mL distilled water. Filter sterilize. Add 1 mL of each to 225 mL mBPWp (at 47°C). The concentration in mBPWp will be 10, 10, and 8 mg mL^{-1}, respectively.

A2.1.121 Modified duncan–strong sporulation broth (for *Clostridium perfringens*)

Proteose peptone	15.0 g
Yeast extract	4.0 g
Sodium thioglycolate	1.0 g
$Na_2HPO_4 \cdot 7H_2O$	10.0 g
Raffinose	40.0 g

Dissolve ingredients in 1000 mL distilled water and sterilize by autoclaving for 15 min at 121°C. Adjust to pH 7.8, using filter-sterilized 0.66 M sodium carbonate.

A2.1.122 Modified fecal coliform agar (a differential medium for fecal and nonfecal coliforms)

Tryptose	10.0 g
Proteose peptone No. 3	5.0 g
Yeast extract	3.0 g
NaCl	5.0 g
Lactose	12.5 g
Bile salts No. 3	1.5 g
Aniline blue (water blue)	0.1 g
Agar	15.0 g

Dissolve ingredients except agar in 1.0 L of distilled water by steaming. Adjust pH to 7.4. Add agar, bring to the boil, and continue to heat until the agar dissolves. Do not sterilize. Cool to 45°C and pour into sterile Petri dishes.

A2.1.123 Modified iron milk medium (for stormy fermentation test)

Fresh whole milk	1.0 L
Ferrous sulfate · $7H_2O$	1.0 g

Dissolve ferrous sulfate in 50 mL distilled water. Add slowly to 1 L milk and mix with magnetic stirrer. Dispense 11 mL medium into 16 × 150 mm culture tubes. Autoclave 12 min at 118°C. Prepare fresh medium before use.

A2.1.124 Modified Mr-VP broth (Mr-VP tests for *Bacillus cereus*)

Trypticase	10.0 g
Yeast extract	2.5 g
Glucose	5.0 g
Na_2HPO_4	2.5 g
Agar	3.0 g

Dissolve ingredients except agar in 1.0 L of distilled water by steaming. Adjust pH to 7.4. Add agar, bring to the boil and continue to heat until the agar dissolves. Dispense 10 mL portions to 13 × 100 mm tubes. Sterilize at 121°C for 15 min.

A2.1.125 Modified VP broth (AVP test medium for *Bacillus cereus*)

Proteose peptone	7.0 g
Glucose	5.0 g
NaCl	5.0 g

Dissolve the ingredients in distilled water, and dispense 5.0 mL into test tubes. Sterilize at 121°C for 15 min.

A2.1.126 Mr-VP (buffered glucose) broth (a medium for Mr-VP tests)

Peptone	7.0 g
K_2HPO_4	5.0 g
Glucose	5.0 g

Dissolve ingredients in 1.0 L of distilled water by steaming. Adjust pH to 6.9. Dispense 10.0 mL into test tubes and sterilize at 121°C for 15 min.

A2.1.127 Modified XLD agar (aeromonas media=Ryan's XLD agar) (a selective medium for the isolation of *Aeromonas hydrophila*)

Proteose peptone	5.0 g
Yeast extract	3.0 g
L-Lysine monohydrate	3.5 g
L-Arginine monohydrate	2.0 g
Inositol	2.5 g
Lactose	1.5 g
Sorbitol	3.0 g
Xylose	3.75 g
Bile salts no. 3	3.0 g
Sodium thiosulfate	10.67 g
NaCl	5.0 g
Ferric ammonium citrate	0.8 g
Bromothymol blue	0.04 g
Thymol blue	0.04 g
Agar	12.5 g

Dissolve ingredients except agar in 1.0 L distilled water, mix thoroughly. Adjust pH 8.0. Add agar, bring to the boil and continue to heat until the agar dissolves. Heat to boil. Do not autoclave. Cool to 50°C and add one vial (5.0 mg) of ampicillin selective supplement per liter of media.

Buffered motility-nitrate

A2.1.128 Mrs agar/broth (a selective medium for *Lactobacillus* spp.)

Proteose peptone no. 3	10.0 g
Beef extract	10.0 g
Yeast extract	5.0 g
Glucose	20.0 g

K_2HPO_4	1.0 g
Sodium acetate trihydrate	5.0 g
Tri-ammonium citrate	2.0 g
Manganese sulfate · $7H_2O$	0.2 g
Magnesium sulfate · $4H_2O$	0.5 g
Agar	15.0 g

Dissolve ingredients except agar in 1.0 L of distilled water by steaming. Adjust pH to 6.4. Add agar, bring to the boil and continue to heat until the agar dissolves. Sterilize at 121°C for 15 min. Cool to about 45°C and pour into sterile Petri dishes.

To prepare **Mrs broth**, omit the agar. In **carbohydrate fermentation**, add 1% carbohydrate by adding sterile carbohydrate solution into cooled medium after sterilization.

A2.1.129 Mueller–Hinton agar (a medium for antimicrobial susceptibility test)

Beef extract	300.0 g
Casamine acid	17.5 g
Starch	1.5 g
Agar	17.0 g

Dissolve ingredients except agar in 1.0 L of distilled water by steaming. Adjust pH to 7.4. Add agar, bring to the boil and continue to heat until the agar is dissolved. Sterilize at 121°C for 15 min. Cool to about 45°C and pour into sterile Petri dishes.

A2.1.130 Neomycin blood agar (for the selective isolation of *Clostridium perfringens*)

Add 1 mL of sterile neomycin sulfide solution (100 mg neomycin sulfide dissolved in 10 mL of distilled water, sterilize by membrane filtration) to 100 mL of melted blood agar before pouring into sterile Petri dishes.

A2.1.131 Neutral red chalk lactose agar (a medium for lactic streptococci)

Peptone	3.0 g
Meat extract	3.0 g
Yeast extract	3.0 g
Lactose	10.0 g
Calcium carbonate	15.0 g
Neutral red (1% aq. solution)	5.0 mL
Agar	15.0 g

Dissolve ingredients except agar in 1.0 L of distilled water by steaming. Adjust pH to 6.8. Add agar, bring to the boil and continue to heat until the agar dissolves. Sterilize at 121°C for 20 min. Cool to about 45°C and pour into sterile Petri dishes.

A2.1.132 Neutralizing broth

Pancreatic digest of casein	5.0 g
Yeast extract	2.5 g
Dextrose	10.0 g
Sodium thioglycolate	1.0 g
Sodium thiosulfate	6.0 g
Sodium bisulfite	2.5 g
Polysorbate 80	5.0 g
Lecithin	7.0 g
Bromocresol purple	0.02 g

Dissolve ingredients in 1.0 L of distilled water by steaming. Adjust pH to 7.6. Dispense 7.0 mL of medium into test tubes and sterilize at 121°C for 15 min.

A2.1.133 Nitrate broth (for nitrate test)

Beef extract	3.0 g
Peptone	5.0 g
KNO_3 (nitrite-free)	1.0 g

Dissolve ingredients in 1.0 L distilled water and adjust pH 7.0. Dispense 5 mL portions into 16 × 125 mm tubes. Autoclave 15 min at 121°C.

A2.1.134 Nutrient agar/broth (a general purpose medium for microorganisms)

Yeast extract	3.0 g
Peptone	5.0 g
NaCl	8.0 g
Agar	20.0 g

Dissolve ingredients except agar in 1.0 L of distilled water by steaming. Adjust pH to 7.0. Add agar, bring to the boil and continue to heat until the agar dissolves. Sterilize at 121°C for 15 min. Cool to about 45°C and pour into sterile petri dishes.

To prepare **nutrient broth**, omit the agar.

To prepare **NAMn agar**, dissolve 3.08 g of manganese sulfate in 100 mL of distilled water. Add 1.0 mL of this solution into 1.0 L of nutrient agar, and sterilize at 121°C for 15 min.

A2.1.135 O-nitrophenyl-beta-D-glucopyranoside peptone water (a medium for the detection of beta-galactosidase activity)

O-nitrophenyl-beta-D-glucopyranoside	0.6 g
NaH_2PO_4 buffer (0.01 M, pH 7.5)	100.0 mL

Dissolve at room temperature and sterilize by filtration.

Add aseptically 1 part of O-nitrophenyl-beta-D-glucopyranoside (ONPG) solution to 3 parts of sterile peptone broth (prepared with pH 7.5). Distribute the medium in 2.0 mL amount into sterile test tubes. Check sterility of medium by incubating at 35°C for 24 h.

A2.1.136 Orange serum agar/broth (for acid-tolerant microorganisms; *Bacillus coagulans*, *Geobacillus stearothermophilus*, lactobacilli, yeasts and molds)

Tryptone or trypticase	10.0 g
Yeast extract	3.0 g
Glucose	4.0 g
K_2HPO_4	3.0 g
Orange serum	200.0 mL
Agar	17.0 g

Dissolve ingredients except agar in 800 mL of distilled water by steaming. Adjust pH to 5.5. Add agar, bring to the boil and continue to heat until the agar dissolves. Add 200 mL of orange serum (OS) and mix well. Sterilize at 121°C for 15 min. Cool to about 45°C and pour into sterile Petri dishes.

To prepare **OS broth**, omit the agar.

A2.1.137 Osmophilic agar (for the growth of osmophilic and osmotolerant yeasts and molds)

This medium is prepared by dissolving a Worth agar in a 45° Brix syrup containing 350 g of sucrose and 100 g of glucose in 1.0 L of medium. Sterilize at 108°C for 20 min. Cool to 50°C and pour into sterile Petri dishes.

A2.1.138 Oxidation–fermentation broth (Hugh–Liefson's carbohydrate fermentation medium)

Peptone	2.0 g
NaCl	5.0 g
K_2HPO_4	0.3 g

Bromocresol purple	0.08 g
Carbohydrate	10.0

Dissolve ingredients in 1.0 L of distilled water, except carbohydrates. Adjust pH to 7.0. Dispense 7 mL into tubes containing inverted Durham tubes. Sterilize at 121°C for 10 min.

Sterilize stock carbohydrate solution (50% wt./vol.) by autoclaving or by filtration (0.2 μm pore size). Add carbohydrate solution into base broth to provide 1% (wt./vol.) final carbohydrate concentration. Following carbohydrate would be supplemented separately: adonitol, salicin, rhamnose, glucose, inositol, lactose, mannitol, raffinose, sucrose, xylose, dulcitol, and glycerol.

For halophilic *Vibrio* spp., add NaCl to a final concentration of 3.0%.

To use bromothymol blue or phenol red instead of bromothymol blue.

A2.1.139 PE-2 agar/broth (peptone yeast extract agar/broth) (plate count agar medium, mesophilic or thermophilic spore count)

Yeast extract	3.5 g
Peptone	6.0 g
bromocresol purple (2% ethanol sol.)	2.4 mL
Agar	18.0 g

Dissolve ingredients except agar in 1.0 L of distilled water by steaming. Adjust pH to 7.0. Add agar, bring to the boil and continue to heat until the agar dissolves. Sterilize at 121°C for 15 min. Cool slowly to about 50°C and add 1% carbohydrate.

To prepare **PE-2 broth**, omit the agar.

(The dilution to be created by distributing 20 mL of sample to six Petri dishes was taken into account in the calculation.)

A2.1.140 Pentachloronitrobenzene rose bengal yeast extract sucrose agar (a selective and differential medium for *Penicillium*)

Yeast extract	20.0 g
Sucrose	150.0 g
Pentachloronitrobenzene	0.1 g
Chloramphenicol	0.05 g
Chlortetracycline	0.05 g
Rose Bengal	0.025 g
Agar	20.0 g

Dissolve ingredients except agar in 1.0 L of distilled water by steaming. Adjust pH to 5.6. Add agar, bring to the boil and continue to heat until the agar dissolves. Sterilize at 121°C for 15 min. Cool to about 50°C, and add sterile chloramphenicol and chlortetracycline into medium, mix thoroughly, and pour into sterile Petri dishes.

A2.1.141 Peptone sorbitol bile salt broth (for identification of *Yersinia enterocolitica* from dairy products)

Na_2HPO_4	8.23 g
$NaH_2PO_4 \cdot H_2O$	1.2 g
Bile salts no. 3	1.5 g
NaCl	5.0 g
Sorbitol	10.0 g
Peptone	5.0 g

Dissolve ingredients in 1.0 L distilled water and adjust pH to 7.6, dispense 100 mL into bottles, and sterilize at 121°C for 15 min.

A2.1.142 Phenylalanine deaminase agar (for *phenylalanine deaminase* medium)

Yeast extract	3.0 g
L-Phenylalanine	1.0 g
Na_2HPO_4	1.0 g
NaCl	5.0 g
Agar	12.0 g

Dissolve ingredients except agar in 1.0 L of distilled water by steaming. Adjust pH to 7.3. Add agar, bring to the boil and continue to heat until the agar dissolves. Sterilize at 115°C for 10 min. Cool to about 45°C and pour into sterile Petri dishes.

A2.1.143 Plate count agar/broth (a standard plate count medium)

Yeast extract	2.5 g
Tryptone	5.0 g
Glucose	1.0 g
Agar	15.0 g

Dissolve ingredients except agar in 1.0 L of distilled water by steaming. Adjust pH to 7.0. Add agar, bring to the boil and continue to heat until the agar dissolves. Sterilize at 121°C for 15 min. Cool slowly to about 45°C and pour into sterile Petri dishes.

To prepare **PC agar with antibiotic**, add 100 mg chloramphenicol into mixed media. Sterilize at 121°C for 15 min. Cool to about 45°C and pour into sterile Petri dishes.

To prepare **PC broth**, omit the agar.

A2.1.144 Polymyxin–acriflavidine–lithium chloride–ceftazidime–aesculin–mannitol agar (a selective and differential medium for *L. monocytogenes*)

Colombia blood agar base	39.0 g
Yeast extract	3.0 g
Esculin	0.8 g
Ferric ammonium citrate	0.5 g
Mannitol	10.0 g
Phenol red	0.08 g
Lithium chloride	15.0 g

A2.1.144.1 Selective supplement

Polymyxin B	10.0 mg
Acriflavie hydrochloride	5.0 mg
Ceftazidime	10.0 mg

Dissolve ingredients except agar in 800 mL of distilled water by steaming. Adjust pH to 5.5. Add agar, bring to the boil and continue to heat until the agar dissolves. Add 200 mL orange serum and mix well. Sterilize at 121°C for 15 min. Cool to about 45°C and add selective supplement (dissolve supplements in 2.0 mL of distilled water). Pour into sterile Petri dishes.

A2.1.145 Polypectate gel agar/broth

Sodium polypectate	70.0 g
Peptone	5.0 g
K_2HPO_4	5.0 g
KH_2PO_4	1.0 g
$CCl_2 \cdot 2H_2O$	0.6 g

Heat 500 mL of distilled water in blender, and place all ingredients except sodium polypectate and blend to dissolve. Add polypectate last, in small amounts per addition, with slow stirring to diminish exclude air. Adjust pH to 7.0, and then add 500 mL of the distilled water. Sterilize at 121°C for 15 min, cool to 48°C, and pour into sterile Petri dishes.

To prepare **PG agar**, add 1.5% agar into PG broth.

A2.1.146 Polymyxin pyruvate egg yolk mannitol bromothymol blue agar (for the cultivation of *Bacillus cereus*)

Peptic digest of animal tissue	1.0 g
Mannitol	10.0 g

Sodium pyruvate	10.0 g
N_2PO_4	2.5 g
NaCl	2.0 g
KH_2PO_4	0.25 g
$MgSO_4$	0.1 g
Bromo thymol blue	0.1 g
Agar	18.0 g
Polymixin B	100,000 IU
Egg-yolk emulsion	50.0 mL

Dissolve ingredients except agar, polymyxin B and egg-yolk emulsion in 1.0 L of distilled water by steaming. Adjust pH to 7.4. Add agar, bring to the boil and continue to heat until the agar dissolves. Sterilize at 121°C for 15 min. Cool to about 50°C. Aseptically add supplement: sterile rehydrated contents of one vial Polymixin B supplement and 50 mL of sterile egg-yolk emulsion. Mix well and pour into sterile Petri plates.

Supplement: Rehydrate the contents of one vial aseptically with 5 mL sterile distilled water and mix well. Prepare egg-yolk emulsion as indicated in "Egg-Yolk Emulsion".

A2.1.147 Potato dextrose agar/broth (acidified) (a selective medium for yeasts and molds)

Potato, infusion from	200.0 g
Glucose	20.0 g
Agar	15.0 g

Boil 200 g of peeled potatoes for 1 h in 1.0 L of distilled water. Filter and make up the filtrate to 1 L. Add glucose and mix. Adjust pH to 5.6. Add agar, heat to dissolve agar. Sterilize at 121°C for 15 min.

Acidification: After sterilization, cool to about 50°C and acidify sterilized medium to 3.5 with sterile 10% tartaric acid solution. Mix thoroughly, and pour into sterile Petri dishes. To preserve the solidifying properties of the agar, do not heat the medium after the addition of tartaric acid.

To prepare **PD broth**, omit the agar.

A2.1.148 Potato dextrose agar/broth with antibiotic (a selective medium for yeasts and molds)

Prepare an antibiotic solution containing 500 mg each of chlortetracycline HCl and chloramphenicol in 100 mL of sterile phosphate buffered distilled water and mix. (Not all material dissolves; therefore, the suspension must be evenly dispersed before pipetting into the medium.) Add 2 mL of this antibiotic solution per 100 mL of sterilized and cooled (about 50°C) potato dextrose agar or broth. After being swirled to mix, and pour the medium into sterile Petri dishes.

A2.1.149 Preenrichment medium (an enrichment medium for *Yersinia enterocolitica*)

Na_2HPO_4	7.1 g
NaCl	1.0 g
Potassium chloride	1.0 g
Special peptone	10.0 g
Yeast extract	20.0 g

Dissolve ingredients in 1.0 L of distilled water by steaming. Dispense the medium in 10 mL amount into test tubes. Sterilize at 121°C for 15 min. Cool to about 50°C, and add the following filter-sterilized (0.22 μm pore size) solution to give the final concentrations indicated per L of medium:

Magnesium sulfate · $7H_2O$	10.0 mg
Calcium chloride	10.0 mg

A2.1.150 Preston enrichment broth (a selective enrichment broth for *Campylobacter*)

Nutrient broth	90.0 mL
Lysed horse blood	50.0 mL
Polymyxin B	5000.0 IU
Rifampicin	10.0 mg
Trimethoprim lactate	10.0 mg
Cycloheximide	100.0 mg

Dissolve ingredients except antibiotics in distilled water by steaming. Dispense 10 mL into test tubes. Sterilize at 121°C for 15 min. Cool to about 50°C, and add 5% blood and filter sterilized (0.22 μm filter) antibiotics.

A2.1.151 Pseudomonas cetrimide agar (for the selective isolation and counting of *Pseudomonas aeruginosa*)

Gelatin peptone	20.0 g
Magnesium chloride	1.4 g
Potassium sulfate	10.0 g
Cetrimide	0.3 g
Glycerol	10.0 mL
Agar	13.6 g

Dissolve ingredients except agar in 1.0 L of distilled water by steaming. Adjust pH to 7.2. Add 10 mL of glycerol and boil to dissolve completely. Add agar, bring to the boil and continue to heat until the agar dissolves. Sterilize at 121°C for 15 min. Cool to about 50°C and pour into sterile Petri dishes.

A2.1.152 Pyrazine–amidase agar (for the identification of *Yersinia*)

Tryptic (trpticase) soy agar	30.0 g
Yeast extract	3.0 g
Pyrazine-carboxamide	1.0 g

Dissolve ingredients in 1.0 L 0.2 M Tris-malate (pH 6.0), mix thoroughly. Heat to boiling, dispense 5 mL into tubes. Sterilize at 121°C for 15 min.

A2.1.153 PYR broth

Peptone	20.0 g
Brain heart infusion	3.1 g
Sodium carbonate	2.5 g
Dextrose	2.0 g
Sodium chloride	2.0 g
Na_2PO_4	0.4 g
L-Pyroglutamic acid-beta-naphthylamide	0.1 g

Dissolve ingredients in 1.0 L of distilled water by steaming. Adjust pH to 7.8. Dispense int test tubes in 7 mL amount. Sterilize at 121°C for 15 min. Do not overheat.

A2.1.154 RA-2 agar (for enumeration of heterotrophic microorganisms)

Yeast extract	0.5 g
Peptone	0.5 g
Acid hydrolysate casein	0.5 g
Glucose	0.5 g
Soluble starch	0.5 g
K_2HPO_4	0.3 g
Magnesium sulfate (anhydrous)	0.024 g
Sodium pyruvate	0.3 g
Agar	15.0 g

Dissolve ingredients except agar in 1.0 L of distilled water by steaming. Adjust pH to 7.2. Add agar, bring to the boil and continue to heat until the agar dissolves. Sterilize at 121°C for 15 min. Do not overheat. Cool to about 45°C and pour into sterile Petri dishes.

A2.1.155 Rappaport–Vassiliadis medium

A2.1.155.1 Broth base

Tryptone	5.0 g
NaCl	8.0 g
KH_2PO_4	1.6 g
Distilled water	1.0 L

A2.1.155.2 Magnesium chloride solution

$MgCl_2 \cdot 6H_2O$	400.0 g
Distilled water	1.0 L

A2.1.155.3 Malachite green oxalate solution

Malachite green oxalate	0.4 g
Distilled water	100.0 mL

To prepare the complete medium, combine 1.0 L mL broth base, 100 mL magnesium chloride solution, and 10 mL malachite green oxalate solution (total volume of complete medium is 1110 mL). Adjust the final pH to 5.5. Broth base must be prepared on the same day that components are combined to make a complete medium. Magnesium chloride solution may be stored in dark bottle at room temperature up to 1 year. To prepare solution, dissolve entire contents of $MgCl_2 \cdot 6H_2O$ from newly opened container according to formula, because this salt is very hygroscopic. Malachite green oxalate solution may be stored in dark bottle at room temperature up to 6 months. Dispense 10 mL volumes of complete medium into test tubes. Sterilize at 115°C for 15 min.

A2.1.156 Reinforced clostridial agar/broth

Yeast extract	3.0 g
Peptone	10.0 g
Lab-Lemco meat extract	10.0 g
D-Glucose	5.0 g
Sodium acetate	5.0 g
L-Cysteine hydrochloride	0.5 g
Soluble starch	1.0 g

Dissolve ingredients except agar in 1.0 L of distilled water by steaming. Adjust pH to 7.1. Add agar, bring to the boil, heat until the agar dissolves. Sterilize at 121°C for 15 min. Do not overheat.

To prepare **RC broth**, omit agar.

A2.1.157 Rogosa agar (for selective medium for lactobacilli)

Tryptone or trypticase	10.0 g
Yeast extract	5.0 g
D-Glucose	20.0 g
Tween 80	1.0 g
Potassium dehydrate phosphate	6.0 g
Ammonium citrate	2.0 g
Sodium acetate	25.0 g
Glacial acetic acid	1.32 g
$MgSO_4.7H_2O$	0.575 g
$MnSO_4.4H_2O$	0.14 g
$FeSO_4.7H_2O$	0.034 g
Agar	15.0 g

Dissolve ingredients except agar in 1.0 L of distilled water by steaming. Adjust pH to 5.4. Add agar, bring to the boil and continue to heat until the agar dissolves. Do not sterilize in autoclave. Cool and pour into sterile Petri dishes.

A2.1.158 Sabouraud dextrose agar (a selective medium for yeasts and molds)

Peptone	10.0 g
Glucose	40.0 g
Agar	5.0 g

Dissolve ingredients except agar in 1.0 L of distilled water by steaming. Adjust pH to 5.6. Add agar, bring to the boil and continue to heat until the agar dissolves. Dispense into test tubes and sterilize at 121°C for 15 min. Avoid overheating which could result in a softer agar medium, especially those with a low pH.

A2.1.159 Salmonella-shigella agar (a differential selective medium for *Salmonella* and *Shigella*)

Lab-Lemco powder	5.0 g
Peptone	5.0 g
Lactose	10.0 g
Bile salts	8.5 g
Sodium citrate	10.0 g
Sodium thiosulfate	8.5 g
Ferric citrate	1.0 g

Brilliant green	0.00018 g
Neutral red	0.025 g
Agar	15.0 g

Dissolve ingredients except agar in 1.0 L of distilled water by steaming. Adjust pH to 7.0. Add agar, bring to the boil, heat until the agar dissolves. Do not autoclave. Cool to about 50°C, mix and pour into sterile Petri dishes.

A2.1.160 Sea water agar (a selective medium for halophilic microorganisms)

Yeast extract	5.0 g
Peptone	5.0 g
Beef extract	3.0 g
Agar	15.0 g
Sea water (synthetic)	1.0 L

Dissolve ingredients in synthetic seawater and adjust pH to 7.5. Add agar, bring to the boil and continue to heat until the agar is dissolved. Sterilize at 121°C for 15 min. Cool to 45°C and pour into sterile Petri dishes.

A2.1.161 Seller's differential agar (for gram-negative nonfermentative bacilli: *Pseudomonas aeruginosa* and *Acinetobacter calcoaceticus*)

Yeast extract	1.0 g
Peptic digest of animal tissue	20.0 g
L-Arginine	1.0 g
D-Mannitol	2.0 g
Sodium chloride	2.0 g
Sodium nitrate	1.0 g
Sodium nitrite	0.35 g
Magnesium sulfate	1.50 g
Dipotassium phosphate	1.0 g
Bromo thymol blue	0.04 g
Phenol red	0.008 g
Agar	15.0 g

Dissolve ingredients in 1 L of distilled water (except agar) and heat to dissolve ingredients completely. Adjust pH to 6.7. Add agar, heat until boil to dissolve the agar. Sterilize at 121°C for 10 min. Cool the medium to 50oC and add 22 mL (0.15 mL or two drops per 7 mL medium in tube) of 50% sterile dextrose solution.

A2.1.162 Semisolid brucella medium (a differential medium for *brucella* and *campylobacter*)

Tryptone	10.0 g
Peptamin	10.0 g
Glucose	1.0 g
Yeast extract	2.0 g
NaCl	5.0 g
Sodium bisulfide	0.1 g
Cystine hydrochloride	0.2 g
Glycine	10.0 g
Agar	1.6 g

Dissolve ingredients except agar in 1.0 L of distilled water. Adjust pH to 7.0. Add agar, heat to boiling until the agar dissolves. Distribute 10 mL into test tubes and sterilize at 121°C for 15 min. Cool tubes in upright position.

A2.1.163 Shigella broth

A2.1.163.1 Base

Tryptone	20.0 g
K_2HPO_4	2.0 g
KH_2PO_4	2.0 g
NaCl	5.0 g
Glucose	1.0 g
Tween 80	1.5 mL

Dissolve ingredients in 1.0 L distilled. Adjust pH to 7.0 and sterilize at 121°C for 15 min.

A2.1.163.2 Novobiocin solution

Dissolve 300 mg novobiocin in 1.0 L distilled water. Sterilize by filtration through 0.45 μm membrane. Add 2.25 mL concentrate to 225 mL base.

A2.1.164 Sulfide, indole, motility agar

Casein peptone	20.0 g
Ferrous ammonium sulfate	0.2 g
Meat peptone	6.6 g
Sodium thiosulfate	0.2 g
Agar	3.0 g

Dissolve ingredients except agar in 1.0 L of distilled water by steaming. Adjust pH to 7.3. Add agar, bring to the boil and continue to heat until the agar dissolves. Dispense 7 mL into tubes. Sterilize at 121°C for 15 min. Cool to about 50°C, and add 100,000 IU of polymyxin B sulfate. Pour into sterile Petri dishes.

A2.1.165 Simmons citrate agar (a citrate test medium)

Ammonium dihydrogen phosphate	1.0 g
K_2HPO_4	1.0 g
NaCl	5.0 g
Sodium citrate	2.0 g
$MgSO_4.7H_2O$	0.2 g
Bromothymol blue	0.08 g
Agar	15.0 g

Dissolve ingredients except agar in 1.0 L of distilled water by steaming. Adjust pH to 6.8. Add agar, bring to the boil and continue to heat until the agar dissolves. Dispense into test tubes and sterilize at 121°C for 15 min.

A2.1.166 Skim milk (5%) (a medium for the propagation of microorganisms souring milk and milk products)

Dissolve 5.0 g of skim milk in 100 mL distilled water. Dispense into test tubes and sterilize at 121°C for 5 min.

A2.1.167 Sodium dodecyl sulfate polymyxin sucrose agar

Proteose peptone	10.0 g
Beef extract	5.0 g
Sucrose	15.0 g
NaCl	20.0 g
Sodium dodecyl sulfate (sodium lauryl sulfate)	1.0 g
Bromothymol blue	0.04 g
Cresol red	0.04 g
Agar	15.0 g

Dissolve ingredients except agar in 1.0 L of distilled water by steaming. Adjust pH to 7.6. Add agar, bring to the boil and continue to heat until the agar dissolves. Sterilize at 121°C for 15 min. Cool to about 50°C, and add 100,000 IU of polymyxin B sulfate. Pour into sterile Petri dishes.

A2.1.168 Sodium citrate brain heart infusion broth (for isolation of *Brucella* spp.)

Sterilize 1.0 L of BHIB at 121°C for 15 min. Cool to about 50°C and add aseptically filter-sterilized solutions of sodium citrate to give 1% and 10 mL of 1% serum albumin.

A2.1.169 Sodium lactate agar

Trypticase	10.0 g
Yeast extract	10.0 g
K_2HPO_4	0.25 g
Agar	15.0 g

Dissolve ingredients in 1.0 L of distilled water by steaming. Adjust pH to 7.0. Add agar, bring to the boil and continue to heat until the agar dissolves. Sterilize at 121°C for 20 min. Cool to about 50°C and pour into sterile Petri dishes.

A2.1.170 Sorbitol–MacConkey agar

Peptone or gelysate	17.0 g
Protease peptone no. 3 or polypeptone	3.0 g
Sorbitol	10.0 g
Bile salts, purified	1.5 g
NaCl	5.0 g
Neutral red	0.03 g
Crystal violet	0.001 g
Agar	13.5 g

Dissolve ingredients except agar in 1.0 L distilled water, mix thoroughly. Adjust pH 7.1. Add agar, bring to the boil and continue to heat until the agar dissolves. Sterilize at 121°C for 15 min.

A2.1.171 Spirit blue milk fat agar

Pancreatic digest of casein	10.0 g
Yeast extract	5.0 g
Agar	20.0 g
Spirit blue	0.15 g

Dissolve ingredients (except agar) in 1 L of distilled water. Mix thoroughly. Add agar and heat with frequent agitation and boil for 1 min to completely dissolve the agar. Sterilize at 121°C for 15 min. Cool to 50°C. Aseptically add 30 mL Lipase Reagent (3%) or other lipid source and mix thoroughly and pour into sterile Petri dishes.

A2.1.172 Sporulation broth (for *Clostridium perfringens*)

Polypeptone or tryptose	15.0 g
Yeast extract	3.0 g
Soluble starch	3.0 g
Magnesium sulfate	0.1 g
Sodium thioglycolate	1.0 g
Na_2HPO_4	11.0 g

Dissolve ingredients in 1.0 L of distilled water by steaming. Adjust pH to 7.8. Distribute into test tubes. Sterilize at 121°C for 15 min. Before use, exhaust in flowing steam for 20 min.

A2.1.173 Spray's fermentation medium (for *Clostridium perfringens*)

Tryptone	10.0 g
Neopeptone	10.0 g
Sodium thioglycolate	0.25 g
Agar	2.0 g

Dissolve all ingredients except agar in 1 L distilled water by heating and adjust pH to 7.4. Add agar and heat with agitation to dissolve the agar. Dispense 9 mL portions into 16 × 125 mm tubes. Autoclave 15 min at 121°C. Before use, heat in boiling water or flowing steam for 10 min. Add 1 mL of 1% carbohydrate from sterile carbohydrate solution to 9 mL base.

A2.1.174 Staphylococci medium No. 110 (a selective medium for *Staphylococcus aureus*)

Peptone	10.0 g
Yeast extract	2.5 g
Lactose	2.0 g
Gelatin	30.0 g
Mannitol	10.0 g
NaCl	75.0 g
K_2HPO_4	5.0 g
Bromothymol blue	0.02 g
Agar	15.0 g

Dissolve ingredients except agar in 1.0 L of distilled water by steaming. Adjust pH to 7.0. Add agar, bring to the boil and continue to heat until the agar dissolves. Sterilize at 121°C for 15 min. Cool to about 45°C and pour into sterile Petri dishes.

A2.1.175 Starch agar (a medium for starch hydrolysis test)

Yeast extract	3.0 g
Peptone	5.0 g
Potato (soluble) starch	10 g
Agar	15.0 g

Dissolve ingredients except agar in 1.0 L of distilled water by steaming. Adjust pH to 6.8. Add agar, bring to the boil and continue to heat until the agar dissolves. Dispense into test tubes and sterilize at 121°C for 15 min. Cool to about 50°C and pour into sterile Petri dishes.

To prepare **starch ampicillin agar** (a selective medium for *Aeromonas hydrophila*) add ampicillin (30 mg L^{-1}) into a cooled medium (at 50°C) after sterilization.

A2.1.176 Streptomycin thallous acetate actidione agar (for the isolation of *Brochothrix thermosphacta* from food samples)

Peptone	20.0 g
Yeast extract	2.0 g
Dipotassium hydrogen phosphate	1.0 g
Magnesium sulfate	1.0 g
Agar	13.0 g

Dissolve ingredients in 1 L of distilled water (except agar) and heat to dissolve ingredients completely. Adjust pH to 7.0. Add agar, heat until boil to dissolve the agar. Sterilize at 121°C for 10 min. Cool the medium to 50o̲C and add following selective supplements.

A2.1.177 Streptomycin thallous acetate actidione selective supplement (g per L of medium)

Streptomycin sulfate	500.0 mg
Thallous acetate	50.0 mg
Cycloheximide	50.0 mg

A2.1.178 Sucrose agar/broth

Sucrose	10.0 g
Yeast extract	3.0 g
Peptone	5.0 g
Agar	15.0 g

Dissolve ingredients in 1.0 L of distilled water by steaming. Adjust pH to 6.9. Add agar, bring to the boil and continue to heat until the agar dissolves. Sterilize at 115°C for 20 min. Cool to about 50°C and pour into sterile Petri dishes.

To prepare **sucrose broth**, omit agar.

A2.1.179 Sulfite indole motility medium (an indole, motility, and H_2S test medium for anaerobes)

Yeast extract	3.0 g
Peptone	10.0 g
Tryptone	10.0 g
Glucose	1.0 g
L-Lysine hydrochloride	10.0 g
L-Cystine	0.2 g
Ferric ammonium citrate	0.2 g
Sodium citrate	2.0 g
NaCl	5.0 g
Gelatin	80.0 g
Bromocresol purple	0.02 g
Agar	4.0 g

Dissolve ingredients except agar in 1.0 L of distilled water by steaming. Adjust pH to 7.3. Add agar, bring to the boil and continue to heat until the agar dissolves. Dispense 5.0 mL into test tubes and sterilize at 121°C for 15 min.

A2.1.180 Tellurite cefixime sorbitol macconkey agar

Peptone	20.0 g
Bile salts no. 3	1.5 g
Crystal violet	0.001 g
Sorbitol	10.0 g
Sodium chloride	5.0 g
Neutral red	0.03 g
Agar	15.0 g

Dissolve ingredients in 1.0 L distilled. Adjust pH to 7.1 and sterilize at 121°C for 15 min. Add 10 mL of following supplement after cooling of medium to 50°C.

A2.1.180.1 Supplement

Cefixime	0.05 mg
Potassium tellurite	2.5 mg

Dissolve ingredients in 10 mL of distilled water and sterilize by autoclaving or filter sterilizing.

A2.1.181 Tellurite mannitol glycine agar/broth (for *Staphylococcus aureus*)

Lithium chloride	10.0 g
Lab-Lemco' powder	5.0 g
Yeast extract	5.0 g
Lithium chloride	5.0 g
Mannitol	20.0 g
Sodium chloride	5.0 g
Glycine	1.2 g
Sodium pyruvate	3.0 g

Dissolve ingredients in 1.0 L of distilled water by steaming. Adjust pH to 6.9. Add agar, bring to the boil and continue to heat until the agar dissolves. Sterilize at 115°C for 20 min. Cool to about 50°C and add 10 mL telluride solution (1%) to the medium at 50°C–55°C. Mix well and pour into sterile Petri dishes.

To prepare Tellurite mannitol glycine **broth**; omit agar.

A2.1.182 Tetrathionate broth (a selective enrichment medium for *Salmonella*)

Proteose peptone	5.0 g
Bile salts	1.0 g
Sodium thiosulfate	30.0 g
Calcium carbonate	10.0 g

Dissolve ingredients in 1.0 L of distilled water by steaming. Adjust pH to 8.0. Heat to boiling. Dispense into sterile test tubes. Cool to about 50°C and add 2 mL of iodine solution to each 100 mL of base. Add 1 mL of 1/1000 solution of brilliant green per 100 mL base. Sulfathiazole (0.125 mg per mL of medium) may be added to prevent excessive growth of products. Do not autoclave after addition of iodine solution.

A2.1.182.1 Iodine solution

Iodine	6.0 g
Potassium iodide	5.0 g

Dissolve ingredients in 20 mL of distilled water.

A2.1.183 Thermoacidurans agar (a medium for *Bacillus thermoacidurans*)

Yeast extract	6.0 g
Proteose peptone	6.0 g
Glucose	6.0 g

K_2HPO_4	4.7 g
Agar	20.0 g

Dissolve ingredients except agar in 1.0 L of distilled water by steaming. Adjust pH to 5.0. Add agar, bring to the boil and continue to heat until the agar dissolves. Sterilize at 121°C for 15 min. Cool to 50°C, and pour into sterile Petri dishes.

(The dilution to be created by distributing 20 mL of sample to six Petri dishes was taken into account in the calculation.)

A2.1.184 Thioglycolate agar (a medium for anaerobic, microaerophilic and aerobic microorganisms)

Fluid thioglycolate medium	1.0 L
Agar	20.0 g

Suspend the agar in *FT* medium. Heat to boiling, and continue to heat until the agar dissolves. Sterilize at 121°C for 15 min. Cool to about 45°C, and pour into sterile Petri dishes.

A2.1.185 Thiosulfate citrate bile salt sucrose agar (a selective and differential medium *vibrio*)

Sodium thiosulfate	10.0 g
Sodium citrate.$2H_2O$	10.0 g
Oxgall	5.0 g
Sodium cholate	5.0 g
Sucrose	20.0 g
Pancreatic digest of casein	5.0 g
Pancreatic digest of animal tissue	5.0 g
Yeast extract	5.0 g
NaCl	10.0 g
Iron citrate	1.0 g
Thymol blue	0.04 g
Bromothymol blue	0.04 g
Agar	14.0 g

Dissolve ingredients except agar in 1.0 distilled water. Adjust pH to 8.6. Add agar, bring to the boil and continue to heat until the agar dissolves. Do not autoclave. Cool to 45°C, and dispense into sterile Petri dishes.

A2.1.186 T_1N_1 agar/broth (a medium for cultivation of *Vibrio cholerae*)

Trypticase (pancreatic digest of casein)	10.0 g
NaCl	10.0 g
Agar	15.0 g

Dissolve ingredients except agar in 1.0 L of distilled water. Adjust pH to 7.2. Add agar, bring to the boil and continue to heat until the agar is dissolved. Dispense into test tubes, and sterilize at 121°C for 15 min. Allow to solidify in an inclined position (long slant).

To prepare **T_1N_1 broth**, omit the agar.

A2.1.187 Toluidine blue deoxyribonucleic acid agar (a medium for detection of deoxyribonuclease activity of microorganisms)

$CaCl_2$	5.5 g
NaCl	10.0 g
Toluidine blue	0.083 g
Deoxyribonucleic acid	0.3 g
Tris (hydroxymethyl) aminomethane	6.1 g
Agar	10.0 g

Add Deoxyribonucleic acid (DNA) to cold water, and heat slowly while stirring. Add the rest of the ingredients except agar into DNA solution, mix well and adjust pH to 9.0. Add agar, bring to the boil and continue to heat until the agar dissolves. Do not autoclave. Cool to 45°C and pour into sterile Petri dishes.

A2.1.188 Tomato juice agar (a medium for yeasts and other aciduric microorganisms)

Tomato juice	20.0 g
Yeast extract	10.0 g
Glucose	10.0 g
K_2HPO_4	0.5 g
KH_2PO_4	0.5 g
Magnesium sulfate	0.2 g
Peptone	10.0 g
NaCl	0.01 g
Ferrous sulfate	0.01 g
Manganese sulfate	0.01 g
Agar	15.0 g

Dissolve ingredients except agar in 1.0 L of distilled water. Adjust pH to 6.7. Add agar, bring to the boil and continue to heat until the agar dissolves. Sterilize at 121°C for 15 min and pour into sterile Petri dishes.

A2.1.189 Transport media

Transport media are employed in the safe collection, transportation, and preservation of microbiological samples. They are formulated with minimal nutrients to increase microbial survival without replication. A relatively high pH minimizes the extraction of bacteria due to acid formation.

A2.1.190 Campylobacter thioglycolate medium with antibiotics (use as transport media for sample containing *campylobacter*)

Pancreatic digest of casein	17.0 g
Pancreatic digest of soybean meal	3.0 g
Glucose	6.0 g
NaCl	2.5 g
Sodium thioglycolate	0.5 g
Agar	1.6 g
Cystine	0.25 g
Sodium sulfite	0.1 g
Amphotericin B	2.0 mg
Cephalothin	15.0 g
Trimethoprim	5.0 g
Vancomycin	10.0 g
Polymyxin B	2500.0 IU

Dissolve ingredients except agar in 1.0 L of distilled water by steaming. Adjust pH to 7.4. Add agar, bring to the boil and continue to heat until the agar dissolves. Sterilize at 121°C for 15 min. Aseptically add presterilized antibiotic solution and dispense 7.0 mL into test tubes.

A2.1.191 Cary and blair transport medium (use as a transport medium for samples containing microorganisms)

Sodium thioglycolate	1.5 g
Na_2HPO_4	1.1 g
NaCl	5.0 g
Agar	5.0 g

Dissolve ingredients except agar in 1.0 L of distilled water by steaming. Add agar, bring to the boil and continue to heat until the agar dissolves. Cool to about 50°C and add 9 mL of 1.0% aq.

calcium chloride solution. Adjust pH to 8.0. Dispense 7.0 mL into test tubes and sterilize at 121°C for 15 min.

A2.1.192 Stuart transport medium (use as a transport medium for bacteria, fungi, or parasites)

Sodium thioglycolate	1.0 g
Sodium glycerophosphate	10.0 g
Calcium chloride	0.1 g
Methylene blue	0.002 g
Agar	3.0 g

Dissolve ingredients except agar in 1.0 L of distilled water by steaming. Adjust pH to 7.3. Add agar, bring to the boil and continue to heat until the agar dissolves. Dispense 5.0 mL into test tubes and sterilize at 121°C for 15 min.

A2.1.193 Swab amies transport medium (a general transport medium)

Charcoal	10.0 g
Na_2HPO_4	1.15 g
NaCl	3.0 g
KH_2PO_4	0.2 g
Potassium chloride	0.2 g
Sodium thioglycolate	1.0 g
Calcium chloride	0.1 g
Magnesium chloride	0.1 g
Agar	4.0 g

Dissolve ingredients except agar in 1.0 L of distilled water by steaming. Adjust pH to 7.3. Add agar, bring to the boil and continue to heat until the agar dissolves. Dispense into test tubes and sterilize at 121°C for 15 min.

A2.1.194 Swab Stuart's transport medium (use as a transport medium for bacteria, fungi or parasites)

Sodium glycerophosphate	10.0 g
Calcium chloride	0.1 g
Mercaptoacetic acid	1.0 g
Agar	7.5 g

Dissolve ingredients except agar in 1.0 L of distilled water by steaming. Adjust pH to 7.4. Add agar, bring to the boil and continue to heat until the agar is dissolved. Dispense the medium in 5.0 mL amount into test tubes and sterilize at 121°C for 15 min.

A2.1.195 Transport medium (use as a transport medium for bacteria, yeasts, or molds)

Charcoal	10.0 g
Sodium thioglycolate	1.0 g
Sodium glycerophosphate	10.0 g
Calcium chloride	0.1 g
Methylene blue	0.002 g
Agar	3.0 g

Dissolve ingredients except agar in 1.0 L of distilled water by steaming. Adjust pH to 7.4. Add agar, bring to the boil and continue to heat until the agar dissolves. Dispense 5.0 mL into test tubes and sterilize at 121°C for 15 min.

A2.1.196 Tributyrin agar (a test medium for lipolytic activity)

Peptone	5.0 g
Yeast extract	3.0 g
Tributyrin (glyceryl tributyrate)	10.0 g
Agar	15.0 g

Dissolve ingredients except agar in 1.0 L of distilled water and adjust pH to 7.5. Add agar, heat until the agar dissolves. Sterilize at 121°C for 20 min.

Lipolytic test. Prepare 1/10, 1/100, 1/1000, and 1/10000 homogenates of the melted butter in quarter-strength Ringer's solution and within 10 min, to transfer 1 mL quantities of each dilution to separate sterile Petri dishes. Add 16 mL of the medium (cooled to 50°C) into Petri dishes, mix. Incubate plates at 30°C for 3 days. After incubation, the medium appears opaque but a zone of clear medium surrounds lipolytic colonies.

A2.1.197 Trimethylamine N-oxide broth (an identification medium for *Campylobacter*)

Nutrient broth	25.0 g
Yeast extract	1.0 g
New Zealand agar	2.0 g
Trimethylamine N-oxide	1.0 g

Dissolve ingredients in 1.0 L of distilled water by steaming. Dispense 4.0 mL into test tubes and sterilize at 121°C for 15 min.

A2.1.198 Triple sugar iron agar (a medium for differentiation of microorganisms by three sugar fermentation and H_2S production)

Beef extract	3.0 g
Yeast extract	3.0 g
Peptone	15.0 g
Proteose peptone	5.0 g
NaCl	5.0 g
Lactose	10.0 g
Sucrose	10.0 g
Glucose	10.0 g
Ferrous ammonium sulfate	0.2 g
Sodium thiosulfate	0.2 g
Phenol red	0.024 g
Agar	13.0 g

Dissolve ingredients except agar in 1.0 L of distilled water by steaming. Adjust pH to 7.3. Add agar, bring to the boil and continue to heat until the agar dissolves. Dispense 8 mL into test tubes and sterilize at 121°C for 15 min. Cool in slanted position.

A2.1.199 Trypticase agar/broth (a medium for differentiation of motility test and fermentation reactions)

Tryptone	20.0 g
Phenol red	0.02 g
Bile salts no. 3	1.5 g
K_2HPO_4	1.5 g
Agar	15.0 g

Dissolve ingredients except agar in 1.0 L of distilled water by heating while stirring. Adjust pH to 7.2. Sterilize at 121°C for 15 min.

For use in fermentation studies, add carbohydrate in the desired concentration (usually 10.0 g L^{-1}) and adjust pH if necessary. Dispense into tubes, filling them half full. Sterilize at 121°C for 15 min. Store at room temperature.

To prepare **trypticase novobiocin agar**, prepare a stock solution of novobiocin in distilled water. Add enough filter-sterilized novabiocin solution to equal 20.0 mg L^{-1} of cooled medium (at 50°C). Inoculate medium using a needle by stabbing the needle to one-half the depth of the agar column. After incubation, look for carbohydrate fermentation and motility.

To prepare **trypticase broth**, omit agar.

A2.1.200 Trypticase peptone glucose yeast extract agar/broth (a medium for *Clostridium perfringens*)

Trypticase or tryptone	10.0 g
Peptone	5.0 g
Yeast extract	20.0 g
Glucose	4.0 g
Sodium thioglycolate	1.0 g
Agar	15.0 g

Dissolve ingredients except agar in 1.0 L of distilled water by steaming. Adjust pH to 7.3. Add agar, bring to the boil and continue to heat until the agar dissolves. Sterilize at 121°C for 15 min. Cool to about 45°C and pour into sterile Petri dishes.

To prepare **TPGY broth with trypsin**, after steaming TPGY broth to drive off oxygen and cooling, add the trypsin to the TPGY broth immediately before inoculating to give a final concentration of 0.1%. (Prepare a 1.5% aq. solution of trypsin. Sterilize by filtration through a 0.45 μm filter, and refrigerate until use.)

A2.1.201 Trypticase (tryptone) soy agar/broth

Trypticase peptone	17.0 g
Phytone	3.0 g
NaCl	5.0 g
K_2HPO_4	2.5 g
Glucose	2.5 g

Dissolve ingredients except agar in 1.0 L of distilled water by steaming. Adjust pH to 7.3. Add agar, bring to the boil and continue to heat until the agar dissolves. Sterilize at 121°C for 15 min.

To prepare **TS broth**, omit the agar. Dispense 8 mL into test tubes and sterilize at 121°C for 15 min.

To prepare **TS−polymyxin broth**; Dissolve 500,000 units polymyxin B sulfate in 33.3 mL distilled water. Filter-sterilize add 0.053 mL sterile 0.15% polymyxin B solution to 8 mL TS broth, and mix thoroughly.

For **Trypticase soy-blood agar** (without polymyxin); Cool sterilized agar to 50ºC. Add 5 mL defibrinated sheep blood to 100 mL agar. Mix and dispense 20 mL portions to 15 × 100 mm petri dishes.

A2.1.202 Trypticase (Tryptic) soy agar/broth (a general-purpose medium for microorganisms)

Tryptone	17.0 g
Soy (phytone) peptone (soytone)	3.0 g

Glucose	2.5 g
NaCl	5.0 g
K_2HPO_4	2.5 g
Agar	15.0 g

Dissolve ingredients except agar in 1.0 L of distilled water by steaming. Adjust pH to 7.3. Add agar, bring to the boil and continue to heat until the agar dissolves. Sterilize at 121°C for 15 min.

To prepare **TS broth**, omit the agar.

To prepare **TS ampicillin agar** (a selective enrichment media for *A. hydrophila*), add 30 mg of ampicillin into cooled (50°C) 1.0 L of TS medium and pour into sterile Petri dishes.

To prepare **TS blood agar**, add 5% sheep blood into agar medium.

To prepare **TS polymyxin broth**, add 0.1 mL of 0.15% polymyxin B sulfate solution (add 33.3 mL of sterile distilled water into 50 mg polymyxin B sulfate with a sterile syringe to give a 0.15% solution).

To prepare **TS agar magnesium sulfate-NaCl medium**, add 30.0 g $MgSO_4.7H_2O$ into 1.0 L of cooled (50°C) TS agar medium.

A2.1.203 Tryptone broth (indole test medium)

Tryptone	10.0 g
NaCl	5.0 g

Dissolve ingredients in 1.0 L of distilled water and mix by steaming. Adjust pH to 7.5. Filter the solution through filter paper. Dispense 5.0 mL into test tubes. Sterilize at 121°C for 15 min.

A2.1.204 Tryptone glucose extract agar (cultivating and counting microorganisms in water and dairy products)

Beef extract	3.0 g
Tryptone	5.0 g
Glucose (dextrose)	1.0 g
Bromocresol purple	0.04 g
Agar	15.0 g

Dissolve ingredients except agar in 1.0 L of distilled water by steaming. Adjust pH to 7.0. Add agar, bring to the boil and continue to heat until the agar dissolves. Sterilize at 121°C for 10 min. Cool to 50°C and pour into sterile Petri dishes.

A2.1.205 Tryptone glucose yeast extract agar/broth (see plate count agar)

A2.1.205.1 Tryptone (trypticase) soy (TS) agar/broth (general purpose nonselective medium for growth of a wide variety of microorganisms)

Tryptone or trypticase	17.0 g
Soya peptone (soytone) or phytone	3.0 g
K_2HPO_4	2.5 g
NaCl	5.0 g
Glucose	2.5 g
Agar	15.0 g

Dissolve ingredients except agar in 1.0 L of distilled water. Adjust pH to 7.3. Add agar, bring to the boil and continue to heat until the agar dissolves. Sterilize at 121°C for 15 min, pour into sterile Petri dishes.

To prepare **TS broth**, omit the agar.

To prepare TS broth blood agar; cool medium to 47°C and add 5% sheep blood. Pour into sterile Petri dishes.

To prepare **TS-ferrous sulfate-NaCl agar/broth** add 0.035 g and 20.0 g ferrous sulfate into ingredients of TS ingredients.

A2.1.206 Tryptone (Trypticase) broth (general purpose media)

Tryptone or trypticase	10.0 g
Glucose	5.0 g
K_2HPO_4	1.25 g
Yeast extract	1.0 g
Bromocresol purple (2% alcoholic sol.)	2.0 mL

Dissolve ingredients in 1.0 L of distilled water, and heat if necessary. Add bromocresol solution and dispense 10 mL into test tubes, and sterilize at 121°C for 15 min.

A2.1.207 Tryptose sulfite cycloserin agar/broth (a selective medium for *Clostridium perfringens*)

Tryptose	15.0 g
Soyprotein	5.0 g
Yeast extract	5.0 g
Sodium metabisulfite	1.0 g
Ferric ammonium citrate	1.0 g
Agar	15.0 g

Dissolve ingredients except agar in 1.0 L of distilled water. Adjust pH to 7.6. Add agar, bring to the boil and continue to heat until the agar dissolves. Sterilize at 121°C for 15 min, pour into sterile Petri dishes.

D-Cycloserine solution: Dissolve 1 g D-cycloserine (white crystalline powder) in 200 mL of distilled water. Sterilize by filtration and store at 4°C until use.

Final medium: For pour plates, add 20 mL of D-cycloserine solution to 250 mL base. To prepare prepoured plates containing egg yolk, also add 20 mL of 50% egg yolk emulsion. Mix well and dispense 18 mL into 15 × 100 mm petri dishes. Cover plates with a towel and let dry overnight at room temperature before use.

To prepare Tryptose sulfite cycloserine (**TSC**) **broth**, omit the agar.

To prepare **TSC dextrose broth**, add 1.0 g dextrose into medium.

To prepare **egg yolk free TCD sulfite agar**, omit the egg yolk.

A2.1.208 Trypticase soy yeast extract agar

Trypticase soy agar	40.0 g
Yeast extract	6.0 g

Dissolve ingredients in 1.0 L distilled water, adjust pH to 7.3 and sterilize at 121°C for 15 min.

A2.1.209 Tyrosine agar (for tyrosine test)

Beef extract	3.0 g
Peptone	5.0 g
Tyrosine	5·0 g
Agar	15 g

Dissolve ingredients except agar in 1.0 L of distilled water by steaming. Adjust pH to 8.6. Add agar, bring to the boil and continue to heat until the agar dissolves. Sterilize at 121°C for 15 min.

A2.1.210 Urea agar (a urea test medium)

Peptone	1.0 g
Glucose	1.0 g
NaCl	5.0 g
KH_2PO_4	2.0 g
Urea	20.0 g
Phenol red	0.012 g
Agar	15.0 g

Dissolve ingredients except agar in 1.0 L of distilled water by steaming. Adjust pH to 6.8. Add agar, bring to the boil and continue to heat until the agar dissolves. Dispense into test tubes and sterilize at 121°C for 15 min.

A2.1.211 Universal preenrichment *broth*

Casein enzymic hydrolysate	5.0 g
Dextrose	0.50 g L^{-1}
Disodium phosphate	7.0 g L^{-1}
Ferric ammonium citrate	0.10 g L^{-1}
Magnesium sulfate	0.250 g L^{-1}
Monopotassium phosphate	15.0 g L^{-1}
Proteose peptone	5.0 g L^{-1}
Sodium chloride	5.0 g L^{-1}
Sodium pyruvate	0.20 g L^{-1}

Dissolve ingredients in 1.0 L of distilled water by heating and adjust pH to 6.3. Dispense into screw-capped tubes and sterilize at 121°C for 15 min

A2.1.212 Veal infusion agar/broth

Veal, infusion	500.0 g
Proteose peptone no. 3	10.0 g
NaCl	5.0 g
Agar	15.0 g

Dissolve ingredients in 1.0 L distilled and heat with agitation to dissolve agar. Adjust pH to 6.8 and dispense 7 mL portions into tubes. Sterilize at 121°C for 15 min. Incline tubes to obtain 6 cm slant.

To prepare broth, omit agar.

A2.1.213 Vibrio parahaemolyticus sucrose agar (a selective and differential medium for *Vibrio parahaemolyticus*)

Tryptose	5.0 g
Tryptone	5.0 g
Yeast extract	7.0 g
Sucrose	10.0 g
NaCl	30.0 g
Bile salts No. 3	1.5 g
Bromothymol blue	0.025 g
Agar	15.0 g

Dissolve ingredients except agar in 1.0 L of distilled water by steaming. Adjust pH to 8.6. Add agar, bring to the boil and continue to heat until the agar dissolves. Sterilize at 121°C for 15 min. Cool to about 45°C and pour into sterile Petri dishes.

A2.1.214 *Vibrio vulnificus* sucrose agar

Tryptose	5.0 g
Tryptone	5.0 g
Yeast extract	7.0 g
Sucrose	10.0 g
NaCl	30.0 g
Bile salt no 3	1.5 g
Bromothymol blue	0.025 g
Agar	15.0 g

Dissolve ingredients except agar in 1.0 L distilled water, mix thoroughly. Adjust pH 6.8. Add agar, bring to the boil and continue to heat until the agar dissolves. Boil to dissolve ingredients and cool to 50°C–55°C and dispense into sterile Petri dishes.

A2.1.215 Victoria blue butterfat (or margarine) agar (for the detection of lipolytic activity)

A2.1.215.1 Preparation of victoria blue base

Boil 2.0 g of powdered Victorian blue in 200 mL of distilled water until thoroughly dispersed. Slowly add a 10% solution of NaOH with constant mixing until the color disappears from the solution. Allow the water insoluble precipitate (the basic dye) to settle out. Filter off the precipitated basic dye and wash with distilled water made slightly alkaline with ammonium hydroxide. Dry the dye at 30°C.

A2.1.215.2 Preparation of fat

Obtain separated-fat or margarine fat as already described (Butter-fat agar).

A2.1.215.3 Preparation of dye/fat mixture

Heat 100 g of fat in a conical flask with 100 mL of distilled water and some glass beads. When boiling, slowly add Victorian blue base with constant mixing until the fat is saturated with dye (the fat will be deep red, with some particles of undissolved dye at the bottom of the flask). Boil for 30 min. Separate the fat from the bulk of the water, and filter overnight at 37°C. Separated the filtered fat from any residual water, using a separating funnel, dispense in 30 mL screw-capped bottles in 10 mL amounts and sterilize at 121°C for 15 min.

A2.1.215.4 Preparation of basal medium

The basal medium may consist of yeast extract agar, nutrient agar, or tryptose blood agar base. Whichever basal medium is used the agar content should be increased to 2%, and the pH adjusted to 7.8. Distribute 20 mL into screw-capped bottles (30 mL) and sterilize at 121°C for 20 min.

A2.1.215.5 Preparation of the complete medium

Add aseptically 1 mL of the melted dye/fat mixture to 20 mL of sterile molten basal medium (at 50°C). Emulsify by vigorously shaking for 1 min and pour into sterile Petri dishes. Surface air bubbles can be eliminated by immediately flaming the surface rapidly with a Bunsen flame.

A2.1.216 Violet red bile agar (a selective medium for coliforms and *Escherichia coli*)

Yeast extract	3.0 g
Peptone	7.0 g
Bile salts no.3	1.5 g
Lactose	10.0 g
NaCl	5.0 g
Neutral red	0.03 g
Crystal violet	0.002 g
Agar	15.0 g

Dissolve ingredients except agar in 1.0 L of distilled water by steaming. Adjust pH to 7.4. Add agar, bring to the boil and continue to heat until the agar dissolves. Do not boil for more than 2 min. Do not autoclave. Cool slowly to about 45°C and pour into sterile Petri dishes.

To prepare **VRB glucose agar**, add 1% glucose into melted agar.

A2.1.217 Violet red bile agar with MUG (VRB-2 agar) (a selective medium for *Escherichia coli* H7:0017)

VRBA	1.0 L
MUG (4-methylumbelliferyl-ß-D-glucuronide)	0.2 g

Dissolve ingredients except agar in 1.0 L of distilled water by steaming. Adjust pH to 7.4. Add agar, bring to the boil and continue to heat until the agar dissolves. Do not sterilize. Cool slowly to about 45°C and pour into sterile Petri dishes.

A2.1.218 Vogel–Johnson agar (a selective and differential medium for coagulase positive *Staphylococcus aureus*)

Tryptone	10.0 g
Yeast extract	5.0 g
Mannitol	10.0 g
K_2HPO_4	5.0 g
Lithium chloride	5.0 g
Glycine	10.0 g

Phenol red	0.025 g
Agar	15.0 g

Dissolve ingredients except agar in 1.0 L of distilled water by steaming. Adjust pH to 7.2. Add agar, bring to the boil and continue to heat until the agar dissolves. Sterilize at 121°C for 15 min. Cool to about 50°C and add 20 mL of 1% telluride solution. Mix thoroughly, and pour into sterile Petri dishes.

A2.1.219 Wagatsuma blood agar (an identification medium for *Vibrio parahaemolyticus*)

Peptone	10.0 g
Yeast extract	3.0 g
K_2HPO_4	5.0 g
NaCl	70.0 g
Mannitol	10.0 g
Crystal violet	0.001 g
Agar	15.0 g

Dissolve ingredients except agar in 1.0 L of distilled water by steaming. Adjust pH to 8.0. Do not autoclave. Wash rabbit or human erythrocytes three times in physiological saline, and reconstitute to original blood volume. Add 2 mL of washed erythrocytes to 100 mL of agar cooled to about 45°C and pour into sterile Petri dishes.

A2.1.220 Xylose lysine deoxycholate agar (a selective medium for *Salmonella* and *Shigella*)

Yeast extract	3.0 g
Lysine	5.0 g
Xylose	3.75 g
Lactose	7.5 g
Saccharose	7.5 g
Sodium deoxycholate	2.5 g
Ferric ammonium citrate	0.8 g
Sodium thiosulfate	6.8 g
NaCl	5.0 g
Phenol red	0.08 g
Agar	15.0 g

Dissolve ingredients except agar, ferric ammonium citrate, and sodium thiosulfate in 1.0 L of distilled water by steaming. Adjust pH to 7.4. Add agar, bring to the boil and continue to heat until the agar dissolves. Sterilize at 121°C for 15 min. Cool to about 50°C, and add a 20 mL of sterile aq. solution containing 34% sodium thiosulfate and 4% ferric ammonium citrate. Mix thoroughly, and pour into sterile Petri dishes.

A2.1.221 Yeast extract glucose broth (for bacterial count and inhibit yeasts and molds)

Glucose	30.0 g
Yeast extract	10.0 g
Polypeptone	2.0 g
Acetic acid	5.0 g
Agar	15.0 g
Bromocresol green	0.1 g
Pimaricin	0.1 g

Dissolve ingredients except agar and pimaricin in 1.0 L of distilled water. Heat to dissolve and adjust pH to 6.8. Add agar, heat to dissolve the agar. Sterilize at 121°C for 15 min.

Cool to 50°C and add sodium pimaricin and mix by gently shaking. Pour into sterile Petri dishes.

To prepare **Yeast Extract Glucose broth motility medium**, use only 0.5 g of agar instead of 15 g. Add 0.5% of triphenyl tetrazolium chloride into cooled medium (at 50°C).

A2.1.222 Yeast extract glucose lemco agar/broth (for growth characteristics of lab and *streptococcus*)

Peptone	10.0 g
Lemco (meat extract)	10.0 g
NaCl	5.0 g
Glucose	5.0 g
Yeast extract	3.0 g
Bromocresol green	0.1 g
Agar	15.0 g

Dissolve ingredients except agar in 1.0 L of distilled water by steaming. Adjust pH to 7.0. Add agar, bring to the boil and continue to heat until the agar dissolves. Sterilize at 121°C for 20 min. Cool to about 50°C and pour into sterile Petri dishes.

To prepare **YGL broth**, omit the agar.

To prepare **semisolid YGL**, add 0.3% agar.

To prepare **YES broth**, add 0.5% sucrose into medium instead of glucose (or use other carbohydrates)

A2.1.223 Yeast extract peptone (GYEP) agar (for isolation and counting of *Pediococcus*)

Glucose	10.0 g
Yeast extract	5.0 g
Peptone	5.0 g
Sodium acetate	2.0 g
Tween-80	0.25 g
$MgSO_4.7H_2O$	200.0 mg
$MnSO_2.4H_2O$	10.0 mg
$FeSO_4.7H_2O$	10.0 mg
NaCl	5.0 g
$CaCO_3$	5 g
Agar	15.0 g

Dissolve ingredients in 1.0 L of distilled water by steaming. Adjust pH to 6.8. Add agar, bring to the boil and continue to heat until the agar dissolves. Add milk and mix thoroughly. Sterilize at 121°C for 15 min. Cool to about 50°C and pour into sterile Petri dishes.

A2.1.224 Yeast extract skim milk agar/broth (general nonselective medium for bacteria from milk and milk products)

Yeast extract	3.0 g
Peptone	5.0 g
Fresh whole or skim milk	10.0 mL
Agar	15.0 g

Dissolve ingredients in 1.0 L of distilled water by steaming. Adjust pH to 7.2. Add agar, bring to the boil and continue to heat until the agar dissolves. Add milk and mix thoroughly. Sterilize at 121°C for 15 min. Cool to about 50°C and pour into sterile Petri dishes.

A2.2 Alphabetical listing of stains and reagents

A2.2.1 Preparation of stains

A2.2.1.1 Basic fuchsine stain (for simple stain)

Dissolve 0.5 g basic fuchsine in 20.0 mL 95% ethanol. Dilute to 100 mL with distilled water. Filter if necessary, to remove any excess dye particles.

A2.2.1.2 Brilliant green solution (1% aq.)

Dissolve 1.0 g brilliant green in 100 mL of distilled water.

For enrichment broth of salmonella: Add 2 mL 1% brilliant green solution into 1000 mL sterile distilled water and mix by gently shaking.

A2.2.1.3 Bromocresol purple ethanol solution

Add 2.0 g of bromocresol purple to 10 mL of ethanol, and complete to 100 mL with distilled water.

A2.2.1.4 Bromocresol purple indicator

Add 0.1 g of bromocresol purple to 20 mL of 0.05 N NaOH in a volumetric flask. Heat gently with constant stirring. Add 5 mL of 0.05 N NaOH. This constitutes the stock solution.

To prepare the indicator solution, dilute 1 mL of the stock solution with 9 mL of distilled water.

A2.2.1.5 Bromothymol blue indicator

Dissolve 0.04 g bromothymol blue in 50 mL of ethanol and 50 mL of distilled water.

A2.2.1.6 Capsule stain

Primary stain: Crystal violet (1% aq.).

Decolorizing agent: Copper sulfate (20% aq.).

Capsule staining procedure-1. Prepare thin smears of bacterial culture on a microscope glass slide. Allow the smear to only air-dry. Do not heat-fix as this will cause the capsule to shrink or be destroyed. Apply 1% crystal violet and allow it to remain on the slide for 2 min. With the slide over the proper waste container provided, gently wash off the crystal violet with 20% copper sulfate. Caution: Do not wash the copper sulfate and stain directly into the sink. Blot the slide dry with bibulous paper. Observe with the oil immersion lens.

Capsule staining India Ink procedure-2. Place a drop of India ink onto a clean glass slide. Using a sterile loop, obtain a sample and place it onto the slide and mix it with the drop of India ink. Obtain another sterile glass slide and having laid it at an angle on one end of the first slide, spread out the drop into a film to have a thin layer of the smear-ink mixture. Allow the slide to stand for about 5 mins (air-dry). Saturate the slide with crystal violet for a minute and then tilt the slide to drain excess dye. Allow the slide to dry (air-dry). View the slide using 100x.

Capsule staining procedure-3. Mix equal amounts of India ink and water. Mix a loopful of culture in this dye, spread the smear to cover the surface of slide. Air-dry, do not heat fix. Flood the smear with Liefson's methylene blue stain for 3 min. Wash with water. Background of slide appears dark, the capsule clear unstained rings with small blue cell body.

A2.2.2 Carbolfuchsin (for simple stain)

A2.2.2.1 Stain solution A

Basic fuchsine	0.3 g
Ethanol	10.0 mL

A2.2.2.2 Solution B

Phenol crystals	5.0 g

Dissolve basic fuchsine in alcohol.

Dissolve the phenol crystals in 95 mL distilled water.

Combine two solutions and filter through filter paper. Resultant solution should be removed from light and filtered as necessary to remove crystal formation.

A2.2.3 Crystal violet stain (for simple stain)

Dissolve 0.5 g crystal violet in distilled water, and filter through filter paper.

A2.2.3.1 Simple staining procedure

Prepare a thin smear from culture or sample on a clean glass slide. Stain for 2 min with basic fuchsine stain. Wash the slide with tap water, dry, and examine.

A2.2.4 Hucker modified gram staining (for *campylobacter* spp.)

A2.2.4.1 Crystal violet

Solution A: Dissolve 2.0 g crystal violet in 200 mL ethyl alcohol.

Solution B: Dissolve 8.0 g ammonium oxalate in 800 mL distilled water.

Mix solutions A and B. Let the solution stand overnight or for several days until the dye goes into solution. Filter through filter paper.

A2.2.4.2 Iodine

Iodine	1.0 g
Potassium iodide	2.0 g

Dissolve ingredients in sufficient water to make 300 mL.

A2.2.4.3 Decolorizer

95% ethyl alcohol

A2.2.4.4 0.3% Carbol fuchsine counterstain

Solution A: Dissolve 0.3 g basic fuchsine in 10 mL ethyl alcohol.

Solution B: Dissolve 5.0 mL phenol (melted crystal) in 95 mL distilled water.

Mix solution A and B.

A2.2.5 Gram staining procedure

Prepare a thin smear in a drop of water on a clean glass slide. Air-dry, and gently fix with heat. Stain for 1 min with crystal violet. Wash the slide with tap water. Add iodine for 1 min. Wash the slide with tap water. Decolorize with ethyl alcohol until alcohol wash is clear. Wash the slide with tap water. Counterstain with carbol fuchsine for 10 to 20 s. Wash the slide with tap water, dry, and examine.

A2.2.6 Gram stain (kopeloff and cohen modification)

A2.2.6.1 Methyl violet

Mix 30.0 mL of crystal violet or methyl violet 6B (1% aq.) in 30.0 mL of (5%) sodium bicarbonate solution. Allow the solution to stand at room temperature for 5 min or more before using.

A2.2.6.2 Iodine

Dissolve 2.0 g of iodine in 10 mL of 1.0 N sodium bicarbonate solution and then add 90.0 mL of distilled water.

A2.2.6.3 Decolorizer

Mix 50.0 mL of ethyl alcohol (95%) and 50.0 mL acetone.

A2.2.6.4 Counterstain

Dissolve 0.1 g basic fuchsine in 100.0 mL distilled water.

Modified gram staining procedure. Prepare a thin smear in a drop of water on clean glass slide. Air-dry, and gently fix with the least amount of heat necessary. Flood the smear with crystal violet or methyl violet for 5 min. Flush the smear with iodine solution for 2 min. Drain without blotting, but do not allow the smear to dry. Add decolorizer drop wise until the runoff is colorless (10 s or less). Air-dry the slide. Counterstain with basic fuchsine for 20 min. Wash the excess stain from the slide by short exposure tap water, and then air-dry. If the side is not clear, immerse in xylene. Cell that decolorizes and accept the basic fuchsine stain are Gram-negative. Cells that do not decolorize but retain the crystal violet or methyl violet stain are Gram-positive.

A2.2.7 Gram staining reagents

A2.2.7.1 Crystal violet (also use for simple staining)

Dissolve 0.5 g crystal violet in sufficient water to make 100 mL and filter through filter paper.

A2.2.7.2 Iodine solution

Dissolve 1.0 g iodine and 2.0 g potassium iodide in sufficient water to make 300 mL.

A2.2.7.3 Decolorizer

Acetone	250.0 mL
Isopropyl alcohol	750.0 mL

Mix acetone and alcohol.

A2.2.7.4 Safranin solution

Safranin	0.25 g
Alcohol	10.0 mL

Dissolve safranin in the alcohol. Add 100 mL of distilled water and filter through filter paper.

Gram staining procedure. Prepare a thin smear from culture or sample. Stain for 1 min with crystal violet. Wash the slide with tap water. Add iodine for 1 min. Wash the slide with tap water. Decolorize with ethyl alcohol until alcohol wash is clear. Wash the slide with tap water. Counterstain with safranin for 30 to 60 s. Wash the slide with tap water, dry, and examine.

A2.2.8 **Gray's double dye stain (direct microscopic counting of microorganisms)**

1. Methylene blue (1% aq.)	50.0 mL
Methanol	50.0 mL
Add dye to the methanol.	
2. Basic fuchsine (1% aq.)	25.0 mL
Methanol	25.0 g

Add dye to the methanol.

Before use mix two solutions. Stain slide for a few seconds with mixed stain, dry and examine. Bacteria are stained blue and the background pink. If the slide needs defatting use xylene and wash off with methyl alcohol before staining.

A2.2.9 **Lactophenol-cotton blue (for wet-mounting and staining of molds)**

Saturated solution of cotton	
Blue (soluble aniline blue)	10.0 mL
Lactophenol	10.0 mL

Mix equal parts of lactophenol and cotton blue solution and add 80 mL of distilled water.

A2.2.10 **Lactophenol solution (for wet microscopic preparations of molds)**

Phenol crystals	100 g
Lactic acid	100 mL
Glycerol	200 mL

Dissolve phenol in 100.0 mL distilled water without heat; then add the lactic acid and glycerol.

A2.2.11 Liefson's flagella stains

A2.2.11.1 Solution A

Basic fuchsine	1.2 g
Ethanol (95%)	100.0 mL

Dissolve ingredients with frequent shaking. Store in a tightly stoppered bottle to prevent evaporation.

A2.2.11.2 Solution B

Dissolve 3.0 g tannic acid in 100.0 mL distilled water. If it is intended to keep this for some time as a stock solution, add phenol to a concentration of 0.2% to prevent microbial growth.

A2.2.11.3 Solution C

Dissolve 1.5 g NaCl in 100.0 mL distilled water.

These solutions are stable at room temperature. To prepare the working solution, mix equal quantities of solutions A, B and C, and store in a tightly stoppered bottle in the refrigerator (stable for several weeks) or deep freezer (stable for months). If stored deep-frozen, the stain should be well shaken after thawing since the ethanol tends to separate from the water.

A2.2.11.4 Counter stain

Dissolve 1.0 g methylene blue in 100.0 mL distilled water with frequent shaking.

Flagella staining procedure. Prepare a smear and air-dry. Do not fix Mix solution A, B, and C in equal portions (about 2 mL of each sufficient). Flood the smear with mixed solution and allow for 10 min. Wash in top water. Flood with counterstain for 10–15 min. Wash with tap water. Air-dry and examine. The vegetative cells stain blue and the flagella red.

A2.2.12 Lipid globule stain (Burdon's method)

A2.2.12.1 Solution A

To prepare solution A, dissolve 0.3 g of Sudan black B in 100.0 mL 70% alcohol. After the bulk of the dye has dissolved, shake solution at intervals during the day and allow to stand overnight. Filter if necessary, to remove undissolved dye. Store in a well-stoppered bottle.

A2.2.12.2 Solution B

Dissolve 0.5 g of safranin in 100.0 mL distilled water.

Lipid globule staining procedure. Prepare smear and let dry thoroughly in air. Heat fix with minimal flame. Flood slide with solution A for 10 to 20 min. Drain off the excess stain, and blot dry. Wash the slide for 5–10 s with chemically pure (CP) xylene, and blot dry. Counterstain with solution B for 10–20 s. Wash the slide with tap water, blot dry, and examine.

A2.2.13 Liefson's methylene blue stain (for simple and capsule stain)

Methylene blue	0.3 g
Alcohol	30.0 mL
Distilled water	100.0 mL

Dissolve the methylene blue in the alcohol. Add the distilled water and filter the solution through paper.

A2.2.14 Malachite green stain (0.01% aq.)

Dissolve 0.01 g malachite green in 100.0 mL distilled water, and store in refrigeration.

A2.2.15 Methylene blue indicator stain (for methyl blue reduction test)

Dissolve 0.05 g methylene blue in 100 mL of distilled water.

A2.2.16 Neutral red solution (0.0015% aq.)

Neutral red	0.0015 g
Distilled water	100.0 mL

Dissolve 0.0025 g neutral red in 100.0 mL distilled water.

A2.2.17 Newman lambet stain-modified Levowitz-Weber (for direct microscopic count)

Methylene blue chloride	0.5 g
Ethanol (95%)	56.0 mL
Xylene	40.0 g
Glacial acetic acid	4.0 mL

Add 0.5 g of methylene blue chloride to 56.0 mL of ethanol and 40.0 mL of xylene in a 200 mL flask. Swirl to dissolve, and then let stand for 12–24 h at 4.4°C–7.2°C. Add 4.0 mL of glacial acetic acid, and filter through Whatman No. 42 paper or equivalent. Store the stain in a tightly closed bottle having a cap that will not be affected by the reagents.

A2.2.17.1 Direct microscopic stain for leukocytes

Dry smear prepared with food sample at 45°C–50°C in dry air own. Cool to room temperature. Stand and defat for 2 min with Newman–Labbet stain. Drain off the excess stain by resting the long edge on the surface of absorbent paper. Dry thoroughly, using forced air from a blower or fan. Rinse the dried stained slides by passing through three changes of tap water at 38°C–43°C. Dry rapidly and thoroughly, using forced air from a blower or fan, and examine.

Normal bacterial cells are stained heavily; the background material should be evenly and lightly stained. Plasmolyzed bacterial cells should stain with various degrees of intensity depending on the degree of plasmolysis. Leukocytes stain with the cytoplasm slightly darker than the background and the nuclear areas more deeply stained which allows their differentiation into types.

A2.2.18 North aniline methylene blue stain (direct microscopic counting of microorganisms)

Mix 3.0 mL of aniline oil with 10.0 mL of 95% ethanol, and then slowly add 1.5 mL of HCl with constant agitation. Add 30.0 mL of saturated alcoholic methylene blue solution, and then dilute to 100.0 mL with distilled water. Filter before use and keep in tightly stoppered bottle.

A2.2.19 O-nitrophenyl-beta-D-glucopyranoside test reagents (For beta-D-galactosidase activity)

A2.2.19.1 Reagent A

Dissolve 6.9 g $NaH_2PO_4.H_2O$ in 45 mL of distilled water. Add 3 mL of 30% NaOH and adjust to pH 7.0. Complete volume to 50 mL with distilled water and store in refrigerator.

A2.2.19.2 Reagent B

Dissolve 80.0 g ONPG in 15 mL distilled water at 37°C. Add 5 mL of 1.0 M monosodium phosphate solution (A). Solution should be stored in refrigerator. Prior to its use appropriate portion of the buffered 0.0133 M ONPG solution should be warmed to 37°C.

Galactosidase activity test. Use cultures from TSI agar slant or nutrient agar containing 1.0% lactose. Emulsify a large loopful of growth in 0.25 mL of physiological saline solution. Add one drop of toluene to each tube and shake well (to aid liberation of the enzyme). Allow to stand for 5 min at 37°C. Add 0.25 mL buffered 0.0133 M ONPG solution to suspension, shake the tubes well, and incubate in a water bath at 37°C. Read after 30 min, 1 h, and 24 h of incubation. Positive results are indicated by the development of a yellow color.

A2.2.20 Oxidase test reagent

Dissolve 50.0 mg g of triphenyl-p-phenylenediamine dihydrochloride in 5.0 mL distilled water. Store up to 1 week at 4°C in dark brown container or cover the tube to protect from light.

Oxidase test. Add several drops of reagent on suspect colonies. If the organisms produce oxidase, the colonies will turn pink, red, then black, in that order. Discard plates after 20 min.

A2.2.21 Rose Bengal

Rose Bengal	1.0 g
Calcium chloride ($CaCl_22H_2O$)	0.01 g
Phenol	5.0 g

Dissolve the phenol in 100.0 mL distilled water. Add the other ingredients and stir until dissolved. Filter the solution thoroughly filter paper.

A2.2.22 Safranin staining solution (for simple staining)

Safranin	0.25 g
Alcohol	10.0 mL
Distilled water	100.0 mL

Dissolve the safranin in the alcohol. Add the distilled water and filter the solution through paper.

A2.2.23 Spore stain (Schaeffer–Fulton modification stain; Ashby's)

A2.2.23.1 Malachite Green

Dissolve 5.0 g malachite green in 100 mL distilled water and filter.

A2.2.23.2 Safranin

Dissolve 5.0 g safranin in 100 mL distilled water.

Spore staining procedure. Prepare a smear and air-dry. Heat fix by passing through the flame. Flood the smear with malachite green stain and allow for 1 min. Heat to steaming three or four times for 5 min, then cool. Wash in top water for 30 s. Flood with solution B for 2 min. Wash lightly, dry without blotting, and examine. Spores should stain green and vegetative cells red.

A2.2.24 Wolford's stain (for direct microscopic count)

A2.2.24.1 Solution A

Dissolve 1.0 g of basic fuchsine in 100.0 mL of alcohol.

A2.2.24.2 Solution B

North aniline oil methylene blue stain	10 mL

Make dilute stain for making direct microscopic counts by mixing 2 mL of solution A with 10 mL of solution B and 88 mL of distilled water. Filter before use.

Dilute stain solution may be kept under refrigeration for 3–4 weeks, but should be discarded after that time.

A2.2.25 Preparation of reagents

A2.2.25.1 Acetone alcohol

Mix 700 mL of alcohol and 300 mL of acetone.

A2.2.25.2 Alcohol 70%

Dissolve 700 mL alcohol in sufficient distilled water to make 1000 mL.

A2.2.25.3 Barrit's reagent (Voges-Proskauer reagents)

A2.2.3 Reagent A

Dissolve 5.0 g α-naphthol in sufficient alcohol to make 100 mL.

A2.2.3 Reagent B

Dissolve 40.0 g KOH in sufficient distilled water to make 100 mL.

A2.2.3 Aqueous iodine

Dissolve 5.0 g iodine in sufficient distilled water to make 100 mL.

A2.2.26 Bicarbonate buffer (0.1 M, pH 9.6)

Dissolve 1.5 g of Na_2CO_3 and 2.93 g of $NaHCO_3$ in 1.0 L of distilled water. Store at room temperature for more than 2 weeks.

A2.2.27 Chromic acid cleaning solution

Potassium dichromate	63.0 g 100.0 g
Distilled water	35.0 mL 750.0 mL
Concentrate sulfuric acid	960.0 mL 250.0 mL

Add the potassium dichromate to the water in a 2.0 L flask. Slowly and carefully add the acid. Chromic acid cleaning solution can be used repeatedly until it begins to turn green. Great care should be taken when preparing and handling this cleaning solution, as it is extremely corrosive. *Use protective clothing and eye protection.*

A2.2.28 Citric-phosphate buffer

Citric acid 0.1 M	21.6 mL
Na_2HPO_4 0.2 M	28.4 mL

Dissolve the ingredients and adjust pH to 5.5. Sterilize at 121°C for 20 min.

A2.2.29 Citrate buffer (0.1 M)

Add 46.30 g of sodium citrate dihydrate to the solution. Add 4.36 g of citric acid to the solution. Add distilled water until volume is 1 L.

A2.2.30 Coagulase plasma

Coagulase plasma should contain 0.15% EDTA.

A2.2.31 Formalized physiological saline solution

Dissolve 8.5 g of NaCl in 1.0 L of distilled water. Sterilize at 121°C for 20 min. Cool to room temperature, and add 6.0 mL of formaldehyde solution (36%). Do not autoclave after addition of formaldehyde.

A2.2.32 Gelatin phosphate buffer

Gelatin	2.0 g
Na_2HPO_4	4.0 g

Dissolve ingredients by heating gently in 800.0 mL of distilled water. Adjust pH to 6.2 with HCl. Complete to 1.0 L with distilled water, and sterilize at 121°C for 20 min.

A2.2.33 Glycerin-salt solution (buffered)

Glycerin (reagent grade)	100 mL
K_2HPO_4 (anhydrous)	12.4 g
KH_2PO_4 (anhydrous)	4.0 g
NaCl	4.2 g

Distilled NaCl and bring volume to 900 mL with water. Add glycerin and phosphates. Adjust pH to 7.2. Sterilize at 121°C for 15 min. For double strength (20%) glycerin solution, use 200 mL glycerin and 800 mL distilled water.

A2.2.34 Glycerol formaldehyde solution

Dissolve 5.0 g formaldehyde in sufficient distilled water to make 100 mL and add 25 mL of glycerol and mix.

A2.2.35 Glycine-HCl buffer (0.05 M, pH 1.5)

Glycine	7.5 g
HCl	0.416 g

Dissolve ingredients in 250 mL distilled water. Complete to 1.0 L with distilled water and sterilize at 121°C for 20 min.

A2.2.36 Glycine-NaOH buffer

Glycine	6.01 g
NaOH	2.31 g

Dissolve ingredients in 250 mL distilled water. Complete to 1.0 L with distilled water and sterilize at 121°C for 20 min.

A2.2.37 Hydrogen peroxide solution (H_2O_2; 0.2%)

Dissolve 0.2 mL hydrogen peroxide in distilled water. Complete to 100 mL with distilled water.

A2.2.38 Hydrogen chloride solution (0.1 N)

Dissolve 36.46 g of NaCl in 1.0 L of distilled water and sterilize at 121°C for 15 min.

A2.2.39 Kovac's reagent (for indole test)

p-Dimethyl-amino-benzaldehyde	5.0 g
Amyl or isoamyl or butyl alcohol	75.0 g
Hydrochloric acid (conc.)	25.0 mL

Dissolve the para-dimethyl-amino-benzaldehyde in the alcohol, and then slowly add the hydrochloric acid. Store in the refrigerator.

A2.2.40 Lactophenol picric acid solution (for wet mounting and staining of molds)

Lactic acid	100.0 mL
Phenol	100.0 g
Glycerol	200.0 mL
Picric acid (sat. aq. sol.)	100.0 mL

Dissolve the phenol in the picric acid solution without heat, and then add the lactic acid and the glycerol.

A2.2.41 Lugol's iodine solution (for starch hydrolysis)

Dissolve 10.0 g potassium iodine in 30 mL distilled water. Then add 5.0 g iodine and sufficient distilled water to make up 100 mL, and mix.

A2.2.42 MacFarland standards

McFarland Standards are tubes labeled 1 through 10 and filled with suspensions of barium salts. Make suspensions of barium sulfate as follows:

1. Prepare 1.0% solution of CP sulfuric acid.
2. Prepare 1.0% solution of CP barium chloride.
3. Prepare 10 MacFarland standards as follows:

MacFarland standard no.	$BaCl_2$ 1% (mL)	H_2SO_4 1% (mL)	Bacterial number ($\times 10^6$ mL^{-1})
1	0.1	9.9	300
2	0.2	9.8	600
3	0.3	9.7	900
4	0.4	9.6	1200
5	0.5	9.5	1500
6	0.6	9.4	1800
7	0.7	9.3	2100
8	0.8	9.2	2400
9	0.9	9.1	2700
10	1.0	9.0	3000

Each "MacFarland Standard No" approximates the turbidity of bacterial solutions corresponding to the McFarland Scale number. Thus tube 7 represents the turbidity of bacteria at a concentration of 2.1×10^9 per mL. If you have a culture and wish to quickly determine its approximate population, its turbidity can be visually compared to a set of McFarland Standards. If its turbidity falls between tubes 7 and 8, then the number of bacteria per mL will be between 2.1 and 2.4 billion per mL. The advantage of these standards is that no incubation time or equipment is needed to estimate bacterial numbers.

A2.2.43 Melzer's solution (used to help identify mushrooms)

Depending on the formulation, it consists of approximately 2.50%–3.75% potassium iodide and 0.75%–1.25% iodine, with the remainder of the solution being 50% water and 50% chloral hydrate.

Melzer's is used by exposing fungal tissue (especially mushrooms) or cells to the reagent, typically in a microscope slide preparation, and looking for any of three-color reactions: if fungal tissue contains starch, it will turn navy blue to black when tested with an iodine solution. This is called the amyloid reaction.

A2.2.44 Ninhydrin reagent

Dissolve 3.5 g ninhydrin in 100 mL of 1:1 mixture of acetone and butanol. Store refrigerated.

A2.2.45 Nitrate peptone water

Dissolve 10 g peptone and 2 g KNO_3 in 1.0 L distilled water and adjust pH to 7.0. Sterilize at 121°C for 20 min.

A2.2.46 Nitrate reagents

a. *Potassium hydroxide (40%) creatin solution.* Dissolve 40 g of KOH in a sufficient quantity of distilled water to make 100 mL of solution. When cool, add 0.3 g of creatine. Caution: Reagent breaks down very rapidly at room temperature. Stir the mixture until it dissolves. Store in brown bottles at 4°C for no longer than 3 days or for 21 days at −18°C.
b. *Alpha-naphthol solution.* Dissolve 5 g of alpha-naphthol in 100 mL of 99% isopropyl alcohol. Store in brown bottles in refrigerator.
c. *Antifoam solution.* Mix 1 g of antifoam in 10 mL distilled water.

A2.2.47 Nitrate reduction reagents (for nitrate reduction test)

A2.2.47.1 Solution A (sulfanilic acid)

Sulfanilic acid	7.01 g
Sulfuric acid (concentrate)	286.0 mL
Distilled water	714.0 mL

Slowly add the sulfuric acid to the distilled water. Mix carefully. Add the sulfanilic acid and mix well to dissolve. Store under refrigeration.

A2.2.47.2 Solution B (dimethyl-α-naphthylamine)

Dimethyl-α-naphthylamine	5.01 g
Sulfuric acid (conc.)	286.0 mL
Distilled water	714.0 mL

Slowly add the sulfuric acid to the distilled water and mix carefully. Then add the dimethyl-α-naphthylamine, and mix well to dissolve. Store under refrigeration.

A2.2.48 Peptone water (0.1%)

Dissolve 1.0 g peptone in 1.0 L of distilled water. Adjust pH to 7.0. Dispense in a sufficient quantity to allow for loss during sterilization. Sterilize at 121°C for 15 min.

A2.2.49 Tetramethyl-p-phenylenediamine dihydrochloride (1%)

Dissolve 1.0 g N,N,N',N'-Tetramethyl-p-phenylenediamine · 2HCl in sufficient distilled water to make 100 mL

A2.2.50 Phosphate buffered formalin (40%)

Add 400 mL of formaldehyde (37%–40%) into 600 mL of distilled water. Add 4.0 g sodium phosphate monobasic and 6.5 g sodium phosphate dibasic (anhydrous) into mixture and mix by gently shaking. Adjust pH to 7.2 with 1.0 N NaOH solution and sterilize at 115°C for 10 min.

A2.2.51 Phosphate buffer-3% NaCl

K_2HPO_4	1.15 g
KH_2PO_4	0.2 g
NaCl	30.0 g
KCl	0.2 g

Dissolve ingredients in 1.0 L distilled water. Adjust pH to 7.2 with 1.0 N NaOH solution. Sterilize at 115°C for 10 min.

A2.2.52 Phosphate buffer saline

Dissolve 8.0 g Na_2HPO_4, 2.44 g KH_2PO_4 and 4.25 g NaCl in 500.0 mL distilled water. Adjust pH to 7.2 with 1.0 N NaOH solution and sterilize at 115°C for 10 min.‘

A2.2.53 Phosphate buffer solution

A2.2.53.1 Stock solution

Dissolve 1.8 g KH_2PO_4 in 25.0 mL distilled water. Adjust the pH to 7.2 with 1.0 N NaOH solution and dilute to 1.0 L. Sterilize at 121°C for 15 min, and store in refrigerator.

A2.2.53.2 Dilution

Dilute 1.25 mL of stock solution to 1.0 L with distilled water. Prepare dilution blanks in suitable container. Sterilize at 121°C for 15 min.

0.72 g per mL x 1.25 mL = 0.09 g+1000 mL

A2.2.54 Physiological saline solution (0.85%)

Dissolve 8.5 g of NaCl in 1.0 L of distilled water, adjust pH to 7.0, and sterilize at 121°C for 15 min.

A2.2.55 Quarter strength ringer's solution (diluent and suspending liquid)

Sodium chloride	2.25 g
Potassium chloride	0.105 g
Calcium chloride	0.12 g
Sodium bicarbonate	0.05 g

Dissolve ingredients in 1.0 L of distilled water. Tube in 9.0 mL portions and sterilize at 121°C for 15 min.

A2.2.56 Ringers solution

Sodium chloride	8.6 g
Potassium chloride	0.3 g
Calcium chloride $6H_2O$	0.33 g
Sodium bicarbonate	0.05 g

Dissolve ingredients in 1.0 L of distilled water by steaming. Adjust pH to 7.0 by hydrochloric acid and/or sodium hydroxide. Sterilize by autoclaving at 121°C for 15 min.

A2.2.57 Sodium chlorine solution (0.14 M, pH 9.5)

Dissolve 8.19 g of NaCl in 1.0 L of distilled water, adjust pH to 9.5 by NaOH solution, and sterilize at 121°C for 15 min.

A2.2.58 Sodium hippurate solution

1 g of sodium hippurate is added to 100 mL of distilled water.

A2.2.59 Potassium hydroxide creatine solution

Dissolve 40.0 g of KOH in sufficient distilled water to make 100 mL. Add 0.3 g of creatine. Store for no longer than 3 days in refrigerator.

A2.2.60 Sodium hydroxide solution (NaOH; 1.0 N)

Dissolve 5.2 g of NaOH in 100 L of distilled water and sterilize at 121°C for 15 min.

A2.2.61 Sea water (synthetic)

NaCl	24.0 g
KCl	0.7 g
Magnesium chloride.$6H_2O$	5.3 g
Magnesium sulfate.$7H_2O$	7.0 g

Dissolve ingredients in 1.0 distilled water.

A2.2.62 Silver nitrate

Dissolve 5.0 g silver nitrate in sufficient distilled water to make 100 mL.

A2.2.63 Sodium bicarbonate

Dissolve 10.0 g sodium bicarbonate in 1.0 L distilled water. Sterilize by filtration.

A2.2.64 Sodium citrate solution (2%)

Dissolve 20.0 g sodium citrate in 100 mL distilled water. Sterilize by filtration.

A2.2.65 Sodium hydroxide

Dissolve 40.0 g NaOH in distilled water and to make 1.0 L.

A2.2.66 Sodium deoxycholate solution

Dissolve 0.5 g sodium deoxycholate in in 100 mL distilled water.

A2.2.67 Disodium hydrogen phosphate/potassium dihydrogen phosphate buffer (Na_2HPO_4; 0.15 M and pH 9.0)

$K_2HPO_4.2H_2O$ (Solution A)	1.78 g
Dissolve ingredients in 1.0 L distilled water.	
$KH_2PO_4.2H_2O$ (Solution B)	1.56 g
Dissolve ingredients in 1.0 L distilled water.	

Preparation of a phosphate buffer of 0.15 M and pH 9.0; Mix 100 mL solution A. 5.0 mL Solution B. Dilute to 200 mL with distilled water. pH is adjusted either by adding Solution A (basic) or Solution B (acidic).

A2.2.68 Stabilizer solution

Place 2.5 g of cellulose gum and 10 mL formalin (37% formaldehyde) into boiling distilled water in high-speed blender. With blender running, add the cellulose and formalin. Blend for about 1 min.

A2.2.69 Stabilizer solution alternate

Add 3.0 to 5.0 g pectin and 2.0 mL formalin 40% (37% formaldehyde) into 100 mL distilled water while agitating in high-speed blender. Treat solution with vacuum or heat to remove air bubbles. (If blender is not available, mix dry stabilizer with a small quantity of alcohol to facilitate incorporation with water.) Adjust pH 7.0 to 7.5.

A2.2.70 Trypsin solution (1:250; 5%, wt./vol.)

Add 5 g trypsin to 100 mL sterile H_2O. Refrigerate. Swirl the solution periodically until all of the powder has dissolved. Aliquot the solution into plastic vials (1 mL/vial) and freeze

A2.2.71 Triphenyl-Tetrazolium chloride

Dissolve 0.05 g Triphenyl-Tetrazolium chloride in 100.0 mL distilled water and dissolve. Filter sterilizes before use. Keep in dark bottle in the refrigerator. When deep purple color appears, it should be discarded. Autooxidation can be retarded by addition of ascorbic acid with 0.1%.

A2.2.72 Zinc sulfate solution ($ZnSO_2$; 33%)

Dissolve 33 g $ZnSO_2$ in 100 mL distilled water.

Further reading

Further reading

[1] Abbasi P, Kargar K, Doosti A, Mardaneh J, Ghorbani-Dalini S, Dehyadegari MA. Molecular detection of diffusely adherent *Escherichia coli* strains associated with diarrhea in Shiraz, İran. Arch Ped Infect Dis 2017;5(2):e37629.

[2] Abdul-Hussein ZK, Raheema RH, Inssaf AI. Molecular diagnosis of diarrheagenic *E. coli* infections among the pediatric patients in Wasit Province, Iraq. J Pure Appl Microbiol 2018;12(4):1–12.

[3] Acuff GR. Media, reagents, and stains. In: Vanderzant C, Splittstoesser DF, editors. Compendium of methods for the microbiological examination of foods. 3rd ed. Washington, DC: American Public Health Association; 1992. p. 1093–208.

[4] Al-Alak SK, Qassim DK. Molecular identification of 16S rRNA gene in *Staphylococcus aureus* isolated from wounds and wurns by PCR technique and study resistance of fusidic acid. Iraqi J Cancer Med Genet 2016;9(1):25–30.

[5] Andjelković U, Gajdošik MS, Gašo-Sokač D, Martinović T, Josić D. Foodomics and food safety: where we are. Food Technol Biotechnol 2017;55(3):290–307.

[6] Ann ML. Food microbiology laboratory. Florida: CRC Press; 2005.

[7] Anonymous. Difco manual: dehydrated culture media and reagents for microbiology. 10th ed. Detroit: Difco laboratories; 1985.

[8] Armstrong D. *Listeria monocytogenes* infections. In: Evans AS, Brachman PS, editors. Bacterial infection of humans. 2nd ed. New York: Plenum Medical Book Company; 1991. p. 395–402.

[9] Ashton D, Bernard D. Thermophilic anaerobic sporeformers. In: Vanderzant C, Splittstoesser DF, editors. Compendium of methods for the microbiological examination of foods. 3rd ed. Washington: American Public Health Association; 1992. p. 309–16.

[10] Atlas RM. Handbook of microbiological media. 4th ed. Florida: CRC Press; 2010.

[11] Anonymous. Bacterial agents of enteric diseases of public health concern: *Salmonella* serotype *Typhi, Shigella and Vibrio cholerae*. Manual for the laboratory identification and antimicrobial susceptibility testing of bacterial pathogens of public health importance in the developing world. Geneva: World Health Organization; 2003. p. 103–62.

[12] Bahceci KS, Acar A. Determination of guaiacol produced by *Alicyclobacillus acidoterrestris* in apple juice by using HPLC and spectrophotometric methods, and mathematical modeling of guaiacol production. Eur Food Res Technol 2007;225:873–8.

[13] Baron EJ, Peterson LR, Finegold SM. Diagnostic microbiology. Mosby-Year Book, Inc; 1994.

[14] Baross JA, Lenovich LM. Halophilic and osmophilic microorganisms. In: Vanderzant C, Splittstoesser DF, editors. Compendium of methods for the microbiological examination of foods. 3rd ed. Washington: American Public Health Association; 1992. p. 199–212.

[15] Benenson AS. Cholera. In: Evans AS, Brachman PS, editors. Bacterial infection of humans. 2nd ed. New York: Plenum Medical Book Company; 1991. p. 207–25.

[16] Bennett RW, Lancette GA. FDA, bacteriological analytical manual. 8th ed. Gaithersburg, MD: AOAC International; 1998.

[17] Bennett RW, Lancette GA. Bacteriological analytical manual. 8th ed. Gaithersburg, MD: AOAC International; 1998. Available from: https://www.fda.gov/food/laboratory-methods-food/bacteriological-analytical-manual-bam.

[18] Bennett RW. Detection and quantitation of gram-positive nonsporeforming pathogens and their tosins. In: Pierson MD, Stern NJ, editors. Foodborne microorganisms and their toxins: developing methodology. New York: Marcel Dikker, Inc.; 1986. p. 345–92.
[19] Beuchat LR, Pitt JI. Detection and enumeration of heat-resistant molds. In: Vanderzant C, Splittstoesser DF, editors. Compendium of methods for the microbiological examination of foods. 3rd ed. Washington: American Public Health Association; 1992. p. 251–63.
[20] Bilung LM, Ulok V, Tesfamariam FM, Apun K. Assessment of *Listeria monocytogenes* in Pet. Food Agric. Food Sec. 2018;7:23–8.
[21] Blair JE, Williams REO. Phage typing of *Staphylococci*. Bull. World Health Organ. 1961;24:771–84.
[22] Brackett RE, Splittstoesser DF. Fruits and vegetables. In: Vanderzant C, Splittstoesser DF, editors. Compendium of methods for the microbiological examination of foods. 3rd ed. Washington: American Public Health Association; 1990. p. 919–27.
[23] Bridson EY. The oxoid manual. 8th ed Basingstoke: Oxoid Limited; 1990.
[24] Campbell MS, Wright AC. Real-time PCR analysis of *Vibrio vulnificus* from oysters. Appl Environ Microbiol 2003;69(12):7137–44.
[25] Centre for Food Safety. Microbiological guidelines for food: for ready-to-eat food in general and specific food items. Food and Environmental Hygiene Department, Hong Kong, 2014.
[26] Chay C, Dizon EI, Elegado FB, Norng C, Hurtada WA, Raymundo LC. Isolation and identification of mold and yeast in medombae, a rice wine starter culture from Kompong Cham Province, Cambodia. Food Res 2017;1(6):213–20.
[27] Chen J-Q, Healey S, Regan P, Laksanalamai P, Zonglin HZ. PCR-based methodologies for detection and characterization of *Listeriamonocytogenes* and *Listeria ivanovii* in foods and environmental sources. Food Sci Hum Wellness 2017;6(2):39–59.
[28] Cheyne BM, Dyke MIV, Anderson WB, Huck PM. The detection of *Yersinia enterocolitica* in surface water by quantitative PCR amplification of the ail and yadA genes. J Water Health 2010;08 (3):488–9.
[29] Cliver DO. Foodborne viruses. In: Speck ML, editor. Compendium of methods for the microbiological examination of foods. Washington: American Public Health Association; 1976. p. 462–70.
[30] Cliver DO, Ellender RD, Fout GS, Shields PA, Sobsey MD. Foodborne viruses. In: Vanderzant C, Splittstoesser DF, editors. Compendium of methods for the microbiological examination of foods. 3rd ed. Washington: American Public Health Association; 1992. p. 763–87.
[31] Colins-Thompson DL, Bunning VK. Thermoduric microorganisms and heat resistance measurements. In: Vanderzant C, Splittstoesser DF, editors. Compendium of methods for the microbiological examination of foods. 3rd ed. Washington: American Public Health Association; 1992. p. 169–81.
[32] Cousin MA, Jay JM, Vasavada. Psychrotrophic microorganisms. In: Vanderzant C, Splittstoesser DF, editors. Compendium of methods for the microbiological examination of foods. 3rd ed. Washington: American Public Health Association; 1992. p. 153–68.
[33] Cowman S, Kelsey R. Bottled water. In: Vanderzant C, Splittstoesser DF, editors. Compendium of methods for the microbiological examination of foods. 3rd ed. Washington: American Public Health Association; 1992. p. 1031–6.
[34] Croxen MA, Law RJ, Scholz R, Keeney KM, Wlodarska M, Finlay BB. Recent advances in understanding enteric pathogenic *Escherichia coli*. Clin Microbiol Rev 2013;26(4):822–80.
[35] Dan Li, D, Shen M, Xu Y, Liu C, Wang W, Wu J, et al. Virulence gene profiles and molecular genetic characteristics of diarrheagenic *Escherichia coli* from a Hospital in Western China. Gut Pathogenity 2018;10(35):1–11.
[36] de Almeida KM, Bruzaroski SR, Zanol D, de Melo M, dos Santos JS, Alegro LCA, et al. *Pseudomonas* spp. and *P. fluorescens*: population in refrigerated raw milk. Ciência Rural 2017;47(01):e20151540.

[37] de Oliveira GB, Favarin L, Luchese RH, McIntosh SD. Psychrotrophic bacteria in milk: how much do we really know? Braz J Microbiol 2015;46(2):313–21.

[38] de Souz CMOCC, Abrantes SMP. Detection of enterotoxins produced by *B. cereus* through PCR analysis of ground and roasted coffee samples in Rio de Janeiro, Brazil. Food Sci Technol 2011;31(2):443–9.

[39] Degaute D, Chapusette P, Vanoudenhove JL, Pierret C, Serruys-Schoutens E. Comparison of a biphasic medium plus routine early subculture with a slide blood culture system. Eur J Clin Microbiol 1985;4(5):475–82.

[40] Denny CB, Corlett Jr. DA. Canned foods-tests for cause of spoilage. In: Vanderzant C, Splittstoesser DF, editors. Compendium of methods for the microbiological examination of foods. 3rd ed. Washington: American Public Health Association; 1992. p. 1051–92.

[41] Dierksen KP, Sandine WE, Trempy JE. Expression of Ropy and Mucoid phenotypes in *Lactococcus lactis*. J Dairy Sci 1997;80:1528–36.

[42] Donnelly CW, Brackett RE, Doores S, Lee WH, Lovett J. Listeria. In: Vanderzant C, Splittstoesser DF, editors. Compendium of methods for the microbiological examination of foods. 3rd ed. Washington: American Public Health Association; 1992. p. 637–61.

[43] Donnelly LS, Graves RR. Sulfide spoilage sporeformers. In: Vanderzant C, Splittstoesser DF, editors. Compendium of methods for the microbiological examination of foods. 3rd ed. Washington: American Public Health Association; 1992. p. 317–23.

[44] dos Santos LR, do Nascimento VP, de Oliveira SD, Flores ML, Pontes AP, Ribeiro AR, et al. Polymerase Chain Reaction (PCR) for the detection of *Salmonella* in artificially inoculated chicken meat. Sao Paulo Inst Tropical Med 2001;43(5):247–50.

[45] Doyle MP. Detection and quantitation of foodborne pathogens and their toxins: gram-negative bacterial pathogens. In: Pierson MD, Stern NJ, editors. Foodborne microorganisms and their toxins: developing methodology. New York: Marcel Dikker, Inc.; 1986. p. 317–44.

[46] Dryer JM, Deibel KE. Canned foods-tests for commercial sterility. In: Vanderzant C, Splittstoesser DF, editors. Compendium of methods for the microbiological examination of foods. 3rd ed. Washington: American Public Health Association; 1992. p. 1037–49.

[47] DuPond HL, Mathewson JJ. *Escherichia coli* diarrhea. In: Evans AS, Brachman PS, editors. Bacterial infection of humans. 2nd ed. New York: Plenum Medical Book Company; 1991. p. 239–54.

[48] Dyle MP. Foodborne illness: pathogenic *Escherichia coli*, *Yersinia enterocolitica* and *Vibrio parahaemolyticus*. Lancet 1990;336(1823):1111–15.

[49] Entis P. Membrane filtration systems. In: Pierson MD, Stern NJ, editors. Foodborne microorganisms and their toxins: developing methodology. New York: Marcel Dikker, Inc.; 1986. p. 91–106.

[50] Ercoşkun A. Gıda Maddeleri Tüzüğü. İşçi Sağlığı ve İş Güvenliği Tüzüğü. Hemay-Petek Sağlık Yayınları, Ankara, Yayın No: 2; 1987.

[51] Erkmen O, Erten H, Sağlam H. Fermente Ürünler Teknolojisi ve Mikrobiyolojisi. Ankara: Nobel Academic Publishing Education Consultancy Trade Co. Ltd.; 2020.

[52] Erkmen O. Survival of *Salmonella typhimurium* in Feta cheese during manufacturing and ripening period. MSc. Thesis, METU, Gaziantep, Graduate School of Natural and Applied Sciences, Food Engineering Department, Gaziantep; 1988.

[53] Erkmen O. Behavior of *Staphylococcus aureus* in Turkish feta cheese during manufacture and ripening. J Food Prot 1995;58(11):1201–5.

[54] Erkmen O. *Staphylococcus aureus* porter of individuals working in food processing and marketing in gaziantep region and bacteriophage typing of isolated strains. Food Technol 1996;1(9):53–7.

[55] Erkmen O. Survival of virulent *Yersinia enterocolitica* during the manufacture and storage of Turkish feta cheese. Int J Food Microbiol 1996;33:285–92.

[56] Erkmen O. Behavior of *Staphylococcus aureus* in refrigerated and frozen ground beef and in Turkish style sausage and broth with and without additives. J Food Process Preservation 1997;21:279–88.

[57] Erkmen O. Gıda Mikrobiyolojisi. 5. Baskı, Eflatun Yayınevi, Ankara; 2017.
[58] Erkmen O. Laboratory practices in microbiology. London: Academic Press; 2021.
[59] Erkmen O. Nozokomiyal Stafilokok Enfeksiyonları Ve Bunların Kontrolü Üzerinde Araştırma. Doktora tezi, Gaziantep Üniversitesi, Tıp Fakültesi, Mikrobiyoloji ve Klinik Mikrobiyoloji Anabilim Dalı, Gaziantep; 1994.
[60] Erkmen O, Söylemez Z. Gaziantep Yöresi Yoğurtlarının Kimyasal ve Mikrobiyolojik Analizleri. Türk Hij ve Deney Biyol Derg 1994;51(1):59–62.
[61] Erkmen O, Aydemir S. Beyaz peynir üretimi. In: Erkmen O, Erten H, Sağlam H, editors. Fermente Ürünler Teknolojisi ve Mikrobiyolojisi. Ankara: Nobel Akademik Yayıncılık; 2020. p. 117–56.
[62] Erkmen O, Bozoglu TF. Food microbiology principles into practice. Volume 1: Microorganisms related to foods, foodborne diseases and food spoilage. Chicester: John Wiley and Sons, Ltd.; 2016.
[63] Erkmen O, Bozoglu TF. Food microbiology principles into practice. Volume 2: microorganisms in food preservation and processing. Chichester: John Wiley and Sons, Ltd.; 2016.
[64] Erkmen O, Bozoglu TF. Behaviour of *Salmonella typhimurium* in feta cheese during its manufacture and ripening. Lebensm-Wiss u-Technol 1995;28:259–63.
[65] Erkmen O, Önder AB. Kültür mantarı üretimi. In: Erkmen O, Erten H, Sağlam H, editors. Fermente Ürünler Teknolojisi ve Mikrobiyolojisi. Ankara: Nobel Akademik Yayıncılık; 2020. p. 739–64.
[66] Fakruddin M, Mannan KSB, Andrews S. Viable but nonculturable bacteria: food safety and public health perspective. ISRN Microbiol. Article ID 703813; 2013.
[67] Fayer R, Gamble HR, Lichtenfels JR, Bier JW. Waterborne and foodborne parasites. In: Vanderzant C, Splittstoesser DF, editors. Compendium of methods for the microbiological examination of foods. 3rd ed. Washington: American Public Health Association; 1992. p. 789–809.
[68] FDA. Bacteriological analytical manual online. U.S. Food and Drug Administration, Gaithersburg, MD: AOAC International; 2001.
[69] FDA. Laboratory methods (food). Food and Drug Administration. https://www.fda.gov/food/science-research-food/laboratory-methods-food; 2020.
[70] FDA. Bacteriological analytical manual. 8th ed., Revision A, FDA/Center for Food Safety and Applied Nutrition; 2001.
[71] Fenselau C, Demirev PA. Characterization of intact microorganisms by MALDI mass spectrometry. Mass Spectrom. Rev 2001;20:157–71.
[72] Ferone M, Gowen A, Fanning S, Scannell AGM. Microbial detection and identification methods: bench top assays to omics approaches. Compr Rev Food Sci Food Saf 2020;19:3106–29.
[73] Fleming HP, McFeeters RF, Daeschel MA. Fermented and acidified vegetables. In: Vanderzant C, Splittstoesser DF, editors. Compendium of methods for the microbiological examination of foods. 3rd ed. Washington: American Public Health Association; 1992. p. 929–52.
[74] Flowers RS, D'Aoust JY, Andrews WH, Bailey JS. Salmonella. In: Vanderzant C, Splittstoesser DF, editors. Compendium of methods for the microbiological examination of foods. 3rd ed. Washington: American Public Health Association; 1992. p. 371–422.
[75] Foegeding PM, Ray B. Repair and detection of injured microorganisms. In: Vanderzant C, Splittstoesser DF, editors. Compendium of methods for the microbiological examination of foods. 3rd ed. Washington: American Public Health Association; 1992. p. 321–421.
[76] Foegeding RM. Detection and quantitation of sporeforming pathogens and their toxins. In: Pierson MD, Stern NJ, editors. Foodborne microorganisms and their toxins: developing methodology. New York: Marcel Dikker, Inc.; 1986. p. 393–423.
[77] Froning G, Izat A, Riley G, Magwire H. Eggs and egg products. In: Vanderzant C, Splittstoesser DF, editors. Compendium of methods for the microbiological examination of foods. 3rd ed. Washington: American Public Health Association; 1992. p. 857–73.

[78] FSSAI. Manual of methods of analysis of foods. New Delhi: Food Safety and Standards Authority of India Ministry of Health and Family Welfare Government of India; 2012.
[79] Gilligan PH, York MK. Sentinel level clinical laboratory guidelines for suspected agents of bioterrorism and emerging infectious diseases *Brucella* species. Washington: American Society for Microbiology; 2016.
[80] Goldman E, Green LH. Practical handbook of microbiology. 2nd ed. Northwestern: CRC Press; 2014.
[81] Goma OM, Momtaz OA. 16S rRNA characterization of a *Bacillus* isolate and its tolerance profile after subsequent subculturing. Arab J Biotechnol 2007;10(1):107–16.
[82] Gomes TAT, Elias WP, Scaletsky ICA, Gutha BEC, Rodriguesc JF, Piazzab RMF, et al. Medical microbiology diarrheagenic *Escherichia coli*. Braz J Microbiol 2016;47(1):3–30.
[83] Gonzalez-Rey C, Svenson SB, Bravo L, Rosinsky J, Ciznar I, Krovacek K. Species detection of *Plesiomonas shigelloides* isolated from aquatic environments, animals and human diarrheal cases by PCR based on 23S rRNA gene. FEMS Immunol Med Microbiol 2000;29:107–13.
[84] Gounot AM. Psychrophilic and psychrotrophic microorganisms. Experimental 1986;42:1192–7.
[85] Guarino PA, Gray RJH. Spices and gums. In: Vanderzant C, Splittstoesser DF, editors. Compendium of methods for the microbiological examination of foods. 3rd ed. Washington: American Public Health Association; 1992. p. 961–74.
[86] Halatsi K, Oikonomou J, Lambiri M, Mandilara G, Vatopoulos A, Kyriacou A. PCR detection of *Salmonella* spp. using primers targeting the quorum sensing gene sdiA. FEMS Microbiol Lett 2006;259:201–7.
[87] Hall WH. Brucellosis. In: Evans AS, Brachman PS, editors. Bacterial infection of humans. 2nd ed. New York: Plenum Medical Book Company; 1992. p. 133–49.
[88] Halpin-Dohnalek MI, Marth EH. *Staphylococcus aureus*: production of extracellular compounds and behavior in foods-a review. J Food Prot 1989;52(4):267–82.
[89] Han J-Y, Song W-J, Kang D-H. Optimization of broth recovery for repair of heat-injured *Salmonella enterica* serovar typhimurium and *Escherichia coli* O157:H7. J Appl Microbiol 2019;126:1923–30.
[90] Hanson RS, Phillips JA. Chemical composition. In: Gerhardt P, Murray RGE, Costilow RN, Nester EW, Wood WA, Krieg NR, Phillips GB, editors. Manual of methods for general bacteriology. Washington: American Society for Microbiology; 1981. p. 328–92.
[91] Hao H, Liang J, Duan R, Chen Y, Liu C, Xiao Y, et al. *Yersinia* spp. identification using copy diversity in the chromosomal 16S rRNA gene sequence. PLoS One 2016;11(1):e0147639.
[92] Harmon SM, Geopfert JM, Bennett RW. Bacillus cereus. In: Vanderzant C, Splittstoesser DF, editors. Compendium of methods for the microbiological examination of foods. 3rd ed. Washington: American Public Health Association; 1992. p. 593–604.
[93] Harrigan WF, McCance ME. Laboratory methods in food and dairy microbiology. 3rd ed. London: Academic Press Inc; 1998.
[94] Hartman PA, Deibel RH, Sieverding LM. Enterococci. In: Vanderzant C, Splittstoesser DF, editors. Compendium of methods for the microbiological examination of foods. 3rd ed. Washington: American Public Health Association; 1992. p. 523–31.
[95] Hartman PA, Petzel JP, Kaspar CW. New methods for indicator organisms. In: Pierson MD, Stern NJ, editors. Foodborne microorganisms and their toxins: developing methodology. New York: Marcel Dikker, Inc.; 1986. p. 175–217.
[96] Hartman PA, Swaminathan B, Curiale MS, Firstenberg-Eden R, Sharpe AS, Cox NA, et al. Rapid methods and automation. In: Vanderzant C, Splittstoesser DF, editors. Compendium of methods for the microbiological examination of foods. 3rd ed. Washington: American Public Health Association; 1992. p. 665–746.
[97] Hassanzadazar H, Ehsani A, Mardani K, Hesari J. Investigation of antibacterial, acid and bile tolerance properties of lactobacilli isolated from Koozeh cheese. Vet Res Forum 2012;3(3):181–5.

[98] Hatcher Jr WS, Weihe JL, Splittstoesser, Hill EC, Parish ME. Fruit beverages. In: Vanderzant C, Splittstoesser DF, editors. Compendium of methods for the microbiological examination of foods. 3rd ed. Washington: American Public Health Association; 1992. p. 953−60.

[99] Herrero M, Simó C, García-Cañas V, Ibáñez V, Cifuentes A. Foodomics: MS-based strategies in modern food science and nutrition. Mass Spectrom Rev 2012;31(1):49−69.

[100] Heyndrickx M. The importance of endospore-forming bacteria originating from soil for contamination of industrial food processing. Appl Environ Soil Sci 2011; ID 561975.

[101] Hitchins AD, Hartman PA, Todd ECD. Coliforms-*Escherichia coli* and its toxins. In: Vanderzant C, d Splittstoesser DF, editors. Compendium of methods for the microbiological examination of foods. 3rd ed. Washington: American Public Health Association; 1992. p. 325−3369.

[102] Hosoya K, Nakayama M, Tomiyama D, Matsuzawa T, Imanishi Y, Ueda S, et al. Risk analysis and rapid detection of the genus *Thermoascus*, food spoilage fungi. Food Control 2014;41:7−12.

[103] Hou O, Bai X, Li W, Gao X, Zhang F, Sun S, et al. Design of primers for evaluation of lactic acid bacteria populations in: complex biological samples. Front Microbiol 2018;9:1−11.

[104] Hou T-Y, Chiang-Ni C, Teng S-H. Current status of MALDI-TOF mass spectrometry in clinical microbiology. J Food Drug Anal 2019;27:404−14.

[105] ICMSP. International commission on microbiological specifications for foods. microorganisms in foods. Sampling for microbiological analysis: principles and specific applications. 2nd ed. Toronto, ON: University of Toronto Press; 1986.

[106] Ilyanie HY, Huda-Faujan N, Ida MMY. Comparative proximate composition of malaysian fermented shrimp products. Appl Biol 2020;49(3):139−44.

[107] Jay JM. Microbial spoilage indicator and metabolites. In: Pierson MD, Stern NJ, editors. Foodborne microorganisms and their toxins: developing methodology. New York: Marcel Dikker, Inc.; 1986. p. 219−40.

[108] Jay JM, Loessner MJ, Golden DA. Modern food microbiology. 7th ed. New York: Springer; 2005.

[109] Johnston RW, Tompkin RB. Meat and poultry products. In: Vanderzant C, Splittstoesser DF, editors. Compendium of methods for the microbiological examination of foods. 3rd ed. Washington: American Public Health Association; 1992. p. 821−35.

[110] Kautter DA, Solomon HM, Lake DE, Bernard DT, Mills DC. *Clostridium botulinum* and its toxins. In: Vanderzant C, Splittstoesser DF, editors. Compendium of methods for the microbiological examination of foods. 3rd ed. Washington: American Public Health Association; 1992. p. 605−21.

[111] Kaysner CA, Tamplin ML, Twedt RM. Vibrio. In: Vanderzant C, Splittstoesser DF, editors. Compendium of methods for the microbiological examination of foods. 3rd ed. Washington: American Public Health Association; 1992. p. 451−73.

[112] Keusch GT, Bennish ML. Shigellosis. In: Evans AS, Brachman PS, editors. Bacterial infection of humans. 2nd ed. New York: Plenum Medical Book Company; 1991. p. 593−620.

[113] Keyvan K, Özdemir H. Occurrence, enterotoxigenic properties and antimicrobial resistance of *Staphylococcus aureus* on beef carcasses. Ank Univ Faculty Vet J Med 2016;63:17−23.

[114] Khedid MK, Faid F, Mokhtari A, Soulaymani A, Zinedine A. Characterization of lactic acid bacteria isolated from the one humped camel milk produced in morocco. Microbiol Res 2009;164:81−91.

[115] Kiiyuia C. Laboratory manual of food microbiology for Ethiopian Health and Nutrition Research Institute (Food Microbiology Laboratory). Unido Project (YA/ETH/03/436/11−52); 2003.

[116] Kim A-L, Park S-H, Hong Y-K, Shin J-H, Joo S-H. Isolation and characterization of beneficial bacteria from food process wastes. Microorganisms 2021; 9:X.

[117] Kim J, Enache E, Hayman M. Halophilic and osmophilic microorganisms. In: Salfinger Y, Tortorello ML, editors. Compendium of methods for the microbiological examination of food. 15th ed. New York: American Public Health Association; 2014.

[118] Kingombe CIB, Huys G, Tonolla M, Albert MJ, Swings J, Peduzzı R, et al. PCR detection, characterization, and distribution of virulence genes in *Aeromonas* spp. Appl Environ Microbiol 1999;65(12):5293–302.
[119] Koburger JA, Wei C-I. Plesiomonas shigelloides. In: Vanderzant C, Splittstoesser DF, editors. Compendium of methods for the microbiological examination of foods. 3rd ed. Washington: American Public Health Association; 1992. p. 517–22.
[120] Koch AL. Growth measurement. In: Gerhardt P, Murray RGE, Costilow RN, Nester EW, Wood WA, Krieg NR, et al., editors. Manual of methods for general bacteriology. Washington: American Society for Microbiology; 1981. p. 179–207.
[121] Koneman EW, llen SD, Janda WM, Schreckenberger PC, Winn WC. Color atlas and textbook of diagnostic microbiology. 4th ed. J.B. Lippincott Company; 1992.
[122] Krieg NR, Gerhardt P. Solid culture. In: Gerhardt P, Murray RGE, Costilow RN, Nester EW, Wood WA, Krieg NR, Phillips GB, editors. Manual of methods for general bacteriology. Washington: American Society for Microbiology; 1981. p. 143–78.
[123] Labbe RG, Harmon SM. Clostridium perfringens. In: Vanderzant C, Splittstoesser DF, editors. Compendium of methods for the microbiological examination of foods. 3rd ed. Washington: American Public Health Association; 1992. p. 623–35.
[124] Lake DE, Bernard DT, Kautter DA. Mesophilic anaerobic sporeformers. In: Vanderzant C, Splittstoesser DF, editors. Compendium of methods for the microbiological examination of foods. 3rd ed. Washington, DC: American Public Health Association; 1992. p. 275–89.
[125] Lanctte GA, Tatini SR. Staphylococcus aurerus. In: Vanderzant C, Splittstoesser DF, editors. Compendium of methods for the microbiological examination of foods. 3rd ed. Washington, DC: American Public Health Association; 1992. p. 533–50.
[126] Larkin EP. Detection, quantitation, and public health significance of foodborne viruses. In: Pierson MD, Stern NJ, editors. Foodborne microorganisms and their toxins: developing methodology. New York: Marcel Dikker, Inc.; 1986. p. 439–51.
[127] Lenovich LM, Kongel PJ. Confectionary proucts. In: Vanderzant C, Splittstoesser DF, editors. Compendium of methods for the microbiological examination of foods. 3rd ed. Washington, DC: American Public Health Association; 1992. p. 1007–18.
[128] Li L, Mendis, Trigui N, Oliver JD, Faucher SP. Theimportance of the viable but non-culturable state in human bacterial pathogens. Front Microbiol 2014;5:258.
[129] Liewen MB, Bullerman LB. Toxigenic fungi and fungal toxins. In: Vanderzant C, Splittstoesser DF, editors. Compendium of methods for the microbiological examination of foods. 3rd ed. Washington, DC: American Public Health Association; 1992. p. 811–9.
[130] Lyon WJ. TaqMan PCR for detection of *Vibrio cholerae* O1, O139, non-O1, and non-O139 in pure cultures, raw oysters, and synthetic seawater. Appl Environ Microbiol 2001;67(10):4685–93.
[131] Manero A, Blanch AR. Identification of *Enterococcus* spp. with a biochemical key. Appl Environ Microbiol 1999;65(10):4425–30.
[132] Marriott NG, Gravani RB. Principles of food sanitation. 5th ed. New York: Springer Science-Business Media, Inc.; 2006.
[133] Martinez-Murcia AJ, Collins MD. *Enterococcus sulfureus*, a new yellow-pigmented *Enterococcus* species. FEMS Microbiol Lett 1991;80:69–74.
[134] Mayou J, Moberg L. Cereal and cereal products. In: Vanderzant C, Splittstoesser DF, editors. Compendium of methods for the microbiological examination of foods. 3rd ed. Washington, DC: American Public Health Association; 1992. p. 995–1006.
[135] Meekin TAMc, Olley JN, Ross T, Ratkowsky DA. Predictive microbiology: theory and application. England: Research Studies Press Ltd.; 1993.

[136] Messer JW, Midura TF, Peeler JT. Sampling plans, sample collection, shipment, and preparation for analysis. In: Vanderzant C, Splittstoesser DF, editors. Compendium of methods for the microbiological examination of foods. 3rd ed. Washington, DC: American Public Health Association; 1992. p. 25–49.

[137] Miescier JJ, Hunt DA, Redman J, Salinger A, Lucas JP. Molluscan shellfish: oysters, mussels, and clams. In: Vanderzant C, Splittstoesser DF, editors. Compendium of methods for the microbiological examination of foods. 3rd ed. Washington, DC: American Public Health Association; 1992. p. 897–918.

[138] Mirhosseini SZ, Seidavi A, Shivazad M, Pourseify R. Detection of *Clostridium* sp. and its relation to different ages and gastrointestinal segments as measured by molecular analysis of 16S rRNA genes. Braz Arch Biol Technol 2010;53(1):69–76.

[139] Miri ST, Dashti A, Mostaan S, Kazemi F, Bouzari S. Identification of different *Escherichia coli* pathotypes in north and North-West Provinces of Iran. Iran J Microbiol 2017;9(1):33–7.

[140] Mislivec PB, Beuchat LR, Cousin MA. Yeasts and molds. In: Vanderzant C, Splittstoesser DF, editors. Compendium of methods for the microbiological examination of foods. 3rd ed. Washington, DC: American Public Health Association; 1992. p. 239–49.

[141] Nagpal R, Ogata K, Tsuji H, Matsuda K, Takahashi T, Nomoto K, et al. Sensitive quantification of clostridium perfringens in human feces by quantitative real-time PCR targeting alpha-toxin and enterotoxin genes. BMC Microbiol 2015;5:219.

[142] Nayak R, Stewart TM, Nawaz MS. PCR identification of *Campylobacter coli* and *Campylobacter jejuni* by partial sequencing of virulence genes. Mol Cell Probes 2005;19:187–93.

[143] Nguyen AL. Contribution of intrinsic factors to heat resistance of ascospores of *Byssochlamys*. B. Sci., School of Chemical Engineering. The University of New South Wales Sydney, NSW, Australia; 2012.

[144] Nhung PH, Shah MM, Ohkusu K, Noda M, Hata H, Sun XS, et al.). The dnaJ Gene as a novel phylogenetic marker for identification of *Vibrio* species. Syst Appl Microbiol 2007;30:309–15.

[145] Nickelson II R, Finne G. Fish, crustaceans, and precooked seafoods. In: Vanderzant C, Splittstoesser DF, editors. Compendium of methods for the microbiological examination of foods. 3rd ed. Washington, DC: American Public Health Association; 1992. p. 875–95.

[146] Nowak A, Rygala A, Oltuszak-Walczak E, Walczak P. The prevalence and some metabolic traits of *Brochothrix thermosphacta* in meat and meat products packaged in different ways. Sci Food Agri 2011;92:1304–10.

[147] Sullivan J, Bolton DJ, Duffy G, Baylis C, Tozzoli R, Wasteson Y, et al. Methods for detection and molecular characterization of pathogenic *Escherichia coli*. Dublin: Ashtown Food Research Center; 2007.

[148] Ojha SC, Yean CY, Ismail A, Singh AKB. A Pentaplex PCR assay for the detection and differentiation of *Shigella* species. BioMed Res Int 2013; ID 412370.

[149] Okhuysen PC, DuPont HL. Enteroaggregative *Escherichia coli* (EAEC): a cause of acute and persistent diarrhea of worldwide importance. J Infect Dis 2010;202(4):503–5.

[150] Olson KE, Sorrells KM. Thermophilic flat sour sporeformers. In: Vanderzant C, Splittstoesser DF, editors. Compendium of methods for the microbiological examination of foods. 3rd ed. Washington, DC: American Public Health Association; 1992. p. 299–307.

[151] Orsi RH, Wiedmann M. Characteristics and distribution of *Listeria* spp., including *Listeria* species newly described since 2009. Appl Microbiol Biotechnol 2016;100:5273–87.

[152] Paião FG, Arisitides LGA, Murate LS, Vilas-Bôas GT, Vilas-Boas LA, Shimokomaki LA. Detection of *Salmonella* spp, *Salmonella Enteritidis* and typhimurium in naturally infected broiler chickens by a multiplex PCR-based assay. Braz J Microbiol 2013;44(1):37–41.

[153] Palumbo S, Abeyta C, Stelma Jr. G. *Aeromonas hydrophila* group. In: Vanderzant C, Splittstoesser DF, editors. Compendium of methods for the microbiological examination of foods. 3rd ed. Washington, DC: American Public Health Association; 1992. p. 497–515.

[154] Park DL, Pohland AE. Official methods of analysis of foods for mycotoxins. In: Pierson MD, Stern NJ, editors. Foodborne microorganisms and their toxins: developing methodology. New York: Marcel Dikker, Inc.; 1986. p. 425–38.
[155] Park JY, Jeon S, Kim JY, Park M, Kim S. Multiplex real-time polymerase chain reaction assays for simultaneous detection of *Vibrio cholerae*, *Vibrio parahaemolyticus*, and *Vibrio vulnificus*. Osong Public Health Res Perspect 2013;4(3):133–9.
[156] Peeler JT, Houghtby GA, Rainosek AP. The most probable number technique. In: Vanderzant C, Splittstoesser DF, editors. Compendium of methods for the microbiological examination of foods. 3rd ed. Washington, DC: American Public Health Association; 1992. p. 105–20.
[157] Pérez-Lomas M, Cuaran-Guerrero MJ, Yépez-Vásquez L, Pineda-Flores H, Núñez-Pérez J, Espin-Valladares R, et al. The extended methylene blue reduction test and milk quality. Foods Raw Mater 2020;8(1):240–8.
[158] Pervin R, Rahman AM, Zereen F, Ahmed R, Alam ZM. Isolation, identification and characterization of *Staphylococcus aureus* from raw milk in different places of Savar, Bangladesh. Int J Sci: Basic Appl Res 2019;48(7):1–25.
[159] Pettipher GL, Osmundson ME, Murphy JM. Methods for the detection and enumeration of *Alicyclobacillus acidoterrestris* and investigation of growth and production of Taint in fruit juice and fruit juice-containing drinks. Lett Appl Microbiol 1997;24:185–9.
[160] PHE. UK standards for microbiology investigations. Public Health England, https://www.gov.uk/government/collections/standards-for-microbiology-investigations-smi#bacteriology; 2014.
[161] Power DA, McCuen PJ. Manual of BBL: products and laboratory procedures. 6th ed. Cockeysville: Beckton Dickinson and Company; 1988.
[162] Probert WS, Schrader KN, Khuong NY, Bystrom SL, Graves MH. Real-time multiplex PCR assay for detection of *Brucella* spp., *B. abortus*, and *B. Melitensis*. J Clin Microbiol 2004;42(3):1290–3.
[163] Pusch DJ, Busta WA, Moats WA, Schulza AE. Direct microscopic count. In: Speck ML, editor. Compendium of methods for the microbiological examination of foods. Washington, DC: American Public Health Association; 1976. p. 132–51.
[164] Ray B, Bhunia A. Fundamental food microbiology. Florida: CRC Press; 2014.
[165] Regua-Mangia AH, Gomes TAT, Vieira MAM, Irino K, Teixeira LM. Molecular typing and virulence of enteroaggregative *Escherichia coli* strains isolated from children with and without diarrhoea in Rio De Janeiro City, Brazil. J Med Microbiol 2009;58:414–22.
[166] Reid G. The scientific basis for probiotic strains of *Lactobacillus*. Appl Environ Microbiol 1999;65 (9):3763–6.
[167] Reingold AL. Toxic shock syndrome. In: Evans AS, Brachman PS, editors. Bacterial infection of humans. 2nd ed. New York: Plenum Medical Book Company; 1991. p. 727–43.
[168] Richter RL, Ledford LA, Murphy SC. Milk and milk products. In: Vanderzant C, Splittstoesser DF, editors. *Compendium of* methods for the microbiological examination of foods. 3rd ed. Washington, DC: American Public Health Association; 1992. p. 837–8856.
[169] Rico-Munoz E, Samson RA, Houbraken J. Mould spoilage of foods and beverages: using the right methodology. Food Microbiol 2019;81:51e62.
[170] Roberts D, Greenwood M. Practical food microbiology. Massachusetts: Blackwell Publishing Ltd; 2003.
[171] Ruhanya V. Adsorption-elution techniques and molecular detection of enteric viruses from water. J Hum Virol Retrovirology 2016;3(6):1–6.
[172] Schiemann DA, Wauters G. Yersinia. In: Vanderzant C, Splittstoesser DF, editors. Compendium of methods for the microbiological examination of foods. 3rd ed. Washington, DC: American Public Health Association; 1992. p. 433–50.

[173] Schottroff F, Fröhling A, Zunabovic-Pichler N, Krottenthaler A, Schlüter O, Jäger H. Sublethal injury and viable but non-culturable (VBNC) state in microorganisms during preservation of food and biological materials by non-thermal processes. Front Microbiol 2018;9:2773.
[174] Segmiller JL, Evancho GM. Aciduric flat sour sporeformers. In: Vanderzant C, Splittstoesser DF, editors. Compendium of methods for the microbiological examination of foods. 3rd ed. Washington, DC: American Public Health Association; 1992. p. 291–7.
[175] Shamloo E, Hosseini H, Abdi Moghadam Z, Halberg LM, Haslberger A, Alebouyeh M. Importance of *Listeria monocytogenes* in food safety: a review of its prevalence, detection, and antibiotic resistance. Iran J Vet Res 2019;20(4):241–54.
[176] Sharf JM. Recommended methods for the microbiological examination of foods. 2nd ed. Washington: American Public Health Association, Inc.; 1966.
[177] Sharma R, Garg P, Kumar P, Bhatia SK, Kulshrestha S. Microbial fermentation and its role in quality improvement of fermented foods. Fermentation 2020;6:106.
[178] Siddiquie MD, Mishra RP. Age and gender wise distribution pattern of typhoid causing bacteria *Salmonella* serovars in Mahakaushal Region. World J Pharm Res 2014;3(4):1183–203.
[179] Singh H, Rathore RS, Singh S, Cheema PS. Comparative analysis of cultural isolation and PCR based assay for detection of *Campylobacter jejuni* in food and fecal samples. Braz J Microbiol 2011;42:181–6.
[180] Sirockin G, Cullimore S. Practical microbiology. London: McGraw-Hill Publishing Company; 1969.
[181] Sitamahalakshmi S, Rao PR. Microbial load, microflora and quality of pasteurized milk. Int J Innov Sci Res Technol 2019;4(7):364–74.
[182] Skyberg JA, Logue JM, Nolan LK. Virulence Genotyping of *Salmonella* spp. with Multiplex PCR. 2006;50(1):77–81.
[183] Smith JL, Buchanan RL. Shigella. In: Vanderzant C, Splittstoesser DF, editors. Compendium of methods for the microbiological examination of foods. 3rd ed. Washington, DC: American Public Health Association; 1992. p. 423–31.
[184] Smittle RB, Cirigliano MC. Salad dressing. In: Vanderzant C, Splittstoesser DF, editors. Compendium of methods for the microbiological examination of foods. 3rd ed. Washington, DC: American Public Health Association; 1992. p. 975–83.
[185] Splittstoesser DF. Enumeration of heat resistant mold (*Byssochlamys*). In: Speck ML, editor. Compendium of methods for the microbiological examination of foods. Washington, DC: American Public Health Association; 1976. p. 230–4.
[186] Splittstoesser DF. Direct microscopic count. In: Vanderzant C, Splittstoesser DF, editors. Compendium of methods for the microbiological examination of foods. 3rd ed. Washington, DC: American Public Health Association; 1992. p. 97–104.
[187] Stern NJ, Patton CM, Doyle MP, Park CE, McCardell BA. Campylobacter. In: Vanderzant C, Splittstoesser DF, editors. Compendium of methods for the microbiological examination of foods. 3rd ed. Washington, DC: American Public Health Association; 1992. p. 475–95.
[188] Stevenson KE, Segner WP. Mesophilic aerobic sporeformers. In: Vanderzant C, Splittstoesser DF, editors. Compendium of methods for the microbiological examination of foods. 3rd ed. Washington, DC: American Public Health Association; 1992. p. 265–74.
[189] Succi M, Tremonte P, Reale A, Sorrentino E, Grazia L, Pacifico S, et al. Bile salt and acid tolerance of *Lactobacillus Rhamnosus* strains isolated from Parmigiano Reggiano cheese. FEMS Microbiol Lett 2005;244:129–37.
[190] Swanson KMJ, Busta FF, Peterson EH, Johnson MG. Colony count methods. In: Vanderzant C, Splittstoesser DF, editors. Compendium of methods for the microbiological examination of foods. 3rd ed. Washington: American Public Health Association; 1992. p. 75–95.

[191] Tall A, Teillon A, Boisset C, Delesmont R, Touron-Bodilis A, Hervio-Heath D. Real-time PCR optimization to identify environmental *Vibrio* spp. strains. J Appl Microbiol 2012;113:361–72.

[192] Thoerner P, Kingombe IB, Bogli-Stuber K, Bissig-Choisat B, Wassenaar TM, Frey J, et al. PCR detection of virulence genes in *Yersinia enterocolitica* and *Yersinia Pseudotuberculosis* and investigation of virulence gene distribution. Appl Environ Microbiol 2003;60(3):1810–16.

[193] Thorat VD, Bannalikar AS, Doiphode A, Majee SB, Gandge RS, Ingle SA. Isolation, identification and molecular detection of Brucella abortus from Cattle and Buffalo. Int J Curr Microbiol Appl Sci 2017;6 (10):2853–64.

[194] Truchado P, Gil MI, Larrosa M, Allende A. Detection and quantification methods for viable but non-culturable (VBNC) cells in process wash water of fresh-cut produce: industrial validation. Front Microbiol 2020;11:673.

[195] Vedamuthu ER, Raccach M, Glatz BA, Seitz EW, Reddy MS. Acid-producing microorganisms. In: Vanderzant C, Splittstoesser DF, editors. Compendium of methods for the microbiological examination of foods. 3rd ed. Washington, DC: American Public Health Association; 1992. p. 225–38.

[196] Villalobo E, Torres A. PCR for detection of *Shigella* spp. in mayonnaise. Appl Environ Microbiol 1998;64(4):1242–5.

[197] Weiss A, Domig KJ, Kneifel W. Comparison of selective media for the enumeration of probiotic enterococci from animal feed. Food Technol Biotechnol 2005;43(2):147–55.

[198] Wright A. Food microbiology laboratory mannual. <http://fshn.ifas.ufl.edu/faculty/ACWright/FOS4222.htm>; 2008.

[199] Xu Y-L. Foodomics: a novel approach for food microbiology. Trends Anal Chem 2017;96:14e21.

[200] Zulkifli Y, Alitheen NB, Son R, Yeap SK, Lesley MB, Raha AR. Identification of *Vibrio parahaemolyticus* isolates by PCR targeted to the *toxR* gene and detection of virulence genes. Int Food Res J 2009;16:289–96.

Index

Note: Page numbers followed by "*f*" and "*t*" refer to figures and tables, respectively.

F

G

G

I

J

K

L

U

V

Z

Printed in the United States
by Baker & Taylor Publisher Services